Advanced Technologies Management for Retailing:

Frameworks and Cases

Eleonora Pantano
University of Calabria, Italy

Harry Timmermans
Eindhoven University of Technology, The Netherlands

BUSINESS SCIENCE
Reference

Senior Editorial Director: Kristin Klinger
Director of Book Publications: Julia Mosemann
Editorial Director: Lindsay Johnston
Acquisitions Editor: Erika Carter
Development Editor: Myla Harty
Production Editor: Sean Woznicki
Typesetters: Mike Brehm, Natalie Pronio, Jennifer Romacnchak, and Deanna Zombro
Print Coordinator: Jamie Snavely
Cover Design: Nick Newcomer

Published in the United States of America by
Business Science Reference (an imprint of IGI Global)
701 E. Chocolate Avenue
Hershey PA 17033
Tel: 717-533-8845
Fax: 717-533-8661
E-mail: cust@igi-global.com
Web site: http://www.igi-global.com/reference

Library of Congress Cataloging-in-Publication Data
Advanced technologies management for retailing: frameworks and cases /
Eleonora Pantano and Harry Timmermans, editors.
p. cm.
Includes bibliographical references and index.
Summary: "This book contributes to our understanding of applications of new technologies and their impact on the design and development of point of sale systems and on consumers' behavior"--Provided by publisher.
ISBN 978-1-60960-738-8 (hardcover) -- ISBN 978-1-60960-739-5 (ebook) -- ISBN 978-1-60960-740-1 (print & perpetual access) 1. Retail trade--Technological innovations. 2. Retail trade--Management. 3. Retail trade--Automation--Social aspects. I. Pantano, Eleonora, 1982- II. Timmermans, H. J. P.
HF5429.A386 2011
658.8'7--dc22

2011012091

British Cataloguing in Publication Data
A Cataloguing in Publication record for this book is available from the British Library.

Table of Contents

Detailed Table of Contents

Foreword

Advanced Technologies Management for Retailing: Frameworks and Cases brings together research, insights, and practices of strategic management applications that optimize both the opportunities and challenges of emerging and flourishing technologies in retailing.

In the one hundred plus years since James Cash Penny opened his first "Golden Rule" store in the U.S., which grew to be today's "J. C. Penny" stores, the retail world has changed drastically and immeasurably. His motto of "Honor, Confidence, Service, and Cooperation," is still relevant today, and unlike many retailers, almost forty years after his death, JCPenny is flourishing and operating over 1,100 stores world-wide. The more recent story of Sam Walton is equally inspiring.

It is no longer James and Sam's retail world! Retailers are challenged by changes in customers, completion, cultures, global reach, value-chain operations/options, and the enormous impact of technology. Seemingly invincible Wal-Mart stays current with amazing real-time inventory data-base practices that were unthinkable when Sam opened his first store.

This up to date collection of observations by a multinational body of seasoned researchers and practitioners focuses on three major pieces of the new retail world: specific advances in available technologies, new sources and applications of consumer based knowledge, and the impact of these technology and knowledge advances on the behavior of consumers. The contributions in this edited work come from scholars whose home bases are in Italy, Spain, UK, Croatia, Switzerland, Denmark, USA, Canada, Germany, and Cyprus, but whose experience and research have no national boundaries. Like the authors, the applications described are not limited by national borders but are relevant wherever the value chain of contemporary retailing exists—from Peoria to Paris.

My personal observations are related to the way these technological approaches become intertwined with the strategies of the retailers wherein they are to be applied. First, readers must not lose sight of the necessity of technological leaders to communicate and persuade top managers to consider the value of new options, and the willingness and ability of the owners to fund these important new technologies. Wal-Mart invested billions on it point-of-sale data-base system that allows vendors and corporate managers to access broadly into the current status of local stores and their customer base. Acquiring a lasting strategic advantage is the goal of the technologies, and investors must be convinced that there is a rational basis for the necessary investments.

Second, the obvious implication of the value of this book is that the basic facts of change are changing. Years ago, I told business students that within 20 years they would be managing jobs and products in companies whose products didn't then exist. They did not believe it then and thought it was ridiculous. They do now! Yet, to make the same observation today would be ridiculous. Instead of twenty years, I would have to say five years. Thus is the source of the dynamics and excitement of this book. Retailers have become leaders and change agents, and are no longer simply merchandisers from the warehouses of established practices, but the creators of new and better, not just best, practices. The value-chain vehicle begins where the fabrics of retailing originate to the end users and beyond. Now retailers must offer customer service that often times includes obsolescence protection and buy-back options. The technologies necessary to lead, and even keep up with, the process and mechanisms of the new applications in marketing must themselves evolve. And the research on the consumers must continuously grow and keep up.

Third, marketing technology and strategy are facing a major new constraint: sustainability. Corporate strategy is evolving to acknowledge the need to be part of a sustainable world future, and marketing should take a major lead in that new normalcy. My business students' of years ago would have not believed that the issue of sustainability should have any role in marketing or corporate strategy. But they do now!

Fourth, this technology should not be for the retail giants such as IKEA or Carrefour, but should be developed in such a way as to encourage wide-spread diffusion into various marketing segments for small firms as well as large firms, developing nations as well as economically advanced nations, entrepreneurial enterprises as well as well established corporate firms, industrial firms as well as consumer oriented firms, and so forth. Widespread diffusion of new knowledge will happen when these applications are published in widespread outlets of technical information, not just scholarly journals. It is in this sense that I congratulate the authors and publishers of *Advanced Technologies Management for Retailing.*

Fifth, the future starts soon, and new hardware, software, and amazing new applications will come with increasing frequency. As developer and techno-entrepreneurs become more youthful, social media plays a larger and larger role in marketing applications, and the authors of these pieces become the established leaders in this exciting field, the world will be depending on this gravitas for mentoring and monitoring the values and ethics of this field. Computers are becoming much faster, and storage capacities are growing, and so are the information sources on the Internet alone, to say nothing of electronic data collection at point-of-sale is bordering on a level of privacy violation that seems uncontrollable. In his recent book *Contemporary Issues in Ethics and Information Technology,* Bob Schultz concluded that "Since ethical problems of IT confront us with new situations that cannot be handled in the same way of familiar cases, the only constant is the principles behind the cases." And then adds that the higher level principles regulate the conflicts between lower level principles. My concern is that our marketing advances in IT are the work of people who have less and less interest, education, or incentive to worry about ethical issues. Therefore, I urge the authors and readers of this work to use your leadership to maintain clear standards of ethical practices in this important field.

I believe that *Advanced Technologies Management for Retailing* provides many contributions to the scholarly and practical world of contemporary retailing, *where the consumer must remain the focus and the marketing mix remains enhanced, but unbroken.*

Robert L. Bjorklund
Woodbury University, USA

Robert Bjorklund *has degrees from the University of Sioux Falls and the University of Massachusetts, but his interest in retail marketing began in at an early age in the shoe business. His general business management and academic skills developed later, over a period of over forty years while his career moved between business (marketing, strategy, and human resources) and academia. After completing his PhD, he served as Assistant Professor at the Worcester Polytechnic Institute in Massachusetts and later as Associate Professor at the State University at NY in Utica. Shifting to business practice, he joined Hay Associates, in New York was hired by a client as VP of a consumer products manufacturer and later as VP Marketing in commercial real estate in New Jersey. He then returned to teaching at Rider University and New York University. Currently, Bjorklund is Associate Professor of Management and chair of the Management Department in the School of Business at Woodbury University in Burbank, CA. His current research interest is bank trustee boards and the success or failure of American bank members of the FDIC.*

Preface

In recent years, many studies have focused on the best practices which make stores more attractive and appealing for consumers. The application of innovative technologies at the point of sale is a promising and relatively unexplored field of study, in particular when considering the introduction of digital content and interactive technologies allowing consumers to access products in new ways. Many e-retailers have already exploited the opportunities offered by interactive technologies, such as 3D virtual models, digital products management, and knowledge transfer to consumers, in order to enhance consumers shopping experience. Their use in real stores, however, is still limited.

In this scenario, the development and use of innovative shopping assistants for supporting and influencing consumers during their shopping experience plays a key role for retailers and researchers alike. On the one hand, it improves the consumers in-store experience, on the other it gives marketers useful information on consumer preferences and needs, response to new services and retailing strategies, and on market trends. As a consequence, several current research projects focus on developing new virtual salesperson or existing shopping assistant systems, based on shopping trolleys equipped with digital displays or cameras to scan products barcodes. Adding digital and customized content to these tools can be a powerful means to influence customers' experience. The aim is to support consumers, through a user-friendly interface, by giving them fast and detailed information on products, sales, new arrivals, and so on. Consumers may use the offered information to choose among different alternatives, search for promotions, and calculate the value of purchases, by saving time and enjoying the experience.

The main characteristics are interactivity and multimodality of the interaction, in order to achieve an efficient, flexible, and meaningful feeling of interaction with a human. Therefore, it becomes very useful to deepen our understanding of the advanced technologies in a retailing context and their impact on consumer' behavior. In particular, this book aims to investigate the most useful applications of advanced technologies to retailing, the modality of the interaction between consumers and system, and the main benefits of the effective interaction. It also addresses implications for managing products and for improving consumers' knowledge to influence their subsequent buying behavior. Thus, contributions in this book relate to different fields such as Marketing, Computer Science, Psychology, and Management, to provide an integrated approach to the topic.

THE OBJECTIVES OF THIS BOOK

This volume contains original research that contributes to our understanding of applications of advanced technologies, their impact on the design and development of innovative points of sale, and of consumer behavioural intentions towards these technologies. In particular, Section 1, "*Advances in Technologies*

Management for Retailing," deepens our knowledge of the current developments for improving retailing and consumer in-store experiences. A selection of 6 chapters has been chosen to illustrate the changes in the point-of sale caused by innovative technologies capable of totally modifying store layout, payment modalities, and information transfer from and to consumers. The first chapter provides a detailed review of the current technologies available at the point of sale, by focusing barcode scanning, electronic shelf tags, shelf-checkouts, RFID tags, and fingerprint authentication; subsequent chapters highlight the evolution of the store from the traditional point-of-sale to an innovative point-of-consumer. Chapters 3 and 4 investigate the introduction of a specific technology in the store, respectively the Retail Digital Signage, and an advanced Virtual Shopping Assistant; Chapter 5 "*Information and Communication Technologies in Marketing Channels: Product Considerations*" analyzes the new technologies introduced in the different distribution channels more generally, with emphasis on consumer evaluation of these technologies. The last chapter of this section focuses on the Enterprise Architecture for retailing, in order to underline its benefits for innovating, improving, enriching, and increasing the interaction between business and technology.

Section 2, "*Digital Contents Management for Technology-Based Retailing*," analyses digital contents management for innovative, technology-based retailing in terms of information representation, transferring, and searching. In particular, Chapter 7 highlights an integrated conceptual representation of consumer group knowledge, which includes both the influence of collective variables on the decision making process and the investigation of scientific inquiries concerning the role of advanced technologies in relation to the conceptual representation. The successive chapters focus on the use of Web 2.0 or social media as a powerful tool for customizing digital contents, supporting consumer decision making, and improving consumer consumption experience, with emphasis on how by these technologies might be successfully used for building consumer loyalty, improving relationships with consumers, and creating value for e-retailers in the tourism sector. Furthermore, it is possible to exploit Customer Intelligence to collect a substantial amount of information on consumer needs, attitudes, and preferences towards a certain retailer in order to enhance the consumer-retailer relationship and improve the business decision, as illustrated in Chapter 9. With the same purpose in mind, Chapter 10 provides useful ideas for retailers for collecting and managing a large amount of information on consumers, by outlining which variables are capable of major improvements in information sharing in online settings. In the chapter "*You Never Get a Second Chance to Make a First Impression: Meet Your Users' Expectations Regarding Web Object Placement in Online Shops*," the authors propose a mental model for the development of efficient website interfaces, which can be used for the improving websites devoted to online shops.

The chapters included in the Section 3, "*Impact of Advanced Technologies on Consumers Behavior*," aim to investigate the impact of advanced technologies on consumers behavior in terms of consumer opinions, interaction modalities, and purchasing decisions. In particular, Chapter 12 analyses the online recommendation systems by outlining the strength link between consumer acceptance of these systems and the quality and shopping relevance of the provided recommendations. The various chapters focus on the relationship between Human-Computer Interaction (HCI) and aspects of consumer behavior with emphasis on the electronic retail context (e-tailing), by highlighting how 3D interfaces become a key factor for the success of online retail environments. Chapters 14, 15, and 16 exploit the use of the Technology Acceptance Model for analyzing consumers' intentions to adopt a new technology. In particular, chapter 14 focuses on the introduction of mobile purchasing decision support systems at the point of sale, by evaluating their impact on consumer behaviour. Chapter 15 outlines the direct and indirect effects of perceived usefulness, ease of use, and enjoyment on consumer acceptance of new technology in grocery retailing, by taking into account individual experience with Information Technologies. The last chapter investigates the impact of WiFi on consumer behaviour and attitude towards the new technology.

THE TARGET AUDIENCE

The book should be interesting for scholars of Economics, and in particular of Marketing, Management/Industrial Engineering, scholars of Computer Science, and scholars of Psychology. Researchers in these fields can obtain useful information about the latest results in these fields of study, as well as on the potential use of advanced technologies in retailing.

THE VALUE OF THIS BOOK

Presenting a wide number of technologies linked to consumer behaviour in retail environments (i.e. RFID, Shopping Assistant Systems, smart mirror, and so on), the book undoubtedly has an appeal for scholars in Marketing and Management, as well as in Psychology. In particular, it discusses case studies investigating the phenomenon of consumer-computer interaction, which suggest applying these new technologies to influence buying behaviour. The book adds to current research by enriching the frameworks at the disposal of the scholars interested in this field of study, highlighting how the technologies may be used to influence consumer behaviour in a retailing context.

The book provides:

- Complete analysis from a bibliographic point of view on the application of advanced technologies in retailing, including the research on the consumer acceptance of these technologies, product knowledge transfer, and consumer-computer interaction
- A complete view of consumer behaviour in ubiquitous/pervasive environments
- A complete view of research on the digital contents management for retailing
- A first collaborative approach to retailing and the potential of Computer Science in the generation of advanced systems capable of influencing consumer buying behaviour.

Therefore, this book is unique in its kind because the literature on the topic addresses only some of the subjects described, and almost completely separates consumer behaviour and the field of Computer Science. In particular, the authors emphasize the relevant connections between Retailing, Computer Science, and Psychology. These connections can be exploited to individualize innovative advanced systems for influencing consumers' buying behaviour, based on the current research in retailing, knowledge management, and human-computer interaction. The book fills an important gap, because there does not yet exist a text specifically devoted to the application of advanced technologies to retailing.

Eleonora Pantano
University of Calabria, Italy

Harry Timmermans
Eindhoven University of Technology, The Netherlands

Section 1
Advances in Technologies Management for Retailing

This section deepens our knowledge of the current developments for improving retailing and consumer in-store experiences. A selection of 6 chapters has been chosen to illustrate the changes in the point of sale caused by innovative technologies capable of totally modifying the stores layout, payment modalities, and information transfer from and to consumers.

Chapter 1
Point-of-Sale Technologies at Retail Stores:
What Will the Future be Like?

Richard Clodfelter
University of South Carolina, USA

ABSTRACT

This chapter provides a review and synthesis of information related to technologies available at the retail POS (point-of-sale) checkout. Several POS technologies available to retailers are described, detailing their benefits and drawbacks for both retailers and consumers. The five technologies described and analyzed are barcode scanning, electronic shelf tags, shelf-checkouts, RFID tags, and fingerprint authentication. The extent to which retailers have implemented these available technologies is described, and perspectives on the future implementation of these technologies and emerging trends are also presented. Findings would indicate that there will continue to be innovations in retail technology at POS, and shopper expectations will continue to change. At the same time, retailers will probably remain cautious in deciding if and when to adopt new technologies. They must be convinced that the innovations will deliver sufficient value to offset their expenses.

INTRODUCTION

In the first part of this decade, German-based Metro AG, one of the world's ten largest retailers, unveiled a prototype store of the future. The retailer's future store opened on April 28, 2003 in the town of Rheinberg, Germany, but at the end of May 2008, it was moved to Toenisvorst. The store places emphasis on presenting and testing the newest technology designed to cut costs for retailers and make shopping more exciting for consumers. Innovations used include technologies

DOI: 10.4018/978-1-60960-738-8.ch001

such as the RFID tag, interactive kiosks, wireless communications, smart registers, scales that recognize fruits and vegetables from digital images, and smart checkouts. At the store, customers are able to experience how cutting-edge technologies and innovative services can make shopping a more convenient and exciting experience.

Metro's future store provides a glimpse into the future of retail shopping and how retailers may manage their stores in the next few years. Metro's expectation is that the new technologies will boost sales and lower costs by making processes more efficient. Management expects customer satisfaction to increase because products will be more readily available, service will become more individualized, and shopping will be more convenient. One of Metro's primary goals is to learn exactly what opportunities and threats are offered by those technologies. In fact, acceptance or rejection of the technologies by their customers will be the yardstick as to whether, how, and to what extent these new technologies will be introduced on a mass scale.

This chapter provides a review and synthesis of information related to technologies available at the retail POS (point-of-sale) checkout. POS technologies available to retailers are described, detailing their benefits and drawbacks for both retailers and consumers. The extent to which retailers have implemented these available technologies will also be presented, along with results and findings of relevant research that has been previously conducted. The impact that the technology has had on shopper behavior will also be highlighted. Perspectives on future implementation of these technologies and emerging trends are also presented, and future research opportunities in this area are suggested.

Most retailers agree that the industry is going through a shakeout period and one way to survive and grow is through the implementation of technology. Today, retailers are continually making decisions about whether or not to adapt new technologies available in the field. As they wrestle with the decision on whether or not to adopt these new technologies, they must have a full understanding of what the technology involves, how it will affect the way they do business, and whether or not the new technologies will be accepted by consumers. This review and synthesis of information related to POS technology should provide assistance to retailers as they make these decisions in the future.

CONCEPTUAL FRAMEWORK

New technological advances offer new opportunities to retailers for retailers to better manage their firms and enhance the shopping experience for customers, particularly at the point-of-sale (POS). This chapter is designed to deepen the reader's understanding of how technology is being implemented at point-of-sale (POS) and its impact on both the retailer and the customer.

Recent research (Pantano 2010) suggests that retailing can benefit from the implementation of new technologies. In fact, some researchers (Weber & Kantamneni 2002) have found that a retailer's ability to build and defend a competitive position in the market depends to a large extent on the willingness and capacity of the firm to invest in and use technology. Yet, there is a belief by some practitioners that more technology is preferable to less technology; while some researchers (Sethuraman & Parasuraman 2005) suggest that good technology must be "appropriate." In other words, retailers should adopt the technologies that meet their specific strategic aims. Timmor & Rymon (2007), however, found that there has been little effort to measure customer satisfaction with new technological "improvements" and determine what is the most "appropriate" technologies to be used.

More recently, Gil-Saura et. al. (2009) found that not all technological solutions are equally valued by customers. Even though a technology provides benefits, it may not be appreciated by customers. These researchers stressed the need for retailers to invest in new technologies, but to prioritize implementation based on those technologies valued by customers.

Research conducted by Kim & Kim (2008) found that consumers who enjoy their shopping experiences make more purchases, suggesting that retail managers play an important role in the customer satisfaction process. In fact, Pantano & Naccarato (2010) point out that new retail technologies modify the appearance of stores and customer behavior once they are in the store. For example, store appearance is changed in terms of style, layout, and atmosphere—all are elements which can affect a customer's in-store experience (Puccinelli et. al. 2009).

In another vein, new retail technologies, particularly those at point-of-sale (POS), increasingly are encouraging customer to perform services themselves (Weijters et. al. 2007), and in many cases, customers are being forced to use technology-based self service. Literature on forced adoption suggests that imposing an innovation on consumers results in resistance to the innovation and their developing a negative attitude toward the innovation (Ram & Jung 1991). Moreover, psychological reactance theory (Fitzsimons & Lehmann 2004) suggests that restricted freedom leads to frustration and hostile attitudes toward the source of the restriction—the service provider.

Recently, Reinders et. al. (2008) conducted similar research in a retail setting. They also found that forcing customers to use technology-based self service had severe and adverse negative effects on consumer attitudes toward using the technology and toward the retail store (the service provider). Evidence from other researchers (Robertson & Shaw 2009) would also suggest that consumer dissatisfaction with self-service technologies is widespread. Moreover, for some customers, research (Sun et. al. 2009) indicates that people, not technology, play an extremely important role in the formation of positive long-term relationships between customers and the retail store. Such relationships may vanish when customers are forced to use technology during their shopping experiences.

The Technology Acceptance Model (TAM) has also been extremely important to this area of research. According to TAM, customer intention to make use of technology is based on its perceived usefulness and its perceived ease of use. Complementing this model is research by Muller-Seitz et. al. (2009) which stresses that two criteria specifically impact customer acceptance of new technology. First, customers must be aware of a specific usefulness that outweighs the potential disadvantages of the technology. Secondly, customers must believe that the new technology is secure and safe. Their research refines TAM by integrating the factor "security concerns" into their analysis. The researchers suggested that retailers should not only focus on the benefits a technology offers, but also concentrate on reducing customers risk perceptions when they use the technology.

BACKGROUND, ISSUES, AND PROBLEMS

This chapter will examine five technologies available at retail POS checkouts. Barcode scanning will be reexamined to determine if the technology has offered both customers and retailers the increased productivity and pricing accuracy that were initially promised. Electronic shelf tags will be presented and discussed as a method of guaranteeing pricing accuracy at POS checkouts. Direct customer interaction with POS technology through self-checkouts will be examined, along with a discussion of its benefits and challenges shoppers face when using the technology. RFID tags as a replacement for barcodes will also be described. Finally, new technology for customer payment at POS checkouts through fingerprint authentication will be examined.

Barcoding and Scanning

The technology found most often at retail point-of-sale (POS) checkouts is barcode scanners. In fact, it was only a little over 35 years ago that these scanners made their first appearance at a grocery store in Troy, Ohio in 1974 (Greenfield 1992). Today, almost all retailers use barcode scanning at their customer checkouts. All types of companies, from large retail chains to small independent retailers, are using point-of-sale scanners as an essential part of their retail business operation. In a relatively short period of time, technology was implemented by retailers and accepted by their customers, resulting in a dramatic change in the way customers went through the checkout process as they made their purchases.

The use of barcodes and scanners has provided benefits to both retailers and customers. Retailers have netted substantial savings. They no longer have the expense of placing a retail price on each individual item, and checkout productivity has increased due to cashiers being able to check out customers much faster than in the past. Additional savings for retailers have also occurred because of improved sales and inventory data analysis. For customers, bar codes and scanners have delivered faster store checkouts, more detailed sales receipts, and price reductions resulting from retailers passing along cost savings to them. Customers were also promised more accurate checkouts; however, pricing errors have been found to varying degrees as barcode scanning became widely implemented in retail stores. For some customers, this developed into a mistrust of the technology being implemented.

In the years following the widespread implementation of scanners in retail stores, there were a number of academic, government, and industry studies, as well as media reports which indicated pricing errors were occurring because of the technology. These reports heightened customer mistrust. However, the major problem facing all of these initial studies and reports was that almost all of them involved only a small sample of product price checks, and they were narrowly focused (i.e., small number of items price checked, only a few stores in the sample).

In 1995, the *Examination Procedure for Price Verification* (National Conference on Weights and Measures 1995) was adopted in the United States. These procedures specified sampling practices for selecting products to be price checked and established a 98 percent accuracy rate for measuring the goal for pricing accuracy at retail stores. The perspective was that 100 percent accuracy was probably not attainable because of problems the retailer could not control (i.e., customers examining products and returning them to the wrong shelf or rack).

Upon publication of that document, three national studies related to pricing accuracy at retail stores using scanner technology in the U. S. were conducted. A study by the Federal Trade Commission (Federal Trade Commission 1996) sampled six states and found a 95 percent accuracy rate. During that same period, a study conducted at the University of South Carolina (Clodfelter 1998) sampled ten states and found a 96 percent accuracy rate. Later, in 1998, the FTC conducted a follow-up study (Federal Trade Commission 1998) involving 36 states and found an accuracy rate of nearly 97 percent. Additionally, one other large-scale study which was conducted in New Zealand grocery stores (Garland 1992) that found a similar accuracy rate of about 96 percent.

Study results indicated that pricing errors were occurring at stores where scanners and barcodes are used; however, error rates were not as high as some early studies and reports had initially indicated. These national studies also found that errors in the customer's favor were occurring at about the same rate as errors in the retailer's favor. Moreover, errors were not found to be caused by technology problems, but were due to poor management and operational practices.

Scanner technology can be used more effectively. Retail managers must still make improvements in their operational procedures to ensure a higher pricing accuracy for their stores, as well as their customers. They must identify and implement procedures to reduce the number of pricing errors (i.e., pricing audits conducted on a regular basis, established procedures for handling pricing errors when they occur, performance evaluations of store managers tied to pricing accuracy); otherwise, governments may clamor for a return to item pricing which will be more costly for retailers and negate the benefits provided by the scanners.

Price accuracy translates to customer loyalty and profitability. Pricing errors at the checkout harms both the retailer and the consumer. Retailers lose profits when shoppers are undercharged and see a decrease in customer satisfaction as a result of their being overcharged. In some areas, overcharges can even lead to substantial government fines for many retailers. Above all, pricing errors erode customer trust in the technology being used which can lead to a decrease in sales.

Electronic Shelf Labels

With the advent of product barcoding and the removal of item pricing, the shelf label became the customer's only source of price information which created some problems. Nearly every shopper probably has had doubts about whether the price that flashes on the checkout register matches the price posted on the shelf. However, when they purchase a large number of items, it usually is too difficult to remember the price they saw posted. Some customers feel that the convenience of a faster checkout using technology is worth the problems, even with some errors. They believe that trained cashiers make errors too. As more scanning technology was implemented, customers became more comfortable and some of those fears have been allayed as retailers improved their operational practices, but for other customers fears and distrust remained.

For several decades, there has been technology available to retailers that enables them to eliminate these customer concerns. Retailers can implement ESL (electronic shelf label) systems that automatically coordinate shelf and checkout prices with nearly 100 percent accuracy. ESLs are display units that can be positioned on store shelves next to each product being sold to electronically display product information, such as the retail price. The display units and the store's POS cash registers are both connected to the same price file; therefore, the price that the customer sees on the store shelf is exactly what they pay at the register when they check out. ESLs can also be used to display other types of information, such as a sale price, promotional information, or reminders for customers to purchase related items. By using ESLs, retailers can instantaneously update price changes on the store shelves and at the checkout in one or even multiple stores.

Electronic shelf labels provide an option that eliminates the labor, delays, and potential errors when shelf prices are updated manually. However, ESLs have not had widespread acceptance because of the cost of individual displays and the long payback period to receive a desired return on investment (Webb 2002). ESLs have, however, been popular in the few states in the United States that require products to be item-priced—that is, the retail price must appear on each product. In these states, item pricing regulations are waived if retailers implement electronic shelf labels (Figure 1).

Benefits. The primary benefit that ESLs offer retailers and customers is eliminating pricing error at the scanner by coordinating the prices that exist on the store shelves with prices that exist in the POS checkout's price file. Shoppers are assured that when their purchases are scanned at checkout, the price will match the price in the aisle. One academic study found that the rate of pricing errors at scanners was significantly lower for stores using scanner systems and ESL systems than those stores using scanners alone (Goodstein & Escalas 1995).

Using ESLs also reduces labor costs for the retailer at the store level. In fact, some retail ana-

Figure 1. Example of an electronic shelf label

lysts (Fox 2002) indicate that labor savings are the main reason why ESLs have been embraced more widely outside the United States, where labor costs tend to be higher. ESLs provide the retailer the ability to change prices at the touch of a button, saving hours of labor costs involved in making the changes manually. For example, the ESLs for 40,000 product SKUs can be changed in less than one hour (Falkman 2005).

Finally, ESLs can provide retailers assistance with promotional efforts. The ESL signs can flash when an item is on sale, or the tags can help the retailer promote creatively (i.e., offering temporary price reductions only during certain hours of the day). Electronic shelf labels also offer a great value to retailers that frequently alter their prices and need to convey the updated information quickly to their customers (Kroll 2006).

Drawbacks. Cost remains the primary drawback for widespread implementation of ESLs (Korolishin 2004). Even though the price of the systems dropped fairly rapidly when they made their initial appearance, those costs have stabilized in recent years. For example, a unit for one product was priced at about $12 in 1997, but within five years was down to about $8 each (Fox 2002). Most retailers feel that current prices are still too high.

Another major issue expressed by retailers concerning the implementation of ESL technology is over how rapidly they will receive a return on their investment. In those U. S. states which receive a waiver from item-pricing by using ESLs, capital recovery for the technology takes less than a year (Fox 2002). A similar ROI was also found by one retailer testing the technology in Great Britain (Lewis 2007). Outside of these examples, however, the length of time it takes to achieve an acceptable ROI is a source of intense debate.

Implementation. In addition to guaranteeing pricing accuracy, one of the most important roles that ESLs might play is as a hedge against government legislation requiring item pricing. It is possible that by ensuring price consistency between the store shelf and checkout, the widespread usage of electronic shelf labels could lead to a repeal of item-pricing laws or prevent their enactment elsewhere.

ESLs, however, remain one the great enigmas of retail technology—offering apparently great potential, but still not gaining widespread acceptance. For example, at the beginning of the decade, *Stores* magazine (Fox 2002) predicted that electronic shelf labels "...finally appear ready to deliver." However, that potential still has not materialized.

For most retailers, the slow acceptance of ESLs does not appear to be a reflection on the technology's quality or usability, but a simple matter of economics. Concern that the cost of the technology outweighs its benefits is impeding implementation by most retailers. Cost reductions for the technology will probably continue to occur over time, but will waiting to implement ESLs cause retailers to miss the immediate benefits and the potential competitive advantage that the technology offers?

A number of major retailers (i.e., Safeway, 7-Eleven and BJ's Wholesale Club, and Kohl's Department Store) continue to test ESL technology on an individual store level. And, growth of stores implementing ESLs is expected to increase. In fact, one retail research firm projected the market for ESLs would grow at an annual rate of 14.4

percent in North America between and at a rate of 19.9 percent in Europe (*ChainStore Age* 2005).

Self-Checkouts

Another technological change occurring at the POS checkout places existing technology in the hands of shoppers rather than retail store employees. Self-checkouts are an outgrowth of retailers experimenting with ways of making customer checkout more efficient, while resulting in savings for the store. Self-checkout, the ability of customers to scan, bag, and pay for their merchandise without the assistance of a cashier, has only been available for about 12 years. NCR installed the first self-checkout unit in a Ball's Food store in Kansas in 1998. Recent estimates indicate that ten years later approximately 2.5 billion consumer transactions were processed during a single year at retail stores using NCR's FastLane solution (Schulz 2008). Worldwide, self-checkouts are emerging as a major option for supermarkets, discount stores, and other retailers. Yet, it is critical that retailers balance customer use of the technology with enhanced customer service.

Currently, retailers are testing two types of self-checkout technologies. They include

1. Smart trolleys. These are standard shopping carts/trolleys equipped with an extra touch-screen and scanner. As customers select an item for purchase, they scan the item and the price registers on the cart screen. Shoppers can also see a running total of their purchases, a list of their purchases, and extra information about individual items (e.g., the region where a wine was produced). When customers get to the front of the store, information from their cart is downloaded, and they make the required payment. Metro AG has tested this technology at its Future Store in Germany where the smart trolley is known as a PSA (personal shopping assistant). In the United States, Stop & Shop has tested its own version known as the Cart Buddy.
2. Self-Service Checkouts. Customers shop in the traditional way and scan their purchases themselves at an unmanned checkout, paying by credit card, debit card, or cash. In some stores, shoppers must weigh their purchases so the system can verify that all items have been scanned. With this technology, one supervising cashier monitors several self-checkout lanes. In the United States, self- checkouts have been installed at many of the major retailers, Wal-Mart, Home Depot, Lowes, and Kroger. Although less common in Europe, some retailers, such as Sainsbury's, Tesco, and Metro are testing the technology (Mander 2004) (Figure 2).

More research has been conducted about self-checkouts than other technology being implemented at the POS terminal. One key finding is that customers preferred a checkout with a cashier; however, the majority of shoppers indicated that they wanted to learn how to use self-checkouts (Opara-Nadi 2005). Further, an NCR survey showed that respondents would be 33 percent more likely to frequent a store that offered self-checkout services. Customers also responded that shorter lines (36 percent) and greater speed (32 percent), as well as increased privacy (10 percent) were factors they considered in using self-checkouts (Barr 2001).

According to a Gartner/G2 survey, fifty-six percent of consumers in North America said they have used self-checkout. These findings tend to agree with an ACNielsen U. S. Homescan consumer panel survey of 61,500 households across the United States which found that 61 percent of households have tried self-checkout lanes. The data also suggest that customers are beginning to choose their shopping location based on the availability of self checkout (IBM 2004). Additionally, Marzocchi and Zammit (2006) conducted a survey with customers of a supermarket belonging to a

Figure 2. The introduction of new technologies, such as self-checkouts, change the layout and atmosphere of retail stores

major Italian retail chain in northern Italy. Findings supported the concept that the presence of self-checkouts may lend a competitive advantage and offer the retailer the possibility of expanding its customer base. Additionally, researchers have found that more than half of all grocery store customers purchase fewer than a dozen items, and they are much more likely to patronize a store where they can get in and out in a matter of minutes (Grimes 2004). These findings are critical to retailers as they make decisions about large investment in POS technology and gain a competitive edge in the marketplace.

Other useful findings that would aid retailers in making decisions about implementing self-checkouts were reported in IHL Consulting Group's Sixth Annual Report addressing self-checkout systems in North America. They include

- The average number of items in a self-checkout transaction was 6.7 with an average transaction of $32.85.
- On average, impulse purchases for women drop by 50 percent in a self-checkout environment over a standard checkout. For men, the impulse purchases drop 27.8 percent.
- Twenty-nine percent of self-checkout users use it only when there is a line at the other lanes. This means that if there is no line at a lane with at a register with a cashier, customers will choose to use a cashier rather than self-checkout.
- Speed and convenience were the two most-cited reasons for using self-checkout.
- Ninety-four percent of the respondents said they would use a self-checkout, even if they do not necessarily like it (Sheldon & Buzek 2006).

A study at the Vlerick Leuven Gent Management School in Belgium examined customer perceptions of waiting time for both users and non users of self-checkouts and the effect that this had on the level of satisfaction with the shopping experience. The researchers found that users' attitudes towards self-checkouts were significantly affected by the ease of use of the technology, the perceived usefulness of the technology, the reliability of the technology, and the fun associated with using it. Further, it was noted that people who use self service technology were more likely to have lower perceptions of waiting time (Weijters, et. al. 2005). Since perception of waiting time significantly affects the satisfaction associated with the shopping experience, this indicates the importance of getting customers to try and use self-service technology.

On a different note, a Forrester Research report indicated that shoppers have a love-hate relationship with self-checkouts. The research firm found 44 percent of the respondents felt that it was a great way to checkout of a retail store, while 42 percent would rather have a person handle the chore. For some shoppers, it is a matter of preferring to be served after spending money in the store, and some respondents worried about machines taking away jobs from humans (Lindeman 2005). Further, Licata (2004) found that shoppers who love technology see it as progress. Those who are wary of technology or uncomfortable with change view it as one more chore that they have to do. Research by Gemperlein (2006) indicated that there is a strong correlation between those who regularly use self-checkouts in retail stores and consumers who have gotten their airplane tickets at machines in airport terminals. It would seem that usage of self-checkouts will increase as customers' comfort level with the technology increases.

Little research has been conducted concerning how retailers view the implementation of self-checkouts. In the United States, a study by the Food Marketing Institute (FMI SuperMarket Research 2003) asked retailers who were using self checkouts about their return on investment (ROI). More than 60 percent of the retailers questioned rated it as positive, 33 percent as neutral, and 6 percent as negative. When asked about what factored into their comments, 84 percent listed decreased labor costs, and 58 percent said it freed employees for other customer-service tasks.

Benefits. Self-checkout gives customers an option that addresses one of their biggest concerns when shopping—a faster checkout. The self-checkout not only gives customers control over their checkout experience, but helps allow the store to redeploy staff to help other shoppers make purchase decisions.

Self checkout offers consumers a new level of control over their checkout experiences. They gain the ability to scan items at their own pace which gives them the time to verify prices. Cashiers tend to quickly move the merchandise past the price scanner—which rarely allows the customer the time to see if he/she is being charged the correct amount for each item. Self-checkouts also give customers the ability to handle and bag items some items—such as delicate produce or fresh flowers—however they want. Customers are also offered more privacy. For example, some shoppers can avoid their feelings of self-consciousness in front of cashiers when they purchase items for birth control, incontinence, constipation, hair color, feminine hygiene and dentures.

However, self-checkouts probably do not speed up the checkout process. Trained cashiers can scan and bag goods faster than even the most aggressive or enthusiastic shopper; but, actual checkout speed is only part of the equation for customers. Self-checkout has a psychological effect—as long as the shopper is taking an active part in the checkout process, it seems to go faster. And when there are lines at the other registers, a self-checkout system will always seem faster. Indeed, the perception of speed appears to be the main reason driving shopper acceptance of self-checkouts (Dabholkar, et al 2003).

Customer benefits are not the only factor that prompts retailers to provide more self-checkouts (Roussel-Dupre 2002; IBM 2004). Self-checkouts allow retailers to improve labor allocation, leading to greater efficiency. For example, one cashier can oversee the operation of four to eight self-checkout lanes. The self-checkout has also allowed retailers to move staff away from checkout lanes and improve service in other areas of the store. They want to and need to get cashiers to other areas of the store that cannot be automated. Self-checkouts also ease another employee problem that some stores face. Some retail stores, particularly those operating on a 24/7 schedule, have had problems in getting enough reliable workers to operate conventional checkout lanes. Self-checkouts can help solve that problem, and they also allow retailers to better handle unexpected customer traffic during hours in which increased numbers of customers was not expected.

Drawbacks. Many shoppers are afraid of making embarrassing mistakes; therefore, they have been reluctant to use the technology. Consumers' fear factor is one of the greatest obstacles that retailers implementing self-checkouts must ease.

There are also issues with the technology that slow down the checkout process. Customer mistakes when using the technology can bring the system to a halt, and they must wait for the supervising cashier to intervene. If beer or cigarettes are being purchased, the process also shuts down until the customer's ID is checked. Additionally, supervising cashiers can become overwhelmed with handling the problems of several self-checkout registers at once causing some customers to become so frustrated they have abandoned their shopping carts and headed for the parking lot.

Critics of self-checkouts also believe that retailers are putting too much spin on the argument of customers benefiting from the technology. They suggest that the only reason retailers are implementing self-checkouts is to reduce costs (Bhatnagar 2003). They believe that since customers are slower than professional cashiers, using the argument of a faster checkout is only a sleight-of-hand (Schuman 2004). Some critics have even described self-checkouts as a "gimmick" that does not really save the customer any time. In fact, these critics stress that retailers are shifting work to consumers; new tasks are being added to a customer's hectic shopping routine.

Some critics have also expressed concern that self-checkouts will lead to more theft and shoplifting. There are controls, such as cameras at each terminal and scales that weight each item being scanned to determine if it matches the product weight previously in the computer. For example, if a customer tries to fool the system by scanning a can of tuna and then putting a steak in the bag will hear the unit say, "Please remove the item and scan it again." At the same time, the system would signal the supervising cashier to watch what is going on at that terminal. Moreover, one researcher (Opara-Nadi 2005) did not find an increase in shrinkage at retail stores related to the installation of self-checkout systems.

A factor that may be most important to retailers considering the implementation of self-checkouts is that some customers hesitate to use the technology because they simply prefer to deal with human beings, rather than machines (Mander 2004). Some critics even contend that self-checkouts remove a level of human interaction, which is intrinsic in good customer service.

Implementation. There has been rapid implementation of this technology around the world. Australian retailer Woolworths installed self-checkouts at 70 of its stores in 2008. Approximately 20 percent of their sales transactions go through the self-checkouts, and management indicates that they have received tremendously positive customer feedback (just-food.com 2008). Asda in the UK now offers self-checkouts at 99 percent of its stores, and Tesco claims that that 25 percent of its UK transactions are completed through them (Creasey 2009). In fact, Tesco has opened dozens of its Fresh & Easy grocery stores in the United States all with self-checkouts (Kiviat 2008). And, the chain recently opened a Tesco Express convenience store in Kingley, UK with nothing but self-checkouts. The company is testing whether or not offering nothing but self-checkouts may be more efficient and provide better customer service in small stores. At this test store, the same number of staff is employed; however, they are assigned different responsibilities (just-food.com 2009).

In Northern Ireland, the Henderson Group has also tested self-checkouts in an effort to make the shopping experience for customers as convenient and hassle-free as possible (*Grocer* 2007). In the Netherlands, forty percent of customers at one Albert Heijn supermarket already use self-checkouts (Karaian 2008), and management at Home Depot indicates that about 35 percent of sales are generated by self-checkouts at stores equipped with the technology.

One reason explaining the rapid growth in self-checkouts has been that they have the potential to pay for themselves in about a year and save about 200 employee hours a week (NACS Online 2003). Similar findings have also been reported in a Gartner/G2 research report. Researchers found that the return on investment (ROI) for self checkout systems typically occurred within 12 to 18 months (IBM 2004).

Even though many retailers feel pressured to offer self-checkouts to stay competitive, some chains, such as Target and Publix supermarkets in the United States, feel the technology is counter to their stores' emphasis on customer service. Retailers must make the self-checkout process as easy as possible for shoppers. While implementing new technology to make the checkout process more efficient, they must not alienate customers who would rather not use technology. Yet, consumers today are exposed to numerous self-service experiences on a daily basis, whether on the Internet, at the ATM, at airport self-check ins, or at the fuel pump; implementing self-checkout is a natural evolution for retailers. The goal is the same for all these companies—move customers quickly through their transactions.

RFID

Are there other technologies that can be implemented to speed up the checkout process? Are there procedures that do not transfer the checkout task to customers? For years, advertisements have appeared presenting scenarios of customers pushing shopping carts through a checkout lane and an electronic reader being able to scan the price of every item in the cart as it rolls by—without the shopper having to unload the cart and individually scanning each item. This is not a dream; the technology is already available with which to accomplish this feat. However, neither retailers nor customers should expect this technology to be implemented at the checkout tomorrow. Initially, it took years for consumers to accept product scanning, and it took years to fully equip the majority of retailers with scanners and to develop policies and practices that would help alleviate consumer fears of pricing mistakes. Implementation of this technology will not occur overnight; yet, retail analysts feel it will occur. The pace of acceptance is the unknown variable.

RFID—the technology to accomplish this feat—is already available and in use by many retailers in their stockrooms and warehouses. RFID (radio frequency identification) is a technology where RFID-enabled tags, which are more powerful than conventional barcodes, are placed on merchandise. These RFID tags are basically microchips and antennae from which electronic readers can communicate with the tag from a range of approximately 30 feet (Malone, December 7, 2006). With RFID-equipped merchandise, retailers can follow a tagged item from place to place (i.e., from the stockroom to a store shelf or from the shelf to the POS checkout).

RFID technology has already been in use for decades, initially in military applications, such as tracking material in rugged and fast-moving situations. More recently, it has been used to identify vehicles passing through tollbooths so that motorists can be billed monthly rather than pay each time they pass through the booth. Only during this decade has this technology been considered as a complement and an eventual replacement for barcode technology in the retail industry. Obviously, such technology would provide great convenience for customers and labor savings for retailers by reducing the number of cashiers needed at POS checkouts.

Currently, most RFID applications in retailing have centered around the stockroom and warehouse, where pallets or truckloads are scanned at one time by an electronic reader which identifies RFID tags on the outside of each pallet. Moreover, the same technology can be used on individual items to speed up the checkout process for customers if each item is equipped with an RFID tag (Figure 3).

Benefits. Primarily, inventory management has become more automated and up-to-date when RFID technology is implemented at retail stores (Malone October 24, 2006). RFID-enabled handheld devices also significantly reduce time spent looking for misplaced items, and can signal a visual and/or audible alarm to alert security personnel of items or people entering unauthorized areas (Visich et. al. 2009; Ustundag & Tanyas 2009; *Material HandlingManagement* 2009; *Professional Services Close-Up* 2010). Additionally, RFID can replace electronic article surveillance tags now used to set off an alarm if a shoplifter tries to leave a store without paying for the merchandise (Reda 2005).

Most importantly, RFID technology can speed up POS checkouts. To demonstrate how RFID tags on individual products work at checkouts, NCR installed an RFID-enabled self-checkout at Metro Groups' RFID Innovation Center in Neuss-Norf, Germany (Collins 2004). Further, the use of RFID tags can transform retailers' supply chains and consumers' shopping experiences. RFID scanning is performed 10 times faster than barcode methods and can deliver an 11 percent reduction in labor costs based on the fewer cashiers needed at checkouts. And, the tags have been shown to decrease merchandise theft from 11 to 18 percent in stores equipped with the technology (Beckett & Kavanagh 2005).

Three aspects of RFID technology that make it a particularly attractive alternative to barcodes are that 1) the technology allows information to be read by radio waves from tags without requiring line of sight scanning, 2) the technology allows simultaneous and instantaneous reading of multiple tags in the vicinity of the reader, and 3) each

Figure 3. Example of the size of a RFID tag

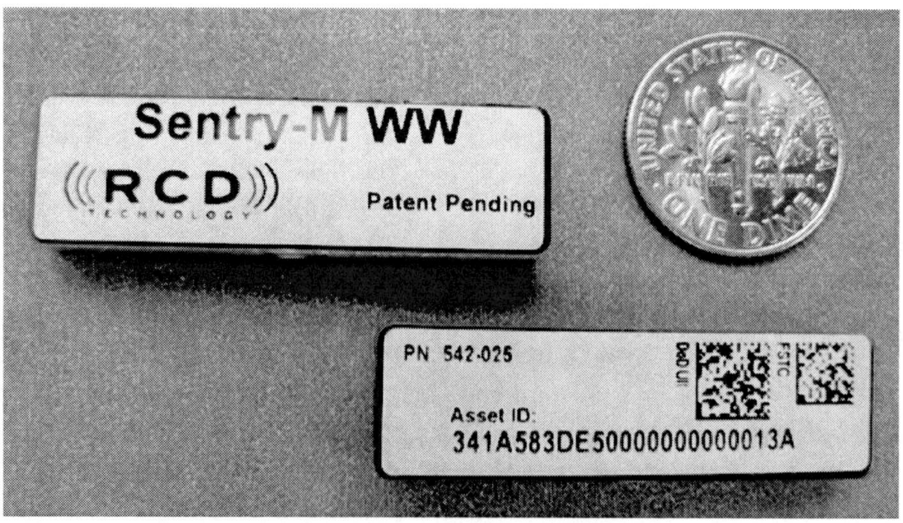

tag can have a unique code that ultimately allows every tagged item to be individually accounted for.

Drawbacks. Some retail analysts project that RFID will ultimately have more value at the item level than it does in cartons/pallets and the movement in the supply chain, but most retailers are not yet ready to make that transition (Reda May, 2006). Costs are the primary concern for retailers. However, as a critical mass is reached, costs will drop. In 2000, the most basic tags cost approximately one dollar each. Within a few years, the costs ranged from $.25 to $.40. Even at those prices, RFID tags were still not yet economical for most retailers to individually tag low-cost items in the store, such as most grocery items. However, some retailers are already using RFID tags on some expensive items, such as electronics and items prone to theft, such as CDs and DVDs.

RFID tags are also a privacy concern for some consumers. If not removed or deactivated, RFID tags can be read by any electronic reader. Will this pose such a large privacy concern for customers that they will reject individually-tagged merchandise? More customer feedback is needed.

Implementation. RFID tags are now making their appearances in retail stores on products from groceries to appliances. Liverpool, an upscale department store chain in Mexico, is already using RFID technology for merchandising and replenishment at two of the 61 stores in the chain. The store uses RFID to count inventory in the backroom and on the selling floor. Through RFID usage, Liverpool has been able to boost its inventory accuracy from 80 to 99 percent (Parks 2007).

In 2005, Mitsukoshi, one of Japan's leading luxury department stores, tested the use of RFID tags in the ladies' shoe department at its flagship store. The results were so impressive that they have moved on to a second product category—designer jeans (Reda March, 2006). Metro stores in Germany rolled out RFID to all of its stores in 2004, enlisting 20 suppliers to put tags on at least some of their products. This increased to 40 vendors in 2006, and store management is discussing with other vendors about which of their products should get tags too (McKinley 2006).

Selexyz (formerly Boekhandels Groep Nederland), the largest bookseller in the Netherlands has also implemented RFID. At each store, the RFID tag allows the store to track books on an individual title basis. The technology can track specific books to each store where the software reconciles the shipping notice with the items that have actually been received, and keeps tabs on each book as they move through the store. In fact, management can use the tags to identify books incorrectly shelved in the wrong section.

Wal-Mart implemented RFID when it required 100 suppliers to incorporate the tags into their systems in order to retain their status as a supplier for the chain. Other top global retailers, including Metro, Target, Carrefour, and Tesco quickly followed in making RFID a prerequisite for its large suppliers (Ayling 2006). However, RFID tests have now slowed and are still only pilots as retailers search for a return-on-investment. Still, Wal-Mart has increased usage of RFID in its stores tenfold, from 100 to 1,000 stores, but RFID installation in its distribution centers is behind schedule. Expansion plans at Target have also been delayed, and have been discontinued at Albertsons, a grocery store chain (Cecere 2007; Rangarajan 2010).

While the incorporation of RFID tags into retailing has not come as quickly as many had initially thought, there continue to be signs that the technology is getting closer to becoming more widely used. Estimates by one research firm, Frost & Sullivan, are that by 2011, retailers will spend more than $4 billion a year on RFID systems and tags which is up from just over $400 million a year in 2004 (*Chain Drug Review* November, 2006; *MultichannelMerchant* 2010).

Yet, widespread implementation of the technology faces several challenges. First, accuracy of the technology must be beyond reproach. The use of RFID for price scanning requires that readers be extremely reliable in providing the correct information. In fact, they need to be more accurate that the standard barcode reader to gain the greatest customer confidence (Dimmock 2003). Secondly, more products must be sent to retailers with RFID tags. There currently are too few individually-tagged items manufactured to make it worthwhile for most retailers to implement RFID scanning at checkouts. Also, retailers and suppliers still need to negotiate who is going to pay to create and affix item-level RFID tags on merchandise.

Once hailed as a technology so significant that it would revolutionize retailing, RFID has come up short in the eyes of many observers. Yet, analysts still believe that the technology will become the norm; however, they expect it to take longer than initially thought (*Chain Drug Review* 2010).

Fingerprint Authentication

New payment methods for customers using new technologies are also being tested by retailers at the POS checkout. One option, using biometric technology, involves fingerprint authentication. Biometrics is a process used to identify or authenticate an individual's identity using any of a series of physical or behavioral characteristics. These characteristics can include, but are not limited to fingerprints, hand or palm geometry, retina and iris scans, facial mapping, signature or writing style, and more recently, DNA maps (Barton, et. al. 2005).

To use fingerprint authentication technology for making payments, customers must first enroll by scanning a fingerprint, entering a search code, and designating accounts from which payments can be made. From that point, they no longer need to write checks or produce a card at the point-of-sale. They pay by touching a fingerprint scanner (*Digital Transaction News* 2007).

This biometric data are then encoded for storage, retrieval, and matching. Sensory devices, such as a fingerprint scanner, are used to collect the data (Prabhakar, Pankanti, and Jain 2003). A reference template, using algorithms, is created from unique characteristics extracted from fingerprints (Bedi & Yang 2009; Bolle, et. al. 2004), and those algorithms cannot be reconverted into an exact copy of the fingerprint (Curiel 2006). Verification occurs when a customer makes a purchase and uses their fingerprint to make the payment. The fingerprint authentication system compares the newly scanned data to a previously stored version. Payment is then accepted or rejected (Figure 4).

Figure 4. Fingerprint authentication technology allows customers to pay for purchases with only the press of a finger

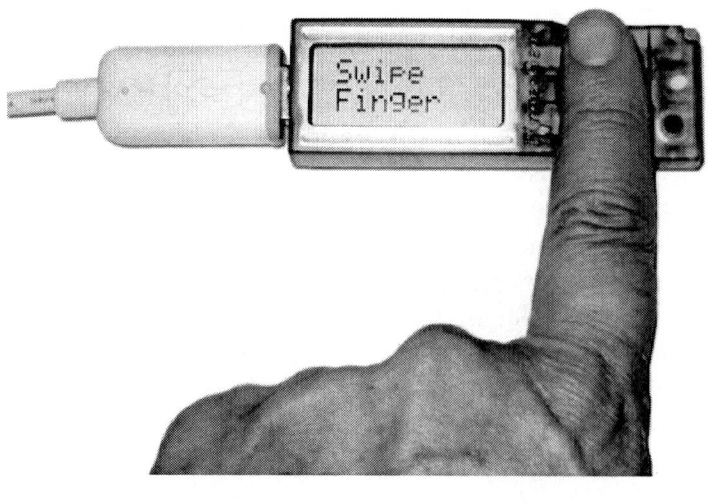

Benefits. From a consumer perspective, fingerprint authentication at point-of-sale checkouts offers several benefits. Once consumers are participating in such a system, they are potentially free from worry about the fraudulent use of their credit cards. They can make payments without carrying any cash or other identifiers, such as credit or debit cards (Wildstorm 2005). With growing consumer concerns about the incidence of bank card fraud and identity theft, a majority of global respondents (Davis and Silver 2004; *Investment Weekly News* 2009) indicated they would accept biometric authentication to verify their identities.

The benefits of fingerprint authentication for retailers, and to a lesser extent, for customers, relate to speed and better information management. The biometric checkout is designed so that the customer has to simply present a finger for identification. Shoppers no longer have to delay checkout lines trying to find credit cards or checks and providing other forms of identification. This almost instantaneous authentication can also replace a loyalty card which enhances the CRM efforts of the retailer (*eWeek* 2006). The potential sophistication of CRM data collected could allow the information to be leveraged into more sales and reduced costs for the retailer.

One of the other attractions of biometric checkout systems is the ability to link authentication with a wide range of payment options (i.e., checking, debit, or credit accounts), including several that offer much lower transaction fees for the retailer. For example, at a typical grocery store, each transaction handled by a fingerprint authentication system costs about ten cents compared to approximately seventy-five cents for a credit card (Clark 2005).

Additionally, employee authentication solutions provider, DigitalPersona Inc., recently integrated fingerprint authentication into the company's software offerings. Such systems potentially can reduce payroll costs related to time and attendance fraud and prevent unauthorized manager over rides (*Corporate IT Update* 2010).

Drawbacks. As with any new technology, fingerprint authentication technology has drawbacks. The primary concern for consumers is protection of their privacy. All organizations must be sensitive to how the public views the handling of the biometric information that it is collecting because many consumers are suspicious of how such technology works and the potential loss of privacy (Curiel 2006). Moreover, civil liberties watchdogs fear this technology gives retailers too much personal information (Elder 2005). Yet, each year, more and more of consumers' personal data are captured and stored by corporate and government interests. Retailers will need to continue to monitor just how much privacy customers are willing to relinquish.

Fingerprint authentication systems also have other drawbacks. Fingerprinting itself suffers from negative associations, courtesy of movies and TV police shows. The perception is that a person's fingerprints are only taken when they are accused of crimes. Moreover, many people simply cannot be fingerprinted. That is the case for some people with thin skin, as well as those who use cleaning chemicals extensively. Individuals who have injured their fingers even slightly can find their prints become either unreadable or altered enough to cause the system to reject the purchase. Many of these issues, however, can be addressed at the point of enrollment. Additionally, there may even be other factors that concern some consumers, such as religious or cultural taboos (Bleicher 2005).

Implementation. In recent years, fingerprint authentication technology has grown in acceptance and use (France and Selormey 2009). In fact, fingerprint verification has become the preferred biometric technology at the point-of-sale over other options such as iris scans and hand geometry because it presents the best combination of a number of factors, including cost, accuracy and size (Oehlsen May, 2005). Moreover, dramatic increases in identity theft and related crimes have probably spurred the increased commercial testing of biometric technology (Linnhoff and Langenderfer 2004; Sraeel 2005).

Many consumers may have already encountered the use of fingerprint authentication technology. One of the simplest examples of the technology is replacing a standard username/password with a fingerprint to log onto a computer. Fingerprint identification has also made its way into portable devices like PDAs—hardware that is vulnerable when lost or stolen. The appearance of fingerprint readers on cell phones has also been released in Japan and Korea, with Europe following closely behind (Kooser 2005). Disney World even uses fingerprint authentication at the park entrance for season pass holders. In the jungles and mountains of Colombia, South America, customers can go to an ATM and press their thumbs on the screen. Then, cash is dispersed. In fact, companies that manufacture automated teller machines have found budding markets for this fingerprint authentication technology all over South America where citizens already are accustomed to the use of fingerprints for general identification (Hannah 2005).

Fingerprint authentication technologies have also been tested by numerous retailers (Clark 2004; Elder 2005; Lucas 2005; *eWeek* 2006; Li and Hwang 2010). Piggly Wiggly, which has 114 grocery stores in the southeastern section of the United States, launched its biometric program in 2005, becoming one of the retail industry's largest commitments to the new technology. Albertson's, Wal-Mart, Target, and Costco have also tested similar systems. Albert Heijn, the Dutch supermarket in the Ahold group, is also investing in fingerprint authentication using Tip2Pay technology (*NACSInsight* 2009).

How rapidly will fingerprint authentication be accepted? Many retailers have a reputation for conservatism in regard to the adoption of new technologies and techniques. Too often, a primary instinct is to wait for a high level of acceptance and proof of success before purchasing and implementing new technologies (McIntosh 2009).

In 2006, the Pay by Touch system (the largest operator of fingerprint authentication systems in retail stores) was being used by more than 2,200 retail locations in the United States (Gores 2006). Usage by consumers, however, remained low. For example, management at one grocery chain where the technology was being implemented declined to say how many customers used the system during its two-year test, but indicated that the number was not significant (McKinney 2008). At another chain, using the Pay by Touch system, only 15 percent of eligible customers enrolled (Clark 2005).

In March 2008, Pay by Touch announced that it would no longer process biometric transactions on behalf of its merchant customers. It was determined that the company could no longer support the fingerprint authentication and payment system as it existed (*The State* 2008). Some other retail stores have continued their tests of fingerprint authentication with a competitive firm, but over 2,200 retailers have ceased their pilot tests—at least temporarily.

What is the future of fingerprint authentication in retail stores? According to the Federal Reserve, check writing in the United States is declining at a rate of 3 to 5 percent a year (Nelson 2005). This change would indicate that there is an opportunity for other payment options at point-of-sale checkouts. Recent survey findings (Clodfelter 2010) from young consumers (ages 18-25) indicated they already rely almost entirely on debit or credit cards when making retail purchases, and they rarely used checks. Consumers seem willing to change how they pay for purchases, and retailers will need to determine how fast consumers will be ready for the next shift in payment options—fingerprint authentication or some other new technology.

Conversely, survey findings (Clodfelter 2010) indicated that consumers may be reluctant to move to fingerprint authentication in retail stores immediately. Paying for products in a store using fingerprint authentication was ranked "last" by respondents. However, they seemed willing to accept the technology when withdrawing money from an ATM, gaining admittance to a business, or gaining access to computer files. As their comfort level with using fingerprint authentication in these areas increases, they may be more willing to accept the technology when making purchases in the future.

The greatest benefit to consumers who use fingerprint authentication systems will be increased safety and security; however, this benefit evaporates if only one or two stores are implementing the technology. Customers would still have to carry credit/debit cards to make purchases at other stores. Wider implementation by retailers will be needed to insure this promised benefit for consumers. More importantly, with a major player like Pay by Touch ceasing operations, the future of fingerprint authentication at retail stores may be more distant than initially predicted.

Increased use of fingerprint authentication systems is likely to reduce the incidence of identity theft, improve consumer convenience by eliminating or reducing password use, and lower prices by reducing fraud costs to retailers. As consumer confidence grows in the large-scale usage of fingerprint authentication, there will likely be a rapid migration toward the technology, but it will likely be customers who make that decision at their own pace.

Implications for Retailers: Solutions and Recommendations

Before making the decision to implement new technologies, retailers must determine its relative advantage. In other words, does the technology provide an advantage in either efficiency or cost-effectiveness? If potential users see no relative advantage in a new technology, they generally will not use it. However, relative advantage alone does not guarantee widespread adoption (Denis, et. al., 2002; Fitzgerald, et. al, 2002). New technologies that are compatible with the intended adopters' values and perceived needs are more readily adopted, particularly if they are perceived as simple to use. Moreover, adoption of new technologies will depend upon retailers being able to convince customers of its benefits—benefits for customers, not just benefits for the retailer (Greenhalgh, et. al. 2004).

For some retailers, installation and use of POS technology started and stopped with barcode scanners. Many retailers have a wait-and-see attitude—let others work out the problems before they make the investment in new technology. Many shoppers are also reluctant to adopt and use new POS technology. In their minds, barcode scanners are not perfect, but will new and different technology, such as RFID scanning, be any more accurate?

Several aspects of customer behavior in relation to adopting technology were revealed by previous research presented in this chapter. Retailers should consider these factors as they make the decision on whether or not to implement new POS technology. Speed—moving through the checkout quickly—is critical for most shoppers. And, perceived speed, as with self-checkouts, may be just as important to customers as real time speed.

Customers are also concerned about how easy new technologies will be to use. No matter which technology (i.e., fingerprint authentication or self-checkouts) that retailers are considering to implement, they must provide opportunities for customers to test or try out the equipment initially with the help of trained store personnel. They must overcome the shopper's fear factor in regards to using the technology.

Pricing accuracy at the point-of-sale checkout should be another critical concern when retailers adopt new POS technologies. Even today, after years of implementation and refinement, errors still occur with traditional barcode scanners. Two of the technologies discussed in this chapter (electronic shelf tags and RFID tags) would almost certainly guarantee that customers would pay the price at checkout that was posted on the store shelf. However, for both technologies, retailers fear the cost is too high to earn their desired return on investment. Particularly in the case of RFID tags, customer checkout would be improved dramatically in real time, and it would not involve the retailer having customers do the scanning, as with self-checkouts. However, due to the expense, neither technology is likely to see widespread adoption soon.

Another observation from the data collected is that retailers seem to more readily adopt technologies where there is a quick return on investment. Barcode scanners and new self-checkout terminals both pay for themselves in approximately one year—primarily through labor savings costs. Retailers can quickly see an impact on their financial bottom line through these investments. These technologies also provide a speedy checkout for shoppers—real or imagined.

The evidence would also indicate that the implementation of self-checkouts will continue at a rapid rate in retail stores, but that may be too fast for some customers. These customers may seek out stores that provide more service from humans. Retailers can use this technology to create a competitive advantage for their stores in the marketplace, but there will also be the opportunity for other retailers to create that competitive advantage through dealing with customers on a more personal level (Jones et. al. 2007)..

Another POS technology discussed in this chapter (fingerprint authentication) also has great promise for speeding up the checkout process. However, in this situation the reluctance to adopt the technology seems to be with consumers, and not retailers. The cost of the technology is not burdensome for most retailers; however, in most pilot tests, there has been extremely low usage from customers. Reasons for them not using the technology again may relate to a "fear factor." That fear may be in relation to using the technology or in giving up more of their privacy. For customers to reach the desired comfort level with this technology may take many more years.

A number of barriers will need to be overcome if the new technologies presented in this chapter are to become an integral part of the retail scene. Installation and acquisition costs of the systems will need to decline. Consumer doubts about accuracy of such systems will need to be eased. And, public concerns about personal privacy will need to be addressed.

FUTURE RESEARCH DIRECTIONS

Retailers must understand to what extent and under what conditions customers will be willing to use POS technology. Research on the adoption of any technological innovation must build on the Technology Acceptance Model (TAM) (Davis, Bagozzi, & Warshaw 1989). The model presents two constructs for forecasting the adoption of computer technology in an organizational setting—perceived usefulness of the technology and the perceived ease of use of the technology. Above all, retailers must realize that some shoppers are technology ready—they have the propensity to embrace and use new technologies for accomplishing goals. At the same time, other customers are not at that same level (Marzocchi and Zammit 2006). Future research studies should incorporate techniques from the TAM model as they attempt to predict customer acceptance of new technological innovations.

Future research will also need to rely on testing new POS technologies in the field. Before widely adopting new approaches, retailers should continue to conduct pilot tests. Much more valuable information can be obtained from customers having actually experienced a technology than having them interact with it only in a laboratory situation.

With all the technologies discussed in this chapter, retailers will need to continue to measure the impact of these innovations on customer behavior.

From the retail perspective, future research will also need to examine the impact of implementing the technology on store traffic, sales, and profitability—in other words, how does it affect return on investment.

Small field tests can also provide valuable information on the costs of installing and maintaining technology, highlight issues of employee and customer acceptance, and provide feedback on how the technology impacts the retailer's competitive position in the marketplace.

CONCLUSION AND SUMMARY

Many retailers fail and others stumble in their efforts to implement technology to enhance the POS/checkout experience for shoppers. In some situations, there may be rejection of the technology by customers. Some retailers have attempted to implement technological innovations before customers are ready to make a change. Timing of the introduction of new technology is critical. From information presented in this chapter, it would appear that most customers today see technology as playing an important role in their shopping experiences, but it must fit their shopping style. Customers must be ready to make a change before new technology will succeed.

There will continue to be innovations in retail technology at POS, and shopper expectations will continue to change. Improved levels of consumer education about the benefits of a technology and more familiarity with it by customers will probably increase adoption levels. At the same time, retailers will probably remain cautious in deciding if and when to adopt new technologies. They must be convinced that the innovations will deliver sufficient value to offset their expenses. New technologies at retail POS checkouts will be implemented in the future; the only question is how quickly that future arrives.

REFERENCES

Ayling, J. (2006, August 3). RFID revolution only just beginning. *just-food.com*.

Barr, V. (2001). Self-scanning checkout picks up speed in mass retail markets. *Stores*, August, 66-70.

Barton, B., Byciuk, S., Harris, C., Schumack, D., & Webster, K. (2005). The emerging cyber risks of biometrics. *Risk Management*, *52*(10), 26–31.

Beckett, H., & Kavanagh, J. (2005). How it works in the real world (use of radio frequency identification tags for the improvement in the efficiency of the staffs). *Computer Weekly*, March 8, 50.

Bedi, H., & Yang, L. (2009). Fair electronic exchange based on fingerprint biometrics. *International Journal of Information Security and Privacy*, (July-September): 76–109. doi:10.4018/jisp.2009100705

Bhatnagar, P. (2003, May 28). Stuck at checkout? DIY lane's open. *CNN/Money*.

Bleicher, P. (2005). Biometrics comes of age: Despite accuracy and security concerns, biometrics are gaining in popularity. *Applied Clinical Trials*, *14*(12), 18–20.

Bolle, R. M., Connell, J. H., Pankanti, S., Ratha, N. K., & Senior, A. W. (2004). *Guide to biometrics*. New York, NY: Springer-Verlag.

Cecere, L. (2007, May 2). Three years of retail RFID pilots. *Forbes*.

Chain Drug Review (2006, November 6). *RFID's day still coming, just later than expected*. (p. 37).

Chain Drug Review (2010, April 26). *RFID may yet prove to be next big thing*. (p. 88).

Clark, K. (2004). Biometrics technology touches down. *Chain Store Age*, *80*(5), 20A.

Clark, K. (2005, February). Pointing to savings, *Chain Store Age*, p. 140.

Clodfelter, R. (1998). Pricing accuracy at grocery stores and other retail stores using scanners. *International Journal of Retail and Distribution Management*, *26*(11), 412–420. doi:10.1108/09590559810246368

Clodfelter, R. (2010). Biometric technology in retailing: Will consumers accept fingerprint authentication? *Journal of Retailing and Consumer Services*, *17*(3), 181–188. doi:10.1016/j.jretconser.2010.03.007

Collins, J. (2004, August 13). Self-checkout gets RFID upgrade. *RFID Journal*.

ComputerWeekly.com. (2002, November 21). *Stores look to self-scanning to cut checkout queues*.

Corporate IT Update *(2010, January11)*. Retail Anywhere integrates DigitalPersona's fingerprint biometrics.

Creasey, S. (2009, May 9). Technology that may seem like science fiction today could become commonplace in the stores of tomorrow. *Grocer*, 46-50.

Curiel, J. (2006, June 25). The last days of privacy. *SFGate.com*, 1-6.

Dabholkar, P. A., Bobbitt, L. M., & Lee, E.-J. (2003). Understanding consumer motivation and behavior related to self-scanning in retailing. Implications for strategy and research on technology-based self-service. *International Journal of Service Industry Management*, *14*(1), 59–96. doi:10.1108/09564230310465994

Davis, D. W., & Silver, B. D. (2004). Civil liberties vs. security. *American Journal of Political Science*, *48*(1), 28–46. doi:10.1111/j.0092-5853.2004.00054.x

Davis, F. D., Bagozzi, R. P., & Warshaw, P. R. (1989). User acceptance of computer technology: a comparison of two theoretical models. *Management Science*, *35*(3), 982–1003. doi:10.1287/mnsc.35.8.982

Denis, J. L., Heber, Y., Langley, A., & Trottier, L. H. (2002). Explaining diffusion patterns for complex health care innovations. *Health Care Management Review*, *27*(3), 60–73.

Digital Transaction News. (2007, November 15). Why Piggly Wiggly decided to roll out biometric payments (pp. 1-3).

Dimmock, B. (2003, June). Will RFID replace the UPC? *Stores*, 88-90.

Elder, J. (2005, June 15). Ready to whip out your finger to pay? *Charlotte Observer*, p. 1A.
eWeek. (2006, January 30). *Consumers resist retail biometrics*.

Falkman, M. A. (2005, July). Future store shows off tomorrow's technology. *Packaging Digest*, 40-43.

Federal Trade Commission. (1996). *A report by the staff of the Federal Trade Commission, Technology Services of the National Institute of Standards and Technology, the States of Florida, Michigan, Tennessee, Vermont, and Wisconsin and the Commonwealth of Massachusetts*. Washington, DC: Federal Trade Commission.

Federal Trade Commission. (1998). *Price check II: A follow-up report on the accuracy of checkout scanner prices*. Washington, DC: Federal Trade Commission.

Fitzgerald, L., Ferlie, E., Wood, M., & Hawkins, C. (2002). Interlocking interactions, the diffusion of innovations in health care. *Human Relations*, *55*(12), 1429–1449. doi:10.1177/001872602128782213

Fitzsimons, G. J., & Lehmann, D. R. (2004). Reactance to recommendations: When unsolicited advice yields contrary responses. *Marketing Science*, *23*(1), 82–94. doi:10.1287/mksc.1030.0033

FMI SuperMarket Research. (2003). Check it out—getting the best of the front end. *FMI SuperMarket Research*, *5*(1).

Fox, B. (2002). Electronic shelf labels draw new interest as prices fall. *Stores*, January.

France, F., & Selormey, D. (2009). Biometrics improving financial accessibility. *Biometric Technology Today*, July-August, 10-12.

Garland, R. (1992). Pricing errors in the supermarket: Who pays? *International Journal of Retail and Distribution Management*, *20*(1), 25–30.

Gathright, A. (2001, October 30). Biometric technology raises hopes, fears and skepticism. *San Francisco Chronicle*, A1.

Gemperlein, J. (2006, June 18). Self-checkout? Just you wait. *Washington Post*.

Gil-Saura, I., Berenguer-Congri, G., & Ruiz-Molina, M. (2009). Information and communication technology in retailing: A cross-industry comparison. *Journal of Retailing and Consumer Services*, *16*, 232–238. doi:10.1016/j.jretconser.2008.11.018

Goodstein, R. C., & Escalas, J. E. (1995). Improving pricing accuracy at the supermarket: Electronic shelving systems and public policy. *Journal of Public Policy & Marketing*, *14*(2), 216–225.

Greenfield, E. (1992). OMR and barcode technologies: empowering education. *Technological Horizons in Education*, *19*(7), 10.

Greenhalgh, T., Robert, G., Macfarlane, F., Bate, P., & Kyriakidou, O. (2004). Diffusion of innovation in service organizations: Systematic review and recommendations. *The Milbank Quarterly*, *82*(4), 581–629. doi:10.1111/j.0887-378X.2004.00325.x

Grimes, W. (2004, April 7). When the cashier is you. *New York Times*, F1.

Grocer. (2007, March 3). Self-checkout tills win for Henderson. *Grocer*, 16.

Hannah, J. (2005). Privacy concerns, expense keep biometrics out of U.S. ATMs. *Information Week* October 12.

IBM. (2004). Satisfy customer demand for self-service and reap the benefits of self checkout. *IBM Retail Store Solutions*, 1-8.

NACSInsight. (2009, June 9). Albert Heijn invests in customer-driven IT.

Investment Weekly News (2009, November 28). Unisys research shows growing global acceptance of biometrics among consumers for protecting identities and personal information. (p. 129).

Jones, P., Williams, P., Hillier, D., & Comfort, D. (2007). Biometrics in retailing. *International Journal of Retail & Distribution Management*, *35*(3), 217–222. doi:10.1108/09590550710735077

just-food.com. (2008, April 21). Woolworths launches self-service checkouts.

just-food.com. (2009, October 22). Tesco on defensive over self-serve stores.

Karaian, J. (2008, June 2). Back to basics. *CFO Europe Magazine*.

Kim, H. Y., & Kim, Y. K. (2008). Shopping enjoyment and store shopping modes: the moderating influence of chronic time pressure. *Journal of Retailing and Consumer Services*, *15*, 410–419. doi:10.1016/j.jretconser.2007.10.003

Kiviat, B. (2008, March 24). The end of customer service. *Time*, 11.

Kooser, A. (2005). Touch and go: Fingerprint readers mean better security—and no more tricky passwords. *Entrepreneur*, *33*(3), 30.

Korolishin, J. (2004). ESL for the little guy. *Stores*, October, 71-72.

Kroll, K. M. (2006). Signs of progress. *Stores*, February, 58.

Lewis, L. (2007). A more able label? *Stores*, September, 54-57.

Li, C., & Hwang, M. (2010). An efficient biometrics-based remote user authentication scheme using smart cards. *Journal of Network and Computer Applications*, (January): 1–5. doi:10.1016/j.jnca.2009.08.001

Licata, P. G. (2004, March 7). Checking yourself out, quite literally. *New York Times*, L13.

Lindeman, T. (2005, April 26). Self-scanning takes off. *Pittsburgh Post-Gazette*, E1.

Linnhoff, S., & Langenderfer, J. (2004). Identity theft legislation: The fair and accurate credit transactions act of 2003 and the road not taken. *The Journal of Consumer Affairs*, *38*(2), 204–216. doi:10.1111/j.1745-6606.2004.tb00864.x

Lucas, P. (2005). Biometrics come into focus. *Collections & Credit Risk*, *10*(3), 18–21.

Malone, R. (2006, December 7). Can RFID invade your privacy? *Forbes*.

Malone, R. (2006, October 24). Smart Store. *Forbes*.

Management, M. H. (2009). University study finds proof of REID benefits. *Materials Handling Management*, *64*(12), 7–8.

Mander, G. (2004). Self-service schemes must meet real needs to win shopper buy-in. *Gartner*, *G2*(June), 1–7.

Marzocchi, G. L., & Zammit, A. (2006). Self-scanning technologies in retail: Determinants of adoption. *Service Industries Journal*, *26*(6), 651–669. doi:10.1080/02642060600850790

McIntosh, D. (2009). Biometrics—a fad or the future? *Biometric Technology Today*, May, 9-12.

McKinley, E. (2006). From pallets to cartons. *Stores*, July, 74-76.

McKinney, M. (2008, April 26). Pay By Touch's backers claim fingers burned. *StarTribune.com*, 1.

MultichannelMerchant. (2010, June 1). *Bakers labels barcoding a success*.

Muller-Seitz, G., Dautzenberg, K., Creusen, U., & Stromereder, C. (2009). Customer acceptance of RFID technology: Evidence from the German electronic retail sector. *Journal of Retailing and Consumer Services*, *16*, 31–39. doi:10.1016/j.jretconser.2008.08.002

National Conference on Weights and Measures. (1995). *Examination procedure for price verification*. Gaithersburg, MD: National Institute of Standards and Technology.

National Petroleum News. (2006). NCR creates "mini" self-checkout solution. *National Petroleum News* March, 27.

Nelson, S. (2005). Finger on the solution. *Stores*, July, 69-70.

Oehlsen, N. (2005). Putting a finger on biometrics. *Credit Card Management*, May, 46-50.

Opara-Nadi, G. (2005). Electronic self-checkout system vs cashier operated system: A performance based comparative analysis. *Capella University*, June.

Pantano, E. (2010). New technologies and retailing: Trends and directions. *Journal of Retailing and Consumer Services*, *17*, 171–172. doi:10.1016/j.jretconser.2010.03.004

Pantano, E., & Naccarato, G. (2010). Entertainment in retailing: The influences of advanced technologies. *Journal of Retailing and Consumer Services*, *17*, 200–204. doi:10.1016/j.jretconser.2010.03.010

Parks, L. (2007). Tracking jeans in Mexico. *Stores*, March, 35-37.

Prabhakar, S., Pankanti, S., & Jain, A. K. (2003). Biometric recognition: Security and privacy concerns. *IEEE Security & Privacy*, *1*(March/April), 33–42. doi:10.1109/MSECP.2003.1193209

Professional Services Close-Up (2010, May 31). Global options group: Bode Technology releases RFID system.

Puccinelli, N. M., Goodstein, R. C., Grewal, D., Price, R., Raghubir, P., & Stewart, D. (2009). Customer experience management in retailing: Understanding the buying process. *Journal of Retailing*, *85*(1), 15–30. doi:10.1016/j.jretai.2008.11.003

Ram, S., & Jung, H.-S. (1991). Forced adoption of innovation in organizations: Consequences and implications. *Journal of Product Innovation Management*, *8*(2), 117–126. doi:10.1016/0737-6782(91)90005-J

Rangarajan, T. S. (2010). Enhancing supply chain management using RFID. *Material Handling Management*, *65*(5), 22–25.

Reda, S. (2005). What you don't know about RFID! *Stores*, September, 26-34.

Reda, S. (2006). Reading the RFID tea leaves. *Stores*, May, 46-50.

Reda, S. (2006). Stepping up the RFID effort. *Stores*, March, 26-30.

Reinders, M. J., Dabholkar, P. A., & Frambach, R. T. (2008). Consequences of forcing consumers to use technology-based self-service. *Journal of Service Research*, *11*(2), 107–123. doi:10.1177/1094670508324297

Robertson, N., & Shaw, R. (2009). Predicting the likelihood of voiced complaints in the self-service technology context. *Journal of Service Research*, *12*(1), 100–116. doi:10.1177/1094670509333789

Roussel-Dupre, S. (2002, February 1). Self-service reality checkout. *Integrated Solutions for Retailers*.

Schulz, D. P. (2008). 10 things you may have missed. *Stores*, March, 20.

Schuman, E. (2004). Self-service need not be self-delusional. *Storefrontbacktalk.com*, August.

Sethuraman, R., & Parasuraman, A. (2005). Succeeding in the big middle through technology. *Journal of Retailing*, *81*(2), 107–111. doi:10.1016/j.jretai.2005.03.003

Sheldon, J., & Buzek, G. (2006, July 20). *Market study: 2006 North American self-checkout systems*. IHL Consulting Group.

Sraeel, H. (2005). Hot on the ID theft trail: Will the real Jane Doe stand up? *Bank Technology News*, *18*(2), 8.

ChainStore Age (2005, February). *Fast growth seen in ESLs, RFID*, (p. 50).

Sun, T., Zixue, T., & Tsai, K. (2009). The role of interdependent self-construal in consumers' susceptibility to retail salespersons' influence: A hierarchical approach. *Journal of Retailing and Consumer Services*, *16*, 360–366. doi:10.1016/j.jretconser.2009.04.002

The State. (2008, March 27). Piggly Wiggly ends Pay By Touch service. *The State*, B7.

Timmor, Y., & Rymon, T. (2007). To do or not to do: The dilemma of technology-based service improvement. *Journal of Services Marketing*, *21*(2), 99–111. doi:10.1108/08876040710737868

Toppo, G. (2006, February 23). Eye scans: A high-tech hall pass? *USA Today*, 12B.

Ustundag, A., & Tanyas, M. (2009). The impacts of radio frequency identification technology on supply chain costs. *Transportation Research Part E, Logistics and Transportation Review*, *45*(1), 29–39. doi:10.1016/j.tre.2008.09.001

Visich, J., Li, S., Khumawala, B., & Reyes, P. (2009). Empirical evidence of RFID impacts on supply chain performance. *International Journal of Operations & Production Management*, *29*(12), 1290–1315. doi:10.1108/01443570911006009

Webb, W. (2002, November 14). On-the-shelf electronic price labels. *EDN*, 36-40.

Weber, M. M., & Kantamneni, S. P. (2002). POS and EDI in retailing: An examination of underlying benefits and barriers. *Supply Chain Management*, *7*(5), 311–317. doi:10.1108/13598540210447755

Weijters, B., Rangarajan, D., Falk, T., & Schillewaert, N. (2007). Determinants and outcomes of customers' use of self-service technology in a retail setting. *Journal of Service Research*, *10*(1), 3–21. doi:10.1177/1094670507302990

Weijters, B., Schillewaert, N., Rangarajan, D., & Falk, T. (2005). *Customers' usage of self service technology in a retail setting*. Vlerick Leuven Gent Management School Working Paper Series 2005/19.

Wildstorm, S. H. (2005). New weapons to stop identity thieves. *Business Week, 3924*, 24.

ADDITIONAL READING

Aggarwal, P., & Cha, T. (1997). Surrogate buyers and the new product adoption process: conceptualization and managerial framework. *Journal of Consumer Marketing, 14*(4-5), 391–201. doi:10.1108/07363769710179419

Alster, N. (2005). A touchy subject. *CFO, 21*(5), 39–42.

Bagozzi, R. P., Davis, F. D., & Warshaw, P. R. (1992). Development and test of a theory of technological learning and usage. *Human Relations, 45*(7), 660–686. doi:10.1177/001872679204500702

Brass, L. (2003). Firms dealing with fingerprint technology see growing demand. *Knight Ridder Tribune Business News,* December 1.

Burke, R. (2002). Technology and the customer interface: What consumers want in the physical and virtual store. *Journal of the Academy of Marketing Science, 30*(4), 411–432. doi:10.1177/009207002236914

Caruso, J. (2004). Biometrics benefits, challenges aired. *Network World* November 1, 17.

Chain Drug Review (2006). Biometic systems next big thing, April 10, 62.

Clark, L. (2006). Spar to trail self-service checkouts. *Computer Weekly*, November 28.

Coupe, K. (2005). Inventory reduction and self-checkout take center stage at the European food-business show. *Retail Technology Quarterly*, July, 42A-43A.

Davis, F. D. (1989). Perceived usefulness, perceived ease of use, and user acceptance of information technology. *Management Information Systems Quarterly, 13*(3), 319–339. doi:10.2307/249008

Deffuant, G., Huet, S., & Amblard, F. (2005). An individual-based model of innovation diffusion mixing social value and individual benefits. *American Journal of Sociology, 110*(January), 1041–1070. doi:10.1086/430220

Duta, N. (2009). A survey of biometric technology based on hand shape. *Pattern Recognition,* (November): 2797–2907. doi:10.1016/j.patcog.2009.02.007

Fiches, M. (2003). Biometrics: The future is today. *Access Control & Security Systems, 46*(7), 20–26.

Gores, P. (2006). One-touch shopping, *JSOnline*, September 9, 1.

Greene, M.V. (2006). Fresh approach to time & attendance. *Stores,* April, 80-82.

Howell, D. (2006). The new customer service. *Chain Store Age, 82*(11), 114.

Langenderfer, J., & Linnhoff, S. (2005). The emergence of biometrics and its effect on consumers. *The Journal of Consumer Affairs, 39*(2), 314–338. doi:10.1111/j.1745-6606.2005.00017.x

Madore, J. (2003). Self-scanning saves time but hits checkout-lane sales. *Newsday*, October 2, 1

Middlemiss, J. (2004). Biometrics add security in insecure times. *Wall Street & Technology,* March 26, 42-46.

Network World (2009). Fingerprints not enough for future government security systems, September 24.

Network World (2009). Voice biometrics haven't caught on just yet, September 4.

Norton, J. A., & Bass, F. M. (1987). A diffusion theory model of adoption and substitution for successive generations of high-technology products. *Management Science, 33*(9), 1069–1086. doi:10.1287/mnsc.33.9.1069

Shulman, R. (2005). Scanning the Horizon. *Progressive Grocer*, *84*(1), 24–26.

ChainStore Age (2007). Checking out self checkout July, 58

Straub, E. (2009). Understanding technology adoption: Theory and future directions for informal learning. *Review of Educational Research*, *79*(2), 625–650. doi:10.3102/0034654308325896

KEY TERMS AND DEFINITIONS

Barcode Scanning: The use of technology to scan UPC codes at retail checkouts to determine a product's price.

Biometrics: Biometrics is a process used to identify or authenticate an individual's identity using any of a series of physical or behavioral characteristics. These characteristics can include, but are not limited to fingerprints, hand or palm geometry, retina and iris scans, facial mapping, signature or writing style, and more recently, DNA maps.

ESL Tags: Display units that can be positioned on store shelves next to each product being sold to electronically display product information, such as the retail price. The display units and the store's POS cash registers are both connected to the same price file; therefore, the price that the customer sees on the store shelf is exactly what they pay at the register when they check out.

Fingerprint Authentication: Technology that uses a customer's fingerprint to authenticate their identity for various purposes—i.e., logging on to computers, entering a business, or making payments for retail purchases.

Radio Frequency Identification (RFID): Is a technology where tags more powerful than conventional barcodes, are placed on merchandise. These tags are basically microchips and antennae from which electronic readers can communicate with the tag from a range of approximately 30 feet. With RFID-equipped merchandise, retailers can follow a tagged item from place to place.

Self-Checkout: A process that involves shoppers using technology to scan, bag, and pay for their merchandise without the assistance of a cashier.

Chapter 2
The Evolution Tornado Retail

Bernd Hallier
EHI Retail Institute, Germany

ABSTRACT

Retail has changed dramatically within a period of the last 60 years. Most obvious for the consumer is the physical outlet, which has developed from a small-size Mum and Papa service store to big hypermarkets as a self-service offer. While the assortment just after World War II consisted only of basic food, during the next decades the product-range exploded due to more and more convenience for the consumers and a broad segmentation of tastes, package-sizes and me-too-choices. The Western part of Europe became an affluent society.

Less seen by the consumers but more by the experts and some dedicated academics is the change of the backstage in retail. The strategic tool to cope successfully with mass-distribution was the introduction of IT-systems in the 1970s. The key to control the flow of the individual product was the product-bar-code used by scanner at the cashier – later also enlarged by chips to control by Efficient Consumer Response (ECR) all the Total Supply Chain.

No attention at all was paid to the evaluation of philosophies offered by the steady upgrade of retail-technologies. While the period 1970/80 was still the push period, when the consumer industry wanted to push the penetration of its products the outlet was the "point of sales" (POS); in the 90ies due to ECR the outlet was rediscovered as the "point of purchase" (POP) with the buying decision and shelf-optimization as a central point; in the last decade big players like WalMart, ALDI, REWE pushed their outlets to be the "point of differentiation" (POD) to gain a Unique Sales Position (USP) in the market. The next big "technological jump forward" will be the intertwining of Facebook, YouTube, and Twitter with new media of retailers. Consumers can gain much more impact onto the listing of products, onto services within a store. It might be the time when the outlets become a "point of consumers" (POC) again.

DOI: 10.4018/978-1-60960-738-8.ch002

INTRODUCTION

Production, distribution and consumers can be defined as a Total Supply Chain, which is one of the most penetrated academic and operational views at the moment – but it mostly describes only the status quo or is a version of technical cooperation between suppliers and distributors (retail industry).

Not yet deeply analyzed is the evolution of production, distribution and consumer as a sustainable evolution over longer time-periods reflecting the introduction of a joint development of those three stakeholders. Innovation-cycles mostly do not emerge separately from one string of those three stakeholders alone, but are in a permanent interaction of influence. Only if this interaction between those three players is successful, then the market will experience a new wave of innovation.

The parameters of this thesis is the history of innovation of the total supply chain in Western Europe; time-cycles in America, in Southern Europe, South East and East Europe are different – for example due to consumer behaviour or simply due to the fact that in the former communist countries modern thoughts could only enter after the lift of barriers between East and West.

The contribution is not aimed to reflect the scientific literature of this academic field but focuses applied science pushed by the author in its 25 years function as a Managing Director of the ISB, DHI and EHI research-institutes and additionally as President of the world leading exhibition for retail-technology "EuroShop". The background of today's EHI are more than 2,000 experts of the Total Supply chain exchanging their know-how and creating standards. Subsidiaries of EHI are among others GS1 Germany (50 percent), Orgainvent (50 percent) and GlobalGAP (100 percent).

BACKGROUND

Innovation Waves

In macro-economics Nikolay Kontratjeff was the first to describe innovation waves/cycles (Kondratjew, 1926); on the company-level it was A. Schumpeter who analyzed the life-cycles (Schumpeter, 1961). In between is the retail industry as a segment of a national/international economy – the first descriptions of 25-year-long cycles starting from the year 1800 were described in 2004. After World War II those innovations had been in roughly 1950 the introduction of self-service/supermarkets, in 1975 the shopping centres and category killers like IKEA, ToysRus and in the year 2000 the internet with B2B and B2C (Bauer and Hallier,1999; Hallier, 2004).

That rough pattern of course can be broken down into more detailed facets for the total retail industry – also as a bench-mark for the individual company performance.

The Mum and Papa Service Stores

For Western Europe a new Chapter of Innovation in retail started after World War II. (ISB, 1988; Hallier, 2001; Hallier, 2004).

The 50ies of the 20th century in Germany for example the characteristics of the food sector had been:

- stores belonging mainly to consumer's cooperatives
- with very atomistic influence onto the suppliers mostly in walking distances of some minutes for the consumers
- organized instore in service concentrating to pack sugar, beans, butter, margarine from bulk-delivery into the quantities demanded by consumers
- with a limited assortment of roughly 200 products

- offered by stores of about 30 to 40 square meters
- in towns quite often split into specialist-shops like milk-shops, fruit and vegetable shops …

As the products were not "pre-packed" of course no real branding could happen for most of the products. Branded goods like "Oetker", "Bahlsen" were in minority. The consumer were guided by the oral advice of the shop-owner/-manager. Due also to small flats and without cooling-opportunities for butter, milk etc. consumers visited daily their neighbourhood-store. The outlet was not only supplying the consumers with products – but was within a limited radius a social communication centre.

The Introduction of Self-Service

In the 50ies and 60ies the traditional service-stores were challenged by "supermarkets" from America (ISB, 1988; Hallier, 2001). The characteristics of those markets is "self-service" which implies:

- products have to be pre-packed either in a depot of the retailer or directly from the supplier.
- The oral "recommendation" of the retailer has to be substituted by design and descriptions on the packed product or by in store advertising either at shelf or special presentation zones.
- In consequence the shelves within the stores had to be accessible
- The selection of products and the action of payment in the cash-zone were split
- In general the cash-register was technically organized into "product-groups" with the result that payment was quick on the one hand side but also with the bad effect for the store not to have product-data.
- Generally the costs for personnel for the retailers decreased due to less man-power
- Self-service became more competitive in comparison to service-stores

According to statistics from the original roots of the EHI Retail Institute the first 80 self-service stores in the Federal Republic of Germany had

Figure 1. Andy Warhol: Coca Cola

a size in-between 35 and 330 square meters in 1952; for 1958 the data of a panel for the size of the assortment shows about 1000 articles – 100 of them in non-food-segment.

On the Way to Mass-Distribution

The phenomenon of the late 60ies and the 70ies is mass-distribution like characterized by Andy Warhol in his Coca Cola (Figure 1) (Hallier, 2004).

Mass-production and mass-distribution are the symbols of democratic consumption: "the president of the USA is drinking Coca Cola – you can drink Coca Cola – Coca Cola is on the shelves of your supermarket!"

Due to the macro-economic recovery of Western Europe and prosperity within the Western World the personal income of consumers increased - followed by more demand – which was reflected by a growth of retail-space and a segmented product-assortment. EHI registered in 1966 still 0.25 million square meters sales-space growing in 1976 to 4.6 million and reaching 1986 about 6,7 million square meters. The potential size of the individual stores seemed never ending: a hypermarket near Cologne/Germany opened with about 35,000 square meters (25 years after the stores-size average of 35 square meters)

The store-size of course also reflected the explosion of the assortment: contrary to 200 articles at the Mum and Papa-store and 1000 articles at the first self-service store now hypermarkets offer on average 40,000 articles. The effect in the relationship between product-range and the consumer as a human being is characterized in the "99 cent" picture of A. Gurski, where within a nearly unlimited assortment only the discount price can be seen- and the consumer only plays the role of an "underdog" in the Figure 2 (Hallier, 2009).

Nevertheless was and is the consumer not only the target, but he is also the driver of the development.

- Due to an increasing mobility of consumers those big-size-markets and Shopping-Centers could be established at the suburbs.
- As consumers live in bigger flats than in the 50ies they can store more products

Figure 2. Andreas Gurski/99 cent

- Also their technical equipment at home (refrigerator) enables them to buy bigger units for longer consumption-periods
- Last but not least they changed their consumer-habits with the trend for more convenience. While in the 50ies they bought for example milk to keep it for three days in a bowl to produce their own yoghurt - now they select between low or high calories, between different fruit-tastes, between other ingredients. Another trend of modern consumption is to have seasonal fruit available all over the year as global supply offers "spring" also when he/she at the consumers's home is experiencing just "autumn".

Additionally the consumer stopped since the 70ies his habit to be loyal like at the Mum and Papa-time: he is using multi-choices! Self-services allowed him to become anonym.

Also his opposite – the retailers – became anonymous. Due to self-service the store-owner could multiply his stores. From atomistic local retailers regional spider-networks/chain stores derived and in the 70ies names of cities/areas became synonyms of the retailer-network: Würzburg-Kupsch, Frankfurt-Latscha, Cologne/Köln – Stüssgen. Those chain-store-operations generated a certain win-win-situation between retail and suppliers: large scale economics. In hot-pursuit of retail expansion and to care for the danger of out-of-stock situation suppliers together with specialized agencies created in the 70ies POS-display-promotions. In the marketing-terminology this period remains as the "push-strategy". Buying much from the supplier created rebates for the retailers which again could be passed on to the consumers to decrease prices and thereby again stimulated increasing demand and growth for those retailers with the lower price.

From POS to Point of Purchase (POP)

The permanent increasing speed of new articles from the producers, from new stores from retailers, also from store-segmentation and store-diversification on retail-level, created the need for data-management not to run out of control of the situation. Consultants and technical suppliers like IBM, Nixdorf started in the beginning of the 70ies to provide information-systems by electronic data-machines; retail got into a strong correlation with the technical development of hardware and software industry. In 1974 for example in Germany EHI and the Branded Goods Association (Markenverband) created as a joint-venture the German Accreditation for bar-coding (European Article Number) which had quickly cross-border-partner organizations. The name "Coorganization" in Germany reflects the willingness of suppliers and retailers to work together. The scanner at the cash-zone of the retailer-outlet brought back to the retailer information about the sales of individual products on store-level. Next steps in the 80ies referred to the optimization of the shelf-space via international work-groups concerned with projects like "Direct Product Profitability (DPP/DPR)" and in the 90ies "Efficient Consumer Response (ECR)" (ISB, 1980; Hallier, 1987; ISB 1989; Heidel, 1990; DHI, 1992; EHI, 1994; Behrends, 2001). Especially the ECR shows that in the center of the total supply chain the consumer's shopping basket was placed again: this meant a swing from the "push-action" to a "push-strategy"! Strategically retail got empowerment by those marketing-tools and gained the dominance over the suppliers (Figure 3) (Hallier, 1995).

The ultimate focus to the purchasing consumer of course is the link between product-data from the scanners in combination with loyalty-cards which provide individual consumer-data; this is a development starting in the USA-market at the 80ies and in Europe with a time-long of twenties years in the beginning of the century.

Figure 3. The empowerment of retail by marketing-tools

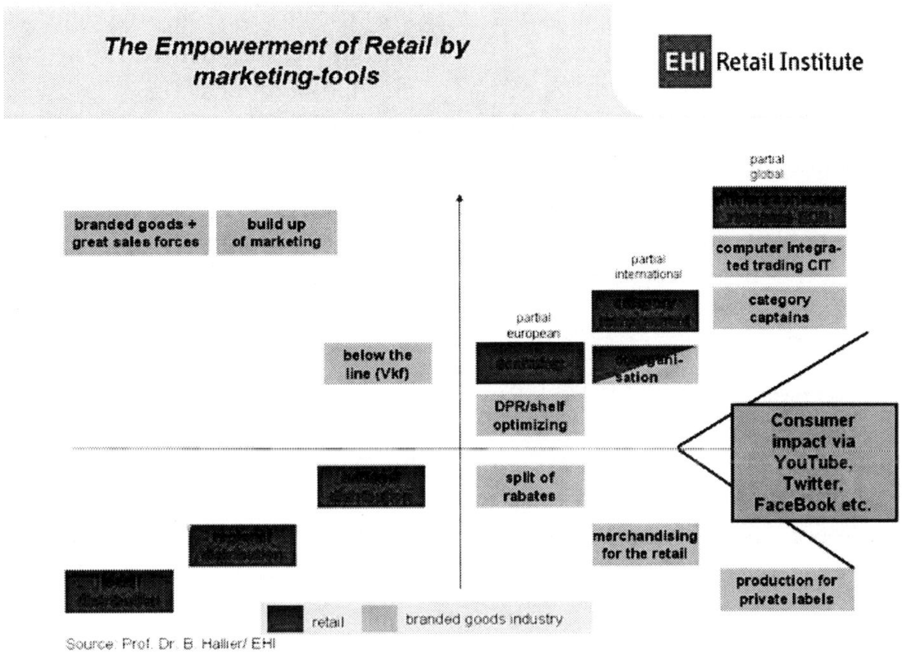

Also politically already the 80ies saw more influence of the consumer towards the total supply chain. One symptom was in Germany the "Green Party" demonstrating against Coca Cola-tins in nature, the waste in general, the burning of garbage in densely populated areas. The German Federal Government reacted by the appointment of a Minister of Environment. He established regulations downsizing one-way-package and to promote multi-trip; he started to add to the term "distribution" the term "redistribution"; he redefined "waste" into "reusable resource". He – Prof. Dr. Klaus Toepfer – initiated the Kyoto-Protocol for sustainability and decrease of CO2-emissions. For his packaging regulations in Germany Toepfer discovered the market-power of the retailers; he threatened to punish them – and forced the retailers by this action to influence the suppliers. DHI/EHI supported the initiative to create more awareness and helped to bridge the political vision and the applicability at retail and suppliers by workshops and the publication of five monographs dealing with this topic (DHI, 1991; EHI, 1993).

In the 90ies another factor of change were several food-scandals like the British Cow Disease. Consumers in Germany were afraid that the meat at the stores came from the UK! Beef at that time (first BSE-crisis in 1994; second crisis in 1996) was an anonymous product – sometimes passing up to 10 different stages of farmers, animal wholesalers, slaughter houses, cutting houses, meat wholesalers, retailers. Again in Germany retailers became pro-actively before national/ regional governments the drivers for innovation for the Total Supply Chain (Figure 4). Organized by EHI in 1994 meat-experts meat suppliers and defined a system of tracking and tracing of meat and organized the control in a joint-venture of EHI and the farmers' CMA (Central Marketing Agency) called Orgainvent. Parallel government introduced ear-marking of cows. In 2002 the EU took over this facultative system as a mandatory EU-regulation (EU 178/2002) (EHI, 1997; EHI, 2001; EHI, 2005).

That tracing and tracking activity demonstrates an important shift for the academic discussions defining "retail-institutionalized" or "process-oriented": the EHI-retailers acted also on behalf of their wholesalers (or their own wholesale-activity) and the suppliers-modern retailers are part of a supply chain. This special role of a driver for innovation for new thinking on behalf of consumers' interests also is reflected by the EHI initiative to stimulate "good agricultural practice" first by "EUREPGAP" and later by the enlarged "GLOBALGAP" as well as lately also by the initiative "Environmental Retail Management" (EHI, 1999; EHI, 2000; 2008; www.european-retail-academy.org/ERM).

Summarizing those trends and initiatives it can be stated that the Point of Purchase (POP) – thinking gives more weight to the consumers than the POS-thinking, but it also shows the gaining influence of Non Government Organizations (NGO) like the EHI Retail Institute, Orgainvent or GlobalGap onto the Total Supply Chain and legislation.

Taking over responsibility not only for the retail-level but for all the Total Supply Chain the backstage becomes involved into business of wholesale, packaging, processing, agriculture, watering and pesticides etc. Retail is drifting strongly away from its definition as "institutionalized retail" towards a definition of "process-oriented retail". Due to the company size of the top-players and the global activities of the retailers the consumers keep the retailers much more reli-

Figure 4. Round table for demand and supply (putting the idea into practice)

able than in the past. Responsibility becomes also part of the competitiveness and communication.

EHI Retail Institute created for the communication with third partners in July 2008 in the internet the environmental platform: www.european-retail-academy.org/ERM and introduced a three years research-program with the title "Environmental Retail Management "(News of August 15th 2008 at www.european-retail-academy.org/ERM). One of the first steps was a theoretical benchmark-project with the title Environmental Retail Flow Chart (News December 19th 2008). Taken all environmental oriented efforts concerning the buying, distributing, marketing of products along the Total Supply Chain as 100 percent – the retailers could judge themselves how much of their efforts in percentage of that total would concern each of the individual steps within the Total Supply Chain.

A comparison of that self-check by several retailers shows that the emphasis by the individual players is in 2009 still very different (Table 1).

Even if on the horizontal level there is the same percentage with two companies the action behind can be very different. For example might on the agricultural level Retailer 1 claim that he is acting environmentally because he is pushing the GlobalGAP-standard, Retailer 2 is promoting integrated production, Retailer 3 organics, Retailer 4 helps Fair Trade, and Retailer 5 is sourcing locally. The conclusion from that Environmental Flow Chart in the beginning of 2010 is that there is a big need to work out methodologies to compare and evaluate the activities of the competitors. And by this claim retailer becomes partner of scientific research. This view is also shared by Peter Jones, Daphne Comfort and David Hillier in the "Case study of the UK's Leading Food Retailers".

Table 1. Environmental flow chart

Level	%	%	%	%	%	%
Agriculture	30	-	25	5	17	-
Processing/ packaging	20	25	10	15	13	-
Building (depots/outlets)	10	-	15	35	20	50
Shopfitting/ processing	20	-	15	5	21	40
POS/Advertising	10	25	20	5	16	5
Consumer Lifestyle	10	50	15	35	13	5
Total	**100**	**100**	**100**	**100**	**100**	**100**

The Point of Differentiation (POD)

Due to the change of the atomistic retail structures to oligopolies on the national or even to a certain degree European level retailers are more and more competitors in one region, in one city. When in Germany in 1980 the first 5 top-ranking food retailers together had a market-share of about 26 percent – now in 2010 their share of the total distribution is together at about 80 percent (Figure 5).

The market power has shifted within the Total Supply Chain from the production towards retail- and most probably will give the consumers more and more influence in the future. It has to be kept in mind that at the moment the annual turnover of the US-retailer WalMart has the equivalent size of the GNP of Switzerland! Under the hypothesis of EHI (based by annual discussions with retailers) that retailers invest about 1 percent of this turnover into the IT-sector of their company, this means an annual investment of WalMart of 3 bln. Euro to follow-up its daily transactions (Table 2).

The problem of oligopolies is to find ways for differentiation in the eyes of the consumers! The ECR-models of the 90ies and the beginning of this century are quite often based on industry-data linked to socio-demographic data of regions. (Hallier, 1997; Hallier, 1999). If assortments are optimized in that way it has the following effects:

- the assortments of the retailers in one area get all the same profile; the customer in the end does not see any differences of the stores of the retail competitors

- according to a standard rule of experience 80 percent of the turnover is covered by 20 percent of the articles. In consequence this means for higher profits on retail-level assortments should be cut down to discounter-sizes.
- If the assortment is the same - then the prices are in the focus; again a factor of discount.

As in the globalizing world with theoretically unlimited free trade there is an over-supply of products the power-swing enables retailers to put pressure onto the suppliers to decrease the production-prices or to increase margins – but the retailer is "mis-using the low price for discounting himself – eroding his profits too! Such a "downswing tsunami" can be watched at the agricultural level in Germany in 2009 – which ended in October 2009 by a subsidy of 280 million Euros from the EU-Commission alone for the milk-farming to give farmers a chance for survival of their production line! Consumers who perhaps had been happy at first about low milk-prices will be confronted later to pay higher taxes to fill that deficit in the EU-budget in the end.

To get out of the downswing of low prices there could be three tactical tools:

Table 2. Potential IT-budget of retail players. 1 Percent of annual turnover invested in electronic data processing for selected companies.

Company	Absolute Amount
WalMart	2 850 million dollar
Carrefour	790 million dollar
Metro	600 million dollar
Auchan	320 million dollar
Coley Myer	180 million dollar
Number 10 Germany	50 million dollar
Number 20 Germany	16 million dollar
Number 1 Turkey	11 million dollar

Figure 5. Concentration in German food retail

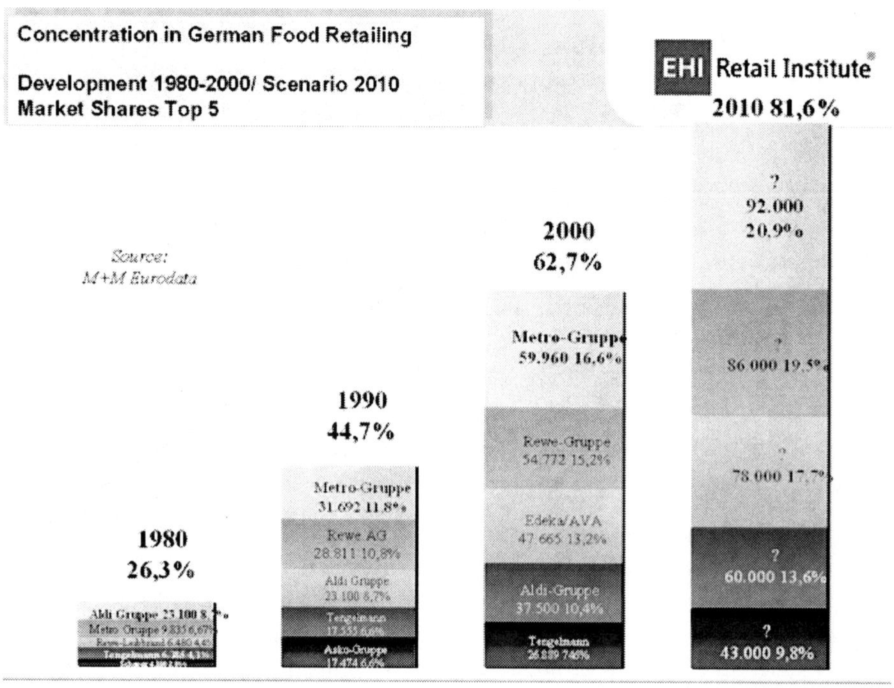

- more exclusive promotions with suppliers with an increase of value
- more private labels
- more branding of retail with soft-factors

At least it can be stated that the situation is not the same everywhere in Europe: for example concerning the margins/profits the UK is much better off than Germany in food-retail; in Croatia the national coop is still market-leader, while in Poland among the top-ten food retailers there is no original Polish company any more since 10 years after the lift of the borders; a company like IKEA created an own life-style-concept and split by this in the awareness of the consumers from its former competitors and is able to establish its own incomparable pricing-strategy

Future Innovation-Concepts

Technology after the introduction of bar-codes and scanners in 1975 now changes roughly from the year 2000 onwards the interactive relationship between retailers to customers again but also retailers to producers. Some of the most important reasons:

- The introduction of the internet is in mail-order-business substituting the printed catalogue. The bankruptcy of Quelle in Germany in 2009 shows that for example this company did not change quickly enough like its competitor Otto-Versand *(Hallier, 2010;Mattmüller, 2010).*
- But internet is also challenging the traditional brick-and-stone business. If there is a speed of 1 percent growth of market-share in the internet-business - then within that 25 years innovation cycle a total of 25 percent is lost for the traditional retail channels. Other (negative or positive) scenarios of growth are of course possible too!
- Retailers therefore change market-concepts for example to increase convenience in the processing of food-preparation towards ready meals in their traditional supermarkets. New vocabularies are invented like "meal-assembling"
- Retailers organize company-crossing loyalty –cards not only in an effort to fix the consumer to their stores but "to catch the fish" in a network of segment-specialists.
- But retailers use those data of course also generally for market-research. It might be questioned in how far they substitute applied scientific work of traditional market-research-agencies in the future.
- A good example for a retail-outlet to test about 50 technical innovations under real buying environment is the Metro Future Stores 1 and 2. *(Hallier, 2004;Hallier, 2009).*
- The second Metro Future Store now is testing mobile shopping concepts by the mobile phones of the consumers – which might have quickly an enlarged market-share (at least already now to be seen in countries like South Korea or even China).
- Other technologies being used are electronic chips which enable RFID or WLAN to optimize the supply flow or availability of products at the shelf.
- The Internet is also changing the retailer-supplier relationship by B2B-paperless communication. Suppliers which are not able to compete in those technologies are getting listed out.
- The Internet and RFID enable also to enforce tracing and tracking of animals or demands for good agricultural practice with decreasing chemicals. In those cases retail becomes an agent on behalf of the consumers or at least segments of them.
- The Internet allows the consumer not only more transparency about product-prices or the above tracing of the production-lines but also enables him to become a "retailer" himself/herself by systems like Ebay.

- Last but not least consumers can gain bigger impact onto retail and suppliers by systems like Facebook, You Tube, twitting. They can evaluate products and stores and use their influence by consumer-networks.
- Retailers on the other hand could use those platform for dialogue, building up initiatives among such communities …
- … even offering "public space" within their stores for those communities. For example it might be thinkable that in evening or on Sundays some space within the store might be available for local "amateur-cooks".
- A retail outlet would get partly event – character as a potential stage for the nearby community

In general technology will gain great influence onto future store concepts – but also delivering many alternative options to segment its marketing-offers. While in the concept of POD there are mainly the two players' suppliers and retailers to determine the product-range, the visual merchandize, the service, now new technology opens the game also for more direct influence of the consumer via modern media. In the terminology of the former chapters the future could perhaps be called POC standing for "point of communication" or in the extreme even "point of customer" as the dialogue with him will be in the focus of retailers and suppliers (Figure 6).

MAIN FOCUS OF THE CHAPTER

EuroShop: Market Place of Innovation

When in 1964 the ISB-Institute für Selbstbedienung and Messe Düsseldorf signed the contract for the promotion of a fair with the name of "EuroShop" the transformation of the traditional service-store in Europe into the self-service stores of the US was in the focus of the intended action.

Exhibition Waves

Starting in 1966 for the first time the exhibition EuroShop had covered in its first innovative "wave" the following topics:

- shopfitting and equipment
- window displays
- refrigerators
- cash registers

Based on this core-business the second wave in the 70ies consisted out of solutions for:

- sales promotion/visual merchandising
- lighting
- stand design/construction

The third wave in the 80ies was concerned with the tools of information

Figure 6. From bilateral to trilateral dialogue

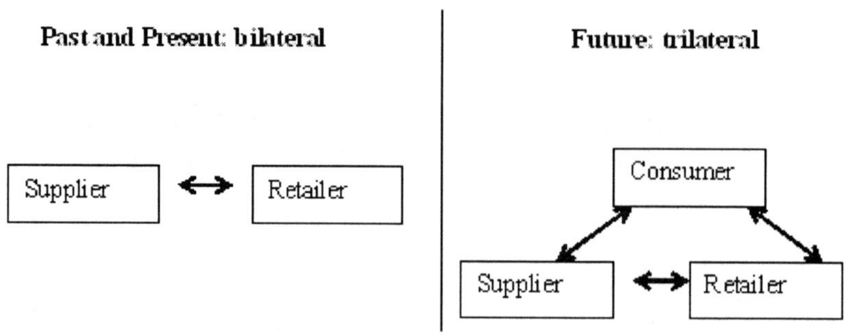

- Merchandising systems
- Information systems
- Communication Technologies

The change from Service to self-service in the 80ies created also problems: theft! The fourth wave of history in 90ies consisted out:

- Security
- Article Surveillance
- Surveillance systems

The new millennium started with an internet hype (new economy) – followed by an internet-baisse. The fifth wave in the new century:

- Internet
- B2B, B2C
- RFID
- WLAN

At store-level the first Metro Future Store in Germany with RFID and WLAN did give a major push to the technical development.

The sixth wave could be detected already in the annual show EuroCIS which is the subsidiary of the EuroShop.

- Mobile technologies like mobile phone/iphone/GPS and new social media like FaceBook, YouTube or LinkdIn will play an important role to influence consumers to visit certain outlets

Parallel to this technological as well as social innovation another strain of development is the ecological wave. EuroShop already reacted to reserve for 2011 special space at the exhibition under the title "Ecopark".

Scope of Penetration

Measuring the impact of EuroShop by the exhibition size in square-meters the number of exhibition, the mix of exhibitions, the number of visiting experts and the international mix visitors the exhibition is an evolution tornado of retail itself.

The exhibition size in square-meters climbed from originally 19,614 in 1966 via 43, 896 in 1984 and up to 106,871 in 2008. Having a German share in 1966 of 88% and 1984 of 96% now only 44% are rented by German exhibitors. This shoes an increase of foreigners from 17 percent in 1966 to 53 percent in 2008.

The number of experts increased from 1966 from 28,762 to 104,766 in 2008 – 59% coming from abroad. New visitor segments come from Russia, India and China. The technological evolution tornado does not stop at national borders – the globalization of technologies and store concepts are reflected by exhibitors and visitors.

CONCLUSION

Seeing the development from POS to POC on the time-axis it can be stated that the trend is from "sales" via "marketing" perhaps even to consumer driven "customization" – which means that the consumer himself takes part via new media in the production-process. The consumer might for example in textiles or shoes select design and materials from electronic catalogues, attach his size or measures by personal electronic sticks and ask for the number of customized products. Beside today's mass-production in future smart niche-production targeting those technical oriented customers might live on good margins (Figure 7).

The POS-period of mass-distribution has changed via the impact of consumer-behaviour (POP) to more segmentation and differentiation (POD) and aiming towards customized activities (POC) with the support of the direct customer-dialogue by the help of new media. If seen in a three-dimensions presentation the development of the new technology enables in this new century retailers to regain the direct retailer/consumer – relationship of the Mum and Papa store

Figure 7. From sales to marketing

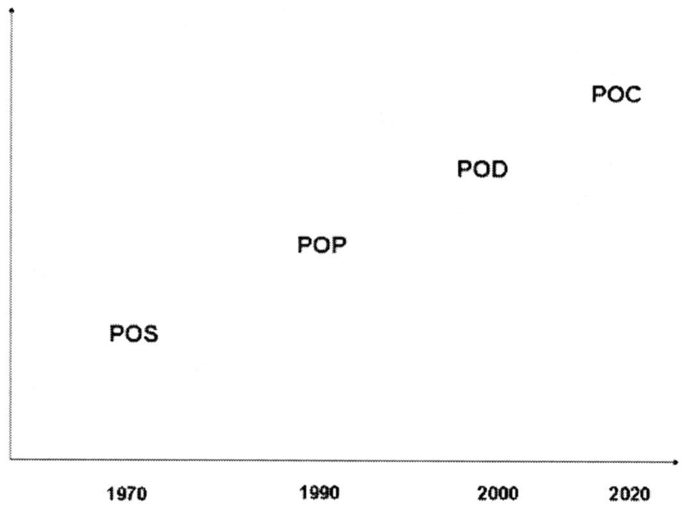

– but the Evolution Spiral (tornado) of retail creates a new Point of Customer (POC) above the old swing of the Mum and Papa store (Figure 8):

While traditionally such a "repetition of the situation" is called in literature "the wheel of retailing" (Savitt, 1984) the three-dimensions-view shows that in history always the parameters change:

- of course the modern shopping centres starting around 1975 in Western Europe are a "repetition" at the concept "all under one-roof" of the start of the department-stores around 1875
- of course the internet-catalogue in 2000 is a "repetition" of the printed catalogue from 1925 but: the consumers, the retailers, the suppliers are on a much higher level! The technological, sociological development of each of the three stakeholders influences each other. Taken this mind-set as a tool of analysis of the situation of the national re-

Figure 8. Evolution tornado retail

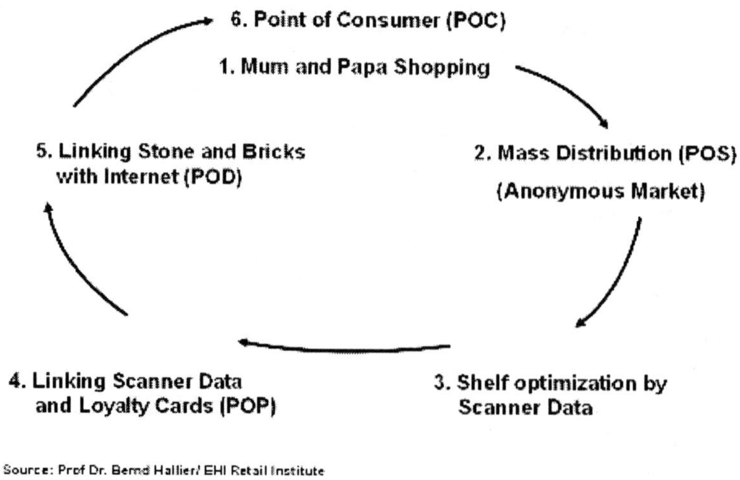

tail of a country it can be even used as a benchmark to evaluate internationally the status quo of that national economy concerning its competiveness, concerning its power of innovation.

It also becomes obvious that the "history of retail outlets" could be analyzed by a broad range of interdisciplinary experts – and could be enlarged within a matrix to different countries/regions. Therefore this contribution is not the final stage but the hopefully starting point of further international scientific cooperation.

REFERENCES

Bauer, H.-J., & Hallier, B. (1999). *Kultur und Geschichte des Handels*. Cologne, Germany: EHI.

Behrends, C. (2001). An der Schnittstelle zwischen Theorie und Praxis. In Hallier, B. (Ed.), *Praxisorientierte Handelsforschung*. Cologne, Germany, Germany: EHI.

DHI. (1991). *Verpackungsverordnung und Entsorgung im Handel*. Cologne, Germany: DHI.

DHI. (1991). *Anforderungen des Handels an Verpackungen*. Cologne, Germany: DHI.

DHI. (1991). *Verpackung und Umwelt im Handel*. Cologne, Germany: DHI.

DHI. (1992). *Auf dem Weg zur: Direkten Produkt-Rentabilität*. Cologne, Germany: DHI.

EHI. (1993). *Mehrwegverpackungen*. Cologne, Germany: EHI.

EHI. (1994). *Scannersysteme- Neue Impulse für Organisation und Marketing*. Cologne, Germany: EHI.

EHI. (1997). *Elektronische Identifikation und Rückverfolgbarkeit von Tieren*. Cologne, Germany: EHI.

EHI. (1999). *Eurep GAP, Verification 2000*. Cologne, Germany: EHI.

EHI. (2000). *Eurepgap, Implementation*. Cologne, Germany: EHI.

EHI. (2001). *Transparenz in der Wertschöpfungskette: Fleisch*. Cologne, Germany: EHI.

EHI. (2005). *Tracing and tracking*. Cologne, Germany: EHI.

Hallier, B. (1987). Sich ins Regal hineinrechnen. *Absatzwirtschaft, 10*, 87.

Hallier, B. (1995). Der Handel auf dem Weg zur Marketingführerschaft. *Absatzwirtschaft, 3*.

Hallier, B. (1997). Wal Mart-Mythos führt zur falschen ECR-Positionierung. In *Dynamik im Handel, 4*. Cologne, Germany.

Hallier, B. (1999). Wird ECR zum Club der Großen? In von der Heydt, A. (Ed.), *Handbuch Efficient Consumer Response*. Munich, Germany.

Hallier, B. (2001). *Praxisorientierte Handelsforschung*. Cologne, Germany: EHI.

Hallier, B. (2004). *EuroShop*. Cologne, Germany: EHI.

Hallier, B. (2009). *Modern stores*. Moscow, Russia: Interexpert.

Hallier, B. (2010). Systemkonkurrenz zwischen stationären Handel und Versandhandel. In Mattmüller, R. (Ed.), *Versand-Handels-Marketing*. Frankfurt, Germany: Deutscher Fachverlag.

Heidel, B. (1990). *Scannerdaten im Einzelhandelsmarketing*. Wiesbaden, Germany: Gabler Verlag.

ISB. (1980). *Physischer Warenfluss*. Cologne, Germany: ISB.

ISB. (1988). *50 Jahre Selbstbedienung, Sonderausgabe*. Cologne, Germany: ISB.

ISB. (1989). *Logistikgerechte Versandverpackungen und Ladungsträger*. Cologne, Germany: ISB.

Jones, P., Comfort, D., & Hillier, D. (2009). *Marketing sustainable consumption within stores: A case study of the UK's leading food retailers*. Sustainability 2009. Retrieved from www.mdpi.com/journal/sustainability

Kondratjew, N. (1926). Die langen Wellen der Konjunktur. *Archiv für Sozialwissenschaft und Sozialpolitik, 56*.

Mattmüller, R. (2010). *Versand-Handels-Marketing*. Frankfurt: Deutscher Fachverlag.

Savitt, R. (1984). The wheel of retailing and retail product management. *European Journal of Marketing, 18*(6/7).

Schumpeter, J. (1961). *Konjunkturzyklen. Eine theoretische, historische und statistische Analyse des kapitalistischen Prozesses*. Göttingen, Germany: Vandenhoeck & Ruprecht.

KEY TERMS AND DEFINITIONS

Evolution Tornado of Retail: It is an enlargement of the Wheel of Retail. While the Wheel's theory is the repetition of situations over a time-period the evolution tornado links the repetition with the technological/sociological upgrade of the development over the time-period.

Point of Consumer (POC): It defines the outlet as an interface of consumer-groups voicing their interests via new social media like Facebook, YouTube, LinkIn etc. The POC might combine shopping, eating/drinking, entertainment.

Point of Differentiation (POD): It is stressing the branding of a retail-outlet – to underline the difference between company A and company B, the difference of the offer C to offer D. The main actor is the retailer – putting his image above the brand-interests of the suppliers; using quite often private labels to demonstrate uniqueness.

Point of Purchase (POP): It is focusing the buying-behaviour of the consumer. The POS-Push action is transformed to a POP-Pull Action. Supplier and retailer are on the same eye's height – acting on behalf of the consumers.

Point of Sales-Marketing (POS): It was developed in the 60ies/70ies to decline out-of-stock situations inside the stores and to push sales by sales-supporting-materials. It was created by the suppliers and used mainly (only) in the interest of the individual supplier.

Chapter 3
Modeling Shopper Responses to Retail Digital Signage

Charles Dennis
University of Lincoln, UK

Andrew Newman
University of Salford, UK

Richard Michon
Ryerson University, Canada

J. Josko Brakus
Brunel University, UK

Len Tiu Wright
De Montfort University, UK

ABSTRACT

This chapter evaluates the impact of digital signage, or digital communications networks (DCNs), on shoppers' perceptions, emotions, and shopping behavior. Digital signage, which consists of screen displays in public spaces showing video, has been little researched to date. The chapter focuses on how consumer shopping behavior can be enhanced by an atmospheric stimulus such as digital signage and the ways in which digital signage can affect consumer perceptions about the images of shopping malls. The chapter reports two interconnected studies. First, a qualitative study is reported with empirical results evaluating how digital signage screens can improve the image of shopping malls and create a favourable shopping atmosphere. The qualitative findings elicit a number of constructs that shoppers use in forming their attitudes towards digital signage. These are supplemented by the views of industry experts and used to form the basis of an attributes list.

The second stage is a quantitative study. The attributes list of digital signage constructs is purified using factor analysis to produce a new scale measuring attitude towards digital signage. This new scale is then applied in developing a new model based on the Elaboration Likelihood Model (ELM). The "direct" route in the ELM suggests that digital signage influences cognition, which then influences emotions, whereas the "peripheral" route is emotion → cognition.

DOI: 10.4018/978-1-60960-738-8.ch003

We predict that these operate in parallel and report a survey of mall consumers (n = 315). Digital signage has a significant, positive, total effect on approach behaviors, mediated by positive affect and (arguably) perception of mall environment. Results extend the Limited Capacity Model of Mediated Message Processing (LCM) from television to digital signage, which predicts the effectiveness of vivid moving visual images as atmospheric stimuli.

In addition to the obvious application to retailers in improving business-to-consumer appeal to shoppers the findings are of use to suppliers of digital signage in business-to-business marketing of their systems to retailers.

INTRODUCTION

This chapter evaluates the impact of digital signage, or digital communications networks (DCNs), on shoppers' perceptions of the retail environment, positive affect, and approach behavior in a shopping mall context. Digital signage consists of 'screen displays located in public spaces showing video material (or private TV channels)' (Clarke, 2003). TV screens have been used in retail environments for some time but since the advent of digital control and flat screens, the use of networks of screens has made digital signage available as an effective, easily controlled communication medium. Referring to digital billboards, i.e. outdoor, the Outdoor Advertising Association of America describes them as: ' ... updated electronically through a variety of methods. Some are networked together, most are operated remotely, and all of them can be updated quickly, sometimes with just the click of a mouse. This ability gives digital [signage] flexibility and nimbleness. This nimbleness gives local businesses a unique and powerful way to reach a large number of geographically targeted consumers very quickly' (OAAA, 2009).

Digital signage content may include, for example, advertisements, community information, entertainment and news. According to POPAI (Point of Purchase Association International), more than 70 percent of purchase decisions are made in store at the point of purchase (Jugger, 1999). Such screen networks go by many names but we use the terminology 'digital signage' here as being most commonly used internationally. Similarly, we use 'shopping mall' (or simply 'mall') as the term becoming more accepted internationally for what has, particularly in Europe, previously been referred to as a 'shopping centre' (or 'shopping center'), i.e. a 'planned retail development ... managed and marketed as a unit' with a 'pedestrian precinct covered from the weather' (Dennis, 2005, adapting from Guy, 1994 and citing Reynolds, 1993).

Digital signage aims to talk to shoppers while they are captive and in the mood to buy. Retailers in countries including the US (Albertson's, Target, Kroger), the UK (Tesco, Asda, Sainsbury) and China (Carrefour) have launched digital signage networks. In addition to pushing merchandise, digital signage also generates hefty advertising revenues. Brand manufacturers pay anywhere from to \$60,000 to \$293,000 for a four-week campaign on Wal-Mart's TV network connecting more than 2,500 stores (The Economist, 2006). Although research figures are sparse, industry insiders estimate that digital signage is currently worth around \$2billion in the US (Computerworld.com 2008).

Digital signage might be considered as contributing to retail atmospherics. Leo J. Shapiro & Associates, the firm that conducts store atmospherics surveys for *Chain Store Age* (Wilson, 2005) categorizes in-store TV among interactive atmospheric elements helping retailers building a competitive advantage. Research indicates that shoppers tend to consider that they would benefit from technological innovations such as electronic shelf-edge displays (a special case of digital signage) and product information kiosks (which parallel digital signage) (Burke, 2002).

This chapter examines the effect of a digital signage network in contributing to retail atmospherics by influencing shoppers' perceptions of the overall retail environment and approach/avoidance responses. The research takes place in shopping mall environments, which differ from the retail store in not being aimed primarily at promoting a single retailer. Rather, digital signage in the mall environment is similar to the outdoor digital billboard, where it is often used to display breaking news, community information and promote a range of local retailers (OAAA, 2009). As outlined by Underhill (2004), the mall is a store of stores, and better provision of information and perceptions of the atmosphere and environment of a mall should enhance shoppers' experiences such that they are likely to stay longer and spend more money (Wright, Newman and Dennis, 2006).

Digital signage is thus an important tool for retail atmospherics, with particularly important potential for shopping malls. Yet there is a paucity of scholarly research into digital signage (Newman, Dennis and Zaman, 2006). This chapter therefore sets out to address this research gap. The chapter aims to make a theory contribution by eliciting the essential elements of shoppers' evaluations of digital signage. There are important implications for mall owners and for retailers, as, if digital signage is found to enhance perceptions of a mall environment, research demonstrates that shoppers transfer perceptions of the mall environment to the store images of individual retailers (Chebat, Sirgy and St-James, 2006), which may significantly impact revenue.

BACKGROUND

The Shopping Mall

Shoppers, of course, use the shopping mall as a convenient way to obtain goods and services. Nevertheless, as we demonstrate in this section, consumers also patronize shopping malls for many less-utilitarian purposes. For example, shopping frequency in malls is correlated with (among other variables) recreation (Roy, 1994); and propensity for unplanned purchases is influenced by hedonic as well as utilitarian considerations (Chebat, 1999). Personal life values also influence mall patronage (Shim & Eastlick, 1998). Shoppers patronize shopping centers for walking and exercise (Hangland & Cimbalo, 1997) and as a social and recreation meeting place (Graham, 1988). The shopping mall is considered as a public place for community development among non-shoppers (Lewis, 1990), for the construction of social links (Aubert-Gamet & Cova, 1999), a city within a city (Backes, 1997) and as an ecological habitat for consumers (Bloch, Ridgway, & Dawson, 1994). Enjoyment and entertainment are important benefits of shopping (e.g. Babin, Darden & Griffin, 1994; Sit, Merrilees & Birch, 2003; Yoo, Park & MacUnnis, 1998), valued by consumers, and reflected in their spending (e.g. Donovan, Rossiter & Marcoolyn, 1994; Jones, 1999; Machleit & Mantel, 2001; Sherman and Smith, 1987; Smith & Sherman, 1993). Shopping malls and retailers therefore add entertainment to their offers, enhancing the experiential benefits (Newsom, Collier & Olsen, 2009).

Against this background of substantial hedonic motivations for shopping, many older malls have difficulty competing against more modern ones (Reynolds, Ganesh & Luckett, 2002). The importance of the physical environment has long been recognized (Baker, 1998; Baker, Grewal & Parasuraman, 1994; Baker, Parasuraman, Grewal, and Voss, 2002; Bitner, 1990; 1992; Theodoridis and Chatzipanagiotou, 2009) and has more recently been extended to that of the shopping mall (e.g. Chebat & Morrin, 2007). Ways that malls can compete include improving the environment, making the mall a more pleasurable place to spend time, resulting in customers staying longer and spending more money (Wright, Newman and Dennis, 2006). The next sub-section briefly reviews prior research into ways that marketers can improve the environment and positively influence shoppers by manipulating atmospheric stimuli.

Retail Atmospherics

The capacity to alter in-store behavior through retail atmospherics is well known by retailers and researchers (e.g. Turley & Milliman, 2000). Retail atmospherics can be adapted to enhance the likelihood of triggering particular shopping behaviors. A wide spectrum of shopping behaviors can be influenced in a variety of retail formats. Appropriate music, in particular, has a positive effect on patronage across a range of retail contexts (Garlin and Owen, 2006, in a meta-analysis). Examples include: manipulation of music styles and tempos that impact sales in supermarkets (Herrington & Capella, 1996; Morin, Dubé & Chebat, 2007); impulse buying in department stores (Yalch & Spangenberg, 1990); responses to waiting in banks (Hui, Dubé & Chebat, 1997); sales in wine shops (North, Hargreaves & McKendrick, 1999); and music "fit" on perceptions of an apparel brand (Beverland, Lim, Morrison & Teziovski, 2006).

Other examples of patron responses to retail atmospherics include: increased sales due to effective exterior store windows (Edwards & Shackley, 1992); the effect of lighting on the number of items handled by shoppers and time spent at a display (Summers & Hebert, 2001); store layout on price perceptions (Smith & Burns, 1996); merchandise arrangement on purchase intentions in a wine store (Areni, Duhan & Kieker, 1999); and gender-appropriate scent on perceptions of apparel store environment, merchandise and approach behaviors such as spending (Spangenberg, Sprott, Grohmann & Tracy, 2006). The influence of atmospheric stimuli on shoppers' perception tends to be holistic rather than piecemeal such that aroma, for example has a positive effect only when a store is moderately busy, i.e. not too empty and not too crowded (Michon, Chebat & Turley (2005).

Apart from in-store behavioral response, retail ambiance influences a variety of consumers' emotions and attitudes: the effect of crowding on shopper satisfaction (Machleit, Kellaris & Eroglu, 1994); the mediating effect of the environment on the affective reactions of department store shoppers (Sherman, Mathur & Smith, 1997); the influence of color on furniture store displays (Babin, Hardesty & Sutter, 2003); the impact of the general environment on store image of a card and gift store (Baker, Grewal & Parasuraman, 1994); the effect of facilities and product assortment on consumers' pleasant emotions (Yoo, Park & MacInnis, 1998). Babin & Darden (1995) also observe that the effect of a store atmosphere might be mediated by a consumer's general shopping style thus producing various reactions from different segments of consumers.

In sum, a wide spectrum of shopping behaviors can be influenced by specific atmospheric stimuli in a variety of retail formats. A selection of Turley's and Milliman's (2000) review of 60 experiments and more recent studies is included in Table 1. The reported success of such marketer-manipulated strategies is leading retailers to develop more advanced techniques aimed at making shopping a more enjoyable experience (Pantano & Naccarato, 2010), notwithstanding that shoppers still rely largely on traditional cues such as personnel, products and layout in evaluating their store experiences (Bäckström & Johansson, 2006).

Mall Atmospherics

Compared with store atmospherics, there are fewer studies about how consumers perceive or respond to a mall's environment, particularly with

Table 1. Selected prior research into the effects of specific retail atmospheric stimuli

Study	Stimulus	Influences	Sample (n = usable questionnaires)
Behavioural influences			
Herrington and Capella (1996)	Music	Sales in supermarkets	In-store survey, n=140 (89 aware of background music)
Yalch and Spangenberg (1990)	Music	Sales in department stores	In-store experimental design, n = 86 (foreground music = 33, background music = 32, and control =21)
Hui, Dubé and Chebat (1997)	Music	Responses to waiting in banks	Retail banking video setup, n = 116 undergraduate students, experimental design (4 types of music plus control)
North, Hargreaves and McKendrick, (1999)	Music	Sales in wine shops	In-store 2x2 experimental display; German and French music, and German and French wines, n = 82
Dubé and Morin (2001)	Music	Positive affect and approach behaviours such as spending	In-store survey, n = 110 shoppers aware of background music post-categorized in the low pleasure intensity (48) and high pleasure intensity (62) conditions
Spangenberg, Sprott, Grohmann and Tracy (2006)	Gender-appropriate aroma	Perceptions of apparel store environment, merchandise and approach behaviours such as spending	Pretesting of feminine and masculine scents (n = 300 students, faculty and staff); in-store field experiment (82 males, 99 females) in congruent and incongruent scent conditions.
Chebat and Michon (2003)	Aroma	Perceptions of mall environment, positive affect and spending	Mall intercept, n=145 with aroma (447 control)
Summers and Hebert, (2001)	Lighting	Number of items handled by shoppers and time spent at a display	Field experiment, shoppers observation (n = 2367) in a 2 (stores) x 2 (lighting conditions) experimental design
Babin, Hardisty and Suter (2003)	Color and lighting	Positive affect and purchase intention in an apparel store	209 females from the university community, average age 33.2 years
Sherman, Mathur and Smith (1997)	Social, image, design and ambience	Positive affect and approach behaviours such as spending	Mall exit intercept n=909
Cognitive influences			
Beverland, Lim, Morrison and Teziovski (2006)	Music "fit"	Perceptions of an apparel brand	20 in-depth consumer interviews
Smith and Burns, 1996	Store layout	Price perceptions	Warehouse grocery store intercept before and after manipulation (n = 182), with control (n = 198)
Baker, Grewal and Parasuraman (1994)	General environment	Store image of a card and gift store	N = 297 undergraduates in a laboratory experiment (2x2x2) opposing prestige to various discount conditions
Affective influences			
Machleit, Kellaris and Eroglu (1994)	Crowding	Shopper (dis)satisfaction	1) University bookstore video simulating high and low crowding situations (n = 76 undergraduates) 2) Actual bookstore under various crowding conditions (n = 140) 3) Two grocery stores under various crowding conditions (n = 232 shoppers)
Yoo, Park and MacInnis (1998)	Facilities and product assortment	Shoppers' positive affect	Shoppers intercept (n = 294) in two large Korean department stores
Perception of mall environment			
Finn and Louvière (1996)	Physical environment	Mall image and patronage	Longitudinal mail surveys in 1988 (n=339), 1992 (n=1042), and 1993 (n=848)
Hildebrandt (1998)	Physical environment	Mall image and patronage	Household panel (n = 2105) over a 9-month period
Ruíz (1999)	Physical environment	Mall image and patronage	Door-to-door survey (n = 177)
Wakefield and Baker (1998)	Physical environment	Positive affect and desire to stay longer	Community mall intercept (n = 438)
Chebat, Sirgy and St-James (2006)	Mall image and atmosphere	Mall image influences store image	Video mall simulation (n = 200 shoppers) in an experimental factorial design; store types (2), mall image (2), shoppers SES (2). Dependent variable: self-congruity and store image

respect to specific stimuli (Table 1). Chebat's and Michon's (2003) study is one notable exception, reporting the positive effects of aroma on perceptions of mall environment, positive affect and spending. The limited research available indicates that, similarly to stores, mall atmospherics influence mall image, shopper affect (e.g. Wakefield and Baker, 1998) and patronage (e.g. Dennis et al., 2002; Finn and Louvière, 1996; Hildebrandt, 1988). Interestingly, mall image also influences store image (Chebat, Sirgy and St-James, 2006).

We predict that mall atmospherics will not only contribute to building mall traffic, but also promote sales and additional spending. Based on the environmental psychology approach (Foxall and Soriano, 2005; Mehrabian and Russell, 1974), a shopping-congruent atmosphere is expected to put shoppers in a favorable mood, have them stay longer in the mall (Newman, 2007), and encourage them to spend more.

Digital Signage as an Atmospheric Stimulus

As a relative newcomer to the retail environment, digital signage networks are now found in the marketing toolbox. Despite the growing commercial importance of digital signage, there is little scholarly research into digital signage. The impact of digital signage on consumers' perceptions of the environment and consumers' responses falls in general within the environmental psychology approach (Mehrabian & Russell, 1974) and, more specifically to the digital signage stimulus, the Limited Capacity Model of Mediated Message Processing (LCM), which models how people process television communications, predicting the effectiveness of vivid moving visual images (Lang 2000).

We are able to cite only four published papers on digital signage in scholarly journals. First, Harrison & Andrusiewicz (2004) report on the technical aspects of scheduling ads. The other three are our papers, comprising: (i) a qualitative study by Newman, Dennis & Zaman (2006), which reports on the acceptability of digital signage to shoppers; (ii) a qualitative investigation by Newman *et al.* (2010), that elicits dimensions of shoppers' evaluations of digital signage; and (iii) a quantitative paper by Dennis et al. (2010) that explores the mediating effects of perception and emotion. We draw upon the datasets of studies (i) and (iii) for this chapter.

As outlined in the 'Retail Atmospherics' subsection above, environmental stimuli, including music (e.g. Morrison & Teziovski, 2006); color (e.g. Babin, Hardesty & Sutter, 2003); lighting (e.g. Golden & Zimmerman, 1986); design (e.g. Baker, Parasuraman, Grewal & Voss, 2002); and aroma (e.g. Ellen & Bone, 1998), induce emotions that in turn influence approach/avoidance behavior (Mehrabian & Russell, 1974). Alternatively, Chebat & Michon (2003) suggest that the effects of atmospheric cues on consumers' emotions and behavioral responses are initially mediated by cognition (e.g. Lazarus, 1991). In the case of the digital signage stimulus, the two competing approaches are not necessarily mutually exclusive but are both consistent with the ELM of Petty & Cacioppo (1986), which has been extensively studied in the field of advertising. The appeal can be either rational or emotional. The rational appeal may be more effective when the elaboration likelihood of the communication situation is high, i.e. when shoppers stop to watch the digital signage and perceive specific information. Under these conditions, a consumer's cognitive responses will determine the behavioral outcome – the 'central route'. Alternatively, when the elaboration likelihood is low, i.e. when the digital signage is perceived as background 'wallpaper', perhaps with pleasant scenes, consumers will not process messages cognitively but may still be influenced emotionally and this emotional appeal may still positively influence approach behavior – the 'peripheral route'. Advertising research over past decades has produced many examples of both the rational appeal (e.g. Golden & Johnson, 1983) and the emotional appeal (e.g. Page, Thorson & Heide, 1990) being claimed to be more effective, with other scholars accepting that the two types of appeal are not mutually exclusive (e.g. Puto & Wells, 1984). People tend to rely on feelings as a way of simplifying judgment that operates mainly in the peripheral mode – the "how do I feel about it" or "feelings-as-information" heuristic, in which people equate pleasant feelings as evidence of liking and satisfaction (Pham, 2004). This mechanism tends to operate by default and is only questioned if an alternative explanation for their feelings is made obvious (Gorn, Goldberg, & Basu, 1993). Irrespective of the competing emotion → cognition (Zajonc and Markus, 1984) or cognition → emotion (Lazarus, 1991) theories, marketers can manipulate atmospherics to improve consumers' images of a location and increase spending.

As mentioned in this section above, there is limited research on the effect of digital signage. By implication, in accordance with the LCM (Lang, 2000), which holds that people have limited cognitive resources to process large quantities of information simultaneously and therefore allocate processing resources to those most demanding stimuli that have a high information rate and distinctive features such as movement, color and vividness (Li & Bukovac, 1999), digital signage should act as a more effective atmospheric stimulus, with higher recall of messages than those that are static or less vivid (Taylor & Thompson, 1982). Moving images attract viewers' attention (Reeves & Nass, 1996). These prior findings support the LCM in this context.

Research on store perception suggests that, at least for hedonic products such as perfume, jewelry and gifts, store atmosphere affects perceptions of product quality (Schlosser, 1998). Hedonic benefits are more likely to increase positive affect and loyalty than utilitarian ones (e.g. Chitturi, Raghunathan and Mahajan, 2008) and hedonic aspects are playing more part in shoppers decisions even for discount shopping (Carpenter and Moore, 2009). Hence, we expect that a pleasant atmosphere will be increasingly relevant for all types of shopping.

Relevant background perceptions tend to be perceived only in a general or ambient manner and there is no guarantee that any specific component of the perception of the environment will actually be evaluated individually (Jacoby, 2002). Perception of the environment can be considered as a "package" of interacting components (Jacoby, 2002). Therefore, the atmosphere or background may be manipulated to "prime" people's perceptions and thus change behavior in such a manner that they may or may not be aware of the presence of the stimulus (Dijksterhuis, Smith, van Baaren & Wigboldus, 2005; Mandel & Johnson, 2002). Even though consumers may pay little attention to digital signage, they are likely to perceive it and its content as part of the atmosphere background. This priming is likely to increase choice and spending (Mandel & Johnson, 2002). Accordingly, an atmospheric stimulus such as digital signage may affect other components of the perception of the mall environment, such as the cleanliness. According to the LCM (Lang, 2000), the moving images of digital signage should constitute an effective atmospheric stimulus that may influence consumers' images of the shopping environment, for example providing information. Any means that helps people with fluent decision processes is likely to result in increased liking (Schwarz, 2004). This is consistent with the conventional conceptualization of decision making processes that involves consciously applying rules governing actions in response to stimuli (Berkowitz, 1993).

Alternative evidence can be presented in support of the claim that atmospheric stimuli influence emotion (Donovan & Rossiter, 1982) and that emotion influences cognitive perceptions and approach behavior (Zajonc and Markus, 1984). This argument is consistent with the peripheral route of the ELM (Petty & Cacioppo, 1986) and the "how do I feel about it", feelings-as-information heuristic (Pham, 2004). The process of direct influence of stimuli on emotion may follow the "relatively basic and automatic ... processes" that snap into play before cognitive thinking can get started (Berkowitz, 1993, p. 12; see also Ullman, 1984). In this conceptualization, relevant digital signage content (e.g. pleasant scenes) may positively influence emotion directly.

Prior research on advertising (e.g. Petty and Cacioppo, 1996) and shopping mall environmental psychology (e.g. Chebat and Michon, 2003) has tended to assume that cognition → emotion and emotion → cognition are separate processes. On the other hand, Berkowitz's (1993) work on anger, consistent with Epstein's (1993) Cognitive-Experiential Self-Theory, suggested that the two tend to work in parallel in response to any particular stimulus: a basic, rapid system based on the unconscious emotional response and a more deliberative, sophisticated cognitive process. In product choice decisions, these routes can work together. Consistent with the LCM, in conditions where the resources of cognitive processing are limited (which may be the case when shoppers are coping with parallel tasks such as work or family responsibilities), spontaneous emotional response to a stimulus has a greater effect on choice. On the other hand, when more cognitive processing resources are available (perhaps when on a dedicated shopping-for-pleasure trip), cognition has more impact (Shiv & Fedorikhin, 1999). Clearly, in the case of the shopping mall, both conditions can apply and we therefore predict that digital signage influences both cognition and emotion; and that cognition and emotion both influence consumers' approach behaviors. We consider both positive affect and positive cognitive evaluations, so (again following Shiv and Fedorikhin, 1999) we predict that positive affect and positive cognitive evaluations will act together to increase approach behavior. A stimulus such as digital signage is expected to have both cognitive and affective elements, both of which can influence approach behavior.

We predict that various types of content on digital signage can influence both cognition and emotion; and that, in accordance with the above argument, cognition and emotion can operate in parallel to influence consumers' approach behaviors. A stimulus such as digital signage is expected to have both cognitive cues, such as providing information about stores and products; and affective elements such as pleasant scenes, both of which can influence approach behavior. The research hypotheses are formalised in the next section.

Research Hypotheses

We hypothesize that mall atmospherics will not only contribute to building mall traffic, but also promote sales and motivate additional spending. Based on the environmental psychology approach (Mehrabian & Russell, 1974; Donovan & Rossiter, 1982), a shopping-congruent atmosphere is expected to put shoppers in a favorable mood, encourage them to visit more frequently and to spend more.

We expect that, according to the LCM (Lang, 2000), the moving images of digital signage should constitute an effective atmospheric stimulus that may influence shoppers' images of the shopping environment. Digital signage should constitute an effective marketer-manipulable atmospherics stimulus, increasing positive affect and positive perceptions of the mall environment. We therefore predict that:

H_1 Digital signage will positively influence positive affect and positive perceptions of the mall environment.

The environmental psychology approach has been previously tested in a retail setting (e.g. Chebat & Michon, 2003). The perception of a pleasant shopping environment (Dubé & Morin, 2001) should elicit positive emotions such as pleasure and arousal (Ang, Leong & Lim, 1997) and result in higher spending (Spies, Hesse & Loesch, 1997). In line with this, Puccinelli, (2006) finds that people who are in a good mood have a better perception of products and are willing to spend more money. Similarly, Fedorikhin & Cole (2004) report that positive mood leads to lower perceived risk and influences consumers' choices towards trying new products. Those authors also find that the effect is greater when consumers' constructive processing is higher, which we consider describes the condition of consumers' comparison shopping in a mall. These prior empirical findings and the conceptual considerations in the 'Digital Signage as an Atmospheric Stimulus' section above lead us to postulate that cognition → emotion and emotion → cognition combine to make up the main influences on approach behavior. In other words:

H_2 The effect of digital signage on shoppers' approach behaviors will be fully mediated by positive affect and positive perceptions of the mall environment.

The research hypotheses are summarized in Figure 1.

METHOD

Research Setting

The fieldwork took place at a well-known sub-regional shopping mall near London, UK. The mall consists of a single level with a typical blend of retail provision and services. In addition to the usual apparel and other comparison retailers, the mall also includes a drugstore and a grocery supermarket, making shoppers' trip purposes more typical of shopping in general (Arentze, Opewal & Timmermans, 2005; Dellaert, Arentze, Bierlaire, Borgers & Timmermans, 1998). Most atmospherics research uses simulations rather than real retail environments and student respondents rather than typical shoppers (for an exception, see Spies, Hesse & Loesch, 1997). For this study, we obtained rare access to data from real mall customers in a real shopping environment.

A plasma screen digital signage network was installed especially for the research, which consisted of a one-week familiarization plus two

Figure 1. Schematic representation of hypotheses H_1 and H_2

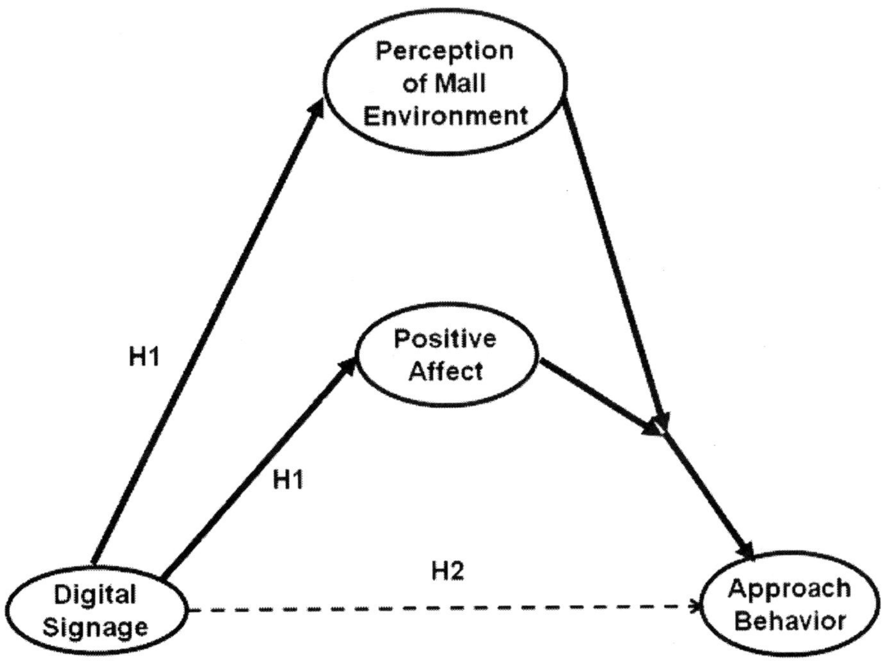

weeks evaluation. The digital signage consisted of twelve 1.2-metre plasma screens distributed around the public areas, plus additional screens placed in the stores of ten participating retailers – making up 22 screens in total. The participating stores were (i) a men's apparel store; (ii) a young women's accessories store; (iii) a children's apparel store; (iv) a children's learning/toy store; (v) a music and media store; (vi) an electronics store; (vii) an optician and optical store; (viii) the drugstore; (ix) a café; and (x) a luxury chocolates store with café.

Qualitative

The first stage consisted of focus groups with shoppers recruited in the mall based on the criterion that they had previously noticed the digital signage. A series of eight group discussions were conducted with subjects varied by age, gender, occupational status and income. Most groups contained six to nine subjects. In total, 51 participants took part including 39 women and 12 men.

A semi-structured discussion guide was used as a basis for group discourse. This contained open-ended questions and cues in order to detect patterns and trends across the groups and encourage spontaneity. Sessions were recorded and later transcribed. Axial coding was used to facilitate the process of listing key facts and recurrent themes. To ensure consistency and systematic analysis, a transcription form based on a standard template was developed. Responses to individual questions were coded and assigned to the particular themes, and the corresponding sections of the text were marked for review. This process facilitated the comparison across groups and reduced the need for repetitive sections in the results. The findings were used to produce a list of attributes of shoppers' evaluations of digital signage.

A second stage of qualitative investigation took the form of interviews with key informants, comprising the managers of the 10 retail stores where the digital signage was installed; the manager of the mall; and the chief executive and marketing manager from the digital signage supplier. The purpose of these key informant interviews was first, to comment on the face validity of the attributes elicited from the focus groups and second, to suggest more dimensions for evaluations of digital signage.

Quantitative

In the quantitative survey, reported shopper spending and other variables were measured using questionnaire responses. Retailer sales data were eschewed in order to avoid anomalies caused by (e.g.) weather, interest rates and competitor advertising that would have confounded the results. The survey instrument was a self-report questionnaire, requiring respondents to rate the digital signage screens and their content, plus the mall and various emotions. The design utilized 5-point Likert-like scales (e.g. very poor to very good). The questionnaire grouped variables into core themes: perception; pleasure-arousal; approach-avoidance; and general demographics.

Broadcast content consisted of two hundred messages sourced partly from promotional material from the mall operator and the retail stores (from both those with, and some without, screens in their stores); and partly new material. The new material on the screens in the mall areas included information about the mall's facilities and public information about external services such as the town theatre and farmers' market. Screens in the participating retail stores mainly carried content specific to those retailers but some also carried selected general information.

Selection of the sample was problematic. Should the study aims have sought to explicate a more representative and random sample, this would have called for a postal survey. Following this route would have placed the study in danger of under-representing the more frequent users of the mall. The findings concern mall shoppers, and the results have implications for mall managers. Therefore, the sample was as representative as practicable of the mall's customers. A convenience mall intercept survey achieved the desired sampling technique (*vis-à-vis* omitting non-customers). Howard (1992); and Hackett & Foxall (1994) use a similar method for comparing shopping malls. This sampling provides high quality data that can be as accurate as other methods (Bush & Hair, 1985). The technique is more likely to select respondents who stay longer, and therefore is weighted so as to be more likely to be representative of mall shoppers' behaviors than would be a true random sample (Nowell & Stanley, 1991).

Researchers intercepted respondents near three coffee shops or in the general mall concourse. Respondents self-completed the questionnaire in order to minimize bias from the interviewer (although a researcher was on hand to help if necessary). Fieldwork spanned two weeks and most opening times, with two or three researchers working at the times when the mall was, respectively, busier and busiest, in order to approximate more representation in the sample at the busiest shopping times. In order to simulate the conditions of a permanent installation (where almost all shoppers would be expected to be aware of the digital signage), respondents were pre-screened based on whether they had seen the screens (92 percent had). The responses consisted of 315 completed questionnaires (after deducting the eight percent of shoppers approached who had not seen the screens). This sample comprised 73 percent females. Females were thus sampled approximately proportionately to their anticipated spending in the mall, based on owners' data. The number classified in the higher socio-economic status of managerial, administrative, professional, supervisory or clerical was 57 percent. This compared, for example, with this mall owner's own data of 63 percent and other typical in-town malls down to 55 percent (Dennis, 2005). The proportion in the older age groups of 45 years and over was 51 percent (in line with the mall owner's expectations of "around 50 percent"). It was therefore considered that the range of socio-economic groupings and age profiles were as representative as practicable of consumers at the mall.

In order to compare the results for the digital signage at the mall with a control location that does not have digital signage, the method followed the approach of McGoldrick and Thompson (1992) and Dennis (2005), by requiring respondents to similarly rate an alternative control shopping location, the one at which respondents shopped most (or next most after this mall). Respondents thus rated both the test mall and the control and the items used are based on the differences between them. For example, for 'pleasure', the rating for 'pleasure' for the control shopping location (1 to 5 scale) was subtracted from the rating for 'pleasure' for the test mall (with 5 added to each value to ensure that all values were positive), making a 9-point scale (which we rescaled 0 to 1 for use in the model). None of the control locations was equipped with digital signage screens.

RESULTS

Qualitative

The discussions commenced with an evaluation of the screens and how these influenced the shopping centre experience. Most subjects said they had observed the screens in more than one location around the centre. Although subjects would have preferred larger screens they acknowledged that the shopping centre was relatively small to accommodate them. The quantity of screens is important to participants as are also their locations within the various shopping areas. Screens located in refreshment and rest areas appeared to attract more notice compared with other locations, such as the main concourse, where they appeared to become less noticeable probably due to the lower dwell time and higher level of visual clutter.

The potential benefit of digital signage for disseminating a range of valuable information was widely acknowledged by all subjects. A typical comment was:

'The screens are a good idea as long as information is useful'.

Advertising was welcomed when it provided local, time-specific information and special offers:

'You don't want to be bombarded but if [there is] a half-price sale on then tell us, don't expect us to have to find it for ourselves. Useful information, sales, special offers, like the [drugstore] do 3 for 2, late night events, video fashion shows or anything like that. ... Or a new range of something that may be coming to stores.'

Community announcements were particularly valued and included issues such as community announcements, special offers and price updates, e.g.:

'Farmer's market, extremely useful. That will actually make quite a difference. Don't want adverts want information.'

'I might be inclined to go in there if there is something that caught my eye. It would make you go there first of all. Farmer's market advert is a community thing. It'd probably be the community thing that would catch my eye before the advertisement for the particular shop. Hoping if you see the community thing you are going to come in and browse the shop. Which might well prove it. If the farmers market comes up and I would stand there and watch it.'

'Or possibly what's on locally at the Playhouse [theatre].'

The shopping times were very sensible like the late night shopping.

The digital signage was viewed as an extension of the visual merchandising efforts and marketing of the centre. One person related an experience abroad:

'They could do with grabbing your attention like when you do go to American Malls you get…I suppose if you heard a noise and looked up and suddenly it says 'our special offer of the week is. Then you might say 'oh' that's a good bargain let's go have a quick look.'

Subjects emphasized simultaneously that in order to sustain the value of information content, regular and timely updates would be important:

'It would be nice if they were updated practically weekly otherwise people would stop looking at them'.

The idea of location and size of screens, *vis-à-vis* their contribution to the surroundings, were introduced into the discussions. Most subjects agreed that, by and large, the locations of screens were poor and a more favorable position would have been where people were willing and able to '*stop for a few moments*'. Comments included, for example:

'The screens are clear but wrong angle'.

and

'The screens are too high'.

Subjects also mentioned that the screens did not grab attention effectively:

'They [screens] are not particularly grabbing, they are boring ... I would make it more flashy, better positioned, bigger and more color'.

'I was thinking why didn't I notice the screens when my attention was drawn to them I could see them everywhere. I think it was because I came in [the mall] with the idea that I was going to go to [the chocolate shop]. Just walked past everything and the screens weren't dynamic enough to arrest my attention'.

Nevertheless, most participants liked the screens:

'Ever so nice, they're lovely' [female subject].

Narratives indicated a preference for screens to be in locations where shoppers are relatively stationary, waiting around, or when the dwell time was greater such as in cafes, lifts and when waiting in queues. Subjects specified specific content that matched their needs and moods with comments such as:

'Music needs to be subtle, calming and not too loud'.

'It depends whether you are actually sitting down in a coffee bar looking at screens or whether you are walking past if you got a sit down audience then it could be a longer ad.',

'The screens to my mind don't really add an awful lot, they are quite small. I think if you are walking by you just don't tend to notice, for me personally I was only sitting down and looked up and saw it. Otherwise you are too busy going to where ever you need to go'.

and

'I only take notice of the screens when I'm sitting down'.

As expected, the subjects' views reflected their age, gender and socio-economic differences. For example, several participants felt that screens were intrusive and somewhat impaired the social dimension of the retail environment. This view was especially evident with the older age groups and in particular those who lived locally:

'They irritate me especially in the coffee shop because I go in the coffee shop to read my paper or talk to a friend. I don't go in to have adverts blaring.'

Where screens had been placed in the busier areas of the concourse, and the content was perceived to be unsuitable for such locations, subjects remarked on the nature of the audio selections – this was further mediated by gender and age. In all the sessions, younger subjects focused with emphasis on the impact of the screen media selection, and in particular the musical content:

'Its enough to put you to sleep ... should play more up to date music. For old people they'd like that because it's relaxing but for us it's boring. We wouldn't stop to watch it because its not eye catching' [teenage subjects].

Significantly, younger age groups viewed the screens as a potential source of entertainment and, when probed, suggested that music videos, movie trailers and celebrity programs were mostly preferred as broadcast material:

'But if there is good music in a shop I'd stay in there, but if its crap music I wouldn't stay in there because it puts me in a really bored mood... R&B, more lively stuff... But that would put off adults... It's got to appeal to like all ages' [teenage subject].

In sum, there were few objections to the digital signage but a minority of the participants considered it to be boring and not attention-grabbing. Despite this and the finding that there were issues with visibility, size and position of the digital signage screens, the consensus of the focus groups was that the digital signage creates an ambience that influences participants' perceptions of the mall environment, giving it a more modern image. The participants reported that the screens add enjoyment and entertainment to their shopping experiences; and provided useful information, particularly community information and information on special events.

A synthesis of these findings was presented to the key informants in the form of proposed questionnaire items and examined by them for face validity. Following the key informant interviews, an attributes list for shoppers' evaluations of digital signage was produced (see Table 2).

Quantitative Results

Models and Measurement Scales

The next step investigated if and how digital signage influences consumers' positive affect, perception of the mall environment and approach behavior. This was done through latent path Structural Equation Modeling (SEM) using SPSS AMOS (Arbuckle, 2006). The SEM analysis illustrates mediation and the indirect effects of the digital signage on consumers' behavioral responses.

Four scales were used in this study: consumers' evaluations of digital signage, perception of the mall environment, positive affect, and approach behavior. All scales were based on multiple-item measurements.

We are unaware of any prior scale for digital signage. This measurement was therefore based first on that previously reported for other stimuli such as aroma (e.g. Ellen & Bone, 1998); and second on the results of the qualitative study reported above. Adapting from Ellen & Bone (1998), we included 'What do you think of the digital signage overall' plus the dimensions of evaluations of digital signage elicited from the qualitative study (5-point scales anchored by 'very poor' to 'very good'). The other three scales were taken from the literature.

Scales were first subjected to exploratory factor analysis (EFA) (ensuring that the scales loaded as distinct variables) before being re-screened through confirmatory analysis and introduced into the structural model. In accordance with Bollen (1989, p. 244) and Kline (2005, p. 314), a limited number of the latent variable indicators were kept in the path analysis, selected for their higher loading in the EFA. Table 3 outlines the measurement scales with selected items, alpha coefficients, factor loadings and sources.

The relationship between latent variables representing consumers' perception of the mall environment, positive affect and approach behavior was investigated in a path analysis (Figure 2). In line with the argument above that cognition → emotion and emotion → cognition tend to work in parallel, we predict that positive affect and positive perception of the mall environment will have a parallel effect on approach behavior. Notwithstanding this, because perception of the mall environment and positive affect are conceptually (and empirically) distinct constructs, we do not combine them into a single variable. Rather, in

Table 2. Initial attribute items for the evaluation of digital signage scale

Evaluation of digital signage item	Source
What do you think of the digital signage overall	Ellen and Bone (1998)
Entertainment	Focus groups
Pleasure	Focus groups
Community information	Focus groups
Information on special events	Focus groups
Easy to see	Focus groups
Sound	Focus groups
Locations of shops	Key informants
Information about the shops	Key informants
Product / price information	Key informants
Brand advertising	Key informants

Table 3. Measurement scales

	Digital Signage	Perception Mall Environment	Positive Affect	Approach Behavior
Explained Variance	18.0%	16.3%	18.0%	11.2%
Alpha	.83	.77	0.82	0.69
Evaluation of Digital Signage very poor/very good				
Community information	.90			
Information on special events	.89			
Entertainment	.87			
What do you think of the digital signage overall	.55			
Perception of Mall Environment (McGoldrick and Thompson, 1992) very poor/very good				
Covered shopping		.80		
Cleanliness		.76		
Availability of good restrooms		.75		
Affective security in the mall and car park		.69		
Affect (Mehrabian and Russell, 1974)				
Unhappy / Happy			.81	
Melancholic / Contented			.81	
Dissatisfied / Satisfied			.80	
Unstimulated / Stimulated			.71	
Approach Behavior (Adapted from Donovan and Rossiter, 1982; Dennis, 2005; and Chebat & Michon, 2003)				
Frequency of visits				.80
Likelihood of revisiting soon, very unlikely / very likely				.69
Spending (non-food)				0.68

Principle components extraction, Varimax rotation

Based on five-point questionnaire scales (except for Approach: Frequency of visits and Spending, which are scale variables)

The top four loading evaluation of digital signage items were retained in the SEM model. Those unused were: (i) pleasure; (ii) easy to see; (iii) sound; (iv) locations of shops; (v) information about the shops; (vi) product / price information; and (vii) brand advertising.

order to demonstrate their parallel effects and neutralize the likely co-linearity problem arising from the parallel effects, we constrain the path coefficients of their effects on approach behavior to be equal. In passing, we speculate that the likely co-linearity arising from the parallel effects may well contribute to the inconsistency of previous research results reported for cognition → emotion *vs.* emotion → cognition directions in previous studies.

The latent variable path analysis outlines the relationships between the stimulus (digital signage) and consumers' responses. The SEM exhibits an excellent goodness of fit (χ^2 = 175, df = 87, CFI =.94, RMSEA =.057). Digital signage has a significant direct influence on perception of the mall environment (Standardized Coefficient =.24, critical ratio (C.R.) = 3.1) and positive affect (Standardized Coefficient =.15, C.R. = 2.2). In turn, perception of the mall environment (Standardized Coefficient=.22, C.R.=3.3) and positive affect (Standardized Coefficient=.20, C.R.=3.3) together have a significant effect on consumers' approach behaviors (it is the unstandardized rather than the standardized coefficients that are constrained equal). The modification indices (MIs) indicate that no additional paths are suggested except for suggested paths in both directions perception of the mall environment → positive affect (MI 37.0, par change 0.38) and positive affect → perception of the mall environment (MI 38.8, par change 0.38). These are the paths that would be expected both from conventional cognition → emotion and competing emotion → cognition paths rather than the hypothesized model, which will be explored later, below. The SEM shows that the influence of digital signage on consumers' behavior is mediated by perception of the mall environment and positive affect. The structural model does not hint at the possibility of any partial mediation (Figure 2). The total effect of digital signage on approach behavior (.08) is significant. Hypotheses H_1 and H_2 are supported.

In following the common shorthand claiming support for our hypotheses, we have not, of course, demonstrated causality, only an association: "data do not confirm a model; they only fail to disconfirm it" (Cliff, 1983). Indeed, Editor (2001) points out that when a model is not disconfirmed, there may be many other competing (but untested) models that are not disconfirmed as well. Accordingly, in order to address such considerations, we have tested alternative models from competing theory frameworks. First, we fitted a conventional environmental psychology cognition → emotion model (Figure 3). This SEM also exhibits an excellent goodness of fit (χ^2 = 131.5, df = 87, CFI =.97, RMSEA =.040). Digital signage has a significant direct influence on consumers' perception of the mall environment (Standardized Coefficient =.23, C.R. = 3.10). Consumers' perception of the mall environment influences consumers' emotions (Coefficient=.47, C.R. = 5.48). In turn, consumers' affect impacts approach behavior (Coefficient=.36, C.R.=3.57). The modification indices (MIs) indicate that no additional paths are suggested. In this alternative conceptualization, testing for mediation (Baron & Kenny, 1986; Iacobucci, Saldanha, & Deng,

Figure 2. Latent path analysis (Testing H_1 and H_2)

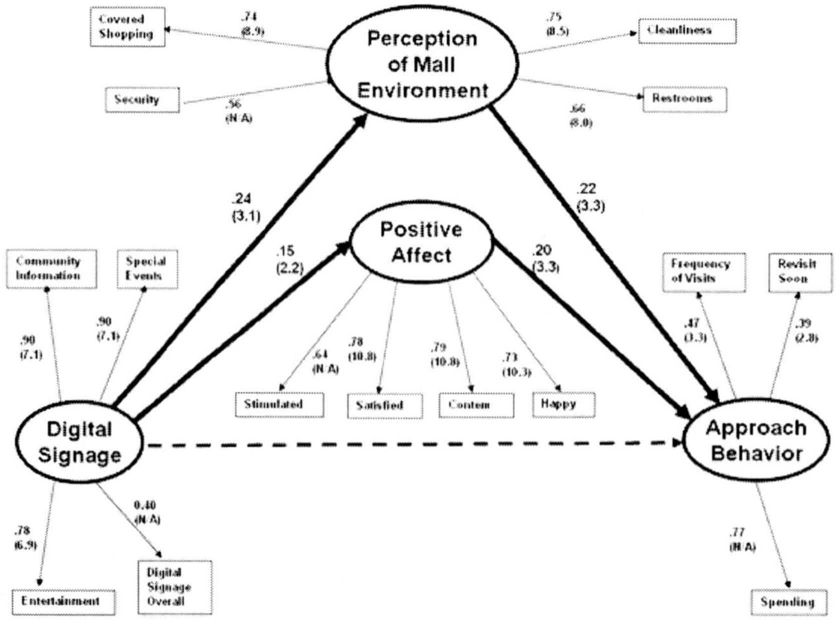

2007), the SEM shows that the influence of digital signage on consumers' behavior is mediated by consumers' perceptions and emotions. The structural model does not hint at the possibility of any partial mediation. Applying the same test also indicates that the effect of digital signage on consumers' emotions is also fully mediated. Nevertheless, we hesitate to confirm this conclusion as there is both conceptual and empirical support for this direct link, illustrated in Figure 2. Later in this section we test an alternative model suggested by the MIs and are unable to disconfirm the direct link stimulus → emotion.

As the literature does not fully resolve the competing theoretical approaches of the direction cognition → emotion *vs*. emotion → cognition, an alternative emotion → cognition model is also fitted to the results for comparison. This SEM emotion → cognition model exhibits a similar goodness of fit to the cognition → emotion model (χ^2 = 147.0, df = 87, CFI =.96, RMSEA =.047) with all paths significant (Figure 4).

Nevertheless, with this model, the MIs suggest the addition of paths that would add elements of cognition → emotion. These are: digital signage → perception of the mall environment (MI = 6.4, par change =.23); and positive affect → approach behavior (MI = 5.2, par change =.36). If these two paths are added to the model, the fit is still good (χ^2 = 130.0, df = 85, CFI =.97, RMSEA =.041) but the path perception of the mall environment → approach behavior becomes non-significant (Figure 5). In this version of the model, the influence of digital signage on approach behavior is not mediated by perception of the mall environment, calling into question this part of H_2. With this modified emotion → cognition model, the standardized total effect of digital signage on approach behavior (.07), emotions (.14) and perception of the mall environment (.23) remains significant. This pragmatic data-driven model (including the path direction emotion → perception of the mall environment) includes a significant path directly from digital signage to emotion (standardized coefficient =.14, C.R. = 2.1). These results lend support to the direct influence of the atmospheric stimulus on positive affect, which is more clearly conceptualized in our originally hypothesized model illustrated in Figures 1 and 2.

In the data-driven model in Figure 5, the path perception of the mall environment → approach behavior is non-significant. For comparison, we also report that, if the constraint in our originally hypothesized model, that positive affect and perception of the mall environment act in parallel

Figure 3. Latent path analysis (cognition → emotion conceptualization)

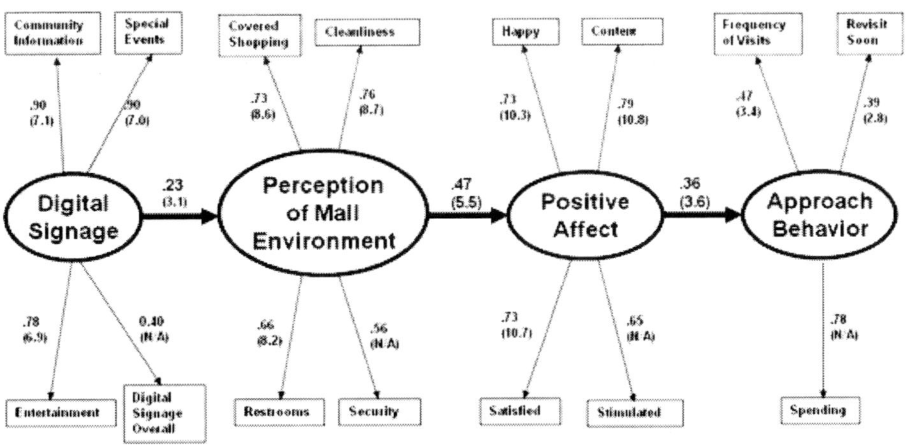

Figure 4. Latent path analysis (emotion → cognition conceptualization)

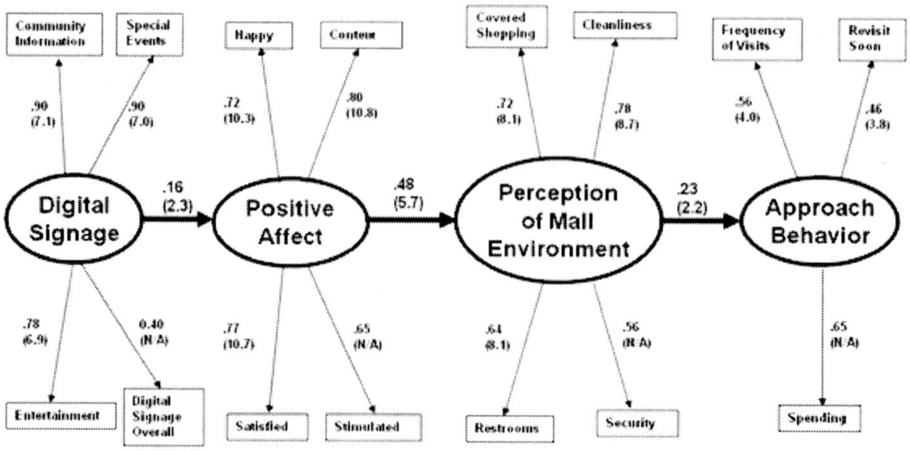

and are constrained equal, is relaxed, the path perception of the mall environment → approach behavior becomes non-significant (standardized coefficient 0.11, C.R. 1.4). These two interpretations would seem to suggest that perception of the mall environment would have no influence (direct or indirect) on consumers approach behavior. This is conceptually inconceivable and inconsistent with the conventional environmental psychology model in Figure 3. We cannot disconfirm that the perception of the mall environment influences consumers approach behavior and therefore, on balance, consider that our originally hypothesized model in Figures 1 and 2 is most preferred. We would most likely attribute the non-significant paths perception of the mall environment → approach behavior in alternative models to the expected co-linearity due to the hypothesized parallel effects of perception of the mall environment and positive affect. In other words, we cannot disconfirm the positive effects of digital signage on perception of the mall environment and positive affect and the mediating effects of perception of the mall environment and positive affect on the relationship between digital signage and consumers' approach behavior (although the mediating effects of perception of the mall environment can be questioned).

The possibility of a direct stimulus → positive affect link is an important one, which we are unable to reject. In terms of the ELM, the cognition → positive affect model in Figure 3 represents the central route, where information provided by the digital signage is valued by consumers and used in their cognitive responses to determine their behavioral outcomes. Nevertheless, shoppers may not always be aware of all the information content on the digital signage (in fact, the majority of shoppers were unable to recall a specific individual store's content featured). In addition to providing information, the digital signage may also act as a background wallpaper constituent of the mall atmospherics. The alternative model with the direct stimulus → positive affect path in Figure 5 (and that part of the originally hypothesized model in Figures 1 and 2) is in accordance with the peripheral route, where processing of the stimulus relies upon the way in which it is presented, rather than cognitive processing. If relevant content on digital signage can influence positive emotions directly, then, in line with the feelings-as-information heuristic, in the absence of any disconfirmation, consumers will interpret the pleasant feelings as representing satisfaction, which should positively influence approach behaviors, as in the models in Figures 2 and 5. In this

Figure 5. Latent path analysis (Alternative modified stimulus → emotion version)

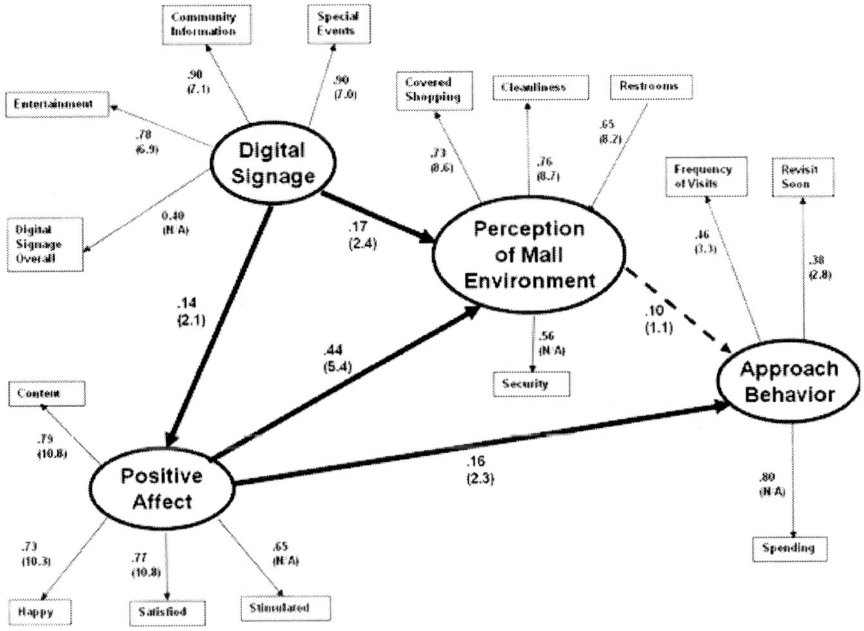

interpretation, non-significance of the perception of the mall environment → approach behavior path would be expected; because consumers use feelings-as-information, they have less need for the cognitive perceptions in taking their shopping decisions.

LIMITATIONS AND FUTURE RESEARCH DIRECTIONS

Before closing, we add a necessary note of caution in the interpretation of our results. First, this study is cross-section and has modeled the effects of evaluations of digital signage rather than its presence or absence. Further experimental studies are called for. We therefore recommend a 'before and after' field experiment. Stage 1 should compare a test mall without digital signage with a control mall. Digital signage should then be installed at the test mall and the two malls compared again. This experiment should demonstrate any causality of the effects of the presence of digital signage.

Second, as mentioned above, Editor (2001) points out that when a model is not disconfirmed, there may be many other competing (but untested) models that are not disconfirmed as well. Accordingly, in order to address such considerations, we have tested alternative models from competing theory frameworks, which we are unable to reject. Notwithstanding, we recommend further research, particularly longitudinal experiments, to evaluate competing models. Specifically, we propose between-subjects and within subjects experiments both in laboratory and field settings, to explore the competing theoretical approaches of the direction cognition → emotion *vs*. emotion → cognition *vs*. parallel routes. Such experiments might compare the effectiveness of different types of digital signage ads, e.g.:

1. **High-cognitive/low emotion**: an ad that contains brief details and price of (say) a tropical island holiday in mainly text form with the logo of the travel company;

2. **High emotion/low cognitive**: an ad that consists (e.g.) of a video of a beautiful tropical lagoon next to a golden sand beach, also with the logo of the travel company; and
3. **High cognitive/high emotion**: an ad that combines the video and text from the first two ads.

Results of a longitudinal experiment such as that proposed should establish causality of the effects of different types of ad content, thus resolving the cognition → emotion *vs.* emotion → cognition *vs.* parallel routes debate. The two experiments together should fill important research gaps, thus helping management with the planning and design of digital signage systems and ads.

More broadly, this research is based on an environmental psychology approach. Consideration needs to be given to broadening the potential theory base, for example to encompass the Theory of Reasoned Action (TRA) (e.g. Ajzen and Fishbein, 1980; Ajzen 1991) and Technology Acceptance (TAM) (Davis, 1989; Venkatesh et al., 2003) families of models.

SOLUTIONS, RECOMMENDATIONS AND CONCLUSIONS

The environmental psychology paradigm linking atmospheric stimuli to perceptions of a retail environment, positive affect and approach behavior is not new to retail atmospherics (although seldom applied for malls). Ample research (e.g. Chebat & Michon, 2003; Dubé & Morin, 2001; Sherman, Mathur & Smith, 1997) has shown that environmental cues will impact consumers' cognition and emotions, and trigger some approach behavior. What is new here is the advent of digital signage in the retail atmospheric toolbox as a stimulus with a substantial effect, as predicted by the LCM. Digital signage has a dual usage: it conveys information when and where shoppers are in the mood to shop (the central route of the ELM), and has an affective or entertainment component (the peripheral route). These results indicate that digital signage is an effective stimulus, adding to positive perceptions of the mall environment, emotions and approach behavior such as spending, as predicted by the LCM. The results therefore extend the applicability of the LCM from television to digital signage.

People who are in a good mood before shopping may have a better perception of the products that they see and consequently spend more (Puccinelli, 2006). Marketers can enhance that process using sensory stimuli that positively influence consumers' moods leading to more spending. This effect is consistent across our parallel model (Figures 1 and 2) and the competing cognition → affect (Figure 3) and alternative affect → cognition (Figure 5) models. Digital signage content can be designed to provide information to shoppers that they consider to be useful (such as community information or store-specific offers). This information will be processed and have a positive effect on consumers approach behaviors (central route). On the other hand, content may also be designed to impact on consumers' emotions directly (e.g. pleasant scenes). This type of atmospheric stimulus need not be processed cognitively but nevertheless, will impact positive emotion and also positive approach behavior (peripheral route, feelings-as-information).

Previous research indicates the effectiveness of only few sensory stimuli significantly associated with increased spending (e.g. aroma, Chebat & Michon, 2003; and music, Mattila & Wirtz, 2001, in a store setting rather than a mall). This study adds a new stimulus, digital signage, as an important tool that mall owners may utilize.

Putting the size of the effect of the digital signage into perspective, we estimate that the digital signage has less influence on approach behavior than do cleanliness, security and helpfulness of staff; but more than the mall's layout, restrooms, or the range of merchandise available. The results demonstrate higher levels of approach behavior (raised by an estimated 1.5 percent), equivalent to increased levels of spend or like-for-like retail sales. Achieving the same result through improvements to design, décor and extensions could cost a regional mall over $20 million. The installation of the digital signage has a manipulable effect on the internal surroundings leading to improvements in consumers' perceptions of the mall. The magnitude of the effect of the digital signage may sound small, but should be viewed in the light of our estimate that, at least in the long term, the improvement in sales figures could be worth around $1 million per year in extra rental income for a regional shopping mall.

In conclusion, regardless of the mechanism (parallel *vs.* stimulus → cognition *vs.* stimulus → emotion), digital signage has a significant, positive, total effect on perception of the mall environment, positive emotion and approach behaviors. The contribution of this chapter is to add digital signage to the short list of successful mall atmospheric stimuli and demonstrate the mediating effects of emotion. The results also extend the Limited Capacity Model of Mediated Message Processing (LCM) from television to digital signage, which predicts the effectiveness of vivid moving visual images as atmospheric stimuli.

ACKNOWLEDGMENT

The authors thank the mall owners for providing access and facilities; and Globecast and How and Why Ltd for a generous research grant.

REFERENCES

Ajzen, I. (1991). The theory of planned behaviour. *Organizational Behavior and Human Decision Processes*, *50*, 179–211. doi:10.1016/0749-5978(91)90020-T

Ajzen, I., & Fishbein, M. (1980). *Understanding attitudes and predicting social behaviour*. Englewood Cliffs, NJ: Prentice-Hall.

Ang, S. H., Leong, S. M., & Lim, J. (1997). The mediating influence of pleasure and arousal on layout and signage effects. *Journal of Retailing and Consumer Services*, *4*(1), 13–24. doi:10.1016/0969-6989(95)00091-7

Arbuckle, J. L. (2006). *Amos (version 7)*. Chicago, IL: SPSS.

Areni, C. S., Duhan, D. F., & Kiecker, P. (1999). Point-of-purchase displays, product organization, and brand purchase likelihood. *Journal of the Academy of Marketing Science*, *27*(4), 428–441. doi:10.1177/0092070399274003

Arentze, T. A., Opewal, H., & Timmermans, H. J. P. (2005). A multipurpose shopping trip model to assess retail agglomeration effects. *JMR, Journal of Marketing Research*, *42*(1), 109. doi:10.1509/jmkr.42.1.109.56884

Aubert-Gamet, V., & Cova, B. (1999). Servicescapes: From modern non-places to postmodern common places. *Journal of Business Research*, *44*(1), 37–45. doi:10.1016/S0148-2963(97)00176-8

Babin, B. J., & Darden, W. R. (1995). Consumer self-regulation in a retail environment. *Journal of Retailing*, *71*(1), 47–70. doi:10.1016/0022-4359(95)90012-8

Babin, B. J., Hardesty, D. A., & Suter, T. (2003). Color and shopping intentions: The intervening effect of price fairness and perceived affect. *Journal of Business Research*, *56*(7), 541–551. doi:10.1016/S0148-2963(01)00246-6

Backes, N. (1997). Reading the shopping mall city. *Journal of Popular Culture*, *31*(3), 1–17. doi:10.1111/j.0022-3840.1997.3103_1.x

Bäckström, K., & Johansson, U. (2006). Creating and consuming experiences in retail store environments: Comparing retailer and consumer perspectives. *Journal of Retailing and Consumer Services*, *13*(6), 417–430. doi:10.1016/j.jretconser.2006.02.005

Baker, J. (1998). Examining the informational value of store environments. In Sherry, J. F. Jr., (Ed.), *Servicescapes: The concepts of place in contemporary markets* (pp. 55–79). Chicago, IL: NTC Business Books.

Baker, J., Grewal, D., & Parasuraman, A. (1994). The influence of store environment on quality inferences and store image. *Journal of the Academy of Marketing Science*, *22*(4), 328–339. doi:10.1177/0092070394224002

Baker, J., Parasuraman, A., Grewal, D., & Voss, G. B. (2002). The influence of multiple store environment cues on perceived merchandise value and patronage intentions. *Journal of Marketing*, *66*(2), 120–141. doi:10.1509/jmkg.66.2.120.18470

Baron, R. M., & Kenny, D. (1986). The moderator-mediator distinction in social psychology research: conceptual, strategic and statistical considerations. *Journal of Personality and Social Psychology*, *51*, 1173–1182. doi:10.1037/0022-3514.51.6.1173

Berkowitz, L. (1993). Towards a general theory of anger and emotional aggression: Implications of the cognitive-neoassociationistic perspective for the analysis of anger and other emotions. In Wyer, R. S., & Srull, T. K. (Eds.), *Advances in social cognition, 6* (pp. 1–46). Hillsdale, NJ: Eribaum.

Beverland, M., Lim, E. A. C., Morrison, M., & Terziovski, M. (2006). In-store music and consumer–brand relationships: Relational transformation following experiences of (mis)fit. *Journal of Business Research*, *59*, 982–989. doi:10.1016/j.jbusres.2006.07.001

Bitner, M. J. (1990). Evaluating service encounter: The effects of physical surrounding and employee responses. *Journal of Marketing*, *54*(2), 69–82. doi:10.2307/1251871

Bitner, M. J. (1992). Servicescapes: The impact of physical surroundings on customers and employees. *Journal of Marketing*, *56*, 57–71. doi:10.2307/1252042

Bloch, P. H., Ridgway, N. M., & Dawson, S. A. (1994). The shopping mall as consumer habitat. *Journal of Retailing*, *70*(1), 23–42. doi:10.1016/0022-4359(94)90026-4

Bollen, K. A. (1989). *Structural equations with latent variables*. New York, NY: John Wiley & Sons.

Burke, R. R. (2002). Technology and the customer interface: What consumers want in the physical and virtual store. *Journal of the Academy of Marketing Science*, *30*(4), 411–432. doi:10.1177/009207002236914

Bush, A. J., & Hair, J. F. (1985). An assessment of the mall intercept as a data collection method. *JMR, Journal of Marketing Research*, *22*(2), 158–167. doi:10.2307/3151361

Carpenter, J. M., & Moore, M. (2009). Utilitarian and hedonic shopping value in the US discount sector. *Journal of Retailing and Consumer Services*, *16*(1), 68–74. doi:10.1016/j.jretconser.2008.10.002

Chebat, J.-C. (1999). Effects of time of day on shopping behavior. *Perceptual and Motor Skills*, *88*(2), 421–422. doi:10.2466/PMS.88.2.421-422

Chebat, J.-C., & Michon, R. (2003). Impact of ambient odors on mall shoppers' emotions, cognition and spending: A test of competitive causal theories. *Journal of Business Research*, *56*, 529–539. doi:10.1016/S0148-2963(01)00247-8

Chebat, J.-C., & Morrin, M. (2007). Colors and cultures: Exploring the effects of mall décor on consumer perceptions. *Journal of Business Research*, *60*, 189–196. doi:10.1016/j.jbusres.2006.11.003

Chebat, J.-C., Sirgy, M. J., & St-James, V. (2006). Upscale image transfer from malls to stores: A self-image congruence explanation. *Journal of Business Research*, *59*(12), 1288–1296. doi:10.1016/j.jbusres.2006.09.007

Chitturi, R., Raghunathan, R., & Mahajan, V. (2008). Delight by design: The role of hedonic vs. utilitarian benefits. *Journal of Marketing*, *72*, 48–63. doi:10.1509/jmkg.72.3.48

Clarke, R. (2003). Question: When is a retailer not a retailer? Answer: When it's a media owner. *The European Retail Digest*, *38*, 20–23.

Cliff, N. (1983). Some cautions regarding the application of causal modelling methods. *Multivariate Behavioral Research*, *18*, 115–126. doi:10.1207/s15327906mbr1801_7

Computerworld.com. (2008, 16 January). *Cisco enters digital signage market*. Retrieved on August 6, 2010, from www.computerworld.com

Davis, F. D. (1989). Perceived usefulness, perceived ease of use, and user acceptance of information technology. *Management Information Systems Quarterly*, *13*(3), 319–340. doi:10.2307/249008

Dellaert, G. C., & Arentze, T, A., Bierlaire, M., Borgers, A. W. J., & Timmermans, H. J. P. (1998). Investigating consumers' tendency to combine multiple shopping purposes and destinations. *JMR, Journal of Marketing Research*, *35*(2), 177–189. doi:10.2307/3151846

Dennis, C., Murphy, J., Marsland, D., Cockett, W., & Patel, T. (2002). Measuring image: Shopping centre case studies. *International Review of Retail, Distribution and Consumer Research*, *12*(4), 353–373.

Dennis, C., Newman, A., Michon, R., Brakus, J., & Wright, L. T. (2010). The mediating effects of perception and emotion: Digital signage in mall atmospherics. *Journal of Retailing and Consumer Services*, *17*(3), 205–215. doi:10.1016/j.jretconser.2010.03.009

Dijksterhuis, A., Smith, P. K., van Baaren, R. B., & Wigboldus, D. H. J. (2005). The unconscious consumer: Effects of environment on consumer behaviour. *Journal of Consumer Psychology*, *15*(3), 193–202. doi:10.1207/s15327663jcp1503_3

Donovan, R. J., & Rossiter, J. R. (1982). Store atmosphere: An environmental psychology approach. *Journal of Retailing*, *58*, 34–57.

Donovan, R. J., Rossiter, J. R., Marcoolyn, G., & Nesdale, A. (1994). Store atmosphere and purchasing behaviour. *Journal of Retailing*, *70*(3), 283–294. doi:10.1016/0022-4359(94)90037-X

Dubé, L., & Morin, S. (2001). Background music pleasure and store evaluation: intensity effects and psychological mechanisms. *Journal of Business Research*, *54*(2), 107–113. doi:10.1016/S0148-2963(99)00092-2

Edwards, S., & Shackley, M. (1992). Measuring the effectiveness of retail window display as an element of the marketing mix. *International Journal of Advertising*, *11*, 193–202.

Ellen, P. S., & Bone, P. F. (1998). Does it matter if it smells? Olfactory stimuli as advertising executional cues. *Journal of Advertising*, *27*(4), 29–39.

Epstein, S. (1993). Emotion and self-theory. In Lewis, M., & Haviland, J. M. (Eds.), *Handbook of emotions*. New York, NY: McGraw Hill.

Fedorikhin, A., & Cole, C. A. (2004). Mood effects on attitudes, perceived risk and choice: Moderators and mediators. *Journal of Consumer Psychology*, *14*(1&2), 2–12. doi:10.1207/s15327663jcp1401&2_2

Finn, A., & Louvière, J. J. (1996). Mall perception, consideration and choice: Anchor store contribution. *Journal of Business Research, 35*, 241–251. doi:10.1016/0148-2963(95)00129-8

Foxall, G. R., & Soriano, M. Y. (2005). Situational influences on consumer's attitudes and behaviour. *Journal of Business Research, 58*(4), 518–525. doi:10.1016/S0148-2963(03)00142-5

Garlin, F. V., & Owen, K. (2006). Setting the tone with the tune, a meta-analytic review of the effects of background music in retail settings. *Journal of Business Research, 59*, 755–764. doi:10.1016/j.jbusres.2006.01.013

Golden, L. G., & Zimmerman, D. A. (1986). *Effective retailing*. Boston, MA: Houghton Mifflin.

Golden, L. L., & Johnson, K. A. (1983). The impact of sensory preferences and thinking vs. feeling appeals on advertising effectiveness. In Bagozzi, R. P., & Tybout, A. M. (Eds.), *Advances in consumer research, 10* (pp. 203–208). Ann Arbor, MI: Association for Consumer Research.

Gorn, G. J., Goldberg, M. E., & Basu, K. (1993). Mood, awareness and product evaluation. *Journal of Consumer Psychology, 2*, 237–256. doi:10.1016/S1057-7408(08)80016-2

Graham, E. (1988, May 13). The call of the mall. *Wall Street Journal*, 7R.

Hackett, P. M. W., & Foxall, G. R. (1994). A factor analytic study of consumers' location specific values, a traditional high street and a modern shopping mall. *Journal of Marketing Management, 10*, 163–178. doi:10.1080/0267257X.1994.9964266

Hangland, A., & Cimbalo, R. S. (1997). Human ethology, age and sex differences in mall walking. *Perceptual and Motor Skills, 85*(3), 845–846.

Harrison, J. V., & Andrusiewicz, A. (2004). A virtual marketplace for advertising narrowcast over digital signage networks. *Electronic Commerce Research and Applications, 3*(2), 163–175. doi:10.1016/j.elerap.2004.02.004

Herrington, J. D., & Capella, L. M. (1996). Effects of music in service environments, a field study. *Journal of Services Marketing, 10*(2), 26–41. doi:10.1108/08876049610114249

Hildebrandt, L. (1988). Store perception and the prediction of performance in retailing. *Journal of Business Research, 17*, 91–100. doi:10.1016/0148-2963(88)90026-4

Howard, E. (1992). Evaluating the success of retail out of town shopping centres. *International Review of Retail, Distribution and Consumer Research, 2*(1), 59–80. doi:10.1080/09593969200000004

Hui, M., Dubé, L., & Chebat, J.-C. (1997). The impact of music on consumers' reaction to waiting for services. *Journal of Retailing, 73*(1), 87–104. doi:10.1016/S0022-4359(97)90016-6

Iacobucci, D., Saldanha, N., & Deng, X. (2007). A meditation on mediation, evidence that structural equations models perform better than regressions. *Journal of Consumer Psychology, 17*(2), 139–153. doi:10.1016/S1057-7408(07)70020-7

Jacoby, J. (2002). Stimulus-organism-response reconsidered: An evolutionary step in modelling (consumer) behaviour. *Journal of Consumer Psychology, 12*(1), 51–57. doi:10.1207/S15327663JCP1201_05

Jones, M. A. (1999). Entertaining shopping experiences, an exploratory investigation. *Journal of Retailing and Consumer Services, 6*, 129–139. doi:10.1016/S0969-6989(98)00028-9

Jugger, S. (1999). The power of packaging. *Admap Magazine*, October.

Kline, R. B. (2005). *Principles and practice of structural equation modeling* (2nd ed.). New York, NY: The Guilford Press.

Lang, A. (2000). The limited capacity model of mediated message processing. *The Journal of Communication, 50*(1), 46–70. doi:10.1111/j.1460-2466.2000.tb02833.x

Lazarus, R. S. (1991). *Emotion and adaptation*. New York, NY: Oxford University Press.

Lewis, G. H. (1990). Community through exclusion and illusion, the creation of social worlds in an American shopping mall. *Journal of Popular Culture*, *24*(2), 121–136. doi:10.1111/j.0022-3840.1990.2402_121.x

Li, H., & Bukovac, J. L. (1999). Cognitive impact ofbanner ad characteristics, an experimental study. *Journalism & Mass Communication Quarterly*, *76*(2), 341–353.

Machleit, K. A., Kellaris, J. J., & Eroglu, S. A. (1994). Human versus spatial dimensions of crowding perceptions in retail environments, a note on the measurement and effect on shopper satisfaction. *Marketing Letters*, *5*(2), 183–194. doi:10.1007/BF00994108

Machleit, K. A., & Mantel, S. P. (2001). Emotional response and shopping satisfaction, moderating effects of shopper attributions. *Journal of Business Research*, *54*, 97–106. doi:10.1016/S0148-2963(99)00093-4

Mandel, N., & Johnson, E. J. (2002). When Web pages influence choice, effects of visual primes on experts and novices. *The Journal of Consumer Research*, *29*(2), 235–245. doi:10.1086/341573

Mattila, A. S., & Wirtz, J. (2001). Congruency of scent and music as a driver of in-store evaluations and behaviour. *Journal of Retailing*, *77*(2), 273–289. doi:10.1016/S0022-4359(01)00042-2

McGoldrick, P. J., & Thompson, M. G. (1992). The role of perception in the attraction of the out of town mall. *International Review of Retail, Distribution and Consumer Research*, *2*(1), 81–98. doi:10.1080/09593969200000005

Mehrabian, A., & Russell, J. A. (1974). *An approach to environmental psychology*. Cambridge, MA: MIT Press.

Michon, R., Chebat, J.-C., & Turley, L. W. (2005). Mall atmospherics: The interaction effects of the mall environment on shopping behaviour. *Journal of Business Research*, *58*(5), 576–583. doi:10.1016/j.jbusres.2003.07.004

Morin, S., Dubé, L., & Chebat, J.-C. (2007). The role of pleasant music and servicescapes: A test of the dual model of environmental perception. *Journal of Retailing*, *1*, 115–130. doi:10.1016/j.jretai.2006.10.006

Netemeyer, R., Bentler, P., Bagozzi, R., Cudeck, R., Cote, J., & Lehmann, D. (2001). Structural equation modelling. *Journal of Consumer Psychology*, *10*(1 & 2), 83–100. doi:10.1207/S15327663JCP1001&2_08

Newman, A., Dennis, C. E., & Zaman, S. (2006). Marketing images and consumers' experiences in selling environments. *Marketing Management Journal*, Fall, 515-599.

Newman, A. J. (2007). Uncovering Dimensionality in the Servicescape: Towards Legibility. *Service Industries Journal.*, *27*(1), 15–28. doi:10.1080/02642060601038601

Newsom, M. K., Collier, D. A., & Olsen, E. O. (2009). Using biztainment to gain competitive advantage. *Business Horizons*, *52*(2), 167–176. doi:10.1016/j.bushor.2008.10.005

North, A. C., Hargreaves, D. J., & McKendrick, J. (1999). The influence of in-store music on wine selections. *The Journal of Applied Psychology*, *2*, 271–276. doi:10.1037/0021-9010.84.2.271

Nowell, C., & Stanley, L. R. (1991). Length-biased sampling in mall intercept surveys. *JMR, Journal of Marketing Research*, *28*(4), 475–480. doi:10.2307/3172787

OAAA. (2009). *Digital billboards, outdoor advertising's glowing future*. Outdoor Advertising Association of America. Retrieved on December 19, 2009 from, http,//www.oaaa.org/marketingresources/digital.aspx

Page, T. J., Thorson, E., & Heide, M. P. (1990). The memory impact of commercials varying in emotional appeal and product involvement. In Agres, S. J., Edell, J. A., & Dubitsky, T. M. (Eds.), *Emotion in advertising* (pp. 255–268). New York, NY: Quorum Books.

Pantano, E., & Naccarato, G. (2010). Entertainment in retailing: The influences of advanced technologies. *Journal of Retailing and Consumer Services*, *17*(3), 200–204. doi:10.1016/j.jretconser.2010.03.010

Petty, R. E., & Cacioppo, J. T. (1986). The elaboration likelihood model of persuasion. In Berkowitz, L. (Ed.), *Advances in experimental social psychology, 19* (pp. 123–205). New York, NY: Academic. doi:10.1016/S0065-2601(08)60214-2

Pham, M. T. (2004). The logic of feeling. *Journal of Consumer Psychology*, *14*(4), 360–369. doi:10.1207/s15327663jcp1404_5

Puccinelli, N. M. (2006). Putting your best face forward, the impact of customer mood on salesperson evaluation. *Journal of Consumer Psychology*, *16*(2), 156–162. doi:10.1207/s15327663jcp1602_6

Puto, C. P., & Wells, W. D. (1984). Informational and transformational advertising, the differential effects of time. In Kinnear, T. C. (Ed.), *Advances in consumer research, 11* (pp. 572–576). Proto, UT: Association for Consumer Research.

Reeves, B., & Nass, C. (1996). *How people treat computers, television and new media like real people and places*. Stanford, CA: Cambridge University Press.

Reynolds, K. E., Ganesh, J., & Luckett, M. (2002). Traditional malls vs. factory outlets, comparing shopper typologies and implications for retail strategies. *Journal of Business Research*, *55*(9), 687–696. doi:10.1016/S0148-2963(00)00213-7

Roy, A. (1994). Correlates of mall visit frequency. *Journal of Retailing*, *70*(2), 139–161. doi:10.1016/0022-4359(94)90012-4

Schlosser, A. E. (1998). Applying the functional theory of attitudes to understanding the influence of store atmosphere on store inferences. *Journal of Consumer Psychology*, *7*(4), 345–369. doi:10.1207/s15327663jcp0704_03

Schwarz, N. (2004). Metacognitive experiences in consumer judgment and decision making. *Journal of Consumer Psychology*, *14*(4), 332–348. doi:10.1207/s15327663jcp1404_2

Sherman, E., Mathur, A., & Smith, R. B. (1997). Store environment and consumer purchase behavior, mediating role of consumer emotions. *Psychology and Marketing*, *14*(4), 361–378. doi:10.1002/(SICI)1520-6793(199707)14:4<361::AID-MAR4>3.0.CO;2-7

Sherman, E., & Smith, R. B. (1987). Mood states of shoppers and store perception, promising interactions and possible behavioral effect. *Advances in Consumer Research. Association for Consumer Research (U. S.)*, *14*(1), 251–254.

Shim, S., & Eastlick, M. A. (1998). The hierarchical influence of personal values on mall shopping attitude and behavior. *Journal of Retailing*, *74*(1), 139–160. doi:10.1016/S0022-4359(99)80091-8

Shiv, B., & Fedorikhin, A. (1999). Heart and mind in conflict, the interplay of affect and cognition in consumer decision making. *The Journal of Consumer Research*, *26*, 278–292. doi:10.1086/209563

Sit, J., Merrilees, W., & Birch, D. (2003). Entertainment-seeking mall patrons, the missing segments. *International Journal of Retail and Distribution Management*, *31*(2), 80–94. doi:10.1108/09590550310461985

Smith, P., & Burns, D. J. (1996). Atmospherics and retail environments, the case of the power aisle. *International Journal of Retail and Distribution Management*, *24*(1), 7–14. doi:10.1108/09590559610107076

Smith, R. B., & Sherman, E. (1993). Effects of store perception and mood on consumer behavior, a theoretical and empirical analysis. *Advances in Consumer Research. Association for Consumer Research (U. S.)*, *20*(1), 631.

Spangenberg, E. R., Sprott, D. E., Grohmann, B., & Tracy, D. L. (2006). Gender-congruent ambient scent influences on approach and avoidance behaviors in a retail store. *Journal of Business Research*, *59*, 1281–1287. doi:10.1016/j.jbusres.2006.08.006

Spies, K., Hesse, F., & Loesch, K. (1997). Store atmosphere, mood and purchasing behavior. *International Journal of Research in Marketing*, *14*, 1–17. doi:10.1016/S0167-8116(96)00015-8

Summers, T. A., & Hebert, P. R. (2001). Shedding some light on store atmospherics. Influence of illumination on consumer behavior. *Journal of Business Research*, *54*, 145–150. doi:10.1016/S0148-2963(99)00082-X

Taylor, S. E., & Thompson, S. C. (1982). Stalking the elusive vividness effect. *Psychological Review*, *89*(2), 155–181. doi:10.1037/0033-295X.89.2.155

The Economist. (2006, March 11). Signs of the times. *The Economist*, *378*(8468).

Theodoridis, P. K., & Chatzipanagiotou, K. C. (2009). Store image attributes and customer satisfaction across different customer profiles within the supermarket sector in Greece. *European Journal of Marketing*, *43*(5/6), 708–734. doi:10.1108/03090560910947016

Turley, L. W., & Milliman, R. E. (2000). Atmospheric effects on shopping behavior, a review of the experimental evidence. *Journal of Business Research*, *49*(2), 193–211. doi:10.1016/S0148-2963(99)00010-7

Ullman, S. (1984). Early processing of visual information. In Gazzaniga, M. S. (Ed.), *Handbook of cognitive neuroscience*. New York, NY: Plenum.

Underhill, P. (2004). *Call of the mall*. New York, NY: Simon & Schuster.

Venkatesh, V., Morris, M. G., Davis, G. B., & Davis, F. D. (2003). User acceptance of information technology: towards a unified view. *Management Information Systems Quarterly*, *27*(3), 425–478.

Wakefield, K. L., & Baker, J. (1998). Excitement at the mall, determinants and effects on shopping response. *Journal of Retailing*, *74*(4), 515. doi:10.1016/S0022-4359(99)80106-7

Wilson, M. (2005). Atmosphere key to well-being. *Chain Store Age*, *81*(10), 94.

Wright, L. T., Newman, A., & Dennis, C. E. (2006). Enhancing consumer empowerment. *European Journal of Marketing*, *40*(9/10), 925–935. doi:10.1108/03090560610680934

Yalch, R., & Spangenberg, E. (1990). Effects of store music on shopping behavior. *Journal of Consumer Marketing*, *7*, 55–63. doi:10.1108/EUM0000000002577

Yoo, C., Park, J., & MacInnis, D. J. (1998). Effects of store characteristics and in-store emotional experiences on store attitude. *Journal of Business Research*, *42*(3), 253–263. doi:10.1016/S0148-2963(97)00122-7

Zajonc, R. B., & Markus, H. (1984). Affect and cognition, the hard interface. In Izard, C. E., Kagan, J., & Zajonc, R. B. (Eds.), *Emotions, cognition, and behavior*. Cambridge, UK: Cambridge University Press.

ADDITIONAL READING

Digital Signage Today. Case studies and articles available from: http://www.digitalsignagetoday.com

Twitter feed available from: http://twitter.com/digsignagetoday.

Harrison, J. V., & Andrusiewicz, A. (2004). A virtual marketplace for advertising narrowcast over digital signage networks. *Electronic Commerce Research and Applications*. 3 (2), 163-175. Available for download from http://www.amazon.com/virtual-marketplace-advertising-narrowcast-networks/dp/B000RR38H8/ref=pd_sim_b_5.

In-Store Marketing Institute. (2009). *New Technologies for Digital Signage Tools that Measure Content Effectiveness, Optimize ROI, and Predict Visual Impact*. In-Store Marketing Institute. Skokie, IL: In-Store Marketing Institute, http://www.instoremarketer.org/.

Kelsen, K. (2010). Unleashing the Power of Digital Signage: Content [th *Screen*. Burlington MA: Focal Press.]. *Strategies*, 5.

Kim, J., Fiore, A. M., & Lee, H. (2007). Influences of online store perception, shopping enjoyment, and shopping involvement on consumer patronage behavior towards an online retailer. *Journal of Retailing and Consumer Services*, *14*(2), 95–107. doi:10.1016/j.jretconser.2006.05.001

Kulviwat, S., Bruner, G. C. II, & Al-Shuridah, O. (2009). The role of social influence on adoption of high tech innovations: The moderating effect of public/private consumption. *Journal of Business Research*, *62*(7), 706–712. doi:10.1016/j.jbusres.2007.04.014

Lundstrom, L.-I. (2008). *Digital Signage Broadcasting: Broadcasting, Content Management, and Distribution Techniques*. Burlington, MA: Focal Press.

Matson, L. (2009). *See It, Hear It, Experience It, Buy It: Increase Sales with Digital Signage, Ambiance Marketing, and Electronic Merchandising*. Innovation Press.

Michon, R., Chebat, J., & Turley, L. W. (2005). Mall atmospherics: The interaction effects of the mall environment on shopping behavior. *Journal of Business Research*, *58*(5), 576–583. doi:10.1016/j.jbusres.2003.07.004

Miller, M. D., & Brunner, C. (2008). Social impact in technologically-mediated communication: An examination of online influence. *Computers in Human Behavior*, *24*(6), 2972–2991. doi:10.1016/j.chb.2008.05.004

Newsom, M. K., Collier, D. A., & Olsen, E. O. (2009). Using "biztainment" to gain competitive advantage. *Business Horizons*, *52*(2), 167–176. doi:10.1016/j.bushor.2008.10.005

Planet, M. (2010). Digital Out-of-Home: Your guide to digital signage application & business trends today. New York, NY: Media Planet. Available for download from http://www.mediaplanet.com.

POPAI. (2010) POPAI G2: Guide to Digital Media in Retail by POPAIdigital. Stoney Stanton: Point of Purchase Advertising International (POPAI). Retrieved 2 August, 2010 from http://www.popai.co.uk.

Ryan, J. (2009). European Digital Signage Survey. Minneapolis: John Ryan a Retail Marketing Agency. Retrieved 2 August, 2010 from http://envirosell.com/downloads/EuropeanDigitalSurveyWIP_V7.pdf.

Schaeffler, J. (2008). *Digital Signage: Software, Networks, Advertising, and Displays: A Primer for Understanding the Business (NAB Executive Technology Briefings)*. Burlington, MA: Focal Press.

Wright, L. T., Newman, A. N., & Dennis, C. E. (2006). Enhancing Consumer Empowerment. *European Journal of Marketing*, *40*(9/10), 925–935. doi:10.1108/03090560610680934

Yoo, W., Lee, Y., & Park, J. (2010). The role of interactivity in e-tailing: Creating value and increasing satisfaction. *Journal of Retailing and Consumer Services*, *17*(2), 89–96. doi:10.1016/j.jretconser.2009.10.003

KEY TERMS AND DEFINITIONS

Digital Signage: Screen displays located in public spaces showing video material. Content typically includes (e.g.) advertisements, community information, entertainment and news. The content showing on digital signage screens is usually updated electronically and they are usually networked together. Most are operated remotely, and all can be updated quickly. Digital signage networks are sometimes referred to as 'digital communications networks' (DCNs).

Elaboration Likelihood Model (ELM): Petty's & Cacioppo's (1986) ELM models the appeal of ads, which can be either rational or emotional. The rational appeal may be more effective when the elaboration likelihood of the communication situation is high, e.g. when shoppers stop to watch digital signage and perceive specific information. Under these conditions, a consumer's cognitive responses will determine the behavioral outcome – the 'central route'. Alternatively, when the elaboration likelihood is low, e.g. when the digital signage is perceived as background 'wallpaper', perhaps with pleasant scenes, consumers will not process messages cognitively but may still be influenced emotionally and this emotional appeal may still positively influence approach behavior – the 'peripheral route'.

Environmental Psychology Approach: Mehrabian's and Russell's (1974) model holds that shoppers' perception's of a pleasant shopping environment should elicit positive emotions such as pleasure and arousal and result in higher spending. People who are in a good mood have a better perception of products and are willing to spend more money.

Limited Capacity Model of Mediated Message Processing (LCM): Lang's (2000) LCM models how people process television communications, predicting the effectiveness of vivid moving visual images.

Retail Atmospherics: Designing the retail store or mall environment to produce emotional effects and/or enhance purchasing behavior.

Shopping Mall: A planned retail development managed and marketed as a unit, usually with a pedestrian precinct covered from the weather. Also referred to as a shopping center/centre.

Chapter 4
The Design of an Advanced Virtual Shopping Assistant for Improving Consumer Experience

Vincenzo Corvello
University of Calabria, Italy

Eleonora Pantano
University of Calabria, Italy

Assunta Tavernise
University of Calabria, Italy

ABSTRACT

The aim of this chapter is to propose the design of an anthropomorphic Virtual Shopping Assistant (VSA), endowed with an advanced system, to be used in the context of an innovative technologically based in-store service. The VSA is able to provide information based on a knowledge management system. It is based on the perceived best human typical seller's characteristics, as well as on the results of psychological studies on consumers' perception of virtual characters. In particular, its interface is anthropomorphic, and thus capable of displaying emotions. This VSA can be used as a mobile application or installed in stores.

INTRODUCTION

In recent years, the demand for new appealing factors capable of improving consumers shopping experience has increased (Diep & Sweeney, 2008; Kim & Niehm, 2009; Pantano, 2010). In fact, these factors influence consumers' satisfaction and loyalty by adding value to goods and services, and improving the interaction with retailers through constant feedback on the products/services provided, as well as consumers preferences (Roussos et al., 2003; Kim et al., 2007; Söderlund & Julander, 2009; Newsom et al., 2009). Technologically advanced devices have been introduced in traditional

DOI: 10.4018/978-1-60960-738-8.ch004

shops in order both to provide a more enjoyable in-store experience (Michon et al., 2006; Chang & Burke, 2007; Pantano & Naccarato, 2010) and to collect information about consumer behaviour (Pantano & Naccarato 2010). Several studies have revealed that consumers show a positive response to the use of new interactive technologies in traditional stores, perceiving the usefulness of advanced technology (Chang & Burke, 2007; Newsom et al., 2009; Kulviwat et al., 2009; Pantano, 2010).

Especially in the e-commerce, the research focused on the development of innovative shopping assistant systems. There are mainly based on the use of virtual agents capable of providing recommendations during consumers' exploration of the e-commerce sites (Ahn, 2010). In fact, many e-retailers have already exploited the opportunities offered by interactive technologies using 3D models (Jin, 2009; Lee & Chung, 2008). In some case, the anthropomorphized interfaces involved are endowed with a wide array of human characteristics, in order to send messages very similar to those transmitted by "real" salesperson (Jin, 2009). The traditional characteristics of the human shopping assistant should therefore be integrated in the virtual one for affecting consumer shopping behavior in the real context of a store, realizing an interface very different from an online virtual seller or a web tool for retailing (Hausman et al., 2009). Despite the volume of research on this topic, the current studies highlight a lack of interactions between the consumer and the virtual seller, as well as a reduced range of service possibilities that the virtual shopper can offer. Moreover, the interfaces utilized are not integrated with a knowledge management system based on consumer preferences, even if information storage is a critical factor in retailing (Pantano, 2010), because the capacity to effectively manage and manipulate information provides a competitive advantage.

For the design of a virtual seller, it is fundamental to take into account some essential parameters related both to technological and psychological factors. In fact, on one hand, the new systems have to be a powerful computational tool capable of providing information based on a knowledge management system; on the other hand, the Virtual Shopping Assistant (VSA) has inevitably to be based on the perceived typical human seller's best characteristics as well as on the results of psychological studies on consumers' perception of virtual characters. This chapter advances the design of a technologically advanced sales system based on virtual agents, proposing the development of a new shopping assistant to be used in the context of an advanced in-store service. In particular, this proposal has an interdisciplinary basis, combining both innovative technological and psychological features. The chapter focuses on some current applications of advanced technologies in retailing, and on the subsequent development of a new technology, which consists of a Virtual Shopping Assistant (VSA) built by using a practice base. In conclusion, the chapter outlines important implications for both marketers and consumers.

THEORETICAL BACKGROUND

Shopping Assistant Systems

A meaningful example of systems created for supporting and influencing of consumers shopping experience are interactive kiosks and shopping assistant systems. The i-PrOS™, created by Charamel GmbH is an interactive kiosk based on a virtual assistant projected onto a screen, capable of catching consumers' attention in a traditional store by delivering multimedia content (Pantano, 2010). The system consists of an interactive screen and a projection foil: consumers interact with the multimedia contents by touch screen sensors, which recognize the finger touch on the surface as a computer mouse-click, getting more information on the products present in the store (e.g. colours, prices, sizes, location, etc.). Despite the quick i-PrOS™ reply to different queries, the system does not have a memory of the users' past actions, thus it cannot suggest new products on the basis of consumers' previous purchases. Usually, these system are located in small stores.

The innovation of this system is the interface, which represents a virtual human. The virtual human is "good looking" and reminds users of a real seller. In this way, consumers have the feeling of really talking to a salesperson. Despite these characteristics, the system does not memorize consumers behavior nor their profile, as a consequence they are not capable of sending recommendations based on their previous purchases.

In shopping assistant systems, the most widely used are the Mobile Shopping Assistant and the IBM Personal Shopping Assistant. The Mobile Shopping Assistant (MSA) created by the METRO Group Future Store is a new application for mobile devices equipped with a camera. It allows the interaction between the mobile and a particular shopping trolley (Pantano & Naccarato, 2010), and consumers can add the products to the shopping basket simply focusing the mobile camera on each good. In particular, the mobile application allows the visualization of the digitally scanned products, the calculation of the total price of the purchases, the localization of specific products in the store, the addition (or removal) of the single item. Furthermore, consumers can ask additional information on each purchase. To date, MSA is used in a large department store in Germany.

The IBM Personal Shopping Assistant is based on a special shopping trolleys available in the store, which consists of a mobile tablet (8.4" display), a Bluetooth® handheld scanner (capable of reading the tag of each product), a cart mount, a charging rack for the storage of the tablet and infared beacon (located throughout the store) for providing products' location information.

The system supports consumers in the choice of favourite items, adding items to the personal shopping list and finding items in the store. Furthermore, it permits graphical visualizations, managing the products in the basket, as well as the products on sale with the exact location in the store.

Given these features, Table 1 analyzes the main characteristics of the current technologies used in stores, comparing Interactive kiosks (as the i-PrOS™) and mobile shopping assistants (such as the ones developed by the METRO Group Future Store). The kiosks provide an interactive interface, which can be used by consumers through the touch screen in order to obtain the desired information through a product knowledge management system (the system has a deep knowledge of the products present in the store), whereas the MSA does not have an interaction with a virtual human as a user friendly interface, but delivers multimodality. Both systems do not provide customized messages through a consumer knowledge management system, which would allow the matching between consumers' requests and the product that better fits the need. However, they have the main common characteristic of supporting consumers during their shopping activity, by involving them in living a new in-store experience. Hence, they need to be integrated with other elements in order to improve the services offer.

VSA Interface

Many researches (Lester et al., 1999; Gulz, 2004; Veletsianos & Miller, 2008; Adamo et al., 2010) focused on the realization of anthropomorphic interfaces have underlined the importance of their humanization in easy-to-use systems for entertainment, communication, and education. In these interfaces, as Bartneck et al. affirm (2005), the displaying of emotions is essential for a natural interaction with humans; moreover, especially expressive virtual faces improve the interaction between the character and the user (Boehner et al., 2007), and their potential is to make the experience more engaging (Gulz & Haakeb, 2006). These animated agents can primarily act as a presenter or guide, as for example "Rea" (Cassell et al., 2000) or "Will" (Churchill et al., 2000). However,

Table 1. The main characteristic of the most used ICTs in retailing

Characteristics	Interactive kiosks	Mobile Shopping Assistant Systems
Interactivity	yes	yes
Multimodality	no	yes
Consumer Knowledge Management System	no	no
Products Knowledge Management System	yes	yes
Personalized product suggestions messages	no	no
User friendly interface and interaction with a virtual human	yes	no
Speed of response	yes	yes
Support of the choice	yes	yes

the realistic modelling of facial movements in anthropomorphic interfaces is one of the most difficult tasks in Computer Graphics, because the most minute changes in facial expression can reveal complex moods and emotions (Bertacchini et al., 2007), and transitions can occur in a fraction of a second (Zhang et al., 2004). Moreover, during conversation human heads change position as well as direction of the glance, and there is an uninterrupted blinking (Cohn et al., 2004). Numerous studies in this field have demonstrated that facial expressions of synthetic agents can be recognized at the same level of those displayed by human faces (Katsikitis, 1997).

For the development of VSA, some basic principles of Human-Computer Interaction (HCI) have been taken into account, although the solution to the problem of a dynamic behaviour is still in progress: fast moving scenes and size makes impossible to load the character on mobile devices. Furthermore, users have to correctly identify the emotions displayed by virtual humans. Moreover, the problem increases when the VSA shows its body, because in human non-verbal communication "language rhythm" has a fundamental role. In fact, important studies of psychology have revealed some forms of rhythmic synchronization in human communication, involving both speech and gesture production. Thus, this synchronization has place through the coordination of a wide range of elements (Bressler & Kelso, 2001); i.e. during talking, people's head, arms and fingers move in a structured temporal organization, synchronized at different levels.

A PRACTICE-BASED APPROACH TO THE DESIGN OF A VSA

The main purpose of the VSA is to change the modes of interaction between the shop organization and the customer in order to: 1) improve the customer's experience and 2) exchange knowledge between the customer and the shop. In order to be effective the VSA needs to be accepted by customers and integrated in their shopping practices. In general, technology acceptance is a widely studied phenomenon (Davis et al., 1989; Lin, 2006; Thong et al. 2006). However the phenomenon of integration of a new technology in the practices of individuals is better understood as a structuring process (Orlikowski 2007; Migliarese & Corvello 2010). Actors actively choose how to use technology, even if the use itself cannot be independent from the characteristics of the system. It is unlikely that the design of a VSA will be successful if it does not take into account the active role of the customer in shaping the practices of interaction that will emerge after the introduction of the new system.

In general there is a well established stream of literature studying the interactions between technology and socio-technical systems and which considers the resulting patterns of action as emergent phenomena. Technology includes patterns of behaviour as a part of its design (DeSanctis & Fulk, 1999; DeSanctis & Scott Poole, 1994; Orlikowski 2000). Therefore, designers try to anticipate the behaviours that users are most likely to adopt and shape technology in order to support those behaviours. Designers' models of users' behaviour are general and ambiguous. Besides, they tend to underestimate the variety of users' behaviour. When the technology introduced is in a *real* context the human system and the technological system influence each other. This problem is often understood in management theory as a problem of alignment. However, alignment is not always achieved. In many cases users find new technological tools difficult to use or not useful. The result is that the technological systems are neglected or, even worse, users must adapt their behaviour to the technology even if it makes their experience worse and their activities more difficult (OECD 2000; Davenport 2005, McAfee 2006).

Several authors have studied the interactions between individuals, organizations and technology during the adoption of a new technology. With collaborative ICT tools, DeSanctis and Poole (1994) proposed the Adaptive Structuration Theory (AST) to study how the *deep structures* embedded in technology affect behaviour. Designers' models of behaviour, once implemented in a technological artifact, are called deep structures. In particular, AST is concerned with appropriation of the structures by workers. Workers can use a technological artifact in ways different than those envisioned by the designers. The same holds for any individual using technology. Orlikowski (2000), instead, observes that what DeSanctis and Poole called structures and she calls practices, emerge from the interaction between organization and technology. They are not embedded in the technology and appropriated by individuals, but to a certain extent created during the interaction. The most visible example of this phenomenon is the modification of the technological artifact by users.

When customers in a shop are invited to use a new tool, they might find some features useful and others useless. They will consider some of the functionalities not worth the cognitive effort needed to learn to use them. Other functionalities will be considered easy enough to be used. Besides, the approach of the customer towards the technology will change according to the specific conditions of each purchase. For example, the way a technological tool is used is expected to change when the purchase is made to satisfy a specific need and under time constraints or to enjoy the shopping experience itself. The use of technological tools such as a VSA, thus, will be characterized by practices shaped by the interactions between technology, contextual conditions and individuals' preferences and decisions.

Summarizing, individuals seldom use technology exactly in the same way as envisioned by designers. Three kinds of reaction are possible:

1. Refusal: the structure embedded in the technology is refused by the individual and the corresponding functionalities are not used;
2. Re-interpretation: the functionalities are used in an original way in order to make them compatible with the individual's emergent practices;
3. Modification: actors directly modify the artifact and, as a consequence, its behavioural content in order to adapt them to the emergent practices.

Magalhães and Silva (2009) propose explicitly taking into account the interactions between routines and technology during the design phase. The two authors consider the interactions between individuals' behaviour and technology as a co-evolution process. This process is emergent; i.e. it is not the result of central design but, instead, it is generated by the complex interactions between decentralized decisions. As a consequence, the result of the use of an artifact is not completely predictable. However, a different approach is possible. Building on the concept of agile software development, Magalhães and Silva (2009) suggest that the implementation of a new technology should go through several short cycles of design and feedback. Steering the processes of co-evolution between technology and organization is proposed as a design strategy. Design and use of technology are not separated processes anymore, but overlap.

The approach of Magalhães and Silva (2009) is particularly suitable for the design and implementation of a VSA. A system of this kind requires acceptance by the user. It can work only if it is accepted by consumers, that is, only if consumers consider it useful and enjoyable. Given its innovativeness it is likely that several adjustments will be required before the VSA shows these features. The progressive introduction of the technology with frequent feed-back and re-design is a promising approach for the introduction of this new technology and for the modification of customers' shopping routines.

The VSA needs to be a flexible and reconfigurable technology. Customers, in fact, show very different purchasing practices. A technology which is only suitable for a minority of these practices is not likely to produce learning processes. Most of the customers are not likely to integrate it in their spontaneous shopping practices.

THE DESIGN OF VSA: TECHNOLOGICAL AND HUMAN FEATURES

The integration of an advanced computational tool with sellers' selected skills allows the development of a new system for supporting purchasing activity. This system, the Virtual Shopping Assistant (VSA), consists of an anthropomorphic computer interface (a virtual human) on a screen or (a virtual face) on a mobile, managing and combining human elements with an information system. In particular, the VSA has technologically advanced features: an anthropomorphic interface, and the basic skills of a real shopper. It represents a new way to influence consumers shopping experience by high customized messages and enjoyable technological elements.

As many researchers affirm, one of the main characteristics making the shopping experience enjoyable is interactivity: in fact, it represents a key feature of the new media (Wu & Chang, 2005), concerns the consumer-to-technology exchange, implies a change in behaviour, and improves users' satisfaction (Chang & Wang, 2008). Another very important feature is multimodality: it concerns the modality of interaction between users and computer that can enhance the human experience in many ways and make it more realistic. Indeed, the virtual seller can show the requested information by texts, images, audio (voice), videos.

However, a necessary characteristic of a VSA is certainly a knowledge management system of consumers. In fact, a virtual seller, as a real salesperson, should have the capacity to remember consumers, their name and their previous purchases. Several studies confirm that consumers appreciate this behavior in a seller (Vignali et al., 2006). In fact, in the online context the salesperson recommendations are capable of strengthening consumers' purchase intentions, by influencing their perceived store quality and credibility (Wang & Doong, in press.).

The system should be capable of adapting the sale tactics in relation to different targets, matching a database related to the consumer behavior (with emphasis on the previous purchases) to a database related to the products available in the store. For example, the VSA can suggest a new good in order to improve sales, or a new version of an old one. Furthermore, the new technology gives retailers the possibility to achieve useful and fast information on consumer response to current marketing strategies (Acquisti et al., 2005). In order to remember previous purchases of a consumer, as well as to suggest new products, the VSA must have deep knowledge of the products of the store (i.e. price, colors, sizes, location) (Vignali et al., 2006; Palamier et al., 2006). This process might be realized in a virtual agent by an *ad hoc* product knowledge management system.

Other important characteristics of a VSA are the following:

1. Customized messages (strictly linked to the consumer knowledge management system): the virtual seller has to be able to produce highly customized messages in relation to the different target consumers and their specific requests and needs. In fact, the system has to provide consumers with personalized product recommendations, based on consumers' personal preferences, in order to facilitate consumption choice. In this way, the system reduces the consumer effort required to choose products (Murray & Häubul, 2009). Furthermore, the VSA should have the possibility to communicate personalized suggestions to consumers, based on their behavior.
2. A user-friendly interface and interaction with a virtual human: the VSA has to seem real. In fact, consumers need to have the feeling of talking with a real person, in order to make the experience more realistic. Moreover, the replies have to be proper and fast, in order to give the feeling of talking with a real salesperson

The human characteristics of a successful seller have always played a key role in the consumer' loyalty, as well as in the purchasing process (Marshall et al., 2003; Williams & Attaway, 2003): patience, a deep knowledge of the products present in the store (Palmatier et al., 2006), the capacity of building a long-term relationship with consumers (Marshall et al., 2003), and different sale tactics in relation to the different targets of purchasers (Sun et al., 2009). Moreover, different researches agree that the fair characteristics of a successful seller, such as a high level of physical attractiveness of the seller, produce a higher level of consumers' satisfaction (Söderlund & Julander, 2009). In particular, a VSA should have characteristics common to a real shopping assistant (Vignali et al.; 2006). It should:

1. identify consumers' needs (both existing and potential) and use this information to influence consumers' buying behaviour. The knowledge system of the consumer profiles (including previous purchases and in-store behaviour) allows the system to know the consumers preferences and suggest new products focusing on this information;
2. stimulate consumers' curiosity. The system should make questions or statements or show attracting images aimed to arouse the consumers' curiosity in order to stimulate their interest;
3. develop relationship with consumers (Marshall et al., 2003; Williams & Attaway, 2003). In fact, when consumers put trust in the salesperson, they are willing to follow his suggestions and prefer to come back to his store for their purchases, in other words the customer relationship enhances the fidelity process and their future buying behaviour;
4. use different sales tactics in relation of the different consumers, because some of them can be more susceptible to salesperson influence, furthermore he has to take care of the different needs and requests (Sun et al., 2009).

Moreover, he/she should provide accurate and fast information: thanks to the knowledge management related to the products and its computational tools, the system can deliver fast answers to the consumers interrogation about several detailed products characteristics, so that consumers have the feeling of talking with a real and informed seller. Finally, the VSA should be patient (the system is always ready to answers to consumers requests, to spend time with them and to wait for their decisions).

Regarding the appearence of the VSA, an anthropomorphic interface might be available also for mobiles forces towards the realization of "Talking Heads", which consists of a virtual head endowed with facial expressions and voice, but without a body. To achieve this goal, Evolutionary Systems Group (ESG), operating at University of Calabria (Italy), has developed an authoring system able to easily produce faces from pictures (Figure 1), displaying facial expressions and talking. The created characters can be used as the interface of the VSA.

This system is related to a parameterized head model with a low number of vertices (131). In particular, the software guarantees:

1. a small number of parameters to animate the facial expressions;
2. slight alterations in a facial expression and the fast passage from one emotion to another without using repetitive pre-built animations;
3. more realistic blinking thanks to the 'noise' function of Perlin (1997), that is a random action of the head and eyelids.

Face3D consists of three components: Face3DEditor for modelling virtual faces, Face3DRecorder for animating the models, and Virtual Theatre for realizing videos (Figure 2). In Face3DEditor it is possible to model the character using pictures or drawings; in Face3DRecorder it is possible to animate the virtual models selecting a facial expression from eight standard ones (neutral, anger, surprise, sadness, fear, joy, disgust, attention), as well as creating and saving small alterations in these expressions. In Virtual Theatre it is possible to realize videos to upload to mobiles or kiosks.

For the creation of the standard facial movements in the system the Facial Action Coding System (FACS) by Ekman, Friesen and Hager (2002) has been used.

Experiments have demonstrated that the characters' standard emotions are easily recognizable in the realized three-dimensional faces.

VSA Architecture

The new system architecture is based on internal software for knowledge management, which is responsible for the efficient matching between the consumer's profile database and the products database; on a multimodal input interaction system, and on the virtual human interface, as shown in Figure 3.

In particular, the steps of the system functioning are as follows:

Figure 1. The production of an anthropomorphized three-dimensional model in the software Face3D (Adamo et al.,2010)

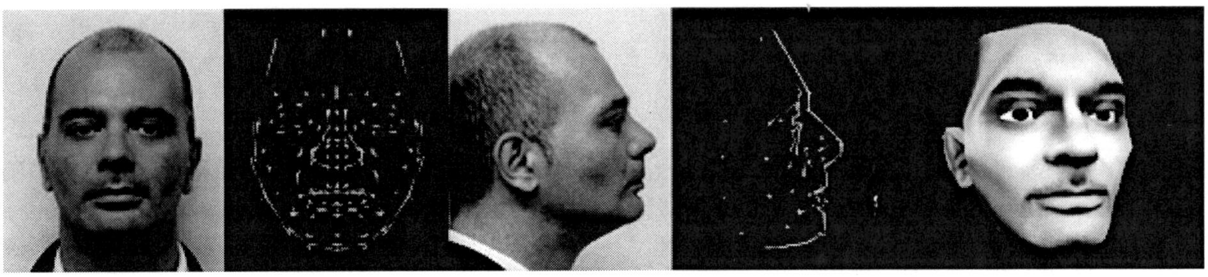

Figure 2. The software Face3D and its three components (Bertacchini et al., 2007)

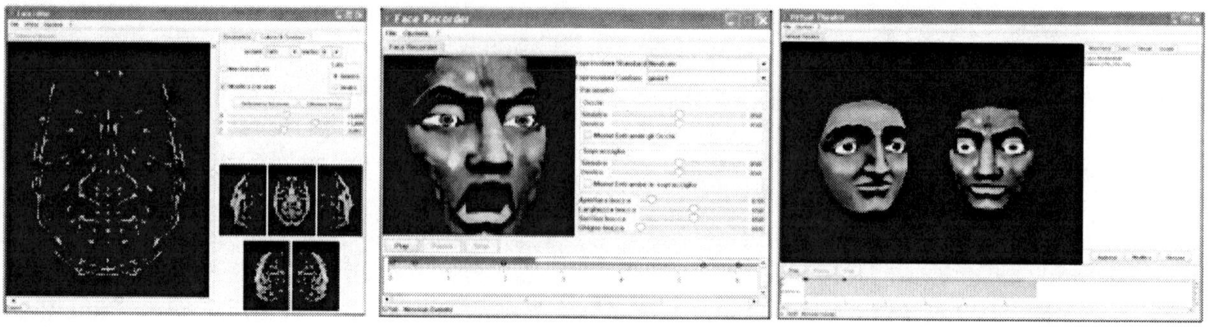

Figure 3. The VSA architecture

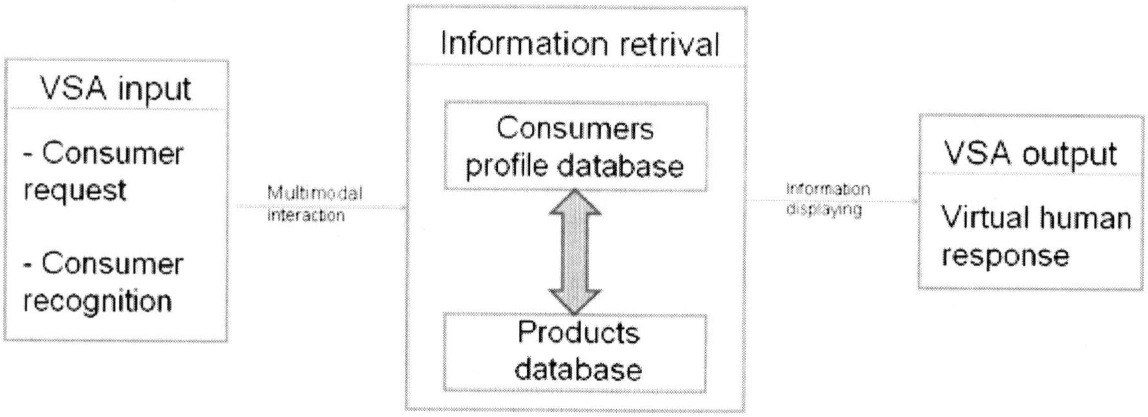

- VSA recognizes consumers. If VSA is installed on a kiosk in the store, a RFID reader system can support this function, by reading a special consumer loyalty card; or fingerprint authentication can be used. In fact, both systems provide the information which allows VSA to recognize the consumer identification data. On the other hand, the VSA can also be installed on a mobile device, therefore, the device directly provides the consumer information;
- consumer sends VSA his/her request, through several input devices (i.e. keybord, voice, mouse click, etc.);
- VSA retrieves the information capable to satisfy the request. In this step the knowledge management systems compare the consumers profile databases with the products database in order to gain the information which best fits the consumer's need. Furthermore, it manages the two different database by adding the consumer choice and removing an items from the products lists;
- VSA displays the final information through the virtual human interface. In this step, the system displays the information by videos or images of products, or by a virtual shopping assistant explanation.

In this way, consumers can save time and achieve detailed information on the products of the store in a new and interactive way. Furthermore, the access to the knowledge related to the product is user-friendly, due to the ease of use of the system, and consumers can access different levels of detail related to each product. In addition, the system can also suggest new products according to consumers' previous purchases.

USAGE SCENARIO

VSA is a mobile application for mobile devices, or it can be installed in a particular kiosk in stores. In the first case, when the consumer is in a store, he/she can start the application and use the mobile device as a shopping assistant. In this way, he/she can interact with the system by using voice, Geographical Position Systems-GPS (systems able to recognize consumer position in the store and his/her movements), mouse-click and writing requests. Furthermore, each product in the store should be equipped with a RFID tag, which memorizes all the product information and relative positions. VSA guides consumers to the store, and then consumer can access the product information by using a panel in the store equipped with RFID tag readers. Information can be displayed in different ways (videos, audios, images, explanations by a virtual human).

In the second case, VSA has similar characteristics to the mobile application (it needs an RFID reader, consumer profile and product database, a knowledge management systems, as well as the virtual human interface), but consumers need a particular loyalty card to interact with the system. In fact, this card memorizes consumer's ID and allows the system to recognize the specific consumers and memorize his/her behavior in the store. Consumers can also use the kiosk input system to interact with the VSA, use its RFID reader (which has similar characteristics to the mobile camera), and visualize the information on the kiosk interface. In addition, a kiosk equipped with a touch screen panel for consumers' requests makes the experience more exciting, improving the consumer-system interaction.

IMPLICATIONS FOR MARKETING

Implications for Marketers

Current research on interactive marketing suggests that companies need to be constantly informed on consumer needs and requests in order to succeed in the changing market (Zahay & Peltier, 2008).

In this contest, two discussion issue emerge from a theoretical standpoint: (1) the reduction of the employees needed, (2) the collection of consistent and fast information on market trends, (3) the providing of high standardized messages.

1. The reduction of the employees needed: consumers avoid the line in the stores and prefer the point of sales who reduce the waiting time, by evaluating positively the shopping mode choice which provides this characteristic (Hsiao, 2009). The introduction of VSA provide a personal and always ready assistant capable of helping them in all the activities involved during the shopping experience. In this way, the number of employees can be reduced, due to the constant availability of a free assistant. As a consequence, the real employees are not requested to memorize a large amount of information on the available products or on frequent consumers, and their work is less stressing and more efficient.
2. the collection of consistent and fast information on market trends: since it is based on the consumer's profile database, the system is able to quickly acquire information on their preferences/needs, creating a personalized communication. Hence, since the database is actively managed and updated, marketers can have constant feedback on the products and services provided to consumers,. In fact, retailers may easily achieve information on the most purchased products, on the frequent consumers demographic characteristics, on the success of the current marketing strate-

gies. These data can be useful to understand the success or failure of such as products and, as a consequence, to develop innovative customized and efficient strategies, capable of both maintaining existing consumers and attracting new ones.

3. The providing of high standardized messages: the VSA is capable of using different sales tactics in relation to different consumers' targets, to manage deep knowledge of products and, as a consequence, to propose highly customized messages.

In addition, marketers can use the information on consumer behaviour to develop new highly customized strategies. Since consumers actively participate in the process of information creation while interacting with the system, marketers can also create highly personalized communication strategies. Thus, taking into account that consumers usually have a strong preference to interact with the seller in order to create the customized solution which best fits their requests, they can actively create the final service. Alternatively, if costumers prefer a standardized offer, the virtual sellers help them avoid using much mental effort, solving effectively the problem of the targeted planned communication

Implications for Consumers

The consumer-system interaction is realistic, due to the exploitation of the multimodal interaction, the user friendly interface (a virtual human), and the fast replies to consumer queries (all elements able to enhance the relationship between consumer and seller, affecting the consumers' satisfaction process). In particular, two aspects of the introduction of a new virtual seller in real contexts merge from a theoretical standpoint:

1. Attractiveness of the new technology: the introduction of advances technologies modifies stores appearance in terms of style, layout and atmosphere. The store environment has an impact on the consumer behaviour in several ways. In fact, the store design and environmental characteristics can modify the consumer's in-store experience (Puccinelli et al., 2009). In particular, the point-of-sale appearance becomes more "futuristic" and oriented to a more technological concept. It can affect consumers approach to the products, providing a new merchandise layout, which can also influence consumer expectations about the search efficiency. The aim of the VSA is to make the store more comfortable for consumers, by providing new efficient services in an engaging way. As a consequence, consumers can spend more time, choose to revisit the store, and prefer it for their purchases. In addition, VSA can attract new consumers, usually less interested in the shopping activities but more in technology (i.e. a younger segment of the population or the male one). Since VSA allow to save time, consumers may evaluate positive this attribute, and as a consequence they might prefer a store which provides this kind of shopping assistance.

In this way, the introduction of advanced technologies on one hand provides useful tools to maintain and interest current consumers, on the other is capable of attracting and exciting new ones.

2. Positive influence of the virtual seller on consumer's behaviour: the enjoyable elements provided in the store influence the consumer's shopping behavior (Diep & Sweeney, 2008; Söderlund & Julander, 2008; Newsom et al., 2009; Roussos et al., 2003) and their attitude to use a new technology (Ha & Stoel, 2009). Indeed, the virtual seller enhances the consumer's shopping experience, influencing their satisfaction as well as their buying behaviour. Furthermore, the possible choice of the VSA can produce more

positive attitudes towards both the system and the shop which proposes it (Reinders et al., 2008).

In particular, the VSA allows consumers to save time, as well as to have a personal assistant capable of supporting the shopping experience by providing useful information on products, promotions and so on, capable to fit their needs and requests. Consumers can exploit the use of VSA to express their personal opinion on current marketing strategies and ask for different services.

In addition, consumers are constantly involved in the in-store service creation and this involvement can positively affect their satisfaction and loyalty to the store.

CONCLUSION

Our analysis underlines the effectiveness of the introduction of a new VSA, based on the integration of real salesperson skills with the main computational tools, in the real context of a store. This virtual seller proposal benefits marketing planning, because it provides crucial information about consumer behaviour, needs and requests, which can be used to develop new customized marketing strategies. In this way, with such information, marketers can identify different response profiles in the market, evaluate and estimate rapidly the consumer response to the introduction of new strategies.

On the basis of our findings, it is possible to realize a prototype to be tested with consumers, in order to obtain statistical data on their emotional and behavioural response by introducing the new virtual seller in the stores. It will be possible to understand if the effectiveness of the system is linked to the category of goods on sales or has the same performance despite the products of the store. Moreover, it is possible to give VSA other basic selling skills of the real salesperson, in order to develop a virtual human as realistic as possible. Thus, the knowledge of the VSA could be appreciated by consumers, due to the increased value which they receive when they interact with a competent seller (Palmatier et al., 2006).

Future studies can focus on what types of retail stores as well as the products VSA might work best in, by carrying out if it is best in department stores, or for goods with a high involvement such as electronics, etc.. Since the high diffusion of iPad devices, further researches could develop new efficient virtual shopping assistant applications focused on these devices, which can be directly connected to a new version of iTunes Stores, capable of delivering "real" products.

Finally, new studies on the consumer perception of the virtual agent's face must be conducted.

REFERENCES

Acquisti, A., & Varian, H. R. (2005). Conditioning prices on purchase history. *Marketing Science*, *24*(3), 367–381. doi:10.1287/mksc.1040.0103

Adamo, A., Bertacchini, P. A., Bilotta, E., Pantano, P., & Tavernise, A. (2010). Connecting art and science for education: Learning through an advanced virtual theater with "Talking Heads". *Leonardo*, *43*(5), 442–448. doi:10.1162/LEON_a_00036

Ahn, H. J. (2010). Evaluating customer aid functions of online stores with agent-based models of customer behavior and evolution strategy. *Information Science*, *180*, 1555–1570. doi:10.1016/j.ins.2009.12.029

Bartneck, C., & Reichenbach, J. (2005). Subtle emotional expressions of synthetic characters. *International Journal of Human-Computer Studies*, *62*, 179–192. doi:10.1016/j.ijhcs.2004.11.006

Bertacchini, P. A., Bilotta, E., Pantano, P., Battiato, S., Cronin, M. D., Di Blasi, G., et al. Tavernise A. (2007). Modelling and animation of theatrical Greek masks in an authoring system. In R. De Amicis & G. Conti (Eds.), *Proceedings of the 5th Eurographics Italian Chapter (EG-IT-07)* (pp. 191-197). Germany: Eurographics Association.

Boehner, K., DePaula, R., Dourish, P., & Sengers, P. (2007). How emotion is made and measured. *International Journal of Human-Computer Studies*, *65*(4), 75–291. doi:10.1016/j.ijhcs.2006.11.016

Bressler, S. L., & Kelso, J. A. S. (2001). Cortical coordination dynamics and cognition. *Trends in Cognitive Sciences*, *5*, 26–36. doi:10.1016/S1364-6613(00)01564-3

Cassell, J., Bickmore, T., & Campbell, L. Vilhja′ lmsson, H., & Yan, H. (2000). Human conversation as a system framework: Designing embodied conversational agents. In J. Cassell, J. Sullivan, S. Prevost, & E. Churchill, (Eds.), *Embodied conversational agents* (pp. 29-63). Cambridge, MA: MIT Press.

Chang, C.-A., & Burke, R. R. (2007). Consumer choice of retail shopping aids. *Journal of Retailing and Consumer Services*, *14*, 339–346. doi:10.1016/j.jretconser.2006.12.003

Chang, H. H., & Wang, I. C. (2008). An investigation of user communication behaviour in computer mediated environments. *Computers in Human Behavior*, *24*(5), 2336–2356. doi:10.1016/j.chb.2008.01.001

Churchill, E., Cook, L., Hodgson, P., Prevost, S., & Sullivan, J. (2000). 'May I help you?': Designing embodied conversational agent allies. In Cassell, J., Sullivan, J., Prevost, S., & Churchill, E. (Eds.), *Embodied conversational agents* (pp. 64–94). Cambridge, MA: MIT Press.

Cohn, J. F., Reed, L. I., Moriyama, T., Xiao, J., Schmidt, K. L., & Ambadar, Z. (2004). *Multimodal coordination of facial action, head rotation, and eye motion*. Sixth IEEE International Conference on Automatic Face and Gesture Recognition (pp. 645-650). Korea: IEEE Computer Series.

Dahlén, M., & Lange, F. (2002). Real consumers in the virtual store. *Scandinavian Journal of Management*, *18*(3), 341–363. doi:10.1016/S0956-5221(01)00012-4

Davenport, T. H. (2005). *Thinking for a living: How to get better performance and results from knowledge workers*. Boston, MA: Harvard Business School Press.

Davis, F. D. (1989). Perceived usefulness, perceived ease of use and user acceptance of information technology. *Management Information Systems Quarterly*, *13*(3), 319–339. doi:10.2307/249008

DeSanctis, G., & Fulk, J. (1999). *Shaping organizational forms*. Thousand Oaks, CA: Sage.

DeSanctis, G., & Scott Poole, M. (1994). Capturing the complexity in advanced technology use: Adaptive structuration theory. *Organization Science*, *5*(2), 121–147. doi:10.1287/orsc.5.2.121

Diep, V. C. S., & Sweeney, J. C. (2008). Shopping trip value: Do stores and products matter? *Journal of Retailing and Consumer Services*, *15*, 399–409. doi:10.1016/j.jretconser.2007.10.002

Ekman, P. Friesen, & W. V., & Hager, J. C. (2002). *The facial action coding system*. London, UK: Weidenfeld & Nicolson.

Gulz, A. (2004). Benefits of virtual characters in computer based learning environments: Claims and evidence. *International Journal of Artificial Intelligence in Education*, *14*, 313–334.

Gulz, A., & Haakeb, M. (2006). Design of animated pedagogical agents-a look at their look. *International Journal of Human-Computer Studies*, *64*, 322–339. doi:10.1016/j.ijhcs.2005.08.006

Ha, S., & Stoel, L. (2009). Consumer e-shopping acceptance: Antecedents in a technology acceptance model. *Journal of Business Research, 62*(5), 565–571. doi:10.1016/j.jbusres.2008.06.016

Hausman, A. V., & Siekpe, J. S. (2009). The effect of Web interface features on consumer online purchase intentions. *Journal of Business Research, 62*(1), 5–13. doi:10.1016/j.jbusres.2008.01.018

Hevner, A. R., March, S. T., Park, J., & Ram, S. (2004). Design science in Information Systems research. *Management Information Systems Quarterly, 28*(1), 75–105.

Hsiao, M.-H. (2009). Shopping mode choice: Physical store shopping versus e-shopping. *Transportation Research Part E, Logistics and Transportation Review, 45*, 86–95. doi:10.1016/j.tre.2008.06.002

Jin, S.-A. A. (2009). The roles of modality richness and involvement in shopping behaviour in 3D virtual stores. *Journal of Interactive Marketing, 23*(3), 234–246. doi:10.1016/j.intmar.2009.04.005

Katsikitis, M. (1997). The classification of facial expressions of emotion: A multidimensional-scaling approach. *Perception, 26*, 613–626. doi:10.1068/p260613

Kim, H., & Niehm, L. S. (2009). The impact of website quality on information quality, value, and loyalty intentions in apparel retailing. *Journal of Interactive Marketing, 23*(3), 221–233. doi:10.1016/j.intmar.2009.04.009

Kim, J., Fiore, A. M., & Lee, H.-H. (2007). Influences of online store perception, shopping enjoyment, and shopping involvement on consumer patronage behaviour towards an online retailer. *Journal of Retailing and Consumer Services, 14*, 95–107. doi:10.1016/j.jretconser.2006.05.001

Kostova, T., & Roth, K. (2002). Adoption of an organizational practice by subsidiaries of multinational corporations: Institutional and relational effects. *Academy of Management Journal, 45*(1), 215–233. doi:10.2307/3069293

Kulviwat, S., Bruner, G. C. II, & Al-Shuridah, O. (2009). The role of social influence on adoption of high tech innovations: The moderating effect of public/private consumption. *Journal of Business Research, 62*(7), 706–712. doi:10.1016/j.jbusres.2007.04.014

Lee, K. C., & Chung, N. (2008). Empirical analysis of consumer reaction to the virtual reality shopping mall. *Computers in Human Behavior, 24*, 88–104. doi:10.1016/j.chb.2007.01.018

Lester, J., Stone, B., & Stelling, G. (1999). Lifelike pedagogical agents for mixed-initiative problem solving in constructivist learning environments. *User Modeling and User-Adapted Interaction, 9*, 1–44. doi:10.1023/A:1008374607830

Lin, J. C., & Lu, H. (2000). Towards an understanding of the behavioural intention to use a website. *International Journal of Information Management, 20*, 197–208. doi:10.1016/S0268-4012(00)00005-0

Magalhães, R., & Silva, A. R. (2009). *Organizational design and engineering*. Working paper 1, INOV Centre for Organizational Design and Engineering (CODE). Retrieved October, 4, 2010, from http://is2.lse.ac.uk/wp/pdf/wp179.pdf

Marshall, G. W., Goebel, D. J., & Moncrief, W. C. (2003). Hiring for success at the buyer–seller interface. *Journal of Business Research, 56*, 247–255. doi:10.1016/S0148-2963(02)00435-6

McAfee, A. P. (2006). Enterprise 2.0: The dawn of emergent collaboration. *MIT Sloan Management Review, 47*(3), 21–28.

Michael, R. W., & Attaway, J. S. (2003). At the interface: The nature of buyer-seller interactions and relationships. *Journal of Business Research*, *56*, 243–246. doi:10.1016/S0148-2963(02)00434-4

Michon, R., Chebat, J.-C., & Turley, L. W. (2006). Mall atmospherics: The interaction effects of the mall environment on shopping behaviour. *Journal of Business Research*, *58*, 576–583. doi:10.1016/j.jbusres.2003.07.004

Migliarese, P., & Corvello, V. (in press). Organizational relations in organizational design and engineering. *International Journal of Organizational Design and Engineering.*

Murray, K. B., & Häubul, G. (2009). Personalization without interrogation: Towards more effective interactions between consumers and feature-based recommendation agents. *Journal of Interactive Advertising*, *23*(2), 138–146. doi:10.1016/j.intmar.2009.02.009

Newsom, M. K., Collier, D. A., & Olsen, E. O. (2009). Using "biztainment" to gain competitive advantage. *Business Horizons*, *52*, 167–176. doi:10.1016/j.bushor.2008.10.005

Oatley, K. (1992). *Best laid schemes: The psychology of emotions*. New York, NY: Cambridge University Press.

OECD. (2000). *Knowledge management in the learning society*. Paris, France: OECD.

Orlikowski, W. J. (2000). Using technology and constituting structures: A practice Lens for studying technology in organizations. *Organization Science*, *11*(4), 404–428. doi:10.1287/orsc.11.4.404.14600

Orlikowski, W. J. (2007). Sociomaterial practices: Exploring technology at work. *Organization Studies*, *28*(9), 1435–1448. doi:10.1177/0170840607081138

Palmatier, R. W., Dant, R. P., Grewal, D., & Evans, K. E. (2006). Factors influencing the effectiveness of relationship marketing: A meta-analysis. *Journal of Marketing*, *70*, 136–153. doi:10.1509/jmkg.70.4.136

Pantano, E. (2010). New technologies and retailing: Trends and directions. *Journal of Retailing and Consumer Services*, *17*(3), 171–172. doi:10.1016/j.jretconser.2010.03.004

Pantano, E., & Naccarato, G. (2010). Entertainment in retailing: the role of advanced technologies. *Journal of Retailing and Consumer Services*, *17*(3), 200–204. doi:10.1016/j.jretconser.2010.03.010

Perlin, K. (1997). *Layered compositing of facial expression*. ACM SIGGRAPH Technical Sketch. Retrived June 25, 2010, from http://mrl.nyu.edu/~perlin/experiments/facedemo

Puccinelli, N. M., Goodstein, R. C., Grewal, D., Price, R., Raghubir, P., & Stewart, D. (2009). Customer experience management in retailing: Understanding the buying process. *Journal of Retailing*, *85*(1), 15–30. doi:10.1016/j.jretai.2008.11.003

Reinders, M. J., Dabholkar, P. A., & Frambach, R. T. (2008). Consequences of forcing consumers to use technology-based self-service. *Journal of Service Research*, *11*(2), 107–123. doi:10.1177/1094670508324297

Roussos, G., Kourouthanasis, P., & Moussouri, T. (2003). Designing appliances for mobile commerce and retailtainment. *Personal and Ubiquitous Computing*, *7*, 203–209. doi:10.1007/s00779-003-0221-3

Söderlund, M., & Julander, C. R. (2009). Physical attractiveness of the service worker in the moment of truth and its effects on customer satisfaction. *Journal of Retailing and Consumer Services*, *16*, 216–226. doi:10.1016/j.jretconser.2008.11.008

Sun, T., Tai, Z., & Tsai, K. C. (2009). The role of interdependent self-construal in consumers' susceptibility to retail sales persons' influence: A hierarchical approach. *Journal of Retailing and Consumer Services*, *16*(5), 360–366. doi:10.1016/j.jretconser.2009.04.002

Thong, J. Y. L., Hong, S., & Tama, K. Y. (2006). The effects of post-adoption beliefs on the expectation-confirmation model for information technology continuance. *International Journal of Human-Computer Studies*, *64*, 799–810. doi:10.1016/j.ijhcs.2006.05.001

Veletsianos, G., & Miller, C. (2008). Conversing with pedagogical agents: A phenomenological exploration of interacting with digital entities. *British Journal of Educational Technology*, *39*(6), 969–986. doi:10.1111/j.1467-8535.2007.00797.x

Vignali, C., Vignali, G., Vranesevic, T., & Vrontis, D. (2006). *Basic selling skills*. UK: Foxwell & Davies.

Wang, H.-C., & Doong, H.-S. (in press). Argument from a spokeperson type: The recommendation strategy of virtual salespersons. *International Journal of Information Management*.

Wu, J. J., & Chang, Y. S. (2005). Towards understanding members' interactivity, trust, and flow in online travel community. *Industrial Management & Data Systems*, *105*(7), 937–954. doi:10.1108/02635570510616120

Zahay, D. L., & Peltier, J. (2008). Interactive strategy formation: Organizational and entrepreneurial factors related to effective customer information systems practices in B2B firms. *Industrial Marketing Management*, *37*, 191–205. doi:10.1016/j.indmarman.2006.10.004

Zhang, L., Snavelvelyn, N., Curless, B., & Seitz, M. S. (2004). Spacetime faces: High resolution capture for modeling and animation. In [ACM Press.]. *Proceedings of, SIGGRAPH2004*, 548–558.

ADDITIONAL READING

Chen, L.-D., & Tan, J. (2004). Technology Adaptation in E-commerce: Key Determinants of Virtual Store Acceptance. *European Management Journal*, *22*(1), 74–86. doi:10.1016/j.emj.2003.11.014

Cheng, H. (in press). Seeking knowledge or gaining legitimacy? Role of social networks on new practice adoption by OEM suppliers. *Journal of Business Research.*

Fishbein, M., & Ajzen, I. (1975). *Belief Attitude, Intention, and Behavior: An Introduction to Theory and Research. Reading.* MA: Addison-Wesley Publishing.

Franco, M., & Mariano, S. (2007). Information Technology Repositories and Knowledge Management Processes: A Qualitative Analysis. *Journal of Information and Knowledge Management Systems*, *37*(4), 440–451.

Kulwait, S., Bruner, G. II, & Al-Shuridah, O. (2008). The role of social influence on adoption of high tech innovations: The moderating effect of public/private consumption. *Journal of Business Research*, *62*, 706–712.

Lee, K. C., & Chung, N. (2008). Empirical analysis of consumer reaction to the virtual reality shopping mall. *Computers in Human Behavior*, *24*, 88–104. doi:10.1016/j.chb.2007.01.018

Oviatt, S. L. (2008). Multimodal interface. In Jacko, J. A., & Sears, A. (Eds.), *The Human-Computer Interaction Handbook: Fundamentals, Evolving Technologies, and Emerging Applications* (pp. 413–433). CRC Press.

Pantano, E., Taversine, A., & Viassone, M. (2010). Consumer perception of computer-mediated communication in a social network. In *proceedings of the 4th International Conference on New Trends in Information Science and Service Science* (pp. 609-614). Korea: AICIT Publisher.

Venkatesh, V. (2000). Determinants of Perceived Ease of Use: Integrating Control, Intrinsic Motivation, and Emotion into the Technology Acceptance Model. *Information Systems Research*, *11*(4), 342–365. doi:10.1287/isre.11.4.342.11872

Yoo, W., Lee, Y., & Park, J. (2010). The role of interactivity in e-tailing: Creating value and increasing satisfaction. *Journal of Retailing and Consumer Services*, *17*(2), 89–96. doi:10.1016/j.jretconser.2009.10.003

KEY TERMS AND DEFINITIONS

Consumer Shopping Experience: The consumers' experience which characterizes the shopping activity. A positive one is capable of influencing the purchasing process and their loyalty to the online/offline retailer.

Emotions: Mental states of readiness that arise from appraisals of events or one's own thoughts. (adapted from Oatley, 1992)

Human-Computer Interaction: A discipline concerned with the design, evaluation and implementation of interactive computing systems for human use and with the study of major phenomena surrounding them.

Practice: An individual's routine use of knowledge for conducting a particular function that has evolved over time under the influence of the individual's history, interests, and actions (adapted from Kostova & Roth 2002).

Shopping Assistant Systems: Systems created for supporting and influencing of consumers shopping experience. They might be based on shopping trolleys or on interactive screens and provide a user friendly interface which can be used by consumers to ask achieve information in an innovative and entertaining way.

Technology Structuration: The process by which people incorporate advanced technologies into their practices.

Virtual Human: It is a virtual agent that is graphically represented by a humanoid avatar.

Virtual Shopping Assistant (VSA): An innovative virtual seller, capable of supporting consumers during their in-store activities and influencing the subsequent decision making and purchasing behavior.

Chapter 5
Information and Communication Technologies in Marketing Channels:
Product Considerations

Irene Gil Saura
University of Valencia, Spain

Marta Frasquet Deltoro
University of Valencia, Spain

Maria Eugenia Ruiz-Molina
University of Valencia, Spain

ABSTRACT

There is broad consensus about the benefits of information and communication technologies (ICT) for distribution channel members. Notwithstanding, it has been pointed out that there is a need to adjust investment to what is strictly necessary. In this vein, this chapter attempts to analyze the level of use and assessment as well as the differences between the different actors' perceptions of their respective suppliers' ICT. In this way, it is intended to discern the major factors which suppliers should take into account when prioritizing the implementation of certain technological solutions, depending on the retail activity.

INTRODUCTION

Organizations must cope with an increasingly changing environment caused by the rapid evolution of technology, among others factors (Porter, 1997). In particular, since information is one of the key success factors in an organization, modern advances in information and communication technologies (ICT) offer new possibilities for business management, representing sources of competitive advantage (Buxmann and Gebauer, 1999).

Regarding marketing channels, diffusion of the use of ICT has involved a deep change in an increasing number of business functions, such as order processing, inventory management, storage,

DOI: 10.4018/978-1-60960-738-8.ch005

transportation and customer service, representing an important source of competitive advantages (Bowerson and Daugherty, 1995). In particular, ICT implementation facilitates the coordination of logistic processes between different channel members (Fen and Luang, 2007).

However, investment in technologies does not always provide the expected returns and therefore, there have been warnings about the dangers of excessive investment in technology (Sethuraman and Parasuraman, 2005), stating that "good" technology is the "appropriate" technology. In order to evaluate this issue, the consideration of the customer point of view on the ICT solutions is crucial (Weinstein, 2002; Frasquet et al., 2008; Gil et al., 2008a; 2008b). Notwithstanding, previous studies have only addressed this issue partially, since they have either focused on a particular technology solution, e.g. self-service technology (Meuter et al., 2000; Dean, 2008), or a single setting, i.e. B2B or B2C.

In this regard, the present chapter pursues a double objective. First of all, we aim to analyze how customers at different levels of distribution channels evaluate the ICT implemented by the sellers. In order to achieve this aim, three samples of manufacturers, retailers, and consumers respond to questionnaires regarding their level of usage of several ICT solutions of their main supplier as well as their assessment of their supplier's technological advancement and alignment. Secondly, we test the existence of significant differences in the use of different technologies across sectors of activity in each of the levels of the marketing channel.

In this way, we expect to infer a series of managerial implications for manufacturers' and retailers' suppliers, as well as for retailers, about the most appropriate technologies for their businesses depending on their level in the marketing channel and the type of product marketed.

THEORETICAL BACKGROUND

ICT is *"a term that encompasses all forms of technology utilized to create, capture, manipulate, communicate, exchange, present, and use information in its various forms (business data, voice conversations, still images, motion pictures, multimedia presentations", etc.)"* (Ryssel et al., 2004:198).

Among the benefits derived from ICT implementation, the literature has pointed out significant improvements in internal processes that ultimately contribute to customer service improvements. In particular, it has been argued that ICT positively affects internal effectiveness through important savings in terms of time, communication costs and personnel expenses; greater reliability, accurateness and error reduction; improvements in inventory management, as well as increases in productivity through the automation of labor intensive tasks (Ellram et al., 1999; Lowson, 2001).

From the consumer point of view, ICT allows service improvements (Lowson, 2001; Gil et al., 2008a; 2008b), providing savings in shopping time through a wider assortment and making it possible to accomplish the required purchases in one shopping trip (Messinger and Narasimhan, 1997).

Additionally, ICT implementation positively affects the relationship value of a distributor with its suppliers and customers (Lewis, 2001; Mentzer and Williams, 2001; Frasquet et al., 2008; Gil et al., 2007; 2008a), involving greater satisfaction (Meuter et al., 2000; Burke, 2002) and increased perceived quality (Gil et al., 2008b). In B2B relationships, ICT creates closer bonds between buyer and seller and is considered a source of commitment (Rebolledo et al, 2005). Additionally a correlation has been observed between buyer satisfaction with the seller's ICT and buyer's future purchase intentions both in B2B (McDonald and Smith, 2004) and in B2C settings (Meuter et al., 2000).

Nevertheless, among the paradoxes of technology it has been emphasized that ICT can integrate but can also generate a feeling of isolation, lack of control or ineptitude (Mick and Fournier, 1998). Additionally, in the context of technology-based service encounters, failures in service design or process may generate customer dissatisfaction (Snellman and Vihtkari, 2003). Finally, it has been also suggested that technology can be a source of satisfaction or dissatisfaction depending on customer attitude towards ICT (Parasuraman and Colby, 2001; Walker et al., 2002) and staff responsiveness (Meuter et al., 2000; Bitner, 2001). In particular, in the business-to-consumer setting, several studies have explored the influence of some consumer characteristics, such as age and gender, on technology acceptance (Davis, 1989; Davis et al., 1989). Regarding age, some researchers provide evidence which suggests that older people may experience more difficulty with the use of computers and other applications of information technology compared to younger individuals and the older customers are, the less likely they are to obtain customer service through a technology-based medium (Dulude, 2002; Burke, 2002; Simon and Usunier, 2007). In contrast, several studies conclude that age has no effect on attitudes towards retail IT solutions (Dabholkar et al., 2003; Eastman and Iyer, 2004; Niemelä-Nyrhinen 2007; Weijters et al., 2007; Dean, 2008).

Concerning gender, while Meuter et al. (2005) find that women are more reluctant to use the retailer's ICT, Odekerken-Schroder et al. (2001) point out that female consumers assess more positively the retailer's technology solutions and this has a higher influence on their loyalty towards the store than for male consumers.

Thus, there is a debate in the literature about the advisability of investing in technology as much as possible, or adjusting investment to what is strictly necessary. Consultants maintain that more technology is always better than lower levels of technological development (Palmer and Markus, 2000), whereas academics state that companies should only adopt technologies that fit the specific strategic aims pursued by the company. Some authors warn about the dangers of "*overengineering*" or of excessive investment in technology (Sethuraman and Parasuraman, 2005).

The impact of ICT in marketing channels has been evaluated by considering, in addition to the investment amount, suitability, as regards technology advancement and alignment between supplier and customer ICT solutions (Powell, 1992; Hausman and Stock, 2003; Wu et al., 2006). Several studies maintain that an intense use of technology contributes to building strong relationships in the value chain, since ICT allow greater speed and quality in information transfer (Clemons and Row, 1993; Booth and Philip, 1998), greater coordination and lower transaction costs (Clemons and Row, 1993) and increased integration between channel members (Wu et al., 2006).

Finally, in retailing it has been suggested that technological solutions should be designed according to the specific requirements of consumers and the characteristics of the store assortment (Burke, 2002; Drennan and McColl-Kennedy, 2003). In this regard, consumers require *high touch* with sales personnel for certain product categories of low frequency purchase, such as home appliances, whereas technical solutions facilitating process automation and time savings are highly valued for frequently purchased goods such as groceries.

Taking into consideration the existing empirical evidence, in the present chapter we aim to contribute to the academic work. We aim to explore the use of a wide catalogue of technology solutions as well as customer assessment in three different settings, manufacturer-wholesaler, wholesaler-retailer and retailer-consumer relationships. To do so, we have considered the following questions:

1. *Which ICT solutions are more intensively used by customers in different marketing channel levels?* We understand that manufacturers, retailers and consumers have different

characteristics and needs in their relationships with their suppliers and, therefore, we expect them to use different ICT solutions provided by their suppliers, and even if they use the same solutions, to differ in the intensity of use of such technologies.

2. *Are there differences in the level of customer ICT use in its relationship with the main supplier depending on the industry?* We aim to analyze whether companies in different activity sectors differ in ICT use in their relationships with their main suppliers.
3. *To what extent are the supplier's technological solutions valued by customers in different levels of marketing channels?* In this way, we try to analyze customer perceptions of the supplier's ICT solutions depending on the customer level in marketing channel (manufacturer, retailer or consumer).
4. *Are there differences in the assessment of the supplier's ICT solutions depending on the industry?* We test if companies belonging to different activity sectors differ in their assessment of their main supplier's ICT solutions.

METHOD

In order to achieve the proposed objectives, quantitative research is performed through a personal survey. Table 1 shows the main characteristics of our research.

The items regarding the intensity of use of the different ICT solutions have been extracted and adapted from Lowson (2001) and Observatorio (2006), whereas those referring to assessment of main supplier's ICT solutions have been extracted from Wu et al. (2006). All items of the questionnaire are measured on a 5-point Likert scale.

The database of manufacturers and retailers is obtained from available secondary information. In particular, the main companies in each activity sector are selected according to their CNAE (National Classification of Economic Activities) and IAE (Tax on Economic Activities) codes obtained from SABI (Iberian Accounting Analysis System), an Informa database that contains the annual reports of the most important Spanish and Portuguese companies in terms of total sum of the assets in the company's balance sheet.

In order to characterize the samples of manufacturers, retailers and consumers, Table 2 and

Table 1. Research characteristics

Universe	Manufacturers of grocery, textile/leather, electronics/home appliances and furniture/wood/decoration,, paper/cardboard/graphical image, automotive, mechanics, chemistry, metal, glass and others Consumers and retailers of grocery, clothing/footwear, electronics/home appliances and furniture/wood/decoration
Geographic scope	Spain (Madrid, Barcelona and Valencia)
Samples sizes	280 manufacturers 304 retailers 400 consumers
Survey method	Personal interview to retailers and manufacturers at their offices Personal interview to final consumers at the store exit
Information collection period	January-March 2007 (manufacturers) October-December 2007 (retailers) September-October 2007 (consumers)
Statistical techniques	Descriptive analysis Analysis of variance (ANOVA)
Statistical software	SPSS version 15.0

3 show their distributions in terms of their classification variables.

As can be observed from Tables 2 and 3, the consumer and the retailer samples are proportionally distributed regarding the industries analyzed.

Table 2. Socioeconomic characteristics of the sample of consumers

Consumers	Number	%
Gender:		
- male	184	46.00
- female	216	54.00
Age:		
- 18-25 years	25	6.30
- 26-35 years	113	28.30
- 36-45 years	156	39.00
- 46-55 years	78	19.50
- 56-65 years	22	5.50
- Over 65 years	6	1.50
Number of stores:		
- grocery	8	15.68
- clothing / footwear	16	31.37
- electronics / home appliances	16	31.37
- furniture / decorations	11	21.57
Length of patronage:		
- 0-5 years	242	60.50
- 6-10 years	119	29.75
- 11-15 years	19	4.75
- Over 15 years	19	4.75

In the manufacturers' sample, the global sample average includes 280 manufacturers, whereas for the comparative analysis between the four analyzed sectors -i.e. grocery, electronics, apparel, and paper/cardboard/graphical image-, we have just considered 43, 38, 40 and 42 companies respectively (58.2% of the total sample).

ICT USE IN MARKETING CHANNELS

In order to answer the first question, i.e. which technological solutions are more intensively used by customers in different marketing channel levels, we calculate the average values of the manufacturers' scores regarding the intensity of ICT use by their corresponding main suppliers (Figure 1).

The results show the use of Internet and e-mail is widely extended in all the industries, with scores exceeding the midpoint of the scale (3) in all cases. In contrast, there still scanty use of radio frequency systems both for identification and transportation.

Additionally, in order to provide an answer to our second research question, we test the significance of the differences in the manufacturer's perception of intensity of ICT use in its relationship with its main supplier across industries. From

Figure 1. Manufacturers' intensity of use of supplier's ICT (industry average)

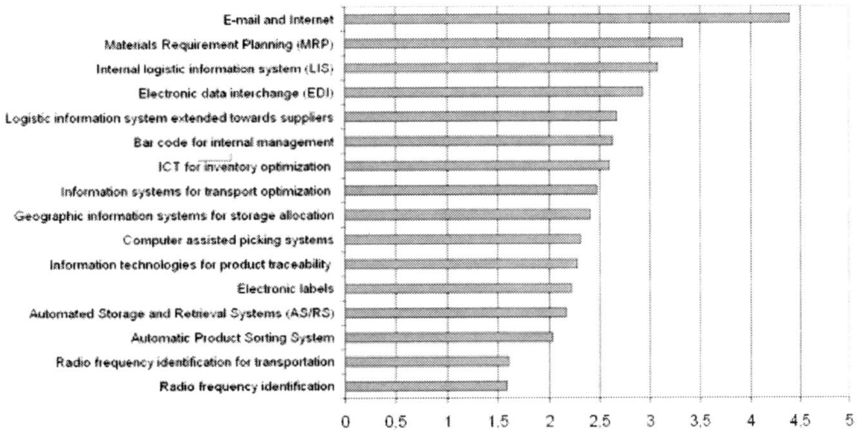

Table 3. Socioeconomic characteristics of the samples of manufacturers and retailers

	Manufacturers		Retailers	
	Number	%	Number	%
% purchases to the main supplier on total purchases of the product:				
- Less than 25%	21	8.1	30	10.0
- 26%-50%	72	27.8	119	39.7
- 51%-75%	73	28.2	74	24.7
- More than 75%	93	35.9	77	25.7
Number of alternative suppliers for the same product:				
- 0-1 suppliers	68	24.3	33	11.0
- 2-4 suppliers	128	45.7	85	28.3
- 5-8 suppliers	45	16.1	100	33.3
- more than 9 suppliers	19	6.8	82	27.3
Industry of the main supplier:				
- grocery	43	15.4	79	26.0
- electronics/ home appliances	38	13.6	87	28.6
- textile/leather	40	14.3	68	22.4
- paper/cardboard/graphical image	42	15.0	70	23.0
- wood	31	11.1	-	-
- automotive	12	4.3	-	-
- mechanics/metal	30	10.7	-	-
- chemistry	28	10.0	-	-
- glass	7	2.5	-	-
- others	9	3.2	-	-
Length of relationship:				
- 0-5 years	33	12.0	72	24.3
- 6-10 years	56	20.4	94	31.6
- 11-15 years	66	24.0	41	13.8
- 16-20 years	45	16.4	43	14.5
- 21-25 years	29	10.5	17	5.7
- Over 25 years	22	8.0	29	9.8

the results shown in Table 4 we infer that there is no industry that clearly uses ICT solutions more intensively than the other activity sectors. However, there are significant differences in Internet and e-mail use, as the electronics and textile manufacturers show a more intense use of ICT solutions in comparison to grocery and furniture/decoration manufacturers. In contrast, furniture and decoration manufacturers use electronic data transmission to a lesser extent, differing significantly from the scores shown by manufacturers of electronics/home appliances.

In a similar vein, we calculate average values for the intensity of use of the main supplier's ICT solutions by retailers and conduct a comparative analysis across sectors (Figure 2).

Similarly to manufacturers, retailers use the different technological applications with unequal intensity, and there is a strong correspondence between the level of use of such technologies by manufacturers and retailers (Pearson's correlation coefficient: 0.915; $p<0.01$).

As for the comparison across sectors, generally, electronics/home appliances retailers declare the most intensive use of ICT solutions their relation with their main supplier. This intensive use of technology refers both to hardware as well as software. In contrast, grocery retailers present the lowest values in comparison to the other retailers, since they never surpass the midpoint of the scale, evidencing their low degree of technological development.

Table 4. Manufacturers' intensity of use of supplier's ICT (by industry)

	Global average	1 Grocery	2 Cloth./footw.	3 Electronics	4 Furnit./deco.	Differences between groups[a]
1. Bar code for internal management	2.63	2.57	2.50	2.47	2.42	-
2. Radio frequency identification	1.59	1.77	1.57	1.31	1.35	-
3. Electronic data interchange (EDI)	2.93	2.93	3.00	3.31	2.00	3-4
4. E-mail and Internet	4.39	4.11	4.60	4.79	3.81	1-3, 2-4, 3-4
5. Internal logistic information system (LIS)	3.07	2.71	3.10	3.12	2.90	-
6. Logistic information system extended towards its suppliers	2.67	2.09	2.95	2.44	2.50	-
7. Materials Requirement Planning (MRP)	3.32	2.97	3.50	3.43	2.65	-
8. Automated Storage and Retrieval Systems (AS/RS)	2.16	2.12	2.55	1.62	1.67	2-3
9. Automatic Product Sorting System	2.03	1.95	2.40	1.68	1.71	-
10. Computer assisted picking systems	2.31	2.19	2.48	1.97	1.87	-
11. ICT for inventory optimization	2.60	2.60	2.70	2.39	2.42	-
12. Information systems for transport optimization	2.47	2.51	2.57	2.13	2.29	-
13. Information technologies for product traceability	2.27	2.71	2.17	2.05	1.93	-
14. Geographic information systems (GIS) for storage allocation	2.40	2.29	2.47	2.03	1.77	-
15. Radio frequency identification (RFID) for transportation	1.61	1.56	1.65	1.47	1.35	-
16. Electronic labels	2.22	2.36	2.20	1.74	2.03	-
In summary, to what extent does your company use technologies for information transmission?	3.28	3.09	3.27	3.47	2.93	-
In summary, to what extent does your company use technologies for information collection?	2.42	2.58	2.72	2.23	2.10	-
In summary, to what extent does your company use technologies for transportation?	2.17	2.45	2.22	1.97	1.97	-
In summary, to what extent does your company use technologies for storage?	2.50	2.34	2.65	2.16	2.23	-

Differences between electronics/home appliances retailers and the other retail sectors are especially large in the use of ICT solutions for information transmission (EDI, email and Internet, internal and external LIS and MRP).

However, no statistically significant differences regarding radio frequency use are observed between the different retailers, for either data identification or transportation, as this technology is still barely developed.

Consumers were asked about their level of use of retailer technological applications that are "visible" for them. For this reason, we have included some items that were not previously considered for ICT use by manufacturers and retailers. The results are displayed in Figure 3 and Table 6.

In contrast to the results obtained for manufacturers and retailers, it is observed that in the relationships between retailers and consumers, Internet and e-mail play a modest secondary role, whereas the intensive use of credit card payment as well as bar codes, scanner and self-service technologies is remarkable.

There are also significant differences regarding the intensity of ICT use across industries. In particular, self-service technologies - i.e. self-weighing and label printing of loose goods, self-scanning check-out - and loyalty programs are

Figure 2. Retailers' intensity of use of supplier's ICT (industry average)

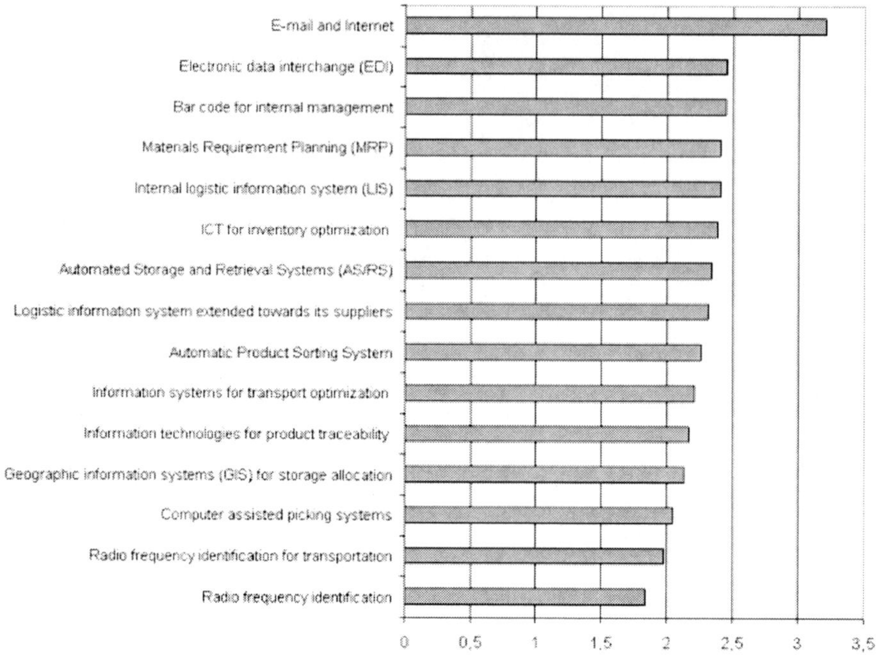

used to a significantly greater extent by grocery retailers in comparison with the other retail sectors.

However, there are no statistically significant differences across retailers as far as the use of bar codes and debit or credit card payment are concerned. Regarding methods of payment, customers of grocery retailers make a more intensive use of store card payment, whereas for electronics retailers, a relatively greater importance of bank transfer payment and mobile payment is observed. Apparel retailers are those showing the lowest scores for cash on delivery.

Furthermore, patrons of electronics/home appliances retailers show higher average intensity of use of communication technologies to contact with the store (telephone, fax and e-mail), security systems, as well as invoicing software in comparison to the other retailers. This sector, together with furniture and decoration retailers (i.e. durable goods retailers), enjoy the highest levels of use of their websites. This result is consistent with the relatively greater percentage of expenditure that these goods represent for consumers and their subsequent greater involvement in the purchase process. Finally, in apparel stores, consumers report a lower level of use of design and office programs.

ICT ASSESSMENT IN MARKETING CHANNELS

Through our third research question we aim to assess to what extent the supplier's technological solutions are valued in different levels of the marketing channel. First of all, we examine manufacturers' assessment of ICT advancement and alignment of their main suppliers (Table 7), where all the scores are below the midpoint of the scale and there are no significant differences across industries.

Regarding the retailer assessment of its ICT advancement and alignment in relation to its main supplier (Table 8), low scores are also observed.

Table 5. Retailers' intensity of use of supplier's ICT (by industry)

	Global average	1 Grocery	2 Cloth./footw.	3 Electronics	4 Furnit./deco.	Differences between groups[a]
1. Bar code for internal management	2.45	2.26	2.51	3.04	2.06	1-3
2. Radio frequency identification	1.84	1.73	1.72	2.09	1.83	-
3. Electronic data interchange (EDI)	2.46	2.13	2.28	3.26	2.33	1-3, 2-3, 3-4
4. E-mail and Internet	3.21	2.52	3.03	4.04	3.44	1-3, 1-4, 2-3
5. Internal logistic information system (LIS)	2.41	2.09	2.31	3.03	2.32	1-3, 2-3, 3-4
6. Logistic information system extended towards its suppliers	2.32	1.95	2.23	2.95	2.25	1-3, 2-3, 3-4
7. Materials Requirement Planning (MRP)	2.41	2.01	2.36	3.01	2.33	1-3, 2-3, 3-4
8. Automated Storage and Retrieval Systems (AS/RS)	2.34	2.04	2.24	2.94	2.23	1-3, 2-3, 3-4
9. Automatic Product Sorting System	2.26	2.03	2.26	2.72	2.09	1-3, 3-4
10. Computer assisted picking systems	2.05	1.83	1.92	2.52	2.01	1-3, 2-3, 3-4
11. ICT for inventory optimization	2.39	2.05	2.44	2.76	2.39	1-3
12. Information systems for transport optimization	2.21	1.97	2.15	2.53	2.24	1-3
13. Information technologies for product traceability	2.17	1.96	2.02	2.57	2.22	1-3, 2-3
14. Geographic information systems (GIS) for storage allocation	2.13	1.89	2.02	2.62	2.07	1-3, 2-3, 3-4
15. Radio frequency identification for transportation	1.98	1.85	1.85	2.22	2.04	-
In general, how intensively does your company use its hardware in the relationship with its main supplier?	2.56	2.28	2.48	3.16	2.42	1-3, 2-3, 3-4
In general, how intensively does your company use its software in the relationship with its main supplier?	2.60	2.22	2.43	3.40	2.49	1-3, 2-3, 3-4

Figure 3. Consumers' intensity of use of retailer's ICT (industry average)

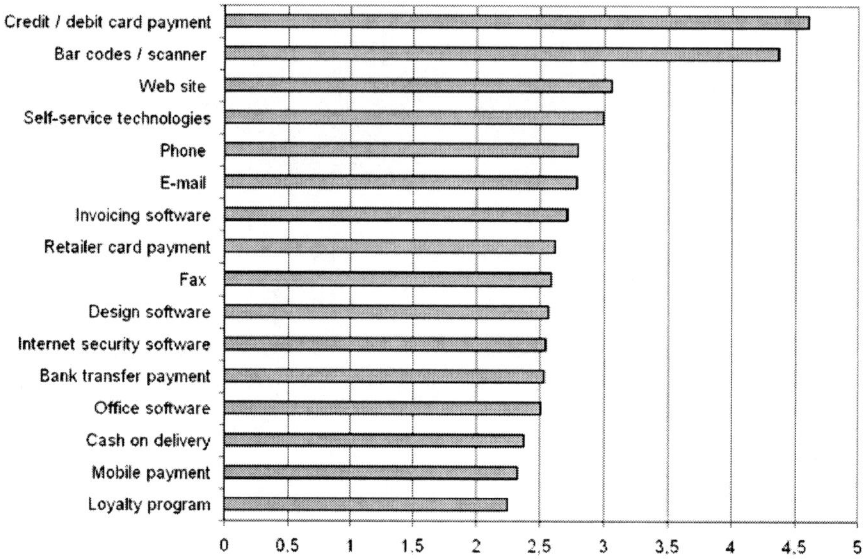

This fact can be understood in the sense that retailers are highly demanding in their relationships with their main suppliers with regards to technological advancement and alignment.

Regarding the fourth research question, consistent with the superior intensity of ICT use by electronics retailers in their relation with their main suppliers, it can be inferred from Table 8 that there is a significantly better ICT assessment in comparison to the other sectors. In contrast to other industries, electronics retailers show scores over the midpoint of the scale for all the items, evidencing the superior technological development of these retailers and their greater coordination with their suppliers for ICT solution developments.

Finally, we analyze customer assessment of retail ICT solutions. According to the results shown in Table 9, consumers' opinions about their retailers' ICT are significantly more favorable than manufacturers' and retailers' assessments of their suppliers' technological solutions. Additionally, there is a more positive assessment of ICT solutions implemented by electronics retailers in comparison with the other retail sectors.

Finally, there are no significant differences across retail sectors as regards the extent to which stores consider customers' opinion on decisions involving IT coordination and development. In this sense, customers suggest that stores pay little attention to their patrons' needs and desires when deciding ICT investments and, therefore, their scores in this item show values lower than the midpoint of the scale.

CONCLUSION

In Spanish marketing channels, the level of use of ICT solutions in the relationships of companies and final consumers with their suppliers is, generally, still low. Differences are observed, however, across types of technology. In reply to our first research question, *Which ICT are more intensively used by customers in different marketing channel levels?*, the use of the e-mail and Internet is common in relationships between manufacturers and their suppliers, as well as between retailers and their suppliers, whereas ICT solutions based on radio frequency technology are still scarcely

Table 6. Consumers' intensity of use of retailer's ICT (by industry)

	Global average	1 Grocery	2 Cloth./footw.	3 Electronics	4 Furnit./deco.	Differences between groups[a]
1. Bar codes / scanner	4.37	4.54	4.31	4.34	4.29	-
2. Self-service technologies	3.00	3.77	2.38	2.77	2.92	1-2, 1-3, 1-4
3. Loyalty program	2.25	2.73	2.03	2.23	1.99	1-2, 1-3, 1-4
4. Credit / debit card payment	4.61	4.46	4.65	4.65	4.70	-
5. Retailer card payment	2.62	3.46	2.48	2.10	2.35	1-2, 1-3, 1-4
6. Bank transfer payment	2.54	2.25	2.20	2.98	2.69	1-3, 2-3
7. Mobile payment	2.33	2.13	2.02	2.70	2.43	1-3, 2-3
8. Cash on delivery	2.38	2.30	2.04	2.56	2.59	2-3, 2-4
9. Phone	2.80	2.47	2.55	3.27	2.89	1-3, 2-3
10. Fax	2.59	2.36	2.29	2.98	2.68	1-3, 2-3
11. E-mail	2.79	2.39	2.36	3.37	2.99	1-3, 1-4, 2-3, 2-4
12. Web site	3.06	2.36	2.70	3.61	3.53	1-3, 1-4, 2-3, 2-4
13. Office software	2.51	2.38	2.15	2.82	2.66	2-3, 2-4
14. Design software	2.57	2.41	2.22	2.87	2.72	2-3, 2-4
15. Internet security software	2.55	2.36	2.25	2.86	2.70	1-3, 2-3
16. Invoicing software	2.72	2.51	2.46	2.99	2.90	2-3

implemented. Regarding relationships between retailer and consumers, the ICT solutions used with greater intensity are methods of payment, bar codes and scanner. Notwithstanding, in view of the significant increase in the number of Internet users in Spain in recent years, e-mail could also be used successfully by retailers as a way of providing customer service and implementing promotional actions.

Table 7. Manufacturers' assessment of supplier's ICT

	Global average	1 Grocery	2 Cloth./footw.	3 Electronics	4 Furnit./deco.	Differences between groups[a]
1. Our partner invests in IT to align their technology with us.	2.32	1.90	2.47	2.22	1.93	-
2. IT advances between my main supplier and my company are well aligned for best supply chain performance.	2.42	2.07	2.37	2.30	2.16	-
3. My company's IT is well aligned with our partner.	2.48	2.40	2.43	2.35	2.51	-
4. In relation to our competitors, our logistic communications are more advanced.	2.42	2.14	2.46	2.47	2.53	-

Table 8. Retailers' assessment of supplier's ICT

	Global average	1 Grocery	2 Cloth./footw.	3 Electronics	4 Furnit./deco.	Differences between groups[a]
1. Our partner invests in IT to align their technology with us.	2.48	2.14	2.28	3.15	2.49	1-3, 2-3, 3-4
2. IT advances between my main supplier and my company are well aligned for best supply chain performance.	2.52	2.14	2.32	3.24	2.50	1-3, 2-3, 3-4
3. My company's IT is well aligned with our partner.	2.45	2.13	2.28	3.15	2.37	1-3, 2-3, 3-4
4. In relation to our competitors, our logistic communications are more advanced.	2.48	2.11	2.36	3.10	2.43	1-3, 2-3, 3-4

With regards to the third question *To what extent are the supplier's technological solutions valued by customers in different levels of marketing channels?*, our results allow us to conclude that assessment of supplier technological solutions is worse in the context of interorganizational commerce (supplier-to-manufacturer or supplier-to-retailer) than in retailer-consumer relationships. This different assessment may have two possible explanations: either final consumers appreciate more than companies the advantages derived from the use of ICT in their relations with their suppliers, or consumers are less demanding than manufacturers and retailers. Further research should be conducted in order to align the seller's technology with buyers' expectations.

These are general conclusions independent of activity sector. In order to refine these results, we considered the second, *Are there differences in the level of customer ICT use in its relationship with the main supplier depending on the industry?*, and fourth research questions, *Are there differences in the assessment of supplier's ICT solutions depending on the industry?*. Follow-

Table 9. Consumers' assessment of retailer's ICT

	Global average	1 Grocery	2 Cloth./footw.	3 Electronics	4 Furnit./deco.	Differences between groups[a]
1.-This STORE invests in technology	3.45	3.20	3.31	3.87	3.41	1-3, 2-3, 3-4
2.-This STORE has the most advanced technology	3.35	3.11	3.21	3.73	3.34	1-3, 2-3, 3-4
3.-In comparison to its competitors, this STORE technology is more advanced	3.28	3.06	3.11	3.67	3.30	1-3, 2-3, 3-4
4.-This STORE considers my opinion as a customer on decisions involving ICT coordination and development in order to improve services and to better satisfy my needs as a client	2.72	2.66	2.68	2.81	2.72	-

ing our results for manufacturers, no significant differences are observed regarding the use and the assessment of their suppliers' ICT solutions. However, for retailers, there is a superior degree of technological development as well as level of demand in electronics/home appliances retailers in comparison to the rest of industries. In contrast, grocery retailers show the lowest levels of use of ICT solutions in their relationships with their suppliers. These differences in ICT use patterns may be due to discrepancies in the assessment of perceived benefits and costs of the use of technology. In this regard, in the case of retailers, these factors may be considered as antecedents of ICT technological development and assessment by the company.

All in all, this evidence is in the line of Sethuraman and Parasuraman (2005), thus supporting the advisability of adapting technology to the company's needs. These needs seem to be determined to a great extent by the level the company occupies in the marketing channel and by the industry it belongs to, i.e. the characteristics and requirements of the product being commercialized by the company.

MANAGERIAL IMPLICATIONS

The results allow us to extract some managerial implications. First of all, from the supplier's point of view, it is desirable to adapt technology to the level of the customer's investment in ICT as well as to the expected degree of technological development. Provided that this adaptation is successful, there will be an acceptable level of technological alignment that may facilitate a long-lasting relationship between seller and customer. Therefore, sellers should prioritize the applications most valued by customers. Thus, we understand that the supplier must assess customer needs derived from the purchase process and the after-sale service as a previous step in its ICT investment decisions.

Secondly, as far as providers of ICT solutions are concerned, more effective communication about the advantages that technology might provide to companies in their relationships with their suppliers is needed, as well as appropriate measures to diminish the disadvantages perceived by marketing channel agents in their use of technological solutions.

Notwithstanding, and in accordance with Curry and Penman (2004), technology in itself may not constitute a source of competitive advantage. In this regard, user's technology acceptance is an issue to be addressed and customer segments should be identified based on their level of training, expertise or interest in technological applications (Parasuraman and Colby, 2001).

For consumers, the wide variety of technological solutions implemented by retailers adds value to the service in terms of convenience, customization, reliability, accurateness and flexibility. Provided that retailers implement training and technical support actions to familiarize consumers with their ICT solutions, technology may constitute a source of mutual benefits for both buyers and sellers.

FURTHER RESEARCH DEVELOPMENTS

The evidence from this study allows us to suggest new further research lines. First, these results should be considered as a first exploratory stage in the assessment of the most suitable technology solutions for marketing channel members depending on their position and the industry they belong to. In this regard, the next step should be to assess the impact of ICT on buyer-seller relationships in terms of customer satisfaction and loyalty. In particular, it may be suitable to carry out research in the framework of the Technology Acceptance Model (TAM) for deep understanding of how retailers and consumers accept the introduction of ICT in the different marketing channels. The TAM is based on the premise that intention to use a technological solution depends on perceived usefulness and ease of use (Davis, 1989; Davis et al., 1989). For instance, this model supports the notion that, due to the biological and psychological aging processes, older people may experience more difficulty with the use of computers and other information technology applications compared to younger individuals (Phang et al., 2006). Therefore, it could be enlightening to conduct an analysis to compare results depending on customer age or computer familiarity.

Alternatively, it might be also interesting to compare the effect of investment in technological solutions on relationship value in the scope of inter-organizational commerce (supplier-manufacturer or supplier-retailer relationships) with the results for retail-consumer relationships, thereby testing for differences in value creation and exchange in both B2B and B2C settings.

Finally, it may be useful to analyze the advantages and disadvantages of different ICT solutions by each type of marketing channel agent, thus facilitating the suppliers' decision making process related to technological solutions to be implemented as well as the design of specific actions to overcome customers' initial reluctance to new ICT solutions and encourage their use of these technologies.

ACKNOWLEDGEMENT

This research has been financed by the Spanish Ministry of Education and Science (Project ref.: SEJ2007-66054/ECON).

REFERENCES

Bitner, M. J. (2001). Service and technology: Opportunities and paradoxes. *Managing Service Quality*, *11*(6), 375–379. doi:10.1108/09604520110410584

Booth, M. E., & Philip, G. (1998). Technology, competencies, and competitiveness: The case for reconfigurable and flexible strategies. *Journal of Business Research*, *41*(1), 29–40. doi:10.1016/S0148-2963(97)00009-X

Bowerson, D. J., & Daugherty, P. J. (1995). Logistics paradigms: The impact of information technology. *Journal of Business Logistics*, *16*(1), 65–80.

Burke, R. R. (2002). Technology and the customer interface: What consumers want in the physical and virtual store. *Journal of the Academy of Marketing Science*, *30*(4), 411–432. doi:10.1177/009207002236914

Buxmann, P., & Gebauer, J. (1999). *Evaluating the use of Information Technology in inter-organizational relationships*. Hawaii Conference on Systems Sciences.

Clemons, E. K., & Row, M. C. (1991). Sustaining IT advantage: The role of structural differences. *Management Information Systems Quarterly*, *15*(3), 275–292. doi:10.2307/249639

Curry, A., & Penman, S. (2004). The relative importance of technology in enhancing customer relationships in banking – a Scottish perspective. *Managing Service Quality*, *14*(4), 331–341. doi:10.1108/09604520410546860

Davis, F. D. (1989). Perceived usefulness, perceived ease of use, and user acceptance of information technology. *Management Information Systems Quarterly*, *13*(3), 319–340. doi:10.2307/249008

Davis, F. D., Bagozzi, R. P., & Warshaw, P. R. (1989). User acceptance of computer technology: A comparison of two theoretical models. *Management Science*, *35*(8), 982–1003. doi:10.1287/mnsc.35.8.982

Dean, D. H. (2008). Shopper age and the use of self-service technologies. *Managing Service Quality*, *18*(3), 225–238. doi:10.1108/09604520810871856

Drennan, J., & McColl-Kennedy, J. R. (2003). The relationship between Internet use and perceived performance in retail and professional service firms. *Journal of Services Marketing*, *17*(3), 295–311. doi:10.1108/08876040310474837

Dulude, L. (2002). Automated telephone answering systems and aging. *Behaviour & Information Technology*, *21*(3), 171–184. doi:10.1080/0144929021000013482

Ellram, L. M., La Londe, B. J., & Weber, M. M. (1999). Retail logistics. *International Journal of Physical Distribution & Logistics Management*, *29*(7/8), 477–494. doi:10.1108/09600039910291993

Feng, C., & Yuang, C. (2007). Application of collaborative transportation management to global logistics: An interview case study. *International Journal of Management*, *24*(4), 623–636.

Frasquet, M., Cervera, A., & Gil, I. (2008). The impact of IT and customer orientation on building trust and commitment in the supply chain. *International Review of Retail, Distribution and Consumer Research*, *18*(3), 343–359. doi:10.1080/09593960802114164

Gil, I., Cervera, A., & Frasquet, M. (2007). Empleo de ICT y efectos relacionales en la cadena logística. *Boletín Económico de ICE*, *2914*, 31–48.

Gil, I., Ruiz, E., & Servera, D. (2008a). Logistic service quality and technology: A comparison between supplier-retailer and retailer-consumer relationships. *International Review of Retail, Distribution and Consumer Research*, *18*(5), 495–510. doi:10.1080/09593960802573385

Gil, I., Servera, D., Berenguer, G., & Fuentes, M. (2008b). Logistics service quality: A new way to loyalty. *Industrial Management & Data Systems Journal*, *108*(5), 650–668. doi:10.1108/02635570810876778

Hausman, A., & Stock, J. R. (2003). Adoption and implementation of technological innovations within long-term relationships. *Journal of Business Research*, *56*(8), 681–686. doi:10.1016/S0148-2963(01)00313-7

Lewis, I. (2001). Logistics and electronic commerce: An interorganizational systems perspective. *Transportation Journal*, 5-13.

Lowson, R. H. (2001). Retail operational strategies in complex supply chains. *International Journal of Logistics Management, 12*(1), 97–111. doi:10.1108/09574090110806253

McDonald, J. B., & Smith, K. (2004). The effects of technology-mediated communication on industrial buyer behavior. *Industrial Marketing Management, 33*(2), 107–116. doi:10.1016/S0019-8501(03)00033-6

Mentzer, J. T., & Williams, L. R. (2001). The role of logistics leverage in marketing strategy. *Journal of Marketing Channels, 8*, 29–48. doi:10.1300/J049v08n03_03

Messinger, P. R., & Narasimhan, C. (1997). A model of retail formats based on consumers' economizing on shopping time. *Marketing Science, 16*(1), 1–23. doi:10.1287/mksc.16.1.1

Meuter, M. L., Bitner, M. J., Ostrom, A. L., & Brown, S. W. (2005). Choosing among alternative service delivery modes: An investigation of customer trial of self-service technologies. *Journal of Marketing, 69*, 61–83. doi:10.1509/jmkg.69.2.61.60759

Meuter, M. L., Ostrom, A. L., Roundtree, R. I., & Bitner, M. J. (2000). Self-service technologies: Understanding customer satisfaction with technology-based service encounters. *Journal of Marketing, 64*, 50–64. doi:10.1509/jmkg.64.3.50.18024

Mick, D. G., & Fournier, S. (1998). Paradoxes of technology: Consumer cognisance, emotions and coping strategies. *The Journal of Consumer Research, 25*, 123–143. doi:10.1086/209531

Observatorio de las Telecomunicaciones y la Sociedad de la Información - Entidad Pública Empresarial Red. (2006). *Diagnóstico tecnológico del sector de comercio minorista*. Retrieved on September 3, 2007, from observatorio.red.es/estudios/documentos/diag_cminoristav1.3.pdf

Odekerken-Schröder, G., De Wulf, K., Kasper, H., Kleijnen, M., Hoekstra, J., & Commandeur, H. (2001). The impact of quality on store loyalty: A contingency approach. *Total Quality Management, 12*(3), 307–322. doi:10.1080/09544120120034474

Palmer, J. W., & Markus, M. L. (2000). The performance impacts of quick response and strategic alignment in specialty retailing. *Information Systems Research, 11*(3), 241–259. doi:10.1287/isre.11.3.241.12203

Parasuraman, A., & Colby, C. L. (2001). *Techno-ready marketing: How and why your customers adopt technology*. New York, NY: Free Press.

Phang, C., Sutanto, J., Kankanhalli, A., Li, Y., Tan, B., & Teo, H. (2006). Senior citizens' acceptance of information systems: A study in the context of e-government services. *IEEE Transactions on Engineering Management, 53*(4), 555–569. doi:10.1109/TEM.2006.883710

Porter, M. E. (1997). Creating advantages. *Executive Excellence, 14*(2), 17–18.

Powell, T. C. (1992). Organizational alignment as competitive advantage. *Strategic Management Journal, 13*(2), 119–134. doi:10.1002/smj.4250130204

Rebolledo, C., Ricard, L., & Prefontaine, L. (2005). The potential of Information Technology in facilitating relationship marketing: The case of large Canadian firms. *Journal of Relationship Marketing, 4*(1/2), 57–72. doi:10.1300/J366v04n01_05

Ryssel, R., Ritter, T., & Gemunden, H. G. (2004). The impact of information technology deployment on trust, commitment and value creation in business relationships. *Journal of Business and Industrial Marketing, 19*(3), 197–207. doi:10.1108/08858620410531333

Sethuraman, R., & Parasuraman, A. (2005). Succeeding in the big middle through technology. *Journal of Retailing, 81*(2), 107–111. doi:10.1016/j.jretai.2005.03.003

Simon, F., & Usunier, J. (2007). Cognitive, demographic, and situational determinants of service customer preference for personnel-in-contact over self-service technology. *International Journal of Research in Marketing, 24*(2), 163–173. doi:10.1016/j.ijresmar.2006.11.004

Snellman, K., & Vihtkari, T. (2003). Customer complaining behaviour in technology-based service encounters. *International Journal of Service Industry Management, 14*(2), 217–231. doi:10.1108/09564230310474174

Walker, R. H., Craig-Lees, M., Hecker, R., & Francis, H. (2002). Technology-enabled service delivery. An investigation of reasons affecting customer adoption and rejection. *International Journal of Service Industry Management, 13*(1), 91–106. doi:10.1108/09564230210421173

Weinstein, A. (2002). Customer-specific strategies-customer retention: A usage segmentation and customer value approach. *Journal of Targeting. Measurement & Analysis for Marketing, 10*(3), 259–268. doi:10.1057/palgrave.jt.5740051

Wu, F., Yeniyurt, S., Kim, D., & Cavusgil, S. T. (2006). The impact of information technology on supply chain capabilities and firm performance: A resource-based view. *Industrial Marketing Management, 35*, 493–504. doi:10.1016/j.indmarman.2005.05.003

ADDITIONAL READING

Ball, D., Coelho, P. S., & Vilares, M. J. (2006). Service personalization and loyalty. *Journal of Services Marketing, 20*(6), 391–403. doi:10.1108/08876040610691284

Dabholkar, P. A., Bobbit, L. M., & Lee, E. (2003). Understanding consumer motivation and behavior related to self-scanning in retailing. *International Journal of Service Industry Management, 14*(1), 59–95. doi:10.1108/09564230310465994

Gil-Saura, I., Berenguer-Contrí, G., & Ruiz-Molina, M. E. (2009). Information and Communication Technology in retailing: A cross-industry comparison. *Journal of Retailing and Consumer Services, 16*(3), 232–238. doi:10.1016/j.jretconser.2008.11.018

Gil-Saura, I., & Ruiz-Molina, M. E. (2009). Customer segmentation based on commitment and ICT use. *Industrial Management & Data Systems, 109*(2), 206–223. doi:10.1108/02635570910930109

Gil-Saura, I., & Ruiz-Molina, M. E. (2009). Logistics service quality and technology investment in retailing. *European Retail Research, 23*(Issue I), 69–82. doi:10.1007/978-3-8349-8203-2_4

Joseph, M., McClure, C., & Joseph, B. (1999). Service quality in banking sector: the impact of technology on service delivery. *International Journal of Bank Marketing, 17*(4), 182–193. doi:10.1108/02652329910278879

Kent, J. L., & Mentzer, J. T. (2003). The effect of investment in interorganizational information technology. *Journal of Business Logistics, 24*(2), 155–175.

Mollá, A., Gil, I., & Ruiz, M. E. (2009). Automation of warehouse management and assortment in the distribution of durable goods: An analysis for the ceramics and building supplies industry in Spain. *Journal of Marketing Channels, 16*(4), 375–390. doi:10.1080/10466690903188054

Srinivasan, S. S., Anderson, R., & Ponnavolu, K. (2002). Customer loyalty in e-commerce: an exploration of its antecedents and consequences. *Journal of Retailing, 78*(1), 41–50. doi:10.1016/S0022-4359(01)00065-3

Weijters, B., Rangarajan, D., Falk, T., & Schillewaert, N. (2007). Determinants and Outcomes of Customers' Use of Self-Service Technology in a Retail Setting. *Journal of Service Research, 10*(1), 3–21. doi:10.1177/1094670507302990

KEY TERMS AND DEFINITIONS

Apps: Abbreviation for applications. An app is a piece of software that can run on the Internet, the computer, or the phone or other electronic device.

Automated Storage and Retrieval System (AS/RS): A high-density rack inventory storage system with un-manned vehicles automatically loading and unloading products to/from the racks.

Electronic Data Interchange (EDI): Computer-to-computer transmission of business information between trading partners based on standard file formats and transaction sets.

Geographic Information Systems or Geospatial Information System (GIS): Mechanism that captures, stores, analyzes, manages, and presents location linked data.

Information and Communication Technology (ICT): *"all forms of technology utilized to create, capture, manipulate, communicate, exchange, present, and use information in its various forms (business data, voice conversations, still images, motion pictures, multimedia presentations", etc.)"* (Ryssel et al., 2004:198).

Logistic Information System (LIS): A system that facilitates shipping, transportation, and warehousing activities. It aims to ensure customer service by getting adequate quantities of the finished product to the proper place in a cost- and time-efficient manner.

Loyalty Program: Structured marketing efforts that reward, and therefore encourage, loyal buying behavior. Examples of loyalty programmes in retailing include loyalty cards, rewards cards, points cards, advantage cards, or club cards.

Marketing Channel: Set of interdependent organizations involved in making goods or services available for use or consumption.

Materials Requirements Planning (MRP): A decision-making methodology used to determine the timing and quantities of materials for purchase.

Radio Frequency Identification (RFID): Technology that uses radio-frequency waves to transfer data between a reader and a tagged movable item to identify, track or locate that item.

Chapter 6
International Fashion Retailing from an Enterprise Architecture Perspective

Torben Tambo
Aarhus University, Denmark

ABSTRACT

International retailing of non-food fashion products, as chain stores impose a particular challenge within Enterprise Architecture (EA) as the same general infrastructure networks, brands, data, business intelligence and applications should work in multiple, semi-compliant geographic regions. Generalised Information System architectures are desirable. A case is presented of a fashion retailing organisation growing from 100 to 2,000 stores with both fluctuant and long-term decision making but also a strong international orientation. The case creates a basis for shaping the wider Enterprise Architecture for retailing. EA has significant potentials for innovation, improvement, enrichment and increased interaction between business and technology. Technology vendors are prepared for EA observations but business seems hesitant: The IT inventory of chain retailing should be managed within EA, positioning the store in large international networks of chains encompassing marketing, supply chain, multi-channel concepts, payment systems and loyalty programs. Conclusively, guidelines are set for a deepening of EA within retail.

INTRODUCTION

The retail industry is by headcount, value adding, square meters and a plethora of other parameters one of the largest industries in many economies. The retail industry is increasingly defining itself into chains (Brun and Castelli 2008), many with a strong international, growth orientation. Gartner (2009) predicts a global annual IT spending with retail on USD 153bn. Lewis (2005) talks about compelling ROI requirements on projects. Several vendor/product-centric proposals for Enterprise Architecture (EA) do exist (IBM, Cisco, SAP, NCR, Oracle), but many have an inclination

DOI: 10.4018/978-1-60960-738-8.ch006

to general or food retailing in single-company, single-country, single-organisation setups, also the detailed EA narrative on 7-Eleven Japan found in Ross et al. (2008). In this chapter, IT and EA within retailing are highlighted within multi-company, non-food, multi-country chains. The application and utilisation of IT is analysed to assess and establish an Enterprise Architecture (EA) for a strengthened support of the development of the chains and the retailing as such and to position various business requirements relatively against each other (Zentes et al 2007).

International retailing, in this case non-food fashion products, as chain stores or branded stores is expanding and imposes a particular case within EA as the same general infrastructure (networks, brands, data) should work in multiple geographic regions with highly different compliance requirements. To create uniformity in global business processes, one overarching information system architecture is desirable, although complicated. Data is increasingly shifting from one-way, batch-oriented, simple price-product data to two-way, real-time communication including rich multi-media content linking head-office with up-to-the-minute business data and reporting. Data drive complex supply chain solutions (Lowson 2001; Wong et al 2006; Choi 2007; Pritchard 2010) within the head-office link logistics services providers and manufacturers to the demands of the stores.

Fiscal requirements (NCR, LS Retail) are critical and include national, regional and local sales tax/VAT with different product-dependent rates and models; specialty printers; local certification processes; handling of multi-currency; volatile currencies; inventory reporting; customs/duty processing down to store level; and excises.

Workforce management (WFM) and optimization (Vedamani 2006, Joshi 2009) ranges from the simple to highly complex local rules for contracts, shifts, breaks, long term schedules, overtime, deviations and job roles; potentially including performance linked compensation.

Payment solutions (Humphrey 2010, SAP 2010) have become increasingly complex with multitudes of payment cards having different properties, different clearing houses, different suppliers of Electronic Funds Transfer (EFT) terminals, increasing number on non-pecuniary payment (Vedamani 2006) and refund options like vouchers, lotteries, smart-phone tags. EFT terminals might be integrated or non-integrated (Tambo 2010b) into the point-of-sales systems each creating complexities and sources of errors; new security debit and credit cards create new requirements on EFT including scanning signatures and biometrical data (Savage 2002).

Loyalty programs (Zentes et al. 2007) introduce several complexities like customer loyalty cards, customer clubs, cross-marketing initiatives giving discounts, earning points and seeking to establish new types of customer relationships. Manufacturer and brand incentives and promotions must also be considered. (Kurata and Yue 2008)

Multi-channel retailing (DMS Retail 2010, Zentes et al. 2007) imposes new and different requirements on the store as transactions are remote and crossing several organisational barriers. Within multi-channel retailing the different entities are expected to loyally contribute and share the benefits.

Most stores in focus have various degrees of multimedia (Cisco 2008) in the form of in-store music, video, information kiosks, surveillance systems and co-sales of media-related products become commonplace. The visual appearance of the store might also be an online service itself. (Visual Retailing 2010)

The research question revolves around how to include an EA for international non-food retail chains with emphasis on placing the right architectural elements at the right levels of the operating organisation and including proposal for management plans of the architectural change processes.

Retail is the end-point of many supply chains (Abernathy et al. 2000). Retail has to comply with standards and regulations in any country of operations. POS and ERP are often detached creating a demand for overcoming "silo thinking" (Tambo 2010b). POS should model exactly – changing - business initiatives and work in any size of store. Retail intelligence starts and ends with POS (Rao 2000). IT service-level-agreements (SLA) on POS needs to secure 100% uptime during opening hours.

BACKGROUND

The elaboration of EA in retailing is providing us with a multi-facetted image. Fundamentals of EA are obviously to be included. Retailing must be understood as a business process and in the context of its formats – the food – non-food segregation is straightforward. Disciplines include IT, HR, tax/vat, accounting, in-store and general marketing. Multimedia also attaches to electronic and visual cross-platform thinking using similar or identical promotion in TV, outdoor, internet, magazines and stores.

EA should govern the store information systems for the performance of the manufacturing, SCM and marketing organisations (Tambo and Koch 2010). New business concepts and augment processes at retail level are critical parameters of competitiveness (Savage 2002). The multimedia requirements and customer loyalty programs of the chain, the city, the mall and the brand more than the store are to be regarded as integrated in the business processes, and changes need support.

Retail Fundamentals

The domain is non-food retailing organised as international chains. Retailing characterised as fixed end-points of supply chains directed towards consumers and based on a number of store formats primarily with that of the specialty store (Palmer and Markus 2000, Cox and Brittain 2004). The speciality store is found numerically dominating in malls and high streets. Chains are organised around a brand with the brand owner in dynamic, but shifting roles as e.g. concept developer, design and purchasing organisation, marketing, and potentially ownership and operations also of the stores of the chain.

Joshi (2009) depicts these fundamental items of information technology in retailing

- Electronic Point-of-sales hardware
- Electronic Point-of-sales software
- Payment systems (EFT ~ Electronic Funds Transfer)
- Automatic Identification and Data Capture (AIDC)
- Infrastructure as databases, servers, networks
- Enterprise Resource Planning (ERP)
- Information Systems for Supply Chain Management (SCM)
- Customer Relationship Management (CRM) systems
- E-commerce
- Emerging Technologies

Vedamani (2006) introduces the 'Merchandise Management Cycle' describing ideal, linked fulfilment processes between the store and the central operating unit. Wong et al. (2005) points out the schism between seasonal and cyclical replenishment; you do not replace summer dresses with summer dresses but with winter coats. Brun and Castelli (2008) further contribute to the criticality of planned replenishment in the fashion chain store.

Retailing Beyond the Store: The Wholesaler

A central organisation is always present in chain stores exemplified with the companies listed in Textilwirtschaft (2008). These companies can be anonymous, associated with the chain, or be a general developer or operator of store concepts. Subsequently, this is named *the wholesale or-*

Figure 1. Sample wholesale, retail and operations design

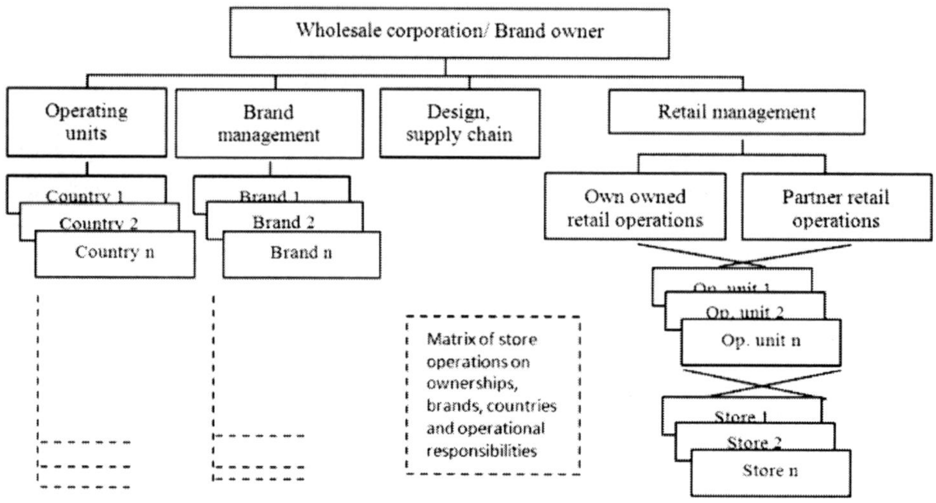

ganisation as the brands in many cases have other distribution channels than the concept stores. The wholesale organisation normally contains the responsibility to develop and support the chain. Including most aspects of design, marketing, sourcing, distribution and retail back office functions (Figure 1).

In most cases, the wholesale organisation supports the chain with IT services or providers – in-house or outsourced or a mix hereof (Ross et al. 2008). Obviously central systems would be managed out of the wholesale organisation such as systems of all parts of the supply chain as well as master data management (MDM) and business intelligence (BI). The wholesale organisation would in most cases offer store concepts including a full line of IT systems: Point-of-sales, security, multimedia, integration services, workforce management, etc. This contributes to the success of chains by creating economies of scale (McIntyre 2009).

Enterprise Architecture

EA, as interpreted in (Bernard 2005), introduces an ideal methodology for comprehension, planning and change from the strategic targets of the enterprise. EA can make a direct line from the top level into the actual implementation and technology artefacts – dependent of the strength of the EA culture of the enterprise. In an outlining of components and artefacts Bernard depict EA as the enterprises commercial, social and technological construct composed by: (1) Goals & initiatives; (2) Products and Services; (3) Data & Information; (4) Systems & Applications; (5) Networks & Infrastructure. In his three-dimensional model (called EA3 cube) this is applied on artefacts, segments and levels-of-detail.

In Doucet et al. (2009) (including Bernard) EA is further elaborated under the conceptualization of Coherency Management (CM) to be offered as a bearing option for aligning business processes and supporting and underlying technology such as information systems by proposing three critical parameters for EA: Alignment, agility and assurance. Alignment addresses the depth and support of business objectives and processes by the technology solution. Agility focuses on speed of change often inhibited by lack of in-depth understanding on technological and social impact of changes. Assurance covers the aspects of compliance: Internal, external, quality, voluntary, governmental, international, moral, social and legal.

Doucet et al. (2009) presents a range of approaches to strengthen the mutual relationships and interactions between business aims and corporate architecture design. In this context, technological systems are not important in a system approach; it can also be management systems, performance measurement systems, legal systems or quality systems. Important is also continuous realignment related to general drift in business focus and challenges. EA addresses the business-led technology of the enterprise. Coherency management addresses the finer coordination between business (management) objectives, governance and drivers and the EA.

Enterprise Architecture on Retailing

EA with international chains will have to address critical issues with this business construct observing both central level, intermediary levels and the store as endpoint.

Organisation and underlying formats (Zentes et al. 2007) are highly affecting the prerequisites of the architecture. International chains do have numerous organizational designs (Figure 1), e.a. spanning from single operating entities, mixed approaches discussed as in the case, into locally oriented designs of franchising concepts of 1 – 4 stores per franchisee. ERP integration is fully relying on the legal organisation, but enterprise integration architectures must come in place (Chen et al. 2008, Chandra and Kumar 2001).

The application framework is shifting towards services requiring a secure delivery platform. In the case below, gift vouchers and customer loyalty cards are mostly issued by external service providers. Payment solutions and various, promotional (Kurata and Yue 2008) and / or non-pecuniary settlements do normally include external services. EFT terminals or integrated EFT functionality have a distinct role. A service oriented approach will be an integral part of an EA.

SCM is shifting from 4 – 6 annual seasonal drops into ongoing replenishment typically within weekly cycles (Brun and Castelli 2008). As laid out by Vedamani (2006) the store is basic in ongoing supply activities. As stated in Wong et al. (2006), seasonal planning is a strong part of this. EA must include SCM activities including product merchandising encompassing product / profile mix with management of categories, variants, brands, gender and prices.

Most authors (Vedamani 2006, Cox and Brittain 2004, Zentes et al. 2007) see HR and Work Force Management (WFM) as important parts of the store operations supported by IT. Most chains do, however, choose a mixed approach including manual, local, national or international solutions. EA must address HR in retailing.

Performance management is commonplace within retailing. Several critical KPI's do exist and must always be supported. A business intelligence framework must be a part of EA consisting of the twofold structure of bidirectional dataflow and performance services. The first mentioned includes Master Data Management (MDM), the latter is an application service embedded in application services delivery framework.

Multimedia or rich media are mostly not included in EA. Most stores do have multimedia in many forms; an EA must address this, especially when it comes to securing uniformity and timeliness of electronic rich media marketing campaigns. Within the case below, certain crossover activities were conducted on goods, represented on internet platforms, and linked to business data.

Alignment over 20 – 40 – 60 countries require an open architecture; SOA can play a role; Enterprise Integration Architecture (EIA) is likely to have a position. Central business targets must be distinguishable from local, regional and national business and compliance requirements.

Agility and assurance (Doucet et al. 2009) are equally critical to fashion retailing as other businesses. As stated, local compliance regulations and fiscalisation seem to change over time. Agility deals with the broad lines of ever changing business and competition environments. Short notices are given on change, but must be balanced with general inertia of business, when it is managing large geographically distributed systems of chains.

Other Enterprise Architecture Frameworks on Retailing

A number of EA's are publicly available mostly vendor-based from either product manufacturers or consultancies, reflecting EA's roots in industry more than research. Subsequently, EA-notes from IBM, Infosys, Cisco, Microsoft, Oracle and SAP are presented. Cisco's Service-Oriented Network Architecture (SONA) is interesting while it from the network infrastructure providers' view seeks to encompass all higher level services of the retail operation. SONA describes retailing EA as a three level model consisting of applications, core common services and a physical infrastructure. Core common services include real time communications, mobility, (data) transport, security, virtualization, application delivery and (network) management, with the first five forming a group of common network services. The applications level is made of:

- Connected retail solutions
- Digital media systems
- Remote expert assistance
- Video surveillance
- Payment systems / EFT
- Workforce optimization / WFM
- Lean store inventory management / SCM

Microsoft offers a range of architectural inputs to both its infrastructure products and its retail systems. Microsoft recognizes the distributed character of retailing.

IBM has a long history with EA in retailing, e.g. the highly operational and pragmatic Retail Application Architecture (Stecher 1993). Additionally, IBM (2009) has specified a Retail Integration Framework and a Store Integration Framework (Gartner 2008) recognizing splits on the organisational level and requirements for integration as part of generalised architectures.

Oracle (2010) and SAP (2010) both provide a very broad spectrum of application services and software solutions for retail and headquarters. The broadness might question the necessity for an EA and potentials of incorporation of mixed strategies (Pollock and Williams 2009), but other tendencies are adverse: New service requirement different agility, media and multimedia services, cost of change and operations and potentials of innovation with competitors using the same technology stack. Microsoft (2005) points also to service orientation and orchestration.

NCR assumes a strong customer and industry adaptability stating "we start with a clean sheet of paper", being strong on full in-store line of products and services, but with no systems for headquarter functions.

Infosys (2008) interprets EA as an ongoing process that needs continuous improvement to secure well described EA services and clear links to relevant governance models.

THE ENDEAVOUR TO GROW INTERNATIONALLY: A CASE

Danish fashion company Bestseller operates 2,000 stores in 40 countries across Europe, Middle East, CIS and North America. The company has grown organically since the first store opened in 1975. Additionally, 2,500 stores are operated in China but these are not included in this case. (Tambo 2010a)

With 100 stores in 1997, Bestseller was facing an inadequate POS strategy. After a market screening NCR was chosen as sole partner for future store POS. 50 Danish stores were "frozen" on older Microsoft Dynamics NAV retail system. In 2003, with 1200 stores at NCR's 1997-software platform NCR was considered irresponsive to requirements for innovation and cost reductions. A screening was done on a host of novel and international POS systems including Fujitsu, WincorNixdorf, IBM and Microsoft. The product and services portfolio in question was broken into the elements:

- Point-of-sales hardware
- Point-of-sales software including necessary back-office software
- Hardware on-site service
- Business intelligence

Point-of-sales software was seen as most critical as it largely sets the scene for the other items; requirements were not explicitly expressed but took more the form of supplier-driven information collection. Implicit requirements included:

- User friendliness of the sales frontend
- Total cost of ownership
- Predefined country localisation packages
- Versatility
- Completeness and one-stop-shopping regarding the requirements above

Microsoft pointed to Icelandic partner Strengur's add-on to Microsoft Dynamics NAV known as Infostore. After many deliberations, Infostore was chosen; a local consultancy with some retail knowledge was also chosen as already being supplier on a number of smaller NAV installations and the original vendor of the 53 frozen NAV systems. IBM SurePOS was chosen as hardware.

Infostore was to replace NCR's system on a country-to-country basis. New countries needed localisations of NCR and for some of these Infostore was chosen; first Russia. Strengur merged with its competitor Landsteinar into LS Retail (2010), which revamped its product portfolio making Infostore the older version. Scepticism with the MS NAV-based systems grew and Microsofts toplevel management intervened. Messages from the stores going from NCR to Infostore complained - 'too complicated', 'we should sell fashion not play bookkeepers'. Plans for rollout of Infostore lost ambitions ending on around 240 stores. From the first Russian implementation the Infostore rollout was characterized as 'crisis management' with a constant chase for scapegoats. Store owners had to pay about EUR 12,000 per installation.

In Slovenia the Infostore system was to be implemented in 20 stores operated by local retail operator Sportina. Sportina persistently claimed that Infostore did not meet Slovenian fiscal requirements. Two years of attempted development of different solutions led to Sportina swapped all POS systems to native Slovenian systems.

In a changed strategy, retail operations in Middle East and CIS were given the right to manage themselves without the central office. In a roll out of retail concepts in new 10 countries such as India, Kasakhstan, Ukraine and Turkey the strategy was outlined to be attracting strong local partners with own POS capabilities and relying on EDI for data exchange.

NCR agreed to prove their innovativeness by offering a new version with 10 years of enhancements and significant system management improvements. Over three years about 230 stores have been converted to this version.

In 2009 a retail operation was to be set up in Canada. The disbelief in the European strategy made the Canadian operating unit choose Retail-1 (GSI 2009) as POS assuming seamless electronic integration with the European operation. The integration has proven difficult and currently, little mutual confidence exists.

Mid-2008 a SOA proposal was made eliminating client-server infrastructure per shop and centralising back office functions as well as introducing a more stand-alone concept around the POS. During 2009 Bestseller was left with 3 – 4 main lines of POS systems covering several versions. Following another intensive market screening for a common POS system Wincor Nixdorf was chosen, but with ongoing discussions on project duration and pace. All stores are owned by partners, legal entities operating independently or with some ownership share from Bestseller, the senior management is looking for "sponsoring" the swap to Wincor Nixdorf to align the POS strategy.

By itself Bestseller is organising first level service, new store openings, training, whereas onsite hardware support is provided by various partners. Support is mostly given in local languages requiring about 15 different languages present in the support organisation. Each shop pays a small yearly service fee of about EUR 200 for full support. New opening pay approx. EUR 12,000. A phone system is routing calls to most linguistically skilled service engineer. Shop owners normally require support during all opening hours including occasional openings. Most failures on service levels are reported by shop managers as a management complaint and a claim on the service fee.

Strategic Business Requirements

In many cases, business requirements have been confused with general pressure for cost reductions even if some requirements have been more persistent. (1) Real-time, online sales figures. (2) Consistent stock value policies, e.g. mean value or FIFO value. (3) Gift vouchers should be interchangeable over larger groups of countries; Bestseller will guarantee for gift vouchers issued by bankrupted partners. (4) Granting certain benefits to loyalty club members. (5) Switch between reporting on colour/size/length breakdown or master product level. (6) Better linkage between stock, goods received, goods sold and returned. (7) Stock transparency between local groups of stores. (8) Automatic replenishment policies driven by point-of-sales on special product groups –the "pull" principle of the supply chain. (9) Simple EFT integration or connectivity across countries. (10) Introducing RFID potentially to replace theft-protection tags. (11) Customer counter/tracker integrated to the reporting of conversion rates. (12) A host of discount principles e.g. bundling, 3-for-2, unbundling, temporal, percentage, absolute. (13) Simple, easy ERP integration knowing partners are using about 20 different ERP systems. (14) Identity management on store employees has sometimes been required, sometimes workforce management.

Lessons Learned

To maintain arms-length principles in trading with concept stores compared to other customers all stores are organised in independent legal entities with a number of stores per entity adapted to the most feasible model of the country in question. Ownership share from Bestseller A/S can be anything from 0 to 100%. Generally, Bestseller owns the stock majority directly or indirectly of about 40% of the stores. The remainder are either individuals with long relationships such as former salesmen or managers or larger companies specialised in retail development and operations.

Co-owners, managers, area managers and core employees in stores are mostly employed on performance based compensation making strong awareness on costs and especially IT costs. Arguments on technological renewal, improved platforms for innovation, stronger systems to business alignment are normally met with strong resentment from local decisions makers. The process of converting from NCR 1997 to LS Retail 2004 ended up in the proof from the business' side that the IT organisation could not justify for increased costs. No systematically mis-alignment measurements were employed during the process (Lankhorst 2009).

Internationalisation

At the time of decision on Infostore it was assumed that localisation packages for NAV would directly be applicable on the full range of country specific requirements. Instead, Infostore needed its own language module, additionally, fiscalisation proved costly. Fiscalisation consists of implementation of several measures to comply with local fiscal requirements. Most expensive were interfaces to specially sealed fiscal printers typically in Eastern Europe. Some countries maintain different tax/vat rulesets for different sizes of enterprises, which increases complexity. Some countries maintain necessity for certain financial flows, subordination of stores to head offices, requirements of physical invoices, disallowing electronic integration, stretching duties into the store, fast rule changing on customs codes and tariffs, etc. Internationalisation on NAV / Infostore accounted for typically 35,000 EUR; too much when a new country starts with a few stores.

Different countries have different statutory requirements on data privacy and anti-competition data exchange; especially Danish Competition Authority imposed harsh restrictions on Danish store data to be anonymous during the projects. Tracking purchases on debit cards are allowed some places, prohibited elsewhere. Categories and other merchandising observations are different from country to country, reflecting different local conditions: Knitwear is of less interest in warm places than in cold. Nonetheless, central governance of reporting was maintained during the projects above.

SOLUTIONS AND RECOMMENDATIONS

EA is ideal for planning and changing retail chains. However, EA must be dedicated to the format of the retail enterprise to assure relevant alignment.

Forming the Foundation for Enterprise Architecture

EA for retail must contain a generic and strategic level with tools for selecting and deselecting services of chain-common nature and services left to lower or local levels, which Bernard (2005) states as mapping of EA Components. The completeness of actor and stakeholder identification must be identified (Towers and Chen 2007) including role and importance; also including electronic interfaces and criticality hereof (Weber and Kantamneni 2002). In Ross et al. (2008) several tools are found for the creation of a corporate EA at the strategic level. Schekkermann (2004) provides guidance in adaption of existing EA frameworks (Figure 2).

Figure 2. Landscape of applications and store impact

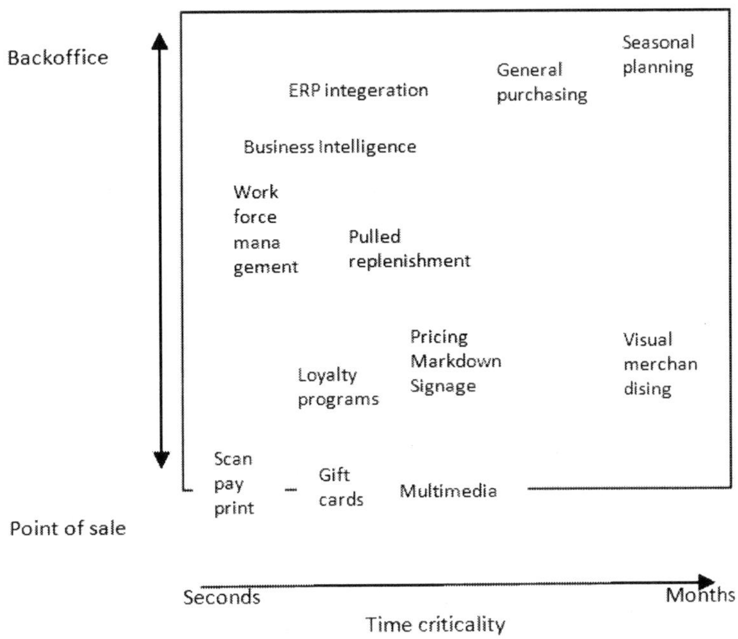

Business relationships (Ford et al. 2003) play a special role as POS vendors seem to form very long relationships over 20 – 30 years; difficulty in changing is a part of this. If internationalization = localization + fiscalisation, NCR has had no limitations in a first line support of the business ambitions within the case. LS Retail did also try to succeed, however with a higher cost impact. Other POS vendors have other issues regarding geographic limitations, internationally operating chains will most likely need a multi-system strategy: Not all POS are suited for all countries; certain countries only fiscally certify a few native POS; partners insisting on host systems; version changes, acquisitions, new markets might require a diverse strategy – EA must contribute in dealing with this.

Enterprise Architecture in Practice

What EA should deal with is essentially the beginning of the story. I.e. a broad comprehension of the strategy, the business, the people and the technology must be established. With Bernard (2005), this figure gives a proposal (Figure 3):

Normally, EA must assume continuous alignment between plans and implementations (Lankhorst 2009). This can be done as management plans (Bernard 2005) or through various portfolio models.

Ross et al (2006) propose a structured and intrinsic approach to secure linkages between enterprise level, business unit level and project level. This includes alignment assessment and coordination activities within a governance framework. Engagement and involvement are key indicators as also discussed in Ford et al. (2003). In retailing, EA would fit ideally into Ross et al.'s model of architecture management practices evolvement, how architecture can traverse a continuum from (1) business silos; (2) standardized technology; (3) optimized core; (4) business modularity.

Recommendable to retail architects would also be Doucet et al.'s (2009) proposal for open minded choice of EA in foundation, extended and embedded modes describing responsibilities, required involvement and business actors' expected contributions.

Methodological critiques must also be in place. EA should drive business and technology simultaneously but business might be excused and ask IT to find a solution. In the case, the IT office occasionally did consulting to retail management to develop the basic architecture of future busi-

Figure 3. Enterprise architecture retailing plan

Future EA Views		**Goals & Initiatives**	**Products & Services**	**Data & Information**	**Systems & Applications**	**Networks & Infra-structure**	**Security Solutions**
High Level View		Strategic Plan	Business Plan	Knowledge Warehouse	Business Systems	Wide Area Network	Security Program
		Versatile, international, but 'simple' retail operations	Appropriate bundles of application services	Products Replenishment strategies Flows and chains Contingencies	A broad application and service delivery platform with point-of-sales and ERP as the cornerstones	Mostly xDSL with VPN Availability measures	General security, anti-fraud framework
Mid Level View		Goals & Initiatives	Business Processes	Information Flows	Support Systems	Local Area Network	System Certification
		Adaptability and compliance with local requirements	Establishing Buying Logistics Marketing Merchandising Selling Reporting	Corporate Master Data → Store logistics → Point of sale → ETL → Business Intelligence → Purchase Planning	Buying, logistics, multimedia, security, workforce mgmt, instore and outstore marketing	Local wired LAN Secure wireless closed area network Support for mobile devices	Robust security infrastructure
Detailed View		Performance Measures	Investment Portfolio	Data Dictionary	Application Inventory	Buildings & Equipment	Data Privacy
		Time performance on point-of-sale level Application performance on rest.	Large, capital intensive, distributed system with strong technical and organizational inter-dependencies	Corporate Master Data Transactions SCM track	Point-of-sales ERP WFM SCM CRM Multimedia Security Management	Point-of-sales hardware, EFT terminal, scanners, sound&light, video panel, kiosks, security, backoffice computers,	Identities, authentication authorizations

ness initiatives. E.g. gift vouchers, intercompany invoices, business intelligence, purchase planning and music appliances.

Growth orientation as a management directive is to be supported by open architectures regarding country adaption and compliance, different operating models and general analysis of the determining business environments. EA could establish an agenda for the growth orientation.

FUTURE RESEARCH DIRECTIONS

As retailing is struggling between offering the customer more and more complex, information-based services, and retail continues to be an area of low margins and staffing considerations "smart" architectures must support the business opportunities the enterprises are seeking.

Other areas for future research might include: Multichannel retailing which is expected to mature and play an increasing role with a series of business, logistical and technological issues.

The decline of multi-brand, mom'n'pop shops will force new entrants to the retailing arena. "Light weight" retailers with experience out of manufacturing or wholesale would require lean store concepts, an issue lightly covered in research.

Most larger retailers are ending the seasonal approaches to store replenishment and switching to continuous weekly or even daily replenishment with direct links between warehouses and stores. Managing stores within a constantly changing portfolio of merchandise is expected to change the store and challenge performance measurement. Further research could support meaningfulness within this level of speed in retailing.

CONCLUSION

EA for retailing must deal with increasingly diverse issues and general frameworks for application services are to be preferred. However, the organisation - including the stores, the store owning entities, the headquarter and occasionally the manufacturers - must insist in a common data platform where a singular and unique Master Data Management function is maintaining integrity. Complex ownership structures and governance within this are issues EA should support. Security is expanding from simple theft and fraud committed by staff to advanced video surveillance and fraud control on payment systems increasing complexity. In this case, other security measures can decrease complexity like biometric login, single sign on, personal tags and corporate wide identities.

Compliance management imposes numerous requirements on a local basis such as fiscalisation, EFT, tax/vat, documentation, odd rules, and ERP-related matters.

Versatility must be embedded within the infrastructure to support wide ranges of multimedia along with or associated to business data.

Supply chain integration links sales, stock levels, replenishment, returned goods with central warehousing and manufacturing using general retail performance management indicators.

Retailing constitutes a highly interesting, but sometimes overlooked, issue in EA except for vendor-centric frameworks, which tend to monopolize or overlook the breadth of requirements. POS has always the primary role. Fiscalisation meets hosts of local and country specific requirements. POS has to be observed in the light of back-office systems: Store ERP, SCM, business intelligence. Organisation is critical to architecture. Analogies to other highly distributed, international, mixed-organisation exist but retailing by magnitude can carry its own practices. Future developments in retailing will benefit strongly from inclusion of broader, business oriented EA with both business strategists and research as strong contributors to this process.

REFERENCES

Abernathy, F. H., Dunlop, J. T., Hammond, J. H., & Weil, D. (2000). Retailing and supply chains in the information Age. *Technology in Society*, *22*, 5–31. doi:10.1016/S0160-791X(99)00039-1

Bernard, S. (2005). *An introduction to enterprise architecture* (2nd ed.). Authors House.

Brun, A., & Castelli, C. (2008). Supply chain strategy in the fashion industry: A model on product, retail channel and brand. *International Journal of Production Economics*, *116*, 169–181. doi:10.1016/j.ijpe.2008.09.011

Chandra, C., & Kumar, S. (2001). Enterprise architectural framework for supply-chain integration. *Industrial Management & Data Systems*, *101*, 209–304. doi:10.1108/EUM0000000005578

Chen, D., Doumeingts, G., & Vernadat, F. (2008). Architectures for enterprise integration and interoperability. *Computers in Industry*, *59*, 647–659. doi:10.1016/j.compind.2007.12.016

Cisco. (2008). *Connected retail architecture: Reuse network services to boost efficiency and profits*. Retrieved September 1, 2010, from https:// www. cisco.com/ /ea Retail.pdf

Cox, R., & Brittain, P. (2004). *Retailing – an introduction* (5th ed.). Prentice Hall / Pearson.

Cunningham, S. J., & Jones, M. (2005). Autoethnography: A tool for practice and education. In *Proceedings of the 6th ACM International Conference on Computer-human interaction.*

Davis, J. M., Kettinger, W. J., & Kuney, D. G. (2009). When users are IT experts too. *European Journal of Information Systems*, *18*, 1.

Doucet, G., Gøtze, J., Saha, P., & Bernard, S. (2009). *Coherency management – architecting the enterprise for alignment, agility and assurance*. AuthorsHouse.

Ford, D., Gadde, L.-E., Håkansson, H., & Snehota, I. (2003). *Managing business relationships*. Wiley.

Gartner (2008). *IBM expands retail SOA foundation to enterprise level*. Retrieved September 1, 2010, from ftp://public.dhe.ibm.com/RE-L03003USEN.PDF

Gartner. (2010). *Dataquest alert: Lead IT spending growth in challenging* 2009. Retrieved September 1, 2010, from http://www.lesechos.fr/medias/2009/0227//300332929.pdf

Grant, D., Hall, R., Wailes, N., & Wright, C. (2006). The false promise of technological determinism. *New Technology, Work and Employment*, 21.

GSI. (2010). *Danish retailer powers North American expansion with retail*, 1. Retrieved September 1, 2010, from http://www.gsi.ca/panda/files/Bestseller_web.pdf

Humphrey, D. (2010). Retail payments: New contributions, empirical results, and unanswered questions. *Journal of Banking & Finance*, *34*, 1729–1737. doi:10.1016/j.jbankfin.2010.04.008

IBM. (2009). *The IBM retail industry framework. Integration, optimization, analytics*. Retrieved September 1, 2010, from http://www-01.ibm.com/software/industry/retail/

Infosys. (2008). *Enterprise architecture organization transformation enables IT governance and shared vision*. Retrieved September 1, 2010, from http://www.infosys.com

Joshi, G. (2009). *Information Technology for retail*. Oxford Univ. Press.

Kurata, H., & Yue, X. (2008). Trade promotion mode choice and information sharing in fashion retail supply chains. *International Journal of Production Economics*, *114*, 507–519. doi:10.1016/j.ijpe.2007.05.021

Lankhorst, M. M. (2009). *Enterprise architecture at work*. Springer Verlag. Lewis, N. (2005). *Enterprise integration: A practical guide for the retail solution architect*. Retrieved September 1, 2010, from http://msdn.microsoft.com/en-us/library/aa479078.aspx

Lowson, R. (2001). Analysing the effectiveness of European retail sourcing strategies. *European Management Journal, 19*, 543–551. doi:10.1016/S0263-2373(01)00068-8

Mathiassen, L., & Nielsen, P. A. (2008). Engaged scholarship in IS research. *Scandinavian Journal of Information Systems, 20*, 2.

McIntyre, C. (2009). Diminishing varieties of active and creative retail experience. *Journal of Retailing and Consumer Services, 16*, 466–476. doi:10.1016/j.jretconser.2009.08.001

NCR. (2010). *NCR corporate website*. Retrieved September 1, 2010, from www.ncr.com

Oracle. (2008). *Oracle on demand for markdown optimization*. Retrieved from http://www.oracle.com / markdown-optimization-data-sheet.pdf

Oracle. (2010). *Oracle and retail*. Retrieved from http://www.oracle.com /us/industries/retail/022189.htm

Palmer, J. W., & Markus, M. L. (2000). The performance impacts of quick response and strategic alignment in specialty retailing. *Information Systems Research, 11*, 241–259. doi:10.1287/isre.11.3.241.12203

Pollock, N., & Williams, R. (2009). *Software and organisations: The biography of the enterprise-wide system or how SAP conquered the world*. Routledge.

Pritchard, S. (2010). Good analysis and use of data can make all the difference. *Financial Times World Retailing, 21*.

Rao, B. P. (2000). Improving retail effectiveness through technology: A survey of analytical tools for physical and online retailers. *Technology in Society, 22*, 111–122. doi:10.1016/S0160-791X(99)00032-9

Retail, D. M. S. (2010). *E-tailing and multichannel retailing*. Retrieved September 1, 2010, from http://www.dmsretail. com/etailing.htm

Retail, L. S. (2010). *LS Retail NAV*. Retrieved September 1, 2010, from http://www.lsretail.com

Ross, J. W., Weill, P., & Robertson, D. C. (2008). *Enterprise architecture as strategy*. Harvard Business School Press.

SAP. (2010). SAP for retail – comprehensive solutions for your business. Retrieved September 1, 2010, from http://www.sap.com/industries/retail/index.epx

Savage, T. (2002). Future trends in retail IT systems. *European Retail Digest, 33*, 17–19.

Schekkermann, J. (2004). *How to survive the jungle of EA Frameworks*. Trafford Publishing.

Stecher, P. (1993). Building business and applications systems with the retail application architecture. *IBM Systems Journal, 32*, 278–306. doi:10.1147/sj.322.0278

Tambo, T. (2010a). Bespoke ERP – a systems view on enterprise management. In *Proceedings of the 19th International Conference on Management of Technology*, Cairo

Tambo, T. (2010b). *IT Retail i et praktisk perspektiv (Lecture notes for retail managers training)*. TEKO-Center.

Tambo, T., & Koch, C. (2010). Enterprise architecture in the supply chain. *In Proceedings of Workshop on Enterprise Architecture Challenges and Responses*. Yonezawa.

Textilwirtschaft. (2010). *Die größten Bekleidungslieferanten in Europa* 2008. Retrieved September 1, 2010, from http://www.textilwirtschaft.de/business/markt/marktfakten/pdfs/64_org.pdf

Towers, N., & Chen, R. (2007). Employing the participative paradigm as a valid approach of supply chain. *International Journal of Retail & Distribution, 36*, 627–637. doi:10.1108/09590550810883478

Vedamani, G. G. (2006). *Retail management – functional principles and practices*. Jaico Publishing House.

Visual Retailing. (2010). *Mockshop – see what you can do*. Retrieved September 1, 2010, from http://www. visualretailing.com

Weber, M. M., & Kantamneni, S. P. (2002). POS and EDI in retailing: An examination of underlying benefits and barriers. *Supply Chain Management: International Journal, 7*, 311–317. doi:10.1108/13598540210447755

Wong, C. Y., Arlbjørn, J. S., Hvolby, H., & Johansen, J. (2006). Assessing responsiveness of a volatile and seasonal supply chain: A case study. *International Journal of Production Economics, 104*, 709–721. doi:10.1016/j.ijpe.2004.12.021

Zentes, J., Morschett, D., & Schramm-Klein, H. (2007). *Strategic retail management – text and international cases*. Gabler.

ADDITIONAL READING

Abbot, H., & Palekar, U. S. (2008). Retail replenishment models with display-space elastic demand. *European Journal of Operational Research, 186*, 586–607. doi:10.1016/j.ejor.2006.12.067

Beaumont, J. R. (1989). Towards an integrated information system for retail management. *Environment and Planning, 21*(3), 299–309. doi:10.1068/a210299

Berman, B., & Evans, J. R. (2010). *Retail Management: A Strategic Approch* (11th ed.). Prentice Hall.

Brito, M. P., Carbone, V., & Blanquart, C. M. (2008). Towards a sustainable fashion retail supply chain in Europe: Organisation and performance. *International Journal of Production Economics, 114*, 534–553. doi:10.1016/j.ijpe.2007.06.012

Choi, T.-M. (2007). Pre-season stocking and pricing decisions for fashion retailers with multiple information updating. *International Journal of Production Economics, 106*, 146–170. doi:10.1016/j.ijpe.2006.05.009

Christopher, M. (1997). Managing Logistics in Fashion Markets. *The International Journal of Logistics Management, 8*(2), 63–74. doi:10.1108/09574099710805673

Christopher, M., Lowson, R. H., & Peck, H. (2004). Creating agile supply chains in the fashion industry. *International Journal of Retail & Distribution Management, 32*(8), 367–376. doi:10.1108/09590550410546188

Cillo, P., & Verona, G. (2008). Search Styles in Style Searching: Exploring Innovation Strategies in Fashion Firms. *Long Range Planning, 41*, 650–671. doi:10.1016/j.lrp.2008.08.001

Coca-Stefaniak, A., Hallsworth, A. G., Parker, C., Bainbridge, S., & Yuste, R. (2005). Decline in the British small shop independent retail sector: exploring European parallels. *Journal of Retailing and Consumer Services, 12*, 357–371. doi:10.1016/j.jretconser.2004.11.007

Davies, M. (2009). Fast Fashion. *Supply Chain Standard* july/august 2009.

Gereffi, G. (1999). International trade and industrial upgrading in the apparel commodity chain. *Journal of International Economics, 48*, 37–70. doi:10.1016/S0022-1996(98)00075-0

Goodwin, D. R. (1990). *The open-to-buy system and retail inventory method: the impact on accurate performance measurement, PhD-Thesis, University of Adelaide Jackson, T., & Shaw, D., (2001). Fashion buying and merchandising management*. Basingstoke: Palgrave Master Series.

Jüttner, U., Christopher, M., & Baker, S. (2007). Demand chain management-integrating marketing and supply chain management. *Industrial Marketing Management, 36*, 377–392. doi:10.1016/j.indmarman.2005.10.003

Lee, C. H., & Rhee, B.-D. (2008). Optimal Guaranteed Profit Margins for Both Vendors and Retailers in the Fashion Apparel Industry. *Journal of Retailing, 84*, 325–333. doi:10.1016/j.jretai.2008.07.002

Levy, M., Grewal, D., Kopalle, P. K., & Hess, J. D. (2004). Emerging trends in retail pricing practice: implications for research. *Journal of Retailing, 80*, xiii–xxi. doi:10.1016/j.jretai.2004.08.003

Liljander, V., Polsa, P., & Riel, A. v. (2009). Modelling consumer responses to an apparel store brand: Store image as a risk reducer. *Journal of Retailing and Consumer Services, 16*, 281–290. doi:10.1016/j.jretconser.2009.02.005

Loebbecke, C. (2009). Furthering Distributed Participative Design - Unlocking the walled gardens. *Scandinavian Journal of Information Systems, 21*(1), 77–106.

McCardle, K. F., Rajaram, K., & Tang, C. S. (2007). Bundling retail products: Models and analysis. *European Journal of Operational Research, 177*, 1197–1217. doi:10.1016/j.ejor.2005.11.009

Moore, C. M., Birtwistle, G., & Burt, S. (2004). Channel power, conflict and conflict resolution in international fashion retailing. *European Journal of Marketing, 38*(7), 749–769. doi:10.1108/03090560410539230

Nair, A., & Kloss, D. J. (2006). An examination of the impact of coordinating supply chain policies and price markdowns on short lifecycle product retail performance. *International Journal of Production Economics, 102*, 379–392. doi:10.1016/j.ijpe.2005.04.009

Newman, A. J., Yu, D., & Oulton, D. P. (2002). New insights into retail space and format planning from customer-tracking data. *Journal of Retailing and Consumer Services, 9*, 253–258. doi:10.1016/S0969-6989(02)00010-3

Oke, A., & Gopalakrishnan, M. (2009). Managing disruptions in supply chains: A case study of a retail supply chain. *International Journal of Production Economics, 118*, 168–174. doi:10.1016/j.ijpe.2008.08.045

Oppewal, H., & Holyoake, B. (2004). Bundling and retail agglomeration effects on shopping behaviour. *Journal of Retailing and Consumer Services, 11*, 61–74. doi:10.1016/S0969-6989(02)00079-6

Patil, R., Avittathur, B., & Shah, J. (2010). Supply chain strategies based on recourse model for very short life cycle products. International Journal of Production Economics Pereira, J.V., (2009). The new supply chain's frontier: Information management. *International Journal of Information Management, 29*, 372–379.

Rajaram, K. (2001). Assortment planning in fashion retailing: methodology, application and analysis. *European Journal of Operational Research, 129*, 186–208. doi:10.1016/S0377-2217(99)00406-3

Sen, A. (2008). The US fashion industry: A supply chain review. *International Journal of Production Economics, 114*, 571–593. doi:10.1016/j.ijpe.2007.05.022

Teller, C., & Reutterer, T. (2008). The evolving concept of retail attractiveness: What makes retail agglomerations attractive when customers shop at them? *Journal of Retailing and Consumer Services*, *15*, 127–143. doi:10.1016/j.jretconser.2007.03.003

Tsay, A. A. (2001). Managing retail channel overstock: Markdown money and return policies. *Journal of Retailing*, *77*, 457–492. doi:10.1016/S0022-4359(01)00055-0

Vaagen, H., & Wallace, S. W. (2008). Product variety arising from hedging in the fashion supply chains. *International Journal of Production Economics*, *114*, 431–455. doi:10.1016/j.ijpe.2007.11.013

Van Weele, A. J. (2010). *Purchasing and Supply Chain Management* (5th ed.). Cengage, Andover.

Wang, W.Y.C., & Chan, H.K., (2009). Virtual organization for supply chain integration: Two cases in the textile and fashion retailing industry. *International Journal of Production Economics*.

Webster, S., & Weng, Z. K. (2008). Ordering and pricing policies in a manufacturing and distribution supply chain for fashion products. *International Journal of Production Economics*, *114*, 476–486. doi:10.1016/j.ijpe.2007.06.010

KEY TERMS AND DEFINITIONS

Chain: A number of stores constructed around the same idea of a brand, an ownership concept, a service, some products. Ownerships might be diverse. The look, the products, the appearance, and/or the attitude of addressing customers would normally be shared across the chain. Chains range from "mini chains" with maybe four or five stores up to thousands of stores.

Electronic Funds Transfer (EFT): Typically a magnetic or smart card reader checking and transferring money from the customer to the store. A keypad is normally built into the EFT. EFT can also just be a functionality within the POS software doing card reading and the financial transaction.

Enterprise Architecture (EA): Is the discipline of strategic planning systems architectures business – technology alignment within private and governmental enterprises. The offset has been Information Technology, which offers relevance in information processing enterprises, but creates scepticism in other enterprises where other governance frameworks also exist.

Fashion: The definition of fashion is often disputed between high-end, more expensive garment products and – used here – general non-food, consumer "wearables", fashion-inspired products within garments, footwear, accessories in almost any price range.

Fiscalisation: Making a system compliant with national tax and excise legislation. Potentially, also requirements of documentation, traceability, validation to avoid tax evasion.

Internationalisation: Making a system ready to meaningful operations in more than one country so no information or technical item, normally being different between countries, are "hard coded" into the system, e.g. currency.

Localisation: Making a system suited for a given language / culture / country in terms of linguistic and verbal support and adaption to a host of other local conditions both technical and people-related.

Loyalty: Retailers and brand owners believe that customers can be more prone to buy his products or visit his store more often than other stores. Loyalty has numerous implementations and legal frameworks different from country to country. Customer Club cards are quite common but vouchers, sweepstakes, mobile phone tags and many other loyalty programs exist.

Point of Sale System (POS): Sometimes ePos for electronic, is the system used by the store employees to enact information processing connected to a sale, typically registering the item and settling a payment. Newer POS offers many options to manage the store and provide external links.

Section 2
Digital Contents Management for Technology-Based Retailing

The main focus of this section is the analysis of digital contents management for a new technology-based retailing in terms of information representation, transferring, and searching. In particular, the selected chapters investigate how it is possible to collect, exploit, represent, and manage the information for achieving useful data for predicting and influencing consumer behavior.

Chapter 7
Frameworks for a Consumer's Group Knowledge Representation

Massimo Franco
University of Molise, Italy

Francesca Di Virgilio
University of Molise, Italy

Loredana Di Pietro
University of Molise, Italy

Angelo Camillo
Woodbury University, USA

ABSTRACT

This chapter discusses three main objectives: (1) the contribution to the body of literature of consumer behaviour demonstrating that consumer groups' knowledge (i.e., two-person dyads, families, peer or friendship groups, teams, and other social units) is relevant for study by consumer researchers; (2) the development of an integrated conceptual representation of consumer's group knowledge including the influence of collective variables on decision making process; (3) the investigation of scientific inquiries regarding the role of advanced technologies in relation to conceptual representation. The approach introduces a new framework applicable both as a tool for enhancing the understanding of consumer's group knowledge, and as a useful guide to future research on consumer knowledge as a whole. The content discussed herein attempts to establish the building block toward the development of a theory of consumer's group knowledge. The study offers direction toward a potential path that could evolve into an established theory regarding consumer's group knowledge in the marketplace.

DOI: 10.4018/978-1-60960-738-8.ch007

INTRODUCTION

For both researchers and practitioners, consumer's knowledge is a critical factor in creating competitive success over time. Competitive advantage, particularly in the last two decades, is no longer measured solely by the amount of resources that are readily accessible or by material production. Having exclusive access to consumer's knowledge can become an integral part of the firm's core competencies. It is commonly acknowledged that intangible resources such as consumer's knowledge can lead to firms being more flexible and thus more competitive in Business System (de Vita, Mercurio & Testa, 2007). To remain economically viable, a firm must constantly meet the needs and wants of consumers while providing superior services and products more effectively than its rivals. Consumer's knowledge though allows firms to strategically place themselves in a better position than their competitors in providing these services. Constant access to new knowledge provides the firm with flexibility to create new products or services or innovate on current products, services, and processes.

Thus, understanding what consumers know has been a topic of enduring interest for scholars of marketing science (Deshpande, 1983; Peter & Olson, 1983; Wilkie & Moore, 2003).

A growing body of literature published over the last three decades attests to the importance of consumer knowledge as an area of investigation that deserves merit and continuum in the field of scientific research. This same literature, however, also suggests that consumer knowledge is extremely complex and that traditional measures may tap only a portion of its richness (Myers, Greyser & Massy, 1979; Pfeffer, 1993; Summers, 2001; McInnis, 2004, Levy, 2006). Consumer's knowledge, in fact, includes several schools of thought (consumer culture theory school, transformative consumer research school, social cognition school, motivation research school and behavioural decision theory school) and several disciplinary approaches.

An analysis of published literature reveals that over the past several years the emphasis was substantively on attitude toward choices among a set of close substitutes. Particular emphasis was evident in assessing the practical importance and the impact of marketing mix elements, specifically, price, promotion, and, (mass) advertising. Theoretically, instead, the dominant paradigm has been borrowed from the field of psychology (cognitive and social in particular) and, to a lesser extent, economics. Nevertheless, the evolution of consumer's knowledge shows that a "theory of consumer" (Teas & Palan, 1997; Summers, 2001) is still not available and scholars' struggles swing from a very ambitious goal of building a comprehensive theory to a less aspiring one of developing theories.

Within the scope of consumer's knowledge, the first part of this chapter discusses how scholars develop the knowledge of a single consumer and the knowledge of a group of consumers in their field of research through published literature on consumer knowledge and on organizational behaviour. In analysing the context it becomes evident that there is a gap in the consumer's knowledge literature. Although published literature recognizes the importance of the consumers as a source of knowledge, is fails to recognize the importance of a "group of consumers" as well. Indeed scholars have not generally included a group of consumers as a level of analysis.

Over the years, several articles published in top tiers management journals have focused primarily on intra - individual behaviours, especially cognitive processes measured accurately by studying the performance of individuals in purchasing goods for consumption (Bagozzi, 2000). There have been only a few analytical attempts to determine if interpersonal interactions of a consumer's group do influence consumer behaviour (Bagozzi, 2000; Briley & Wyer, 2002; Thomas-Hunt, Ogden & Neale, 2003; Cummings, 2004).

Hence, augmenting organizational behaviour literature with research on expertise, effects, and conceptual development, we illustrate a theoretical proposal for improving the current consumer's knowledge representation through integration of one important perspective as our level of analysis: the group (Thomas-Hunt, Ogden & Neale, 2003; Cummings, 2004). We construct a conceptual representation of consumer's group knowledge which includes important collective variables of a group and incorporates the direct effects of these variables on decision making process of consumer's group. We also present a depiction of the decision making process of a consumer's group as a problem – solving process including a continuous flow of reciprocal individual and collective interactions among environment factors, cognitive and affective process and behavioural actions. This innovative approach adopts a sort of "human capital interpretation" (Ratchford, 2001) in which the human capital is a consumer's group and our objective is to investigate its dynamics relevant to "consumer's knowledge models".

Successively, we discuss the possible role of advanced technologies and its impact on our theoretical representation which is an evolution of consumer's knowledge models for the development of a theory of consumption as a fruitful ground.

Technologies are a worthwhile opportunity to improve the efficiency of purchase decision making, and consequently, if well applied, the accuracy of consumer knowledge (Thomas-Hunt, Ogden & Neale, 2003; Chi, 2009; Pantano, 2010). Today the development and use of advanced technologies for supporting and influencing consumers during their shopping experience plays a key role for both retailers and researchers (Chi, 2009). We present three scientific inquiries about how advanced technologies could be quite effective in helping a group of consumers make decisions about a purchase. We close the chapter by discussing limitations and future researches of this work.

This effort constitutes an initial step as a building block yet far from completion of what is considered a theory of consumer's knowledge. Nevertheless we present some useful insights and posit that it might be fruitful in the development of such a "group knowledge theory".

BACKGROUND

Knowledge is a renewable, reusable and accumulating resource of value to the organization when applied in the production of products and services. Knowledge however cannot, as such, be stored in computers; it can only be stored in the human brain. Knowledge has the highest value, the most human contribution, the greatest relevance to decisions and actions; it has sense of power and the greatest dependence on a specific situation or context (Poston & Speier, 2005; Tanriverdi, 2005; Wasko & Faraj, 2005).

Knowledge must exist before information can be formulated and before data can be processed and measured to form information. As such, raw data does not exist if thought or knowledge processes that lead to its identification and collection have already influenced even the most elementary part of data.

It is argued that knowledge which exists, when articulated, verbalized, and structured, becomes information which in turn, when assigned a fixed representation and standard interpretation, becomes measurable data (Alavi & Leidner, 2001).

Critical to this argument is the fact that knowledge does not exist outside an agent (a knower); it is indelibly shaped by one's needs as well as one's initial accumulation of knowledge. Knowledge is thus the result of cognitive processing triggered by the inflow of new stimuli.

One of the reasons that knowledge is such a difficult concept is because this process is recursive, expanding and often discontinuous. According to Grover and Davenport (2001), many cycles of generation, codification and transfer are concurrently occurring in businesses. These cycles feed on each other. Knowledge interacts with information to increase the state space of possibilities and provide new information, which can then facilitate generation of new knowledge. The knowledge process acts on information to create new information that allows for greater possibilities to fulfill old or possibly new organizational needs. This process is often discontinuous, where new needs and their fulfillment mechanism could be created.

In resource-based perspective of knowledge, data are raw numbers and facts. Information is processed data and knowledge is information combined with human thoughts. As mentioned earlier knowledge is indeed the result of cognitive processing triggered by the inflow of new stimuli. Information is converted to knowledge once it is processed in the mind of individuals, and the knowledge becomes information once it is articulated and presented to others. A significant implication of this view of knowledge is that for individuals to arrive at the same understanding of information, they must share the same knowledge framework. Understanding the content and structure of human knowledge is an ongoing challenge to cognitive scientists.

In the following section we discuss the frame structure approach to the topic of consumer knowledge and describe the content and organization of consumer knowledge extrapolated from published literature.

Within the scope of this scenario, consumer knowledge can be defined as the subset of the total amount of information stored in memory that is relevant to product purchase and consumption (Blackwell, Miniard & Engel, 2001). Numerous studies have been conducted in the field of marketing to investigate consumer knowledge and to explore its influence on the individual differentials in consumer behaviour, and to identify the consequences of those differentials on the behavior of the firm. Furthermore researches published in consumer knowledge literature show overwhelming concern about individual processes, attitudes, information processing, judgment, and choice.

The standard depiction of a consumer knowledge structure shows a network of concepts that are linked to each other without any restrictions placed upon membership to the network (Rulke, & Galaskiewicz, 2000). Possible types of associations include the target concept's characteristics, related products, product uses, attitudes and other summary evaluations, as well as purchase-related associations including store and price information, and second-hand memories from such sources as advertisements and word-of-mouth.

Marks and Olson (1981), describe knowledge structures as containing factual knowledge, evaluations, affect, purchase criteria, and decision rules. Russo and Johnson (1980) developed a five-level classification scheme based upon presumed stages in a brand choice process. A factor analysis suggested that three factors were operative: knowledge of product attributes, knowledge centred on situational usage that would distinguish experts from novices, and personal knowledge.

While all these types of aspects have been proven to exist, many other aspects of consumer knowledge have been neglected (Bagozzi, 2000). One of these aspects is exactly the study of a "consumer's group". Consumer's group are in so far assumed to be simply an additive function of the individual activities or, more commonly, are totally ignored. The individualistic bias limits generalization of theoretical propositions and empirical findings across consumer knowledge because a significant proportion of consumer information processing is done not by individuals acting alone but by two or more persons in interaction (Granbois, 1968).

In fact a research study conducted in the late 60s by Granbois (1968) about the in-store behaviour of shoppers identifies some differences between group and individual shopping. The research found that shopping parties of at least three persons deviated more from their original purchase plans (they bought either more or less than originally planned) than did either single shoppers or two-party groups. Furthermore two or more people shopping together were almost twice as likely to buy more than planned than if they had shopped alone.

Over the last three decades studies on consumer's group have been rare. Witt (1969) discusses the influence of small, informal social groups on member brand choice. Rudd and Kohout (1983) drawing on small groups research, sociology, and consumer information processing research, compare information acquisition depth and decision time across individuals, ad hoc cross-gender dyads, and married couples, while Ward and Reingen (1990) analyse how social group's structure influences cognitive structure and how shared cognitive structure influences choice. This perspective is applied to how a group (with several subgroups) makes a consumer decision with consequences for the entire group.

A single pioneer study of great significance was conducted by Ratchford (Ratchford, 2001). The study was about the theoretical and empirical investigation of the "human capital interpretation" in consumer behavior. Within this context "human capital" refers to the accumulated and embodied knowledge, skills, and expertise, which have been acquired and preserved by consumers. In essence it is a variable accumulation of knowledge which could reflect all those qualities of consumers that affect their capacities to reach objectives. In this study, the term knowledge is used as synonymous of ability, attribution, capability, competence, experience, interpretation, intuition, know-how, persuasion, skill, and tradition. Also within this context, a great importance is given to human resources "consumers" which, according to Ratchford (2001) contribute to: (1) better rational choices that have direct effect on the purchase; (2) more effective interactions for the best possible purchase deals; (3) better informed consumers decision-making processes. This is one of the first studies that focused not on one individual consumer but on multitude of consumers.

Following Ratchford's study, we have focused specifically on the role of group's consumers (Bagozzi, 2000; Briley & Wyer, 2002; Thomas-Hunt, Ogden & Neale, 2003; Cummings, 2004). There have been also several analytical attempts to determine if such group interactions do in fact influence consumer behavior (Grier and Deshpandé, 2001). These perspectives suggest that, while knowledge is "owned" at the individual level, the integration of this knowledge to a collective level is necessary. This integration of knowledge typically takes place in groups. Marketing men have conceded that such group factors (as dimension, social class, and ethnic groups) all play some role in consumer decision making (Briley & Wyer, 2002; Okhuysen & Eisenhardt, 2002).

In sum the lack of theoretical and empirical research on consumer group knowledge led us to the development of our conceptual representation. Our study starts by considering the group variables as an important step in understanding consumer's group knowledge. Group variables are seen as a major determinant in attitude formation and attitude change, as well as for other phenomena of importance to the collective level. Subsequently, we analyse the influence of these group variables on the most important process at the base of consumer knowledge: the decision making process. Given the organizational behaviour orientation of this text, we emphasize consumer's group decision making process regarding a purchase.

CONSUMER'S GROUP KNOWLEDGE REPRESENTATION

In response to growing demands for efficiency and flexibility, organizations shift from individual to group-based structures. Groups bring assets, adding knowledge and creativity, increasing the understanding and acceptance of ideas (Tosi, Mero, & Rizzo, 2000). Numerous studies have demonstrated benefits for groups that engage in information exchange and communication within the group (Keller & Staelin, 1987; Gruenfeld, Mannix, Williams & Neale, 1996; Rulke & Galaskiewicz, 2000). Though successful groups take advantage of the perspectives, talents, and ideas of different members, a well-designed group also creates a common understanding of the purchase context through the sharing of knowledge.

The most common definition of group is "a collection of two or more interacting individuals with a stable pattern of relationships between them who share common goals and who perceive themselves as being a group" (Davenport, 1999). This definition can be applied also to consumer's groups conceived as two or more people in social interaction who share common goal: purchase decision making. The final purpose of a consumer's group is to satisfy a need across buying consumer goods. Cartwright and Zander (1968) suggested that the following statements can be applied to a consumer's group: "when a set of people constitutes a group, one or more will characterize them: (a) they engage in frequent interaction; (b) they define themselves as members; (c) they are defined by others belonging to the group; (d) they share norms concerning matters of common interest; (e) they find the group to be rewarding; (f) they pursue interdependent goals; (g) they have a collective perception of their unity; (h) they tend to act in a unitary manner.

The work of Allen (1977) indicates that people prefer to turn to other people rather than interpreting documents to obtain information. Allen (1977) found that engineers and scientists were approximately five times more likely to turn to a person for information than to an impersonal source such as a database or file cabinet. When information is held by multiple members, not only more people within the group possess the information, but group members who possess the information may also provide retrieval cues to each other to aid the introduction of the knowledge and decision making (Ward & Reigen, 1990; West, Garrod & Carletta, 1997; Bagozzi, 2000). More recently, this same tendency has been found even for people with ready access to the Internet and their firm's extensive intranet (Cross & Sproull 2004).

Published literature categorises two macro - types of groups: formal and informal group (Levine & Moreland, 1990). In this study we use the concept of informal groups which better adapts to consumer's groups. From the standpoint of consumer knowledge, informal social groups are generally more important to the marketer, since their less clearly defined structures provide a more conducive environment for the exchange of information and influence about consumption – related topics (Thomas-Hunt, Ogden & Neale, 2003). One key factor in the formation of informal groups is that membership is voluntary; it is not dictated by the organization, rather encouraged by an expression of common interests. Certainly, sometimes the interests that bind individuals together are far more disperse. Groups may develop out a common interest in participating in sports, or as for consumer's groups, in going shopping together. Friendship groups, for example, consist of people with natural affinities for one another. They sit together, take breaks together, and even do things together. Friendship groups provide opportunities for satisfying the social needs of individuals that are so important to their well being; therefore those types of informal groups are an important part of social life. In these and related ways, informal groups can provide their members with social satisfactions, security, and a sense of belonging. For example, each of the following illustrates common usage of informal consumer's group concept: two friends "discussing via chat on a social network (as facebook) about the purchase of an MP3", a family "making a purchase of a bedroom with the assistance of virtual salesperson of IKEA", and a music band "making a purchase of an electronic guitar on e-bay". In this point of view, we discuss that knowledge of the consumer's group lead to richer content and can help explain why consumers do what they do.

The term "group dynamics" refers to the mechanisms of relationship that take place within a

Figure 1. Consumer's group knowledge representation

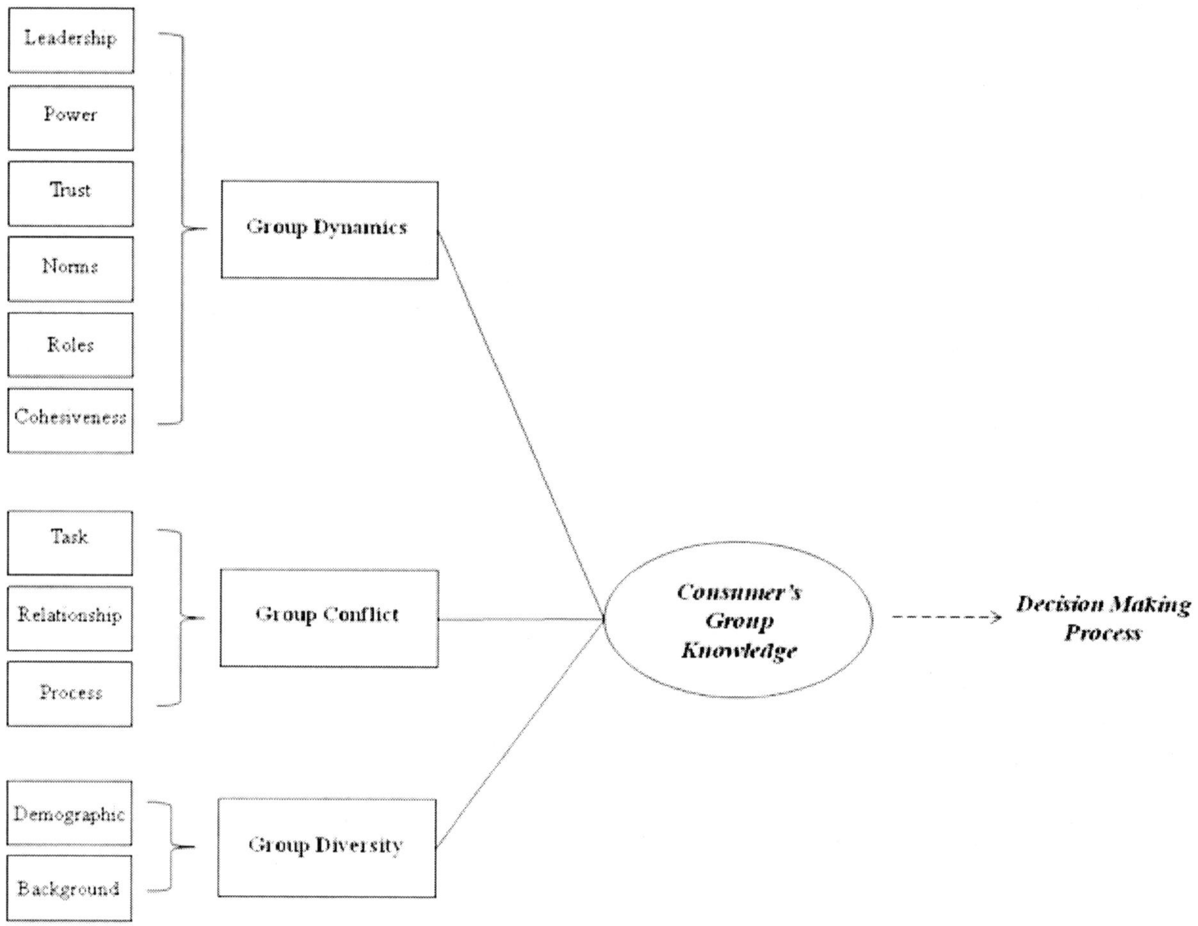

group (Shaw, 1981; Stewart, Manz & Sims, 1999). In general, researchers have found group dynamics to be important for acquiring information (Keller & Staelin, 1987; Gruenfeld, Mannix, Williams & Neale, 1996; Rulke & Galaskiewicz, 2000), for disseminating useful knowledge (Schermerhorn, Hunt & Osborne, 2008), for solving complex problems (Rulke & Galaskiewicz, 2000) and for influencing decision making process (Yoon et al., 2009).

The primary purpose of this chapter is to explore the influence of informal groups dynamics on the decision making process of their members and to develop an integrated conceptual representation of consumer's group knowledge. In addition we enhance our representation by including two other important group variables: group's diversity and intragroup conflict. Figure 1 describes the group dynamics and group's variables included in our representation.

One of the most important group dynamics is leadership. Published theoretical and empirical literature in behavioural sciences fully supports the notion that groups, and particularly certain individuals within the group, influence member behaviour (Rulke & Galaskiewicz, 2000). Within the scope of this study we define leadership as "the process by which an individual of a group influences others" (Bass, 1990) in order to facilitate their purchase decision. In other words, subordinates accept influence from leaders because they respect, like, or admire them as well as because they hold positions of authority (for example a father of a family).

Leadership is connected directly to power. Power is defined in organizational science as the ability to get someone to do something you want done or the ability to make decisions in the way you want them to (Knights & Willmott, 2007). In our representation we consider power as the participation in the making of important decisions which involve severe deprivation of values. This interpretation is consistent with our intention of using power as a weighting function for group decision-making. A unique property of power within this concept is that power is a relationship between two persons and not an absolute attribute of a single individual. This definition of power is closely analogous to autonomy or the ability to achieve one's goals without interference from others. In terms of dependency (Thibaut & Kelley, 1959), it is suggested that power may derive from the ability to help others achieve goals that they otherwise would not be unable to meet, thereby creating dependency in others.

Another variable of group's dynamics is trust. Trust has been defined as a state of a positive, confident, though subjcctive expectation regarding the behaviour of somebody or something in a situation which entails risk to the trusting party (Currall & Judge 1995). Although the concept of trust has been viewed at different levels (group, organization, society), here we focus on trust among group members, which is defined as the extent to which a person is confident in, and willing to act on the basis of words, actions, and decisions of another. The trust literature provides considerable evidence that trusting relationships lead to greater knowledge exchange and to better decisions. Trust groups are defined as groups with close, interpersonal ties and positive, amiable pre-existing relationships among members (Dirks & Ferrin, 2001). When trust exists, people are more willing to give useful knowledge and are also more willing to listen and to absorb others' knowledge (Levin 1999). By reducing the need to verify information, trust also makes knowledge transfer less costly. For example, Levin (1999) found that strong trusting ties usually helped improve knowledge transfer between scientists and engineers. Instead affect-based trust is typically found to be important in the context of social relationships for informal group of consumers.

Groups have norms and values that explain much of a group's decision making. Norms, values, and standards of behavior guide people's judgments and decisions and have often been conceptualized in terms of individualism and collectivism (Postmes, Spears & Cihangir, 2001). In our representation we consider trust as a collective concept, that it is often reflected in a disposition to think of oneself as a member of a group or collective and to evaluate one's own attributes and behavioral outcomes in relation to others. Subsequently norms became the beliefs, moral rules, and values, which guide members to decide a product between several alternatives. Group norms code describes acceptable purchase and influence consumer's group decision making.

Every member of a group plays a certain role within that group. Roles are coherent sets of behaviours that tend to be adopted by the different members of a group, partly as a matter of personal inclination but also as a response to the expectations of others (Fisher, Hunter & Macross, 2001). We presuppose that different roles may emerge and affect the relative participation of members to the decision making process within the group. They also include relationship roles, such as the group encourager, as well as task roles related to the practical aspect of the group such as decision making process.

The nominal definition of cohesiveness is the total field of forces which act on members to remain in the group (Leana, 1985). Two classes of this concept are distinguished and conceptualized: (1) the attractiveness of the group for its members and (2) the extent to which the group mediates goals for its members when cohesiveness is biased in terms of interdependency. Cohesiveness is the last definition included in the group dynamics of our representation. Dirks and Ferrin (2001) highlight the importance of cohesiveness for the success of knowledge transfer because it increases the effort in reconstructing the transferred information and in applying them in terms of knowledge. Group members' desire to remain in their group and have confidence in decisions of others members can create a group mind-set. This mind-set leads group members to make purchase decisions that minimize the risk of negative outcomes for both themselves and others, thanks to a sense of responsibility each member feels for the group. This is because negative outcomes of group-relevant purchase decisions (e.g., failure, disgrace and embarrassment) or the possibility to have more unfavourable consequences for some members can erode group cohesiveness. This may force group members to be closer together. For these reasons, feelings of group cohesiveness may increase cautiousness and cause a stronger tendency to avoid purchase making decisions that could have negative consequences for one-self and other members (Aaker & Lee, 2001).

We expand our representation and add two group's variables to the group's dynamics. The first variable is group conflict. Group conflict may be defined as "a tension between group members due to real or perceived differences" (Jehn & Mannix, 2001). An abundance of recently published literature (Jehn, Rupert & Nauta, 2006; Franco, Di Virgilio & Di Pietro, 2006) has indicated that some types of conflicts may actually be less detrimental (and even beneficial) to group decision making. Therefore, a helpful contribution to the development of our conceptual model is the distinction between different types of conflicts, being (Jehn & Mannix, 2001) relationship, task, and process conflict.

Relationship conflict is a perception of interpersonal incompatibility and typically includes tension, irritation and hostility among group members (De Dreu & Van Vianen, 2001). Task conflict is generally task oriented, depersonalized, and is focused on judgmental differences on the best solution in key decision areas (Jehn & Mannix, 2001). Process conflict is the conflict about how tasks should be accomplished by the group, including the distribution of responsibilities and the delegation of tasks and resources among their members. These three types of intragroup conflicts have different consequences both, negative or positive on group decision making.

Relationship conflict has been theorized to have negative effects on several group outcomes (Jehn, Rupert & Nauta, 2006) as decision making process. Literature suggests that relationship conflict promotes inefficiency and ineffectiveness, leads to a loss of perspective regarding the objects, inhibits individuals' cognitive functioning in assessing new information provided and processing complex information, encourages stereotype listening, and induces the freezing out of iconoclasts from important discussions (Jehn, Rupert & Nauta, 2006). Moreover, Jehn e Mannix (2001) found that relationship conflict diminished decision creativity and quality and decrease knowledge, which hinders the completion of organizational tasks. Time is often spent on interpersonal aspects of the group rather than on technical and decision-making tasks.

On the other hand, several researches show that task conflict has positive impact on group outcomes. Also interaction techniques which force group members to disagree and debate the merits of different alternatives produce superior decisions. Jehn and Mannix (2001) noted that task conflict contributes to decision quality because the synthesis that emerges from the contesting of the diverse perspectives is generally superior to the individual perspectives themselves. Task conflict appears to be positively related to the increase of constructive interpretation of information and to the selection of alternatives to make decisions (Jehn & Mannix, 2001).

There is no consensus in the empirical literature concerning the impact of the process conflict on group decision making. Jehn and Mannix (2001) found that high levels of process conflict had a dysfunctional effect in the group's performance. However it is almost impossible to identify a positive or negative influence of process conflict on group decision making.

Group diversity is another important group's variable part of the knowledge process. Lawrence (1997) suggested that diversity can be studied across at least four different categories of variables: visible demographic attributes (such as gender); relational attributes (such as organizational tenure); status attributes (such as marital status); and personal attributes (such as personal beliefs and perceptions). Jehn and Bezrukova (2004) consider group diversity along six demographic dimensions: age, gender, race, and tenure with the company, level of education, and functional background.

In this context however we adopt the perspective of other authors (Pfeffer 1983; Shaw & Barret-Power, 1998) who suggest that there are two macro - types of diversity which is a better fit to our use of informal group: demographic (or primary) diversity and background (or secondary) diversity.

The term demographic diversity refers to the degree to which a unit (e.g. a working group or organization) is heterogeneous with respect to demographic attributes. Attributes classified as demographic generally include immutable characteristic such as age, gender and ethnicity; attributes that describe individuals' relationship with group, such as group position (leader, follower, etc...).

Background diversity is referred to a difference in the amount of knowledge accumulated in a group, for example the specialization in different problem-solving domains found in Benbasat and Weber (1996) between actors, or when the members of a group have dissimilar belief structures, priorities, assumptions about understandings of alternatives, based on previous training and experience (Shaw & Barret-Power, 1998).

Scholars examining diversity in groups have primarily focused on the consequences of demographic diversity for processes such as communication and decision making (Jehn & Mannix, 2001; Franco, Di Viriglio & Di Pietro 2007). The consistently negative effects of demographic diversity on group processes are likely the result of heightened member emphasis on social categories rather than project-relevant information. Therefore, we posit that demographic diversity should not increase the value of intragroup knowledge and of decision making process.

Background diversity has been hailed as a competitive advantage because minority views "can stimulate consideration of non-obvious alternatives in decision making groups" (Shaw & Barret-Power, 1998). In fact homogeneity limits the variety of views within a group and may decrease the numerous alternatives to purchase a final product. Accordingly, we assume that background diversity should increase the value of intragroup knowledge and of decision making process.

The need for a theoretical representation was born out of the determination that little theoretical research is available on consumer's group knowledge. Conceptual papers are rapidly declining despite the fact that they are critical to the development of knowledge. We therefore hypothesize that an empirical bias could be the main reason behind the lack of conceptual papers. Scholars may imply that papers are only viewed as scientific if they have an empirical component. This bias creates a paradox: scholars tend to focus their efforts on empirical studies that are very often considered inadequate from a theoretical perspective when submitted to an academic outlet. This becomes one of the main reasons for rejection and, in turn, drives many scholars to use rhetorical contortions in order to include theoretical justifications in their empirical studies (Summers, 2001; McInnis, 2004, Levy, 2006, Mari, 2008).

CONSUMER'S GROUP DECISION MAKING PROCESS

In our consumer's group knowledge representation we assume that all group's variables have influence on the decision making process. Studies in this area have focused on individual cognitive processes and the results have shown relatively little concern with how others may influence these decisions or with the possibility that decisions should be studied from a group as well as from an individual perspective (Bagozzi, 2000; Briley & Wyer, 2002; Thomas-Hunt, Ogden & Neale, 2003; Cummings, 2004).

Research has found also that the characteristics of prior knowledge possessed by group members and how information is distributed within the group affect the decision making process (Levine, 1999). The literature about small groups also indicates that group decision making depends not only on information resources available to the group, but also on the processes or structures which groups use to exploit these resources. Stasser, Vaughan & Stewart (2000) further found that as a piece of information was distributed across more individuals within the group, the retrieval of this information became more likely and thus facilitated group decision making.

When decision making is conceived as a cycle of interactions between individual members such that each person is seen to give off and to receive attempts to influence others, the perspective is an interpersonal one. Here, group members act or react in coordinated ways, but no notions of collective concepts are incorporated. Rather, analyses are confined to individual characteristics (e.g., personal beliefs) and individual actions. By contrast, when group decision making is seen as a social process of joint formation of goals and intentions, the perspective is a plural subject one. Here members jointly construct mutual understandings and shared volitional commitments to make a group decision and consciously come to see their actions in this way. Knowledge does not originate solely from the mental processes of an individual; it also originates from the social or collective environment being the group dynamics (Bagozzi & Dholakia, 1999). Such collective concepts as group or social identity become a central variable in knowledge management. Our research is based on the group dynamics or collective approach.

"Literature published over the years concurs that the decision making of a single consumer consists of a multi-step process (Mihal, Sorce & Comte, 1984; Dean & Sharfman, 1996; Knights & Willmott, 2007)." The first step is identifying the need of a product. During this process the consumer could distort, omit, ignore, and/or discount information that provides important cues regarding the existence of product need. This, of course, is problematic because a need cannot be satisfied if it is never recognized. The next step is making a pre-decision purchase; that is a decision about how to make a decision. By assessing the type of need in question, consumer may opt to make a decision. The next step is to individuate possible products to satisfy a need. Because all these possibilities may not be equally feasible, the successive step calls for evaluating alternative products. In the final step, consumer makes a purchase choice. After several alternatives are evaluated, the one that is considered acceptable is chosen.

In this chapter, we consider a different prospective concerning the process of decision making; we present the perspective from a consumer's group point of view and not that of a single consumer as per common research on consumer knowledge. We view consumer decision making as a problem – solving process and assume consumers in general have common goals that they seek to achieve or satisfy. We argue that bringing people together may increase the amount of product knowledge and information available for making good purchase decisions. In other words, the group may be replete with resources available to the members. An additional benefit is that group decisions are likely to enjoy greater acceptance than individual

Figure 2. Consumer's group decision making process representation

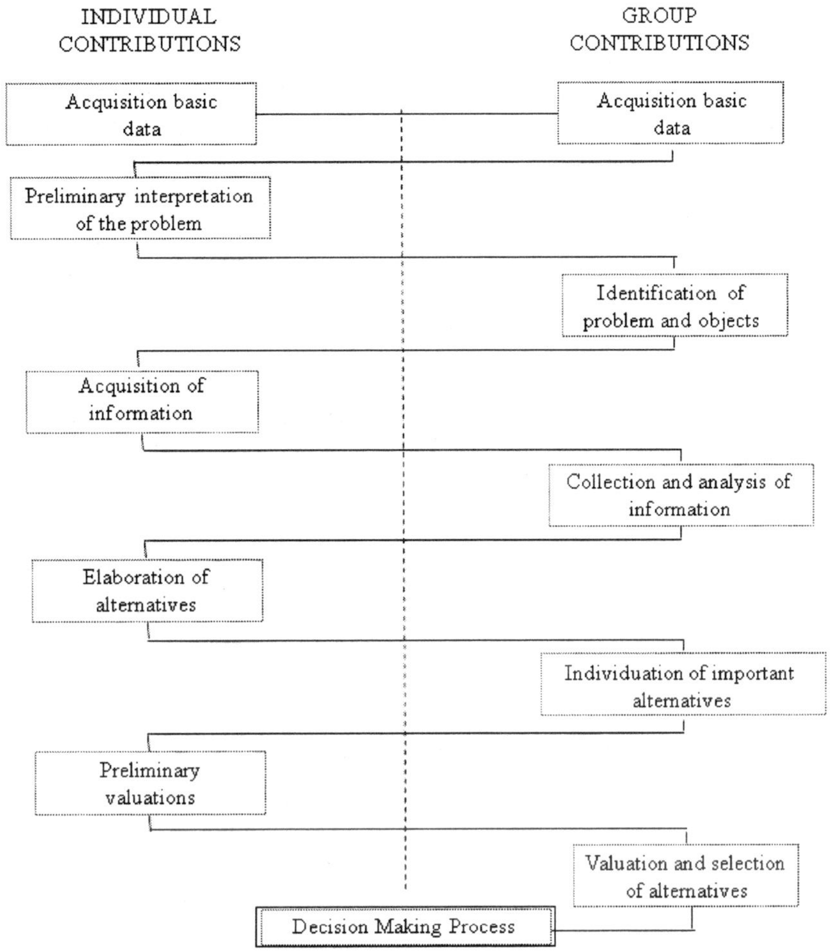

decisions. People involved in making decisions may be expected to understand those decisions better and be more committed to carrying them out than decisions made by someone else.

In our representation we follow the thoughts according to Franco (1991), that consumer's group problem solving is a continuous flow of reciprocal, individual, and collective interactions among environmental factors, cognitive and affective processes and behavioural actions. We divided this stream into separate stages at individual and collective level to simplify the analysis and to facilitate the understanding (Figure 2).

When an individual or group decision initiates, following the general and collective acquisition of basic data, each individual typically begins with a preliminary interpretation of the problem (e.g. set of sub - goals organized into a goal hierarchy, relevant product knowledge) however a problem identification is realized only at group level and serves as a decision frame through which the decision maker views the objects to reach.

Gradually, cognitive processes allow acquiring information at individual level in order to collect and analyse them as an identity. After individual elaborations of alternatives, the group individuates the most important ones. These several alternatives are evaluated and the one that is considered acceptable is chosen. Consequently, in the final step of the process, the group makes a purchase choice.

Such a collective perspective, often advocated but rarely implemented, promotes a broader view of how a groups make decisions than that suggested by traditional studies of information processing.

THE ROLE OF ADVANCED TECHNOLOGIES IN OUR REPRESENTATION: SOME SCIENTIFIC ENQUIRIES

Communication and information technologies are adding new capabilities for rapid and inexpensive consumer input at all stages of the knowledge process. And at each stage of advanced technologies development process consumer's knowledge has experienced enormous improvements (Nambisan, 2002).

Several approaches have been applied to the study of consumer knowledge in the advanced technology environment. A group of researchers applied intention theories to investigate Internet adoptions (Gefen & Straub, 2000; Chen, Gillenson & Sherrell, 2002). Some investigated website characteristics that render a higher quality Internet store (Aladwani & Palvia, 2002, Liang & Lai, 2002). Many were interested in studying types of Internet usage (Ranganathan & Ganapathy, 2002). Others investigated individual characteristics of consumers that affect online purchasing decisions (Bellman, Lohse & Johnson, 1999). In addition to these topics, some researchers have devoted their efforts to categorising products that can be sold on the Internet successfully (Gefen & Straub, 2000), while others have investigated how shoppers transform themselves from non-innovators to innovators (Nambisan, 2002).

The advanced technologies applied to retailing are usually based on pervasive environments and mobile and ubiquitous computing (Pantano & Naccarato, 2010). Retailers have deployed kiosks, interactive displays, handheld shopping devices, computer-enabled grocery carts, and special shopping trolleys to assist with store navigation, provide detailed product information, offer personalized product recommendations and promotions, and expand the available selection of merchandise (Chang & Burke, 2007). Currently, new researches on the technology applications to retailing focus on the development of a smart mirror (Pantano, 2010). Only few prototypes are available on a limited number of stores across the world. The smart mirror consists of integrated software and a hardware system which recognizes consumer by a web cam and reproduces graphically him/her while wearing the product in the store. This system allows consumers to visualize how they look in any frame of the store or to see their new contact lens, and simulate the effect produced by the chose good (Pantano, Taversine & Viassone, 2010). Another recent innovation in retail are virtual worlds - computer-generated physical spaces - represented graphically in three dimensions, that can be experienced by many users, so called avatar (Kohler, Matzler & Fuller, 2009). They provide companies with a representational-rich-mediated environment that facilitates direct and rich interactions with consumers. In fact, the playful environment of virtual worlds has been described as engines of creation that provide the freedom to experiment and lead to unprecedented rates of innovation. Kohler, Matzler, and Fuller (2009) suggest also incorporating the latest technological advances into open innovation practice, namely the emerging technology of virtual worlds. The technology of virtual worlds could further enrich existing web-based consumers' integration methods, by allowing real time, media-rich and highly interactive collaboration between sellers and their consumers.

All those technologies allow obtaining and cataloguing dates and information on a single consumer to increase the consumer knowledge and to influence positively consumers buying behaviour. However, in this study we ask three important scientific questions regarding the role of advanced technologies and their integration in our representation of consumer's group knowledge. Hence we posit that:

1. **Advanced technologies** will be able to understand the consumer's group knowledge
2. **Advanced technologies** will interpret group's variables and their influence on consumer's group decision making process
3. **Advanced technologies** can be integrated in our conceptual representation of consumer's group knowledge?

In considering advanced technologies the core question we argue that it is not *if* and *how* advanced technologies can play a role in our representation but what we want technologies to be capable of doing. With the rapid advancement of IT (Information Technologies) and CT (Communication Technologies) we answer to the question that it is not what technology in general can do for researchers and practitioners, instead what we want technologies to do for all players in the market place.

Nevertheless the task of technologies' integration is compelling. In fact several studies have investigated the impact of the information environment of technologies on decision making process. The results indicate that more information is not always better. A 1977 study by Jacoby, Szybillo and Busato-Schach (1977) show that consumers who are given additional product information feel more satisfied and less confused, but they actually make poorer purchase decisions. Keller and Staelin (1987) find an inverted U-shaped relationship between the amount of information available and decision effectiveness. Walczuch and Lundgren (2004) point out that two or three summary attributes would need to be used because of the limitations in consumers' ability to combine many attributes into an overall rating. Similarly, more choice alternatives do not necessarily improve the objective quality of decisions. An increased number of alternatives contribute to task complexity and too many choice options can produce negative effects.

Despite the limited contribution, other studies indicate that in the evaluation process of the different alternatives, these particular technologies are capable to support consumers. In fact, they have the possibility to easily and rapidly gain detailed and complete information on products and services. Furthermore, consumers can immediately compare different proposals (Berg, van den Arentze & Timmermans, 2009; Pantano, 2010). Haubl and Trifts (2000) demonstrate that in a customizable electronic shopping environment, use of a recommendation agent or a comparison matrix generally leads to an increase in the quality of consumers' consideration sets, as well as enhanced decision quality.

However, at the present time it is very difficult to evaluate the role of technologies on consumer's group because no one is able to learn and evaluate group's variables and understand how to help a group of consumers as of yet. Actually the use of technologies applied to this topic is still in its infancy and is very limited.

For many years firms have been relying on their own intuition or qualitative and empirically based information about organizational behaviour for guidance needed to efficiently sell to consumers. This approach carries heavy financial burdens due to rather expensive investment, especially in human resource. Therefore for the near future the use of advanced technologies must be developed summarising much of this information in a way that gives firms ready access to a wealth of information regarding the consumer's group knowledge. Advanced technologies could be quite effective in helping a group of consumers make decisions about a purchase. The status – quo of the use of technology in group knowledge may lead us to think that the introduction of digital content and advanced technologies that understand group variables may be almost unrealistic. However these technologies would be fascinating to all players allowing consumers to facilitate the decision making process through a user-friendly interface, by giving information related to products, promotions, new arrivals and collecting at the same time information about consumer behaviour and group's variables. To achieve this one must consider two main characteristics which are the interactivity and the multimodality, in order to achieve an efficient, flexible and meaningful feeling of human-computer interaction.

FUTURE RESEARCH DIRECTIONS

Given the nascent nature of the study phenomenon, there may be many exciting opportunities that lay ahead for new research. This study contributes to the development of a program of research on consumer's group knowledge. Because only limited research on consumer's group knowledge exists in the field of consumer research, this study can act as a catalyst for future scientific enquiries in this important area.

The issue of how to effectively design and deploy advanced technologies in this group approach is most certainly an additional future research direction. It has become clear that the digital environment offers many opportunities for firms to interact with consumer's group along with the entire knowledge process. A future research direction is to answer to our scientific inquiries regarding the role of technologies and to understand how these technologies can be successfully applied so to interpret the function of the group's variables. Our study could be tested by a group of stakeholders/experts on this topic (e.g. Delphi methodology).

Future studies can expand on the proposed representation by observing other group's dynamics and by distinguishing also between the different levels (high and low) of group dimensions by examining the differential effects on decision making process.

Replete information on group make-up, group dynamics of professional groups, special interest groups, and other group's types are readily available on all search engines of major internet providers. Many studies on consumer behaviour are now conducted over the internet by using these types of groups. They are categorized by industry, by interest, by profession, by demographics and by other characteristics. With the application of E-commerce and M-commerce studying these groups' behaviour becomes useful in researching consumer group knowledge.

Future studies should compare and contrast firms that use this theoretical representation of consumer's group knowledge with those that do not determine the differential impact on consumers knowledge. At a larger scale a comparison could be made in terms of the impact on sales, consumer loyalty, and development of new products.

It is appropriate to note that important implications for policy and marketing decision making may emerge if future research findings indicate substantial differences in the nature of consumer information processing across types of decision making groups. For instance, further research is needed to explore the effect of the type of decision making unit on other information acquisition variables, such as the content and sequence of the information acquired. In addition, research analysing the information processes of various decision making units could be fruitful. It is hoped that the present chapter will help move consumer research into the realm of social information processing and decision making regarding the consumer's group knowledge.

Future research could analyse and develop training system for employees and companies to gather data from the consumers with respect to variables governing the dynamics within a group of consumers. Generally firms have only few opportunities to gather knowledge and sell the company's product or service. This is a concept more clearly understood by firms that sell highly priced products or services. Under these circumstances the firm should be highly trained so that when given this unique opportunity with a group of consumers it will successfully help the group decision making. Providing the right attitude towards the group of consumers and the right techniques to help the consumers, the firm should have appropriate practices and training on group variables to gather knowledge from the consumers.

This type of research, which provides a rich understanding of the consumer's group, are nonetheless expensive due to high cost of training of employees and salespeople. To understand the group's variables and their influence on a group decision making process, expensive ethnographic and qualitative research techniques are necessary.

Another possible problem is that potential group disagreement over important matters may breed ill will and relationship conflict. Therefore, we may expect that groups will not make purchase decision because of members' intimidation by group leaders. However this may indicate the true nature of human behaviour.

Finally another obvious drawback is that groups are likely to waste time. The time spent socialising before making a purchase decision may be a drain on the group and be very costly to organizations.

CONCLUSION

We recapitulate that the representation we propose will accomplish one important goal with respect to consumer knowledge. Using the human capital interpretation and the group's theory, and guided by the theoretical approaches from related research in consumer behavior and consumption patterns, this chapter provides an analytical framework to explore the consumer's group knowledge investigating the influence of group variables on decision making process.

Our study shows that there are some important potential applications of consumer's group knowledge representation in the study of consumer behavior, with an outline of the major theoretical approaches to these applications. Gaining and utilising consistent knowledge by consumers is not a simple or straightforward task. It is a highly involved and multidimensional process, which is seldom complete or errorless. Furthermore, different elements of this process may separately or jointly exert varying, and sometimes, conflicting influences on the normally complicated decision making process for consumers.

Practitioners can use this conceptual representation to evaluate knowledge of a group of consumers and better target future knowledge management interventions towards those groups most likely to benefit. In fact, the outcome of this study is of benefit to both, the consumers and the firms. From a better understanding of the consumer's group variables, a company will have greater understanding of the true needs and expectations of consumers. The firm learns from its consumers about the knowledge that will assist in product innovation and improvements of processes. Since the firm has a better understanding of the consumer's group variables it will be able to improve service and thus achieve consumers' satisfaction and retention. All that leads to increased sales and the acquisition of a new group of user.

To understand interactions within a group of consumers as a source of knowledge can help the firm also attain a competitive advantage in product and service innovation. Therefore, firms do not need to wait for the time consuming marketing research efforts to stay tuned to the changing nature of the market and need not remember the frequency of knowledge acquisition. It's necessary only to change the level of analysis.

REFERENCES

Aaker, J. L., & Lee, A. Y. (2001). I seek pleasures, we avoid pains: The role of self regulatory goals in information processing and persuasion. *The Journal of Consumer Research*, *28*, 33–49. doi:10.1086/321946

Aladwani, A. M., & Palvia, P. (2002). Developing and validating an instrument for measuring user-perceived Web quality. *Information & Management, 39*(6), 467–476. doi:10.1016/S0378-7206(01)00113-6

Alavi, M., & Leidner, D. E. (2001). Knowledge management and knowledge management systems: Conceptual foundations and research issues. *Management Information Systems Quarterly, 25*(1), 107–136. doi:10.2307/3250961

Allen, T. (1977). *Managing the flow of technology*. Cambridge, MA: MIT Press.

Bagozzi, R. P. (2000). On the concept of intentional social action in consumer research. *The Journal of Consumer Research, 27*, 388–396. doi:10.1086/317593

Bagozzi, R. P., & Dholakia, U. (1999). Goal setting and goal striving in consumer behavior. *Journal of Marketing, 63*, 19–32. doi:10.2307/1252098

Bass, B. M. (1990). *Handbook of leadership. Theory, research a managerial application*. New York, NY: The Free Press.

Bellman, S., Lohse, G. L., & Johnson, E. J. (1999). Predictors of online buying behaviours. *Communications of the ACM, 42*(12), 32–38. doi:10.1145/322796.322805

Benbasat, I., & Weber, R. (1996). Research commentary: Rethinking diversity in Information System research. *Information Systems Research, 7*(4), 389–399. doi:10.1287/isre.7.4.389

Blackwell, R. D., Miniard, P. W., & Engel, J. F. (2001). *Consumer behavior*. Fort Worth, TX: Harcourt College.

Briley, D. A., & Wyer, R. S. (2002). The effect of group membership salience on the avoidance of negative outcomes: Implications for social and consumer decisions. *The Journal of Consumer Research, 29*(3), 400–415. doi:10.1086/344426

Cartwright, D., & Zander, A. (1968). *Group dynamics: Research and theory*. New York, NY: Harper & Row.

Chang, C. A., & Burke, R. R. (2007). Consumer choice of retail shopping aids. *Journal of Retailing and Consumer Services, 14*, 339–346. doi:10.1016/j.jretconser.2006.12.003

Chen, L., Gillenson, M. L., & Sherrell, D. (2002). Enticing online consumers: The technology acceptance perspective. *Information & Management, 39*(8), 705–719. doi:10.1016/S0378-7206(01)00127-6

Chi, Y. L. (2009). A consumer-centric design approach to develop comprehensive knowledge-based systems for keyword discovery. *Science Direct. Expert Systems with Applications, 36*(2), 2481–2493. doi:10.1016/j.eswa.2007.12.061

Cross, R., & Sproull, L. (2004). More than an answer: Information relationships for actionable knowledge. *Organization Science, 15*(4), 446–462. doi:10.1287/orsc.1040.0075

Cummings, J. N. (2004). Work groups, structural diversity, and knowledge sharing in a global organization. *Management Science, 50*(3), 352–364. doi:10.1287/mnsc.1030.0134

Currall, S. C., & Judge, T. A. (1995). Measuring trust between organizational boundary role persons. *Organizational Behavior and Human Decision Processes, 64*, 151–170. doi:10.1006/obhd.1995.1097

Davenport, H. (1999). Groups and teams. In Brook, S. I. (Ed.), *Organisational behaviour*. London, UK: Financial Times Pitman publishing.

De Dreu, C. K. W., & Van Vianen, A. E. M. (2001). Managing relationship conflict and effectiveness of organizational teams. *Journal of Organizational Behavior, 22*, 309–328. doi:10.1002/job.71

de Vita, P., Mercurio, R., & Testa, F. (2007). *Organizzazione aziendale: Assetto e meccanismi di relazione*. Torino, Italy: G. Giappichelli Editore.

Dean, J. W., & Sharfman, M. P. (1996). Does decision process matter? A study of strategic decision-making effectiveness. *Academy of Management Journal, 39*(2), 368–396. doi:10.2307/256784

Deshpande, R. (1983). Paradigms lost: On theory and method in research in marketing. *Journal of Marketing, 47*, 101–110. doi:10.2307/1251403

Dirks, K. T., & Ferrin, D. L. (2001). The role of trust in organizational settings. *Organization Science, 12*(4), 450–467. doi:10.1287/orsc.12.4.450.10640

Fisher, S. G., Hunter, A., & Macrosson, W. D. K. (2001). A validation study of Belbin's team roles. *European Journal of Work and Organizational Psychology, 10*(2), 121–144. doi:10.1080/13594320143000591

Franco, M. (1991). *I piccoli gruppi strutturati. Impatto sui modelli organizzativi aziendali*. Torino, Italy: G. Giappichelli Editore.

Franco, M. (2007). I livelli dell'attore organizzativo: gruppo. In de Vita, P., Mercurio, R., & Testa, F. (Eds.), *Organizzazione aziendale: Assetto e meccanismi di relazione*. Torino, Italy: G. Giappichelli Editore.

Franco, M., Di Virgilio, F., & Di Pietro, L. (2006, April). *A qualitative analysis of conflict types: An intragroup conflict*. Paper presented at the meeting of 10th International Workshop on Team working, EIASM, University of Groningen, Groningen, The Netherlands.

Franco, M., Di Virgilio, F., & Di Pietro, L. (2007, April). *Diversity in work groups: Effects on the level of conflict*. Paper presented at the meeting Workshop on Organisational Design, European Institute for Advanced Studies in Management, Bruxelles, Belgium.

Gefen, D., & Straub, D. (2000). The relative importance of perceived ease of use in IS adoption: A study of e-commerce adoption. *Journal of the Association for Information Systems, 1*, 1–30.

Granbois, D. H. (1968). Improving the study of customer in-store behavior. *Journal of Marketing, 32*, 28–32. doi:10.2307/1249334

Grier, S. A., & Deshpandé, R. (2001). Social dimensions of consumer distinctiveness: The influence of social status on group identity and advertising persuasion. *JMR, Journal of Marketing Research, 38*(2), 216–224. doi:10.1509/jmkr.38.2.216.18843

Grover, V., & Davenport, T. H. (2001). General perspectives on knowledge management: Fostering a research agenda. *Journal of Management Information Systems, 18*(1), 5–21.

Gruenfeld, D. H., Mannix, E. A., Williams, K. Y., & Neale, M. A. (1996). Group composition and decision making: How member familiarity and information distribution affect process and performance. *Organizational Behavior and Human Decision Processes, 67*(1), 1–15. doi:10.1006/obhd.1996.0061

Haubl, G., & Trifts, V. (2000). Consumer decision making in online shopping environment: The effects of interactive decision aids. *Marketing Science, 19*(1), 4–21. doi:10.1287/mksc.19.1.4.15178

Jacoby, J., Szybillo, G. S., & Busato-Schach, J. (1977). Information acquisition behavior in brand choice situations. *The Journal of Consumer Research, 3*, 209–216. doi:10.1086/208669

Jehn, K., Rupert, J., & Nauta, A. (2006). The effects of conflict asymmetry on mediation outcomes. *The International Journal of Conflict Management, 17*(2), 96–109. doi:10.1108/10444060610736594

Jehn, K. A., & Mannix, E. A. (2001). The dynamic nature of conflict: Longitudinal study of intragroup conflict and group performance. *Academy of Management Journal*, *44*, 238–251. doi:10.2307/3069453

Keller, K. L., & Staelin, R. (1987). Effects of quality and quantity of information and decision effectiveness. *The Journal of Consumer Research*, *14*, 200–213. doi:10.1086/209106

Knights, D., & Willmott, H. (2007). *Introducing organizational behaviour and management*. Thomson Learning.

Kohler, T., Matzler, K., & Fuller, J. (2009). Avatar-based innovation: Using virtual worlds for real-world innovation. *Technovation*, *29*, 395–407. doi:10.1016/j.technovation.2008.11.004

Lawrence, B. S. (1997). The black box of organizational demography. *Organization Science*, *8*, 1–22. doi:10.1287/orsc.8.1.1

Leana, C. R. (1985). A Partial test of Janis, groupthink model: Effects of group cohesiveness and leader behavior on detective decision making. *Journal of Management*, *11*(1), 5–17. doi:10.1177/014920638501100102

Levin, D. Z. (1999). *Transferring knowledge within the organization in the R&D arena*. Unpublished doctoral dissertation, North-western University, Evanston.

Levine, J. M., & Moreland, R. L. (1990). Progress in small group research. *Annual Review of Psychology*, *41*, 585–634. doi:10.1146/annurev.ps.41.020190.003101

Levy, S. J. (2006). How new, how dominant? In Lusch, R. F., & Vargo, S. L. (Eds.), *The service-dominant logic of marketing. Dialog, debate, and directions*. Armonk, NY: M.E. Sharpe.

Liang, T. P., & Lai, H. J. (2002). Effect of store design on consumer purchase: An empirical study of online bookstores. *Information & Management*, *39*(6), 431–444. doi:10.1016/S0378-7206(01)00129-X

Mari, C. (2008). *Consumer motivation. Foundations for a theory of consumption*. Naples, Italy: Edizioni Scientifiche Italiane.

Marks, L. J., & Olson, J. C. (1981). Toward a cognitive structure Conceptualization of product familiarity. In Kent, M., & Arbor, A. (Eds.), *Advances in consumer research*. MI: Association for Consumer Research.

McInnis, D. J. (2004). Where have all the papers gone? Reflections on the decline of conceptual articles. *ACR News*, 1-3.

Mihal, W. L., Sorce, P. A., & Comte, E. T. (1984). A process model of individual career decision making. *Academy of Management Review*, *9*(1), 95–103. doi:10.2307/258236

Myers, J. G., Greyser, S. A., & Massy, W. F. (1979). The effectiveness of marketing's R&D for marketing management: An assessment. *Journal of Marketing*, *43*(1), 17–29. doi:10.2307/1250754

Nambisan, S. (2002). Designing virtual customer environments for new product development: Toward a theory. *Academy of Management Review*, *27*(3), 392–413. doi:10.2307/4134386

Okhuysen, G. A., & Eisenhardt, K. M. (2002). Integrating knowledge in groups: How formal interventions enable flexibility. *Organization Science*, *13*(4), 370–386. doi:10.1287/orsc.13.4.370.2947

Pantano, E. (2010). Advanced technologies and retailing: Trends and directions. *Journal of Retailing and Consumer Services*, *17*(3), 171–172. doi:10.1016/j.jretconser.2010.03.004

Pantano, E., & Naccarato, G. (2010). Entertainment in retailing: The influences of advanced technologies. *Journal of Retailing and Consumer Services, 17*(3), 200–204. doi:10.1016/j.jretconser.2010.03.010

Pantano, E., Taversine, A., & Viassone, M. (2010). Consumer perception of computer-mediated communication in a social network. In *Proceedings of the 4th International Conference on New Trends in Information Science and Service Science* 2010, IEEE Computer series.

Peter, P. J., & Olson, J. C. (1983). Is science marketing? *Journal of Marketing, 47*, 111–125. doi:10.2307/1251404

Pfeffer, J. (1983). Organizational demography. *Research in Organizational Behavior, 5*, 299–357.

Pfeffer, J. (1993). Barriers to the advance of organizational science: Paradigm development as a dependent variable. *Academy of Management Review, 18*(4), 599–620. doi:10.2307/258592

Postmes, T., Spears, R., & Cihangir, S. (2001). Quality of decision making and group norms. *Journal of Personality and Social Psychology, 80*(6), 81–99. doi:10.1037/0022-3514.80.6.918

Poston, R. S., & Speier, C. (2005). Effective use of knowledge management systems: A process model of content ratings and credibility indicators. *Management Information Systems Quarterly, 29*(2), 221–244.

Ranganathan, C., & Ganapathy, S. (2002). Key dimensions of business-to consumer websites. *Information & Management, 39*(6), 457–465. doi:10.1016/S0378-7206(01)00112-4

Ratchford, B. T. (2001). The economics of consumer knowledge. *The Journal of Consumer Research, 27*, 397–411. doi:10.1086/319617

Rulke, D. L., & Galaskiewicz, J. (2000). Distribution of knowledge, group network structure, and group performance. *Management Science, 46*(5), 612–625. doi:10.1287/mnsc.46.5.612.12052

Russo, J. E., & Johnson, E. J. (1980). What do consumers know about familiar products? In Olson, J. E., & Arbor, A. (Eds.), *Advances in consumer research*. MI: Association for Consumer Research.

Schermerhorn, J. R. Jr, Hunt, J. G., & Osborn, R. N. (2008). *Organizational behavior* (8th ed.). New York, NY: John Wiley E Sons Inc.

Shaw, J. B., & Barret-Power, E. (1998). The effects of diversity on small work group processes and performance. *Human Relations, 51*, 1307–1325. doi:10.1177/001872679805101005

Shaw, M. E. (1981). *Group dynamics: The psychology of small group behaviour*. New York, NY: McGraw-Hill.

Stasser, G., Vaughan, S. I., & Stewart, D. D. (2000). Pooling unshared information: The benefits of knowing how access to information is distributed among group members. *Organizational Behavior and Human Decision Processes, 82*(1), 102–116. doi:10.1006/obhd.2000.2890

Stewart, G. L., Manz, C. C., & Sims, H. P. (1999). *Team work and group dynamics*. USA: John Wiley & Sons.

Summers, J. O. (2001). Guidelines for conducting research and publishing in marketing: From conceptualization though the review process. *Journal of the Academy of Marketing Science, 29*(4), 405–415. doi:10.1177/03079450094243

Tanriverdi, H. (2005). Information technology relatedness, knowledge management capability, and performance of multibusiness firms. *Management Information Systems Quarterly, 29*(2), 311–334.

Teas, K. R., & Palan, K. M. (1997). The realms of scientific meaning framework for constructing theoretically meaningful nominal definitions of marketing concepts. *Journal of Marketing*, *61*(4), 52–67. doi:10.2307/1251830

Thibaut, J., & Kelley, H. H. (1959). *The social psychology of groups*. New York, NY: John Wiley.

Thomas-Hunt, M. C., Ogden, T. Y., & Neale, M. A. (2003). Who's really sharing? Effects of social and expert status on knowledge exchange within groups. *Management Science*, *49*(4), 464–477. doi:10.1287/mnsc.49.4.464.14425

Tosi, H. L., Mero, N. P., & Rizzo, J. R. (2000). *Managing organizational behavior*. Blackwell Business.

van den Berg, P. E. W., Arentze, T. A., & Timmermans, H. J. P. (2009). ICT and social interaction: Modelling communication mode choice and its effect on travel behavior. *Proceedings of the IATBR Conference*. Jaipur, India, in press.

Walczuch, R., & Lundgren, H. (2004). Psychological antecedents of institution-based consumer trust in e-retailing. *Information & Management*, *42*, 159–177.

Wasko, M. M., & Faraj, S. (2005). Why should I share? Examining social capital and knowledge contribution in electronic networks of practice. *Management Information Systems Quarterly*, *29*(1), 35–57.

West, M. A., Garrod, S., & Carletta, J. (1997). Group decision-making and effectiveness: Unexplored boundaries. In Cooper, C. L., & Jackson, S. E. (Eds.), *Creating tomorrow's organizations: A handbook for future research in organizational behavior*. John Wiley & Sons.

Wilkie, W. L., & Morre, E. S. (2003). Scholarly research in marketing: Exploring the 4 eras of thought development. *Journal of Public Policy & Marketing*, *22*(2), 116–146. doi:10.1509/jppm.22.2.116.17639

Yoon, C., Cole, C. A., & Lee, M. P. (2009). Consumer decision making and aging: Current knowledge and future directions. *Journal of Consumer Psychology*, *19*, 2–16. doi:10.1016/j.jcps.2008.12.002

ADDITIONAL READING

Ahmed, P. K. Lim, K.K. Loh, A.Y.E. (2002). *Learning through knowledge management*. Woburn: Butterworth Heinemann.

Aldrich, H. (1999). *Organization Evolving*. London: Sage Publications.

Argote, L., McEvily, B., & Reagans, R. (2003). Managing knowledge in organizations: an integrative framework and review of emerging themes. *Management Science*, *49*(4), 571–582. doi:10.1287/mnsc.49.4.571.14424

Assael H. (1998). *Consumer behaviour and marketing action*. South-western College Publishing.

Babin, B., & Darden, W. (1995). Consumer self-regulation in a retail environment. *Journal of Retailing*, *71*(1), 47–70. doi:10.1016/0022-4359(95)90012-8

Bagozzi, R. P. Gürhan-Canli, Z. & Priester J.R. (2002). *The social psychology of consumer behaviour*. Buckingham, Philadelphia: Open university press.

Bales, R. F. (1950). A set of categories for the analysis of small group interaction. *American Sociological Review*, *15*, 257–263. doi:10.2307/2086790

Bellenger, D. N., & Korgaonkar, P. K. (1980). Profiling the recreational shopper. *Journal of Retailing*, *56*, 77–92.

Bettman, J. R. (1979). *An information processing theory of consumer choice*. Reading, MA: Addison-Wesley.

Cleveland, J. N. Murphy, K.R. Barnes-Farell, J.L. & Banks. C.G. (1998). *Analysing appraisal as goal-directed behaviour: An alternative approach to performance appraisal research and practice*, Unpublished manuscript, Department of Psychology, Colorado State University, Ft. Collins.

Daft, R. L. (1994). *Management*. Fort Worth: Dryden.

Davenport, T. H., & Prusak, L. (1998). *Working knowledge*. Boston: Harvard Business School Press.

Fieldler, F. E. (1978). Leadership effectiveness. *The American Behavioral Scientist*, *24*, 619–632. doi:10.1177/000276428102400503

Hansen, M. T. (1999). The search-transfer problem: the role of weak ties in sharing knowledge organization subunits. *Administrative Science Quarterly*, *44*(1), 82–111. doi:10.2307/2667032

Haugtvedt, C. P. Herr, P.M. & Kardes, F.R. (2008). *Handbook of consumer psychology*. New York: LEA.

Holbrook, M. B. (1999). *Consumer value: A framework for analysis and research*. New York: Routledge. doi:10.4324/9780203010679

Howard, J. A. (1989). *Consumer Behavior in Marketing Strategy*. Englewood Cliffs, NJ: Prentice Hall.

Howard, J. A., & Sheth, J. N. (1969). *The Theory of Buyer Behavior*. New York, NY: Wiley.

Huczynski, A. A., & Buchanan, D. A. (1991). *Organizational Behaviour: An introductory text*. Prentice-Hall, Hemel Hempstead.

Johnson, M. D. (1998). *Customer Orientation and Market Action*. Upper Saddle River, NJ: Prentice Hall.

Keller, K. L. (1998). *Strategic Brand Management*. Upper Saddle River, NJ: Prentice Hall.

Mohrman, S.A. Cohen, S.G. & Mohrman, Jr.A.M. (1995). *Designing team-based organizations*. San Francisco: Jossey-Bass Publisher.

Morisetti, J. O. (1966). Group performance as a function of task difficulty and size and structure of group. *Journal of Personality and Social Psychology*, *2*, 34–46.

Nonaka, I. (1994). A dynamic theory of organizational knowledge creation. *Organization Science*, *5*(1), 14–37. doi:10.1287/orsc.5.1.14

Nonaka, I., & Takeuchi, H. (1995). *The Knowledge-creating company: How Japanese companies create the dynamics of innovation*. Oxford: Oxford University Press.

O'Shaughnessy, J. (1987). *Why people buy*. New York, NY: Oxford University Press.

Peter, P. J., & Olson, J. C. (2002). *Consumer behavior and marketing strategy* (6th ed.). New York, NY: McGraw-Hill/Irwin.

Pfeffer, J. (1980). A partial test of the social information processing model of job attitudes. *Human Relations*, *33*(7), 457–476. doi:10.1177/001872678003300702

Schiffman, L. G., & Kanuk, L. L. (1987). *Consumer behavior*. Englewood Cliffs: Prentice-Hall.

Sheth, N. J. Mittal, B. & Newman, B.I. (1999). *Customer behavior: Consumer behavior and beyond*. Fort Worth: Dryden Press.

Shonk, J. H. (1992). *Team-based organization*. Illinois: Business one Irwin, Homewood.

Solomon, M. (1999). *Consumer behavior*. Prentice Hall.

Tsai, W. (2001). Knowledge transfer in intraorganizational network: Effects of network position and absorptive capacity on business unit innovation and performance. *Academy of Management Journal*, *44*(5), 996–1004. doi:10.2307/3069443

Wayne, H. D., & MacInnis, D. J. (2004). *Consumer behavior* (3rd ed.). Boston, MA: Houghton Mifflin.

Willmott, P. (1987). *Friendship network and social support*. London: Policy Studies Institute.

KEY TERMS AND DEFINITIONS

Advanced Technologies: Pervasive environments and mobile and ubiquitous computing.

Conceptual Representation: A graphical representation of the most important variables of consumer group knowledge structure.

Consumer Knowledge: A subset of the total amount of information stored in memory that is relevant to product purchase and consumption.

Consumer's Group Knowledge: The knowledge of consumer's group variables relevant to product purchase decision making on the base of "human capital interpretation".

Decision Making Process: Problem – solving process, a continuous flow of reciprocal individual and collective interactions among environment factors, cognitive and affective process and behavioural actions, which allow group consumers to reach objective and satisfy a need.

Group Diversity: Difference respect to demographic attributes (demographic diversity) and in the amount of knowledge accumulated in a group (background diversity) which influence positively and negatively decision making process of a consumer's group.

Group's Dynamics: The mechanisms of relationship that take place within a group important for influencing decision making process: (1) leadership; (2) power; (3) roles; (4) norms; (5) cohesion; (6) trust.

Intragroup Conflict: A tension between group members due to real or perceived differences that may be detrimental (relationship conflict) or beneficial (task and process conflict) to consumer's group decision making.

Scientific Questions: A process of developing an explanation of questions about the role of advanced technologies on our consumer's group representation. It's a technique for investigating our topic and for acquiring new knowledge.

Chapter 8
Internet Management for Communication–Distribution Interaction as a Means to Maximize Customer Consumption Experience:
The Volagratis Case[1]

Claudia Cacia
University of Salerno, Italy

Lucia Aiello
Università Mercatorum, Italy

Pierpaolo Singer
University of Salerno, Italy

Antonella Ferri
Università Mercatorum, Italy

ABSTRACT

The new online applications described as Web 2.0 or Social Media have a significant effect on consumer behaviour and contribute to an unprecedented customer empowerment. This issue has an important influence in many fields of activity, especially ones of the Web's most successful areas, i.e. tourism. The purpose of this chapter is to recommend the Internet as a "point of synergy" in the "promo-distribution" process of tourism. The Internet is a tool that customizes digital content to support consumers' decision making and maximizes customer consumption experience. Despite the lack of extensive research, the corporate world seems to embrace the Web 2.0 concept, which so far appears to be a promising tool for building customer loyalty and strong consumer-relationships that create value for e-retailers. The research tests the models proposed in a case study, which could be considered a core contribution, both in terms of practical implications for corporations, and in academic research.

DOI: 10.4018/978-1-60960-738-8.ch008

INTRODUCTION

Current organization trends are moving in a direction capable of communicating with customers through emotions, in order to involve them both in the purchase process and to satisfy their needs. Indeed, "consumption is an experience deriving from the interaction between a subject – the consumer – and an object – a product, an event, an idea, a person, a place, or any other thing within a given context" (Addis et al, 2001). In this perspective, a reconsideration in an interactive key of customers relations becomes thus imperative for firms to grasp opportunities of development and ensure survival. Accordingly, the wide diffusion of new technologies in communication and business has changed how consumer and product/store knowledge has to be managed and represented digitally. This issue has an important influence in many fields of activity, and this is particularly true for tourism, one of the most successful areas on the Web. Our research prospects an original conceptualization of marketing strategies and operational policies in order to examine and reconstruct the dynamics of "integrated" marketing when identifying potential interaction between communication and distribution for tourism. The main contribution of this study has been the elaboration of the "point of synergy" model, a tool for understanding how tourism industry organizations can create competitive advantage by applying, optimizing and coordinating new technologies (Barney, 1991). The "point of synergy" model has great innovation potential and is a relevant example of knowledge "contamination" where elements of communication are integrated with distribution by means of typical technology tools. In the model, new technologies constitute integration of intelligence, exchange, cross-fertilization, and links between elements, with a role of strengthening development and competitive synergies, knowledge development and relative benefits. Thus, through the Internet users achieve new extremely involving shopping experiences. The Internet facilitates the building, maintaining and strengthening of stakeholder and external relationships (Bennett et al, 2001) in order to create a virtuous circle of "customer satisfaction, attractiveness and business value" (Singer & Cacia, 2009). Besides, in modern systems the customer oriented perspective ends up prevailing over the production system, enables greatly "access to the customer" and establishing lasting relations with him. Said relations seem to represent the prerequisite for ensuring an organization's' competitive advantage. Customers no longer fall within a generic definition of public, passively assisting a show performed by the firm, but they live a gratifying experience if they finding the suppliers organized according to their personal taste. This highlights the role and the opportunities presented by interactive technologies to customer consumption experience. Evolutionary trends in terms of marketing could lead to more and more customized products and processes. All communication vehicles tend to become distribution instruments and vice versa: therefore, means of communication such as Community and Social Networks could become effectively means of distribution. Finally, the research tests the model proposed, i.e. a case study of the Bravofly group, potentially a core contribution in terms of practical implications for corporations.

1. THE COMMUNICATION AND DISTRIBUTION CHAIN IN TOURISM

According to Siano (2001), the communication mix depends on the intersecting of two elements: "communication object" and "final recipients of communication flows". The different communication areas can be classified by these elements. The main factors affecting communication are: the nature of the market, the product, and the firm's financial assets. Consequently, communication can be defined as a circular process involving many factors, regardless of specific areas of communication (Watzalawick, 1967; Merten et al., 1977; Rogers et al., 1981). Generally, the goal of business communication is to inform the external environment of the potential tangible and intangible assets of the company. Therefore, the communication system has a crucial influence in making firms more attractive and interesting for consumers. By improving corporate image firms would be in a position to create a more transparent and efficient relational system by breaking down information symmetries; thus achieving marked competitive advantage. It is necessary to ensure that what is communicated coincides with the facts. For Singer (2010) in the communication process every subject has a dynamic role; whereby communicational relations follow several multi-directional flows, especially as concerns tourism. The mechanisms and the channels of communication are those common to all products, although the subject, timing, costs, degree of trust and continuity in the communicative relationship may change (Wong, 2005). The subjects in the tourism industry play an active role in communication and distribution processes, which can be both strategic and operational. To date, distribution has evolved in complex ways: previous traditional sales channels have been threatened by large organizations. Consumers have been transformed even more significantly, paying more attention to choice, implementing more meticulous research concerning tourism products and by seeking more guarantees from suppliers (Cooper, 2008). Current trends of tourism distribution show the transition from a linear to a circular chain. The "novelty" of the circular chain lies in the fact that each party has both a direct and indirect relationship with the end consumer and can also activate the distribution channel. Generally, this relation is activated by alternative channels with which suppliers reach consumers in an interactive or integrative way. Alternative channels are for the most part virtual systems, such as the Internet, GDS (Global distribution System) and TV-Systems, or cooperative entities. The linear chain is a rigid form of distribution that does not fit in with the principles of flexibility offered by new company tools (Ferri et. al, 2009). New tools have reduced the "distance" between producers and potential customers, due to their interactivity and multimodality. Customers are also involved and may assume different roles, from consumer up to assembler of their own tourist package. The planning of distribution policies are much influenced by end customers who often have greater power in determining the distribution channel to put in place. In tourism distribution, strategies are mainly influenced by the complexity of tourist needs, the evolution of tourism's statutory framework and further, by changes in distribution channels. These trends push tourism providers to reach consumers directly, enabling suppliers to cut costs by extending the short and reducing the long distribution channel,

2. THE PROMO DISTRIBUTION CHAIN AND THE ROLE OF THE INTERNET AS A "POINT OF SYNERGY"

New marketing perspectives consider the close integration between strategies, tools and goals. Distribution and communication are closely linked levers that tend to use the same tools to a greater extent in tourism both by virtue of the characteristics inherent to the tourism product combined with new technologies (Marcussen, 2001). The contact point between communication and distribution concerns the encoding of this relation in terms of its representing a basic element of the relationship (Singer et al., 2010). The link between communication and distribution highlights the need to integrate both strategic and operative policies. Based on these considerations, the promo-distribution chain is the result of the overlapping of circular communication flows and distribution.

Figure 1. The promo-distribution chain in tourist activities

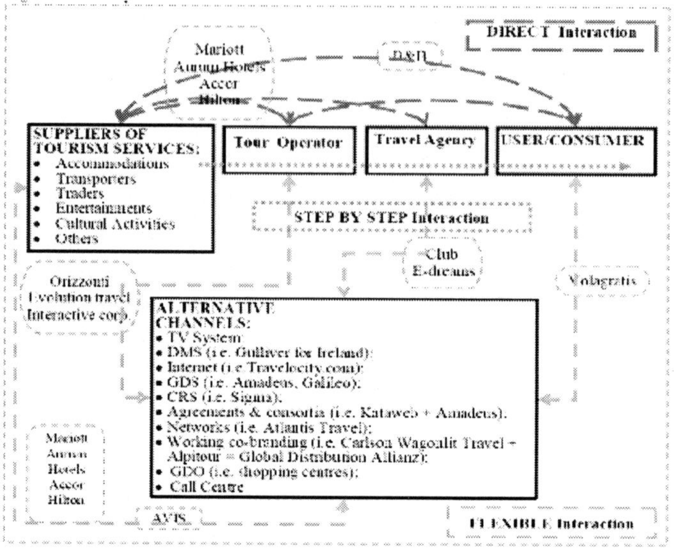

In a previous study the Authors identified three types of interaction:

- step by step (the traditional interaction between operators whereby customers receive information and buy products through the long chain);
- direct (the interaction is bidirectional and occurs directly between all actors: for example Hilton promotes and sells through the website);
- flexible (communication and distribution is achieved through alternative channels) (Singer et. al., 2010).

Figure 1 show how the distribution strategy may affect the reputation or how the intranet can influence the choice of distribution strategy and enhance the relations between service providers. Furthermore, the model highlights the impact of financial communication on the distribution strategy and on the choice of channel. Therefore, the Internet if suitably organized for e-commerce could play a crucial role in distribution strategy. The model highlights the needs for businesses, especially tourism firms, to be aware of the importance of implementing an integrated policy both in delivery and communication processes. Many researchers have noted the potential of the World Wide Web in business, and advocated incorporating the internet into the tourism industry (Burger, 1997). It is a valuable tool for both suppliers and consumers for disseminating information, communication and online purchasing (Law, 2009).

The Internet is a significant element in distribution processes, thanks to its efficiency in terms of cost, time and customer attractiveness (Buhalis, 1999; Urban, 2003). Representing an embedded high-quality system it is capable of influencing, just like price or brand, consumer preference and behavior (Singer, 2002). Consequently, technology plays a significant role in tourism, both as a facilitator of growth, and as an enabling factor to increase and ensure positive experiences for the tourist (Werthener, 2001). Businesses, including customer-oriented and information intensive tourism enterprises, are increasingly adopting e-business models to achieve their organizational goals (Law, 2009). Using the Internet has become vital for businesses to strengthen their customer relationships and to gain a larger market segment (Law, 2009). Dalgety (2002) attributes this widespread growth to technological improvements, cultural change stemming from more sophisticated Internet users and the rapid increase in online travel services. As a result of these societal and technological trends, many tourism-related operators have reorganized their distribution strategies using a multichannel focus (Prideaux, 2002). Accordingly, many firms have moved their business online in order to achieve competitive advantage by reducing costs and increasing customer satisfaction as well as enhancing business collaboration/partnerships, and extending service offerings (Peppard, 1993; Gratzen, 2003; Chen et al, 2005). A number of authors have attempted to identify and classify the potential benefits and implications of doing business on the internet (Verity and Hoff, 1994; Berthon et al., 1996; Quelch and Kein, 1996). Unquestionably, technology significantly influences tourism, both in its role as a facilitator of growth, and as an enabling factor to increase and ensure positive experiences for the customer (Stipanuk, 1993). Other authors have highlighted the opportunities that the Web as a natural all-embracing medium can provide in terms of supplying customers with everything they could possibly want (Ghosh 1998). In this respect, customers benefit from such a channel by gaining in real time gratification of their requirements, greater choice, multi-sensory, accurate and up to date information, and an easy to use interface (Pollock, 1995; Corvi & Bonera, 2005). Likewise, tourism organizations can reduce costs and become more competitive by serving consumers through the Internet (Buhalis, 1998). Therefore, the relationship between communication and distribution becomes even more pronounced (Buhalis, 2008); in reality, the Internet is integrated within the business model as a support, communication and sales tool for tourism products. Thus, a complex product such as tourism can be represented on a website where its multi-media and hyper-textual potential can be fully exploited. At the same time, the advent and development of automated reservation systems, such as Computer Reservation System (CRS), and later, Global Distribution Systems (GDS) have rendered the supply of tourism services global (Go, 1992; Sheldon, 1997). As a consequence, the structure of distribution channels (Benjamin et. al, 1995) has been completely changed. The efficiency and reliability of GDS enable suppliers to distribute and manage their reservations globally, by bridging consumer needs with tourism supply. Hence, great synergies are achieved, where globalization drivers stimulate GDS developments and vice versa (Hopper, 1990). Accordingly, technologies provide a boost to generate relationships between businesses, allowing the creation of an expanded network which can pool information about the market; this also enables supply and demand needs to overlap (Werthner et al., 1999). Consequently, building a successful strategy for Internet applications has become a key element in achieving competitive advantage (Barney, 1991; Boardman, 2005). One way to obtain competitive advantage, often emphasized during recent years, is to provide superior customer value delivery (Woodruff, 1997). In particular, such advantage is deeply affected by technology as it determines the relative cost positioning or differentiating of organizations (Hitt, 1996; Kriebel, 1988; Devaraj et al., 2003). Obviously, this affects every aspect of how business is conducted, changing internal processes as well as external relationships, modifying and restructuring entire economic sectors (Timmers, 1998; Wirtz, 2001; Porter, 2001). The process is also clear from the steady growth that characterizes the digital segment of travel, one of the most significant on the Web, and constantly on the increase. Consequently, unprecedented implications can be drawn for the future of tourism and consumer behaviour (Carrasco et al., 2006; Axhausen, 2003). The Internet enables the distribution of multimedia information; as textual data, graphics, pictures, video, and sounds are easily accessible, the Internet represents the flagship of the IT revolution and has instituted an innovative platform for efficient, lively and timely exchange of both ideas and products. Therefore, the web provides exceptional opportunities for the industry, bridging the gap between customers and suppliers and empowering closer interaction. The global connectivity of businesses and customers reduces traditional barriers in space and time and enables new value constellations that are richer in form than the conventional value chain (Akkermans, 2001). Customers can now search for personalized travel, visit the pages of the relevant tour operators and select the most appropriate tour packages. Through the Internet, individuals can make their thoughts and opinions easily available to a global community of Internet users (Dellarocas, 2003), and a growing number of users actively take advantage of this opportunity. At the same time enterprises have the advantages of building up customer databases, since customers frequently make all their travel arrangements through the same site. The outcome is that the Internet provides infrastructure for inexpensive delivery of multimedia information promotion and distribution for both principals and destinations (Hawkins et. al, 1995). ITs also assist the provision of customized products in order to meet the needs of individual consumers, and as a consequence, they are expected to become instrumental in differentiating tourism supply. The Internet will thus increase the attraction of both institutional and individual consumers, whilst increasing their productivity and efficiency (Buhalis, 1996). The ability of e-retailers to use this framework effectively will increasingly determine their future competitiveness. The increased use of e-commerce enables consumers to communicate directly with tourism organizations in order to request information and purchase products. Consumers empowered by home computing can access information about tourism products and organizations instantly, inexpensively and interactively. They can also effectuate the purchase prior to the experience itself, so the information search becomes a critical factor in the purchasing process (Detlor et al, 2003; Wang, 2008). Consequently, companies should integrate their distribution and communication policies in order to make their offer readily available to consumers and to maximize the latter's consumption experience. Moreover, recent trends indicate that a growing number of internet users are changing their consumer behaviour. As e-Marketer suggests, even more internet users access consumer generated content (CGC), a number expected to grow to 101 million by 2011. Searching for travel-related information is one of the most popular online activities (Pew Internet et al., 2006). The growing tendency of online travel referrals for planning trips has been reported by several travel-related studies (Bonn et al., 1999; MacKay et al., 2005). Consequently, the potential success of e-commerce for tourism products is linked to the ability to manage information exchange processes and, to particular forms of interactive communication (Hoffman, 1996). Notwithstanding potential benefits, the web does not guarantee profitability and may even worsen firm competitive position and industry attractiveness (Porter, 1985; Strassmann, 1990). Only if certain prerequisites are satisfied: long term planning and strategy, innovative business processes re-engineering, top management commitment, and training throughout the hierarchy, can the web be fruitful (Buhalis, 1998). It goes without saying that the Internet is not merely an innovative communication tool through which to transfer content previously conveyed by other media, but demands new, appropriate communication and customer relation strategies, due to its contextual product distribution function. Furthermore, the phases of planning and developing strategies need to consider the role of the Internet as a "point of synergy" between communication and distribution. Therefore, it is quite appropriate to speak of promo-distribution-integrated policies. The Internet is in a core position, representing as it does both a tool of communication, distribution and simultaneously their integration. The net is also a tool for comparing

Figure 2. Internet as point of synergy

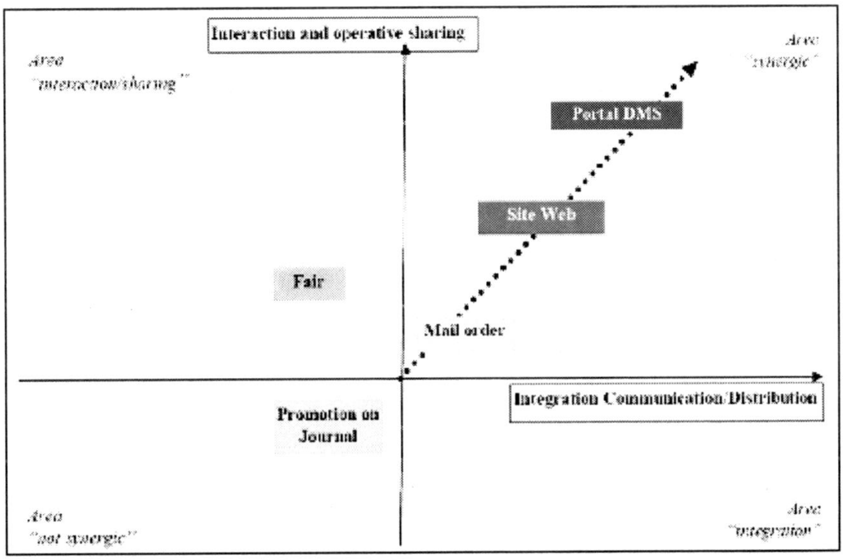

competitors in real time; the role of the Internet as point of synergy is illustrated in Figure 2, where four areas have been identified (Singer et al., 2010).

The first is the "interaction/sharing" area, where we find tools that enable the interaction and sharing only of operational policies (i.e. trade fairs enable different TOs to share space and costs to advertise their products, their primary aim not being to implement integration). The second area "no synergy", is characterized by the lack of interaction and sharing of the two levers (advertising in newspapers). The third area "integration" refers to any tool allowing integration between communication and distribution, but not operational interaction and sharing. The last area "full synergy" where conditions of operational sharing and policy integration are created (destination portal). The Internet, the perfect synthesis between distribution and communication, is organized without a specific role or priority aims. Consequently, it can be said that within the promo-distribution chain, the Internet acts as a point of synergy between the two levers.

3. RELATIONSHIP MARKETING AND INTERNET TO ATTRACT E-CUSTOMERS

In the current competitive landscape, firms tend to create and manage a series of relationships and approaches to company analysis, involving not only the organization but also management and marketing (Ford, 1990). The study of economic events moves from the analysis of a company's capacity through internal and external relationships to create networks, consistent in formal and informal relations, and essential for survival (Gummesson, 2008; Golinelli, 2010). Consequently, distribution and communication policies cannot ignore the relationships companies activate and maintain with their targets. Several authors have developed paradigms with a various marketing focus:

- relationship marketing (RM), addressing attention to relations and customers satisfaction (Peterson et al., 1992) and retention (Buchanan, 1990, Gale, 1994; Gordon, 1999; Kotler et al., 1999; Arussy, 2009);

- experiential marketing, involving customers in sales experience (Schmitt, 1999);
- viral marketing, spreading adverting between individuals or groups and encouraging positive word of mouth (Howard, 2005; Scott, 2007).

Gummesson (2008) defines relationship marketing as an economic, cultural and social phenomenon, with two fundamental dimensions: the value society (value creation as the best result) and the network society (net structure of organizations and of society). Various authors have considered RM both as a tool to improve coordination between "value chains" and as a means of creating synergies between companies (Marcati, 2007). As suggested by Webster (1996), marketing activities are increasingly oriented to customer satisfaction and delivering superior value. RM originates from this approach. Companies have to develop steady and lasting relations with customers, as the key to gaining competitive advantage (Cristopher et al., 1991; Lorenzoni et al., 1999; Kozinets 1999). Customer orientation enhances operational and marketing strategies through three important actions: provide more and better services to customers, establish trust, and build relationships (Reicheild et al., 1990). The market orientation approach allows a) the compiling of market information on current and future customer needs and b) the strengthening of internal and external company relations, thanks to a customer-centred policy. Reicheild (1990) suggested that companies should be customer oriented, competitive and profit-driven. Accordingly, differentiating business from that of competitors and satisfying consumers becomes imperative. Closely aligned with RM, is customer relationship management (CRM) defined as "the combination of strategies and technologies that empowers relationship programs, reorienting the entire organization to a concentrated focus on satisfying customers" (Kurtz, 2008). This approach represents a gradual change of focus: from the quality of products (Total Quality Management), to processes to reduce costs (Business Process Reengineering) to the satisfaction, retention and customer loyalty (Customer Satisfaction and Customer Loyalty) (Fairnet, 2002). On the other hand, CRM owes its raison d'être to RM. Grönroos (1997) defines RM as: "the process of identifying and establishing, maintaining, enhancing and when necessary, terminating relationships with customers and other stakeholders, at a profit, so that the objectives of all parties involved are met ... by mutual exchange and fulfilment of promises". The aim of relational marketing is to build and maintain a loyal customer base, in order to give profit to the organization, to this end, the company must focus on attracting and retaining customers and on strengthening relations with them (Berry, 1983). Over the years the complexity of the competitive environment has prompted companies and users to look in other directions, changing the main objective of RM into attracting, building, maintaining, and consolidating relationships with audiences. The intervention of new technologies has changed the way of managing RM, making the use of WWW essential for competing on the global market. The Internet represents a virtual network, an innovative and eclectic means of relational marketing, minimizing costs and maximizing benefits. Gummesson (1994) has identified thirty marketing relationships (30R) grouped in:

1. Market relationships (traditional and special i.e. triad between supplier/customer/competitor);
2. Non-market relationships: divided into mega-relationships (social relationships) and nano-relationships (between human resources);
3. E-relationships, social relationships, environmental relationships: the e-relationship includes all the relationships and interaction on the net or IT based. These relationships

Figure 3. IT advantages and attracting power

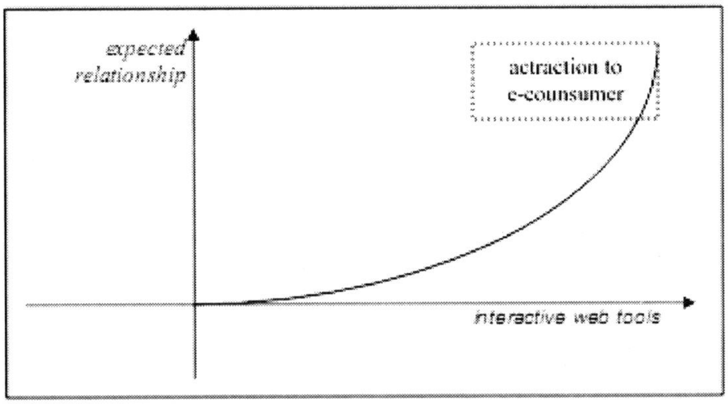

occur in the market-space (from marketplace to market-space) and the electronic channel improves consumer ability to access products and services as well as giving them control of relationships with companies. The strength of E-relationships lie in their capacity to share information and knowledge in real time, in direct and indirect modality; thus reducing the costs/sacrifices of the relationship and extending the control capacity of the parties involved (Pride, 2005). E-relationships require a highly developed web site, as described by Chaffey (2009) in the 5th level model, allowing full interactivity for customers relations. In the light of these considerations, the web and other new IT tools extend the number of potential relationships (Giacomazzi et al., 2008). Figure 3 shows the relations between web tools and expected relationships. In short, IT both facilitates the buying process and attracts greater audiences than traditional instruments.

Andrew (1990) defines attractiveness as a quality that causes an interest or desire in something or someone. Firm attractiveness (Ferri, et al., 2010) achievable by the internet is its capacity to satisfy consumers and to build a corporate web personality, evoked by its design and brand (Aiello et al., 2009). A company is attractive if it strengthens its relationship with market across the web, transforming the target into a loyalty market. Therefore, loyalty intensity depends on strength appeal (Johansen et al., 1996). For Gummesson (2008) loyalty intensity is a possible indicator of ROR (return on relationship) that expresses the impact of relationship management on revenues, costs, return on capital employed and profits. He suggests that the life cycle of a customer relationship is based on the belief that suppliers are able to meet the needs of their clients and fulfill promises. An improvement in perceived quality by customers will increase their satisfaction, loyalty and firm profitability. According to Gummesson it is possible to construct a sequence of interdependent events, called the service profit chain, as follows: good external quality → customer satisfaction → permanence of clients → high income. However, customer satisfaction is not sufficient to guarantee future profits but must remain over time (Gummesson, 2008). The company must plan and implement retention policies based on product customization. Below we suggest the importance of the internet and e-relationship in activating a virtuous circle.

4. CIRCLE OF "ATTRACTIVENESS-CUSTOMER SATISFACTION-BUSINESS VALUE"

To provide superior customer value delivery is a prerequisite when establishing and maintaining long-term customer relationships. Superior customer value calls for customer satisfaction, a factor of increasing importance in highly competitive e-markets. Many authors are focusing on the most appropriate application of ICT in the tourism sector in order to increase the added value of the tourism offer (García-Crespo et al., 2008), which implies the creation of highly customized tourist packages for users. These technologies have a pervasive effect on the creation, production and consumption of tourist products (Stamboulis et al., 2003). It has been demonstrated that information technology can add value to goods and this value is evidenced in premium prices (Nault et al., 1995). Value creation in electronic information networks such as the Internet is grounded on either increased effectiveness in the supply chain or new customer values in the products and services delivered (Methlie, 2000; Bharadwaj, 2000). The Internet is undoubtedly a fitting arena for tourism businesses to develop value-added services for customers (Nysveen et. al., 2003). Value-added services are (Nysveen et al., 2002) "services giving access to various forms of information about the tourism products offered on a website". Many fundamental issues have to be dealt with in order to gain competitive advantage through superior customer value:

- What exactly produces customer value?
- What kind of value added-services do customers prefer on a Web site?
- How can firms build customer loyalty and therefore create business value?

Unquestionably, to build customer loyalty, companies need to shift the focus of e-business from e-commerce (transactions) to e-service (all ties and encounters occurring before, during, and after transactions). Thereby, to deliver superior service quality, Web company managers must first understand how customers perceive and evaluate online customer service. This involves defining what e-service quality is, identifying its underlying dimensions, and determining how it can be conceptualized and measured (Zeithaml et al., 2002). The first formal definition of Web site service quality (e-SQ) was provided by Zeithaml (et al., 2002). In their terms, e-SQ can be defined as the extent to which a Web site facilitates efficient and effective shopping, purchasing, and delivery of products and services (Zeithaml et al., 2002). It is clear from this definition that the meaning of service is comprehensive and includes both pre and post-Web site service features. Academic research has identified a number of criteria that customers use in evaluating Web sites, these include:

- Information availability and content (Li, et al., 1999; Swaminathan, et al., 1999; Hoque, 2000; Wolfinbarger et al., 2001)
- Ease of use or usability (Davis 1989; Davis et. al., 1989; Swaminathan, et al., 1999): A site's search functions, download speed, overall design, and organization are among the key elements that affect usability (Jarvenpaa et al., 1997; Lohse et al., 1998; Nielsen 2000; Novak, et. al., 2000; Montoya-Weiss et al, 2003).
- Privacy/security (Privacy and security are key evaluative criteria in online services (Culnan et al., 1999; Hoffman et al., 1999; Quelch et al., 1996; Montoya-Weiss et al. 2003).
- Graphic style (Hoffman et al. 1996; Lynch et al. 2000; Novak et al. 2000; Montoya-Weiss et al. 2003).
- Fulfilment: Wolfinbarger and Gilly (2001) found that reliability/fulfilment ratings were the strongest predictor of customer satisfaction and quality and the second

strongest predictor of web loyalty (Palmer, et al., 1999; Wolfinbarger & Gilly, 2001).

In terms of shopping online versus offline, online buyers perceive benefits in obtaining information directly from a site rather than having to go through salespeople in an offline store (Zeithaml et al., 2002). In terms of information content, the ability to search price and quality information increases satisfaction with both the experience and product purchased and improves intentions to revisit and repurchase from a Web site (Lynch et al., 2000). Privacy is fundamental and involves the protection of personal information-not sharing personal information collected about consumers with other sites (as in selling lists), protecting anonymity, and providing informed consent (Friedman, et al., 2000). Security, on the other hand, involves protecting users from the risk of fraud and financial loss from the use of credit card or other financial information (Montoya-Weiss et al. 2000). In recent years other criteria such as access, responsiveness, and personalization have become important (Bart, 2005; Vesanen et al., 2006; Miceli et. al., 2007). By using product bundling or personalization for the specific user's tasks, a firm can also add value to services and products. Personalization is a method to give customers information and/or services on a Web site that suits the individual visitor and also includes the relevant adaptation of products sold at the Web site (Nysveen et. al., 2003). Personalized Web sites are simply dynamic Web sites where each consumer can get personally tailored information through user profiles and identification (Thorbjørnsen, 2002). User profiles are information about individual interests, preferences, and demographics, stored in a database. Data can be obtained through the company's existing consumer databases by asking the consumer for profile data on the Web site, or by logging consumer behaviour on the net (Thorbjørnsen, 2002). This makes possible several personalized applications. Above all, personalization turns into the main element of customer attractiveness. Thanks to personalization firms can influence customer perception. According to Varki and Colgate (2001) value perception has a positive effect on customer satisfaction, thus supporting the positive effect of value-added services on customer satisfaction. In the virtual space it is possible to customize the buying process to the needs of each customer. The Internet enables an integrated booking system to be offered to the customer who wishing to arrange his entire tour

Figure 4. The virtuous circle of "A-CS-BV"

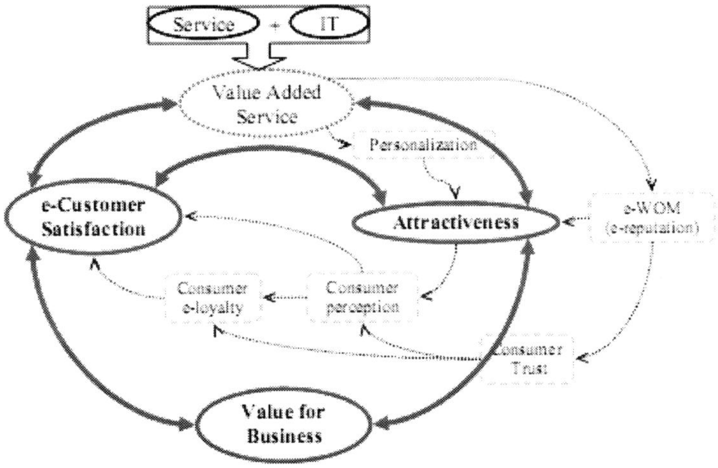

on his own, can select each part of the vacations just as if filling a shopping basket at a virtual department store (Corvi & Bonera, 2005). In the light of the relationship marketing view, a tourism organization can build customer loyalty and switching costs that no other industry player can achieve. According to Bloch (1996) Internet shopping malls will be the future industry magnets. Thanks to ICT, companies can improve consumer experience, reduce costs and improve efficiency and services. This establishes a bidirectional relation between consumer experience, customer satisfaction and attractiveness for the business (Newman, 1973; Woodside, 1989; Cornin, 1992). However, companies cannot only attract customers and increase customer satisfaction, but must also create value for business. Consequently, if the company is able to do this it can also create lasting value (Stiglitz & Weiss, 1981). Value creation comes from a firm's ability to continually adapt to the markets in which it sells and from which it receives resources; adaptation takes place within relationships where communication is central and a priority (Golinelli, 2010). Through the communication process, firms should seek to maximize customer trust (Marcus et al., 1997; Ryan, 2005).

Consequently, firms and tourism organizations can enhance their performance by empowering their strategic marketing and management efforts through undertaking all their functions using advanced ITs. This enables them to improve their networking and ultimately to improve their virtuality. Although, due to the fundamental lack of trust between consumers and businesses on the Web (Hoffman et al, 1999), to increase trust with customers, the Web site (e-retailer) has to take several remedial actions. This is essential because "trust is a crucial variable that determines outcomes at different points in the process and serves as a glue that holds the relationship together" (Singh et al., 2000). In the e-commerce context, customers who do not trust an e-business will not be loyal to it even though they are generally satisfied with the e-business (Anderson et al., 2003; Dwyer et al, 2007). Therefore, e-satisfaction is likely to result in stronger e-loyalty when customers have a higher level of trust in the e-business enabling firms to create business value (Marcus & Wallace, 1997; Bushman et al., 2001). In addition, to maximize the virtuous circle, firms have to reduce perceived customer risk (Guatri & Massari, 1992; Singer et al. 2009). Perceived risk is the consumer's perception of the uncertainty and concomitant adverse consequences of buying a product or service (Chen et al., 2003). Broydrick (1998) maintains that removing risk is an important means of enhancing perceived customer value. Some of the factors that influence perceived risk are e-retailer reputation, perceived quality, and product price (Chen, 2003). As suggested in both classical economics (Heiman et al., 1996) and consumer-research literature (e.g., Bolton et al., 1991; Richardson et al., 1994; Teas et al., 2000), consumers use signals or extrinsic cues (i.e. advertising) to infer product quality and refine their choices. Accordingly, the electronic word of mouth (eWOM) is an important source of risk-reducing information, largely because its independent nature reveals true characteristics of the product (Murray, 1991; Heiman et al., 1996). eWOM refers to interpersonal communication between consumers concerning their personal experiences with a firm or a product (Richins, 1983) and has a significant impact on customer purchase decisions (Bone, 1995; Chen et al., 2009), especially in tourism, due to the intangibility of services. Consumers tend to rely on e-WOM to reduce perceived risk and uncertainty (Bansal et al., 2000), thus it can have a significant influence on travel-related decisions. As it is one of the most important external information sources for travel planning (Fodness et al., 1997) it also contributes to the construction of firm image, and e-retailer should not neglect its importance. Consumers perceive e-retailers with a good reputation as being more trustworthy and credible than those with a poor reputation (Hendrix, 1999). Accordingly, this factor can boost business attractiveness, customer satisfaction and loyalty – thanks to the "consumer reviewing" process - and therefore, foster performance and business value (Bernstein, 1988). Recent trends show a growing use of any internet-based interactive applications that enable consumers and firms to provide interactively access to hypermedia content (machine interactivity, such as personalized Web sites) and to communicate through the medium (person interactivity, such as customer communities) (Hoffman et al., 1996). Personalized Web sites in addition to customer communities and social network are the most common applications on Web today and both appear to be promising tools for building customer loyalty, strong consumer-relationships (Armstrong et al., 1996; Peppers et al., 1997; Holland et al., 2001) creating value for e-retailers.

5. INTERNET TOOLS FOR COMMUNICATION-DISTRIBUTION INTERACTION TO MAXIMIZE CUSTOMER CONSUMPTION EXPERIENCES

The new digital technologies have created new ways and tools for providing new products that meet emerging consumer needs and support corporate communication. The new web enables the application and integration of fresh digital tools. Therefore, to optimize operational policies, managers should be familiar with the rules of using digital media; otherwise excellent planning could fail. The Web 2.0 has led to applications running on the browser, using AJAX technology (data are processed directly on the server, there is no need to install any software on the PC) which can be managed directly on the server interface. As a result, it is easy for companies and potential market to input and update content on a site. A company simply by inserting content, can propose video on multiple platforms free of charge, as well as open a blog, a wiki or a forum (Giacomazzi et. al, 2008).

It is essential to choose the most appropriate digital tools carefully, in line with enterprise goals. Web tools are constructed on the principles of participation, conversation and convergence. This makes consumers, in terms of tools and opportunities, equivalent to businesses. The main tools of Web 2.0 enabling integration between communication and distribution can be clustered into 3 groups (Figure 5). The first group can be defined as "partial integration" and includes: instant messaging (for creating a contact list with which to communicate through an interface at any time); blog (editorial containers or online journal, handled by an author, are shared "diaries" in network. Blogs are often combined with Podcasts -digital audio or video streamed or PC downloadable); e-mailing (for sending messages

Figure 5. Vehicles of integration (partial vs. total integration)

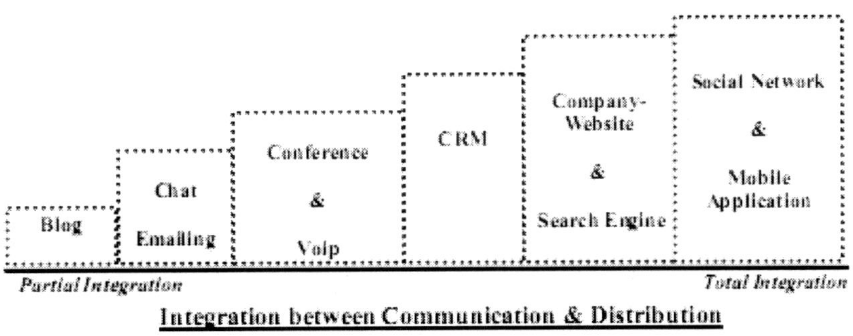

to one or more recipients through e-mail address); wiki (enables knowledge sharing, information exchange and storage). The second group "semi-integration" includes: conferences and VoIP (telephone line through PC, enabling integrated communication and cost savings). The third group "total integration" includes: Customer Relationship Managers (virtual rooms integrating tools for productive activity management, the creation of sales brochures, process analysis, managing relations with the outside world, particularly customers); company's website and search engine; mobile applications; Social Networking (users build personal websites accessible to other users for exchange of personal content and communication). Boyd (2007) defines social network sites as "web-based services that allow individuals to (1) construct a public or semi-public profile within a bounded system, (2) articulate a list of other users with whom they share a connection, and (3) view and traverse their list of connections and those made by others within the system". The nature and nomenclature of these connections may vary from site to site (i.e. MySpace, LinkedIn, and Skyblog), with wikis and tags, that emphasise online collaboration and sharing among users (Facebook allows users to add modules called "Applications" that enhance customer profiles). The strength of the Social Network is in its ability to attract and to link millions of people.

Albeit with the above-mentioned classification, company websites appear as a full integration tool, according to relationship marketing theories, attention shifts to processes and tools able to create, maintain and maximize the consumers relationship. As Gummesson (2008) has suggested, only attention to relationships enables the creation of "ambassador customers" (well satisfied customers that consider suppliers attractive and develop a strong relationship with them or tell others about them) and "loyal customers" (happy customers, but less so than the former, who may increase the business attractiveness and maximize their satisfaction and loyalty (Gummesson, 2008).

As previously mentioned, this relation is functional to achieving a virtuous circle that creates value for businesses. Of all the different tools, social networks enable the realizing and maximizing of the virtuous circle. To date, social networks are not yet a distribution tool, but should become one in order to maximize the positive effects of relational business, especially in tourism, due to its high intangible content, in which an e-retailer's reputation plays a fundamental role. If in the future, should the SN be able to distribute products and services, it will become the best tool for businesses. Indeed, in terms of communication, it already amplifies the effect of e-WOM and, in terms of distribution; it effectively increases the involvement of the consumer buying experience. As a tool accessible to consumers (consumer-controlled, i.e. not self referential), the SN enables a many to many communication (blogs, forums, social networking sites, etc.) without company intercession (Sharkar et al., 2007). Tourism is evidently more than ever, obliged to compete in terms of complex dynamic socio-cultural trends where the norm is greater involvement on the part of consumer tourists claiming more freedom of choice and more customized tourism where they look for new forms of social aggregation (Fabris, 2008).

6. SOLUTIONS AND RECOMMENDATIONS

The gradual empowerment of consumers and of other business stakeholders may be considered the main cause of changes in marketing and online communication in recent years (Cova et. al., 2006). This enhancement process is rendered possible thanks to all the web applications called Web 2.0 and social web. These technologies are creating new dynamics of social interaction that allow users to be active players in the exchange and creation of online content. Consumer-generated content encompasses a variety of media forms and types of web site (Gretzel, 2006). Two modes in which content is created online are consumer reviews and ratings; these are the most accessible and prevalent form of e-WOM (Chatterjee, 2001) that appear to play a progressively more important role in consumer decision-making processes and consumer purchase decisions (Payne, 1991; Haubl, 2000; Chen, 2008). As is evident, the goal of tourism organizations is to investigate the importance of consumer reviews and the role of e-WOM through SNS in customer decision making processes, to effectively maximize the virtuous circle, presented above. The online world is a complete representation of the offline world. As in the offline world, firms should defend themselves from sabotage, whether by competitors or otherwise. Accordingly, the web is starting to move toward total transparency and "truth". It is well known that consumers are influenced by advertising, but will be more motivated, and consider a purchase more reliable if the product or service is recommended by someone else. Various researches highlight the importance of recommendations, especially online where people search for reviews before they make a purchase. Online consumer reviews appear to play an increasing role in consumer decision-making processes. More than 80% of web shoppers use other consumers' reviews when making purchasing decisions. In particular, consumers tend to rely more on consumer reviews when purchasing high involvement products (Park et al., 2007); since travel is a high involvement product, one can expect extensive use of reviews for travel-related decisions. Compete, Inc (2006) found that nearly 50% of travel purchasers visited a message board, forum or online community for their online travel purchasing and one in three said that consumer reviews helped with their purchase decision. Moreover, 25% said they also posted a review on consumer review sites after making their purchase. Clearly, online consumer-generating information is taking on an important role in online traveller decision making. During the consumer decision making process, potential consumers can now access the vast pool of data to evaluate alternatives. Their information search process is facilitated by search engines. As a result, website visitors no longer necessarily enter through a site's home page and brows as they would a brochure (Schipul, 2006). Instead they use search to hunt for, and transport them directly to, specific pieces of information. The web 2.0 has revolutionised the consumer buying process, and has radically altered the origin of information. Consumers are no longer dependent on web site owners to publish the information they seek, as they can increasingly rely on unfiltered, dynamic and topical information provided by their peers. Thus, the services provided by online social shopping networks enable consumers to exchange opinions, recommend and rate products, meet and consult other consumers, and form communities of interest, making the overall shopping experience richer, more enjoyable and more social. Hence, emergent online social shopping websites by bringing the values of recommendations from consumers'online social network facilitate online shopping. Moreover, online social shopping blends two powerful elements of real-world shopping otherwise lost for online consumers: word-of-mouth recommendations from trusted sources and the ability to browse products in the way that naturally leads to discovery. This transformation has prompted a change in the power relationship between e-retailers and e-consumers. Online retailers may be able to increase sales on their site by taking actions to encourage reviewers to reveal more identity-descriptive content about products/services. In this scenario, e-retailers should consider not only the current use of SN, appropriately managing the advantages and disadvantages arising from the same, but should also plan exploitation of its potential opportunities (i.e. the use of SN as a vehicle for distribution), trying to qualify as first movers so as to ensure a sustainable competitive advantage in the long run. Undoubtedly, the use of these advanced technologies can contribute to making tourism products more attractive. Besides, technologies allow the development of on-line applications that can deliver rich multimedia content, blending text, 3D images, sound and video in order to overcome the intangible nature of the product (Pantano, 2009) and consequently, maximize customers' consumption experience. Some SNs are dedicated to particular topics, sharing knowledge or purchasing of products and services, transforming how users research and decide on travel plans (i.e. Yahoo's Trip Planner, TripAdvisors's Inside, VirtualTourist's Trip Planner and others share journals, itineraries and photos) (Vickery et al., 2007). To date, many firms, instead of investing cash in advertising to become known, prefer to allocate these monies to their consumers, so that they themselves bring new customers through word of mouth. Following a broader social media marketing trend, Virgin America is giving away free tickets to influential Twitter users to celebrate the launch of the airline's presence in Toronto. Influencers will also receive free in-flight Wi-Fi, and an invitation to Virgins Toronto launch party (Robles P., 2010). In conclusion, social media has changed the world of marketing forever. Customers, prospects and competitors share information every day through Social Media channels that build relationships, strengthen brands and increase business prospects. And countless people (from start-ups to well-established companies) are creating sizable returns from social media sites such as Twitter, LinkedIn and Facebook. In our view a Social Media Marketing strategy, when well executed, will substantially increase marketing ROI (return of investment). In the following paragraph, we analyze a case of best practices in tourism, explicating proposed solutions and representing opportunities ahead for companies operating in tourism as well as in other sectors.

7. CASE STUDY "BRAVOFLY" [2]

The case study is a research approach, placed between concrete data and methodology paradigms (Lamnek, 2005). Case study research is an essential form of social science inquiry (Yin, 2008). The method is appropriate when investigators either desire or are forced by circumstances: 1. to define research topics broadly and not narrowly; 2. to cover contextual or complex multivariate conditions and not just isolated variables; 3. to rely on multiple and not singular sources of evidence (Yin, 2008). We have investigated on the Tourism's e-retailers by means of qualitative research, exploratory methodology and tools, structured interviews. According to Kerlinger and Lee (2000), "exploratory studies have three purposes: discovering significant variables in the field situation, discovering relations among variables, and laying the groundwork for later, more systematic and rigorous testing of hypotheses". In this direction we have chosen to investigate Volagratis, that is one of the most significant Italian companies of Bravofly' group. In few years, the group, founded by Fabio Cannavale, has recorded high growth rates and has embarked on a path of internationalization, showing high adaptability to context and environmental dynamics. In the implementation of a case study it's important to define the objective, in particular, we have identified the actual scenario of the company; we have individualized the state of the organization's official mission studied; we have found the future perspectives; we have considered the key factors of the firm; we have considered the principal tools of distribution and communication' systems and the role of technology within these systems; we have individualized the confirm/disconfirm of the strategic role of Internet for communication-distribution interaction and the role of e-tools to maximize customers' consumption experience.

The case study analyzed has been organized into two parts. The first phase is: analysis of data on the firm; collection of data and information, interview; elaboration of information; presentation of results. The collection of data and information has been carried out from the firm's web site, articles on firm and interviews. The interviews of Dr. Maria Teresa Rangheri, Marketing Director

Bravofly, Dr. Rosangela Leone, Media Relations and Dr. Antonio Antonaci, Product Manager Bravofly have been led into a first phase by telephone and they are based on ten questions, into a second step they have been led across email, means of reinforcement' to questions asked through phone interview. After, we have realized an interviews elaboration and extrapolation of information and data. The development of the information was based on analysis tools used for removing the information that were not consistent with the research design and extrapolating those which gave confirmation or disconfirmation of the research objectives. The results are synthetic and coherent with the research project and refer to: actual situation of company; company's internet tools to maximize customers' consumption experience. The purpose of the case study is to examine and verify new dimension of tourism, subsequent the introduction of new technologies. This requires the analysis of whether such dimensions are able to defend and reinforce the tourism supply system. The case study on the Volagratis could be considered a contribution to focus on practical implications for tourism corporations. Volagratis is an important e-retailer that has recorded high growth rates and has embarked on a path of internationalization, showing high adaptability to context and environmental dynamics (Singer et al., 2010). Volagratis.it is leading European operators in the online travel industry, founded in May 2004, in Italy, by two partners, in just a few years it has begun a process of international expansion, including the localization of its headquarter in Switzerland (2007). The company's mission is to become the principal web site in Europe for the information and sale of tourism product throughout the policy of highly strategic integration. The main information about the company's turnover are: 2007, 780 thousand air tickets sold, gross revenue 87 million of euro; 2008, 1,230 million air tickets sold, gross revenue 140 million of euro; 2009, 1,8 million air tickets sold, gross revenue 200 million of turnover and 1,8 million air tickets bought by European clients (Lonardi, 2010). The competitive advantages into start activity phases: the online marketing knowledge; the use of technology to simplify the life' client, i.e. the company uses internet to supply a synthesis of information or a telephone booking assistance. The future perspectives for 2010, the projection are 12 million, a growth of 40% and sales to 280 million of euro. For 2011 the company predicts the quotation of Bravofly (Lonardi, 2010). The group is the first in Europe into low-cost' sector and it is increasing into scheduled flight sector. The strategic geographic markets are Italy, France, Germany, Prague and Scandinavia. The company wants to arrive at Turkey. Bravofly is opening two new markets: cruise and vacancy package. The group's objective is the Italian market where the online revenues for these sectors are 200 millions of euros only, while the traditional markets (air tickets) are of 8 millions of euros. The imbalance depends on the scarce offer. The results of the second phase of analysis are examples of descriptive case study and a confirm/disconfirm of theories and model that we have presented. As the online social shopping companies are emerging and getting popular during the recent years, there is rarely scientific research in this filed. The goal of this study is to analyze a case company in order to find out the success factors for an online social company in the fiercer competition. According with our model, Volagratis obtain the synergic point (Singer et al. 2010) between communication and distribution policies applying Internet tools (engineering search, social network, chat, blog, etc.). The group offers a range of multimedia tools (web, phone, TV) integrated into its search engine, available for travellers to look for information, comparable products supply and online booking from their travel and tourism business. Volagratis also offers services to travel agencies, networks and companies. On the group's websites, you can search for and book hotels - thanks to a partnership with leading provider of online hotel reservation (Booking, Venere, Octopus, Travelante, LateRooms, InterContinental Hotels and Expedia) - as well as car rental, free download of destination guides and other information services (Singer et al., 2010). Volagratis is distributed through partner sites including MSN, Yahoo, Lycos, Price Runner, Tiscali, Corriere della Sera, TripAdvisor, SACBO Spa, Virgilio, Banca Intesa, Il Sole 24 Ore, The Online, Momondo.com, lowcost, easyvols.com, Travel Supermarket, etc., so it activates different external relationships with partners in order to put up the customer satisfaction. In fact the company allows consumers: choosing comparable services; consultation and booking of low cost flights; activation of e-relationship. The relentless spread of online agencies moves increasingly towards the use of integrated strategies both in terms of services offer, communication and distribution policies, which is the core of service offer. Therefore, Internet is a tool that has a strategic role into loyalty and customer satisfaction. It is confirmed from the interview of Marketing Director of Bravofly, she affirms "Communication and distribution are integrated functions. We have an area that deals with general marketing of all functions, including communication and distribution" (Singer et al., 2010). Communication and distribution are functions integrated by Internet and the customer is an e-prosumer because he elaborates his product and he compares the different options in real time (Giacomazzi et al., 2008). The advantages of this web tools are: optimization of cost and time, highest interaction between e-retailer and consumer, efficiency in integration policies, customer satisfaction. Besides these resources permit a higher effectiveness and efficiency for the company, e-communication and e-delivery complements the business of customer care (customer service centres are located in Milan and Tunisia) and it is a strategic support for customer service. Internal and external relationship management is accomplished almost exclusively through the web applications (Singer et al., 2010) by which Volagratis tends to raise the perceived value of tourism product offering and, accordingly, to create new sources of competitive advantage. Therefore, it is possible to affirm that this company maximizes the virtuous circle of "attractiveness-customer satisfaction-business value" (Figure 6).

Volagratis uses the net as promo-distribution activities. The Volagratis' product generates an added value for client providing integrate services through e-tools, so it activates the actions

Figure 6. The virtuous circle of "A-CS-BV": Volagratis.com

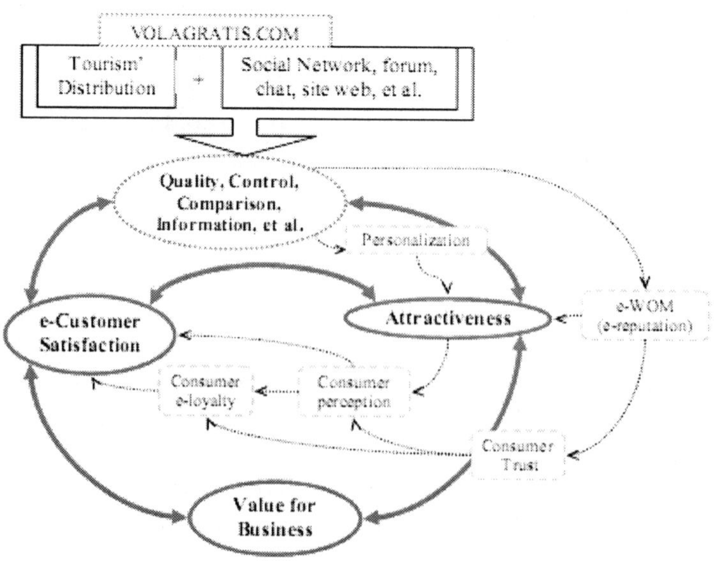

which are directed to quality service through the presence of different opportunities. They create the packaging of product, the control of information and product, the comparison of the different alternative to purchase, the useful information to choose the product more adequate to clients' needs; these actions activate a circle of dependence between "attractiveness-customer satisfaction-business value" (Figure 6). In fact, the online activities are directly and easily controllable in real time and can be changed in a very short time and without cost, so the consumer perceives a high power that influences the level of satisfaction. The power of consumer is based on the capacity to compare in real time the different competitors. Thus the client has the perception to decide the characteristic of product. This requires an high perception of personalized product also through a constant support of human resources (i.e. call-centre). The personalization stimulates the attractiveness that genres e-customer satisfaction, that strengthens the perception of added value for the client, the consumer perception and the consumer e-loyalty. Such results reinforce the role of Internet as a means to maximize customers' consumption experience into management for communication-distribution interaction. Consequently, the consumer through the e-tools can understand and know the mechanisms of experiences of e-consumption, so the client can obtain new abilities or he can strengthen the abilities previously acquired. The e-applications have a strategic importance of the web for effective and efficient management of business dynamic and confirm the Internet as a strategic means to reinforce the e-relationship between e-consumer and e-retailer (Pride, 2005). The Volagratis' web site is developed (5th level of Chaffey, 2009) and this permits a high relationship with e-customers. Therefore, the company, amplifying the number of possible relationships, has built a relationship that improves customer loyalty. So, the Volagratis both facilitates the buying process and can attract greater audiences than traditional instruments. The company' services have a strategic service profit chain, as follows:

- good external quality
- customer satisfaction
- permanence of clients
- high income.

From company perspective e-customer satisfaction and attractiveness influence and determine the value for business through the increase of profit and possible target-product, like Volagratis development's future perspective shows. The value for business includes economic profit, economic value added and shareholder value, and other forms of value such as employee value, customer value, supplier value, channel partner value, alliance partner value, managerial value, and societal value. Many of these forms of value are not directly measured in monetary terms (Swad, 2006). An important aspect for IT is that, in coherence with figure n. 6, the value for business is influenced from e-wom also. It has determined the success of company and a positive e-reputation.

The Figure 7 shows some e-application utilized by the group to stimulate customer satisfaction and affecting consumer behaviours through the e-relationships. Volagratis implements a marketing strategy that applies various levers: the main one is the presence in search engines (SEM - Search Engine Marketing) by adword for Google and overture for Yahoo i.e. SEO (Search Engine Optimization providing free results in thematic areas). In particular, among the main tools for managing internal relationships, the group uses the Calendar Sharing and G-mail, mainly for communication between human resources, and Skype for inter-employees communication and between them and the partners of the group. For interoperability communications, such as that between Marketing Department and IT Department, Project Management tools are useful. These should satisfy many requirements such as the work traceability, estimated time of projects completion, defining and tracking tasks, sharing of design documentation and finally monitoring of work progress. The web is also used for managing external relations, such as Skype and social networks. Skype is both for chatting and for VoIP services, for managing relationships with technology partners while, for communicating with users, taking into account sheer quantity, social networking platforms, including Twitter, Facebook, Slideshare and LinkedIn are preferred. The importance of efficient management of customer relationship is also expressed by the existence on the portal of a flight attendant that supports the user during the various stages of the booking process. The CRM activities are handled via e-mail marketing (newsletters) which is essential for maintaining constant the relationship with customers and for feedback monitoring. E-mail marketing is widely used for the implementation of campaigns aimed at targeting and maintaining customers. Finally, press communication and communication with experts in the field of travel for the dissemination of statistics and user logged analyses as well as consumer behaviour (most frequent bookings, average prices for flights, duration and period of travel, price range and costs) occurs mainly via Slideshare and websites. In conclusion, the analysis of the structure and promo-distribution chain of Volagratis evidences the role of social networks in the processes of management and maintenance of market relationship. Based on the interview of Dr Maria Teresa Rangheri, we have found that the company adheres to several social networks just as in every geographic market one specific social network has taken hold rather than another. Volagratis is on Facebook to satisfy domestic demand (from about four years it has a fan page, in Italy alone the company have acquired more than 6000 fans) and on Twitter to manage the relations with foreign customers (Singer et al., 2010). Into SN there are group' members that attract attention, they are "hub". The hubs are in relationship with other knots' network. Therefore, SN enables both information sharing with specific target audience and direct contact with customers which stimulate a positive e-WOM in an extremely viral environment. Volagratis is the first Italian brand that has activated a shop on Facebook; also, it utilizes the first true social shopping platform, recently arrived from U.S., called Wishpot, the wishing well. Linked to Facebook it allows to catalog the images of products you want to buy or offer or organize them in lists. Once created the account and lists you can share it through Facebook, Twitter or other social media with your contacts. Finally consumers can buy directly through the web all the products that are sold with e-commerce in Italy and worldwide. Thanks to Open Graph (the last platform presented in April by Facebook) Volagratis incorporates some social

Figure 7. Vehicles of integration (partial vs. total integration): Volagratis

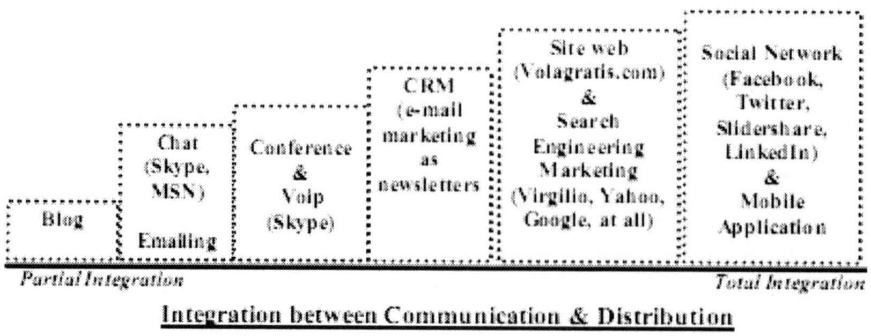

plugging (Like Button, Recommendations) that allows user to interact on Facebook without login. Future opportunities are to be found in mobile applications that should generate dynamic and durable customer's relationships. Consequently, future marketing trends could lead to increasingly audacious and effective customization of processes and products/services, in line with the maximization of customers' consumption experience into an e-relationship approach, that appreciate the circle of "A-CS_BV" proposed in this work.

8. FUTURE RESEARCH DIRECTIONS

The growing importance of Web 2.0 and the effects on consumers and organizations are issues often making headlines and increasingly attracting academic attention. In the face of cruel competition and continually rising customer expectations, e-retailers have necessarily become increasingly interested in their customers and have to address growing attention to factors outside the realm of technology in fever of the behavioural dimensions when conducting an e-business. Many researchers have focused on the ways in which technological applications contribute to changing customer behavioural and to the new challenges facing strategists and marketers (Urban, 2003; McKinsey Quarterly, 2007). However, there is still no systematic research on its importance and its effects on the marketing practice. Assuming the existence of the link between e-satisfaction, e-loyalty and attractiveness, our research findings show that the impact of e-satisfaction can be significantly moderated by variables such as trust and perceived value. At the same time, this study shows that building trusting relationships is an even more difficult challenge that may require e-retailers to go beyond bottom-line profit thinking to differentiate themselves from competitors and to gain competitive advantage. Evidence indicates that customer reviews posted in different forums or online communities, Web-blogs and podcasts are much more powerful as marketing tools than expert product reviews (Gillin, 2007); the influence of blogs and podcasts is increasing because of the rapid expansion of the audiences and contributors (Constantinide, 2008). Currently, there are no reliable data concerning the number of people using SNSs, while marketing research indicates that SNSs are growing in popularity worldwide (comScore, 2007). Nevertheless, information searches across SNs have been confirmed as substantially useful for risk reduction and uncertainty avoidance for tourists (Money et al., 2003). This trend is observable through the utilization of 'social shopping' online platforms where consumers contact other consumers, exchange advice, share their particular needs and interests, rate products and share their experiences in using them (i.e. Buzzillions; Crowdstorm; Ebay Neighborhoods;; Kaboodle; ShopMedia; Stylehive; TheNext;; Trusted Opinion; Viewpoints; Wishpot; Friendster). Consequently, due to the intangibility of tourism products, future research has to understand thoroughly consumer information search preferences, to give e-retailers effective instruments to attract travellers. One of the most critical challenges in customer value co-creation to maximize customer consumption experience will be in creating and maintaining a rewarding innovating experience environment for the customer as opposed to devising the technological infrastructure. Therefore, it is essential for marketers to look to Web 2.0 as a challenge and consider it a new domain of commercial strategy. E-retailers should try to understand the dimensions and the potential consequences of Web 2.0 on marketing practice and business. Understanding the sources of customer value and the consumer motivation in using these applications are the first steps to activate and maximize the virtuous circle presented above. In this scenario, identifying ways to enhance user experience, meeting customer information needs and helping customers become successful will be the keys to future success. As regards further research, we are aware of the necessity to measure and control the model proposed and other potential influencing variables. Finally, ways should be analyzed in the use of mobile applications, for the purpose of generating dynamic and durable customer relationships. Although the case study presented will need to be revised and based on empirical data and benchmarked with other e-tourism companies (i.e. analysis of the Volagratis Business model compared with its main competitors), we trust it provides a contribution that leads to a better understanding of consumer behaviour and customer consumption experience in an online social shopping context.

CONCLUSION

In a new conceptualization of integrated marketing the research prospects how the Internet can play a crucial role in e-distribution strategy if it is properly organized for e-commerce. The chapter has emphasized the e-relationship between e-retailer and e-consumer and proposed the web as an all encompassing medium, which provides customers with everything they could want. Therefore the potential success of e-commerce for tourism product is linked to the ability to manage particular forms of interactive e-communication and e-distribution, in line with the "synergic point" model (Singer et al., 2010). Consequently, e-retailers should provide superior value to customers that allow customer satisfaction and the maintaining of long-term customer relationships. E-satisfaction is likely to result in stronger e-loyalty when customers have a higher level of trust in the e-businesses that allow firms to create business value. By using web personalization for specific user tasks, a firm can also add value to services and products that becomes the main element of customer attractiveness and activates a virtuous circle of A-CS-BV, presented above. To this purpose the evolution in supply and future marketing trends could lead to ever more personalized products and processes. Consequently, tools as Community, SN etc., are key elements to carrying out Social Shopping. E-retailers should use e-tools with cognitive functions, which involve customers while browsing virtual environment and which impact on user attractiveness perception (Pantano, 2010). The implications are that e-tools need to have high human content; no other genre of web services has such high expansion rates as social networks. A key success factor is that social networking services offer modern communication possibilities for people by supporting user interaction and data interchange. Findings from our research show a current snapshot of the social networking sphere attained through an extensive site analysis. Benefiting from new Internet technologies and altered user behaviour, social networking sites have become strategic in new web services that have been emerging with the advent of "Web 2.0". In detail, SNs favour Social Shopping, a method of e-commerce by means of which consumers shop in a social networking environment similar to Facebook or MySpace. Users communicate and accumulate information on products, prices, and deals. Many sites allow users to create customized shopping lists and share them with friends. Some services, such as Groupon, Living Social and Kaboodle even allow users to shop together online to complete the social environment. Social shopping sites may generate revenue not only from advertising and click throughs, but also by sharing information about their users with retailers. SNs have a high impact on consumer behaviour, so companies can use the web to compare prices and share ideas, opportunities, tips and so on. Social shopping is becoming a purchase modality diffuse between e-users, basing their decision to purchase on suggestion communities. In the light of such tendencies, e-retailers have inserted the social aspect into marketing strategy. In the future, firms should learn to co-exist and communicate with an empowered customer insensitive to old-fashioned push marketing and by and large, determined to participate as an equal in the marketing and distribution process. Finally, all indications point to the fact that social media are here to stay: the Internet enables interaction between users and is the perfect point of synergy between e-communication and e-distribution, a means that maximizes customer experience.

REFERENCES

Addis, M., & Holbrook, M. B. (2001). On the conceptual link between mass customization and experiential consumption: An explosion of subjectivity. *Journal of Consumer Behaviour*, *1*(1), 50–66. doi:10.1002/cb.53

Aiello, A., & Cacia, C. (2009, August). Tourism International marketing strategies: the role of quality brand in the valorisations process. In *Proceeding of the 10th Vaasa Conference on International Business*, Vaasa, Finland.

Akkermans, H. (2001). Intelligent e-business: From technology to value. *IEEE Intelligent Systems*, *16*(4), 8–11. doi:10.1109/5254.941352

Anderson, R. E., & Srinivasan, S. (2003). E-satisfaction and e-loyalty: A contingency framework. *Psychology and Marketing*, *20*(2), 123–138. doi:10.1002/mar.10063

Andrew, O., Clore, G. L., & Collins, A. (1990). *The cognitive structure of emotions*. Cambridge University Press.

Armstrong, A., & Hagel, J. (1996). The real value of online communities. *Harvard Business Review*, (May–June): 134–141.

Arussy, L. (2009). Is customer experience relevant in a recession? *Customer Relationship Management*, *13*(4), 12.

Axhausen, K. W. (2003). *Social networks and travel: Some hypotheses. Arbeitsbericht Verkehr- sund Raumplanung, 197, Institut für Verkehrsplanung und Transportsysteme*. Zürich: ETH Zürich.

Bansal, H. S., & Voyer, P. A. (2000). Word-of-mouth processes within a service purchase decision context. *Journal of Service Research*, *3*, 166–177. doi:10.1177/109467050032005

Barney, J. B. (1991). Firm resources and sustained competitive advantage. *Journal of Management*, *17*(1), 99–120. doi:10.1177/014920639101700108

Bart, Y., Shankar, V., Sultan, F., & Urban, G. L. (2005). Are the drivers and role of online trust the same for all websites and consumers? A large-scale exploratory empirical study. *Journal of Marketing*, *69*, 133–152. doi:10.1509/jmkg.2005.69.4.133

Benjamin, R., & Wigand, R. (1995). Electronic markets and virtual value chains on the information highway. *Sloan Management Review*, Winter.

Bennett, A., & Strydom, J. W. (2001). *Introduction to travel and tourism marketing*. JW Strydom.

Bernstein, D. (1988). *Company image*. Guerini e Associati.

Berry, L. (1983). Relationship marketing. In L. Berry, G, L. Shostack, & G. D. Upah (Eds.), *Emerging perspectives on services marketing*. Chicago, IL: American Marketing Association.

Berthon, P., Pitt, L., & Watson, R. T. (1996). The World Wide Web as an advertising medium. *Journal of Advertising Research*, (Jan/Feb): 43–55.

Bharadwaj, A. S. (2000). A resource-based perspective on information technology capability and firm performance: An empirical investigation. *Management Information Systems Quarterly*, *24*(1), 169–196. doi:10.2307/3250983

Bloch, M., Pigneur, Y., & Steiner, T. (1996). *The IT-enabled extended enterprise, applications in the tourism industry*. Retrieved from http://haas.berkley.edu

Boardman, M. (2005). *The language of websites*. London, UK & New York, NY: Routledge.

Bolton, R. N., & Drew, J. H. (1991). A multistage model of consumers' assessments of service quality and value. *The Journal of Consumer Research, 17*, 375–384. doi:10.1086/208564

Bone, P. F. (1995). Word-of-mouth effects on short-term and long-term product judgments. *Journal of Business Research, 32*(3), 213–223. doi:10.1016/0148-2963(94)00047-I

Bonn, M., & Furr, H. S. (1999). Predicting a behavioural profile for pleasure travellers on the basis of Internet use segmentation. *Journal of Travel Research, 37*(4), 330–340. doi:10.1177/004728759903700403

Boyd, D., & Ellison, N. (2007). Social network sites: Definition, history, and scholarship. *Journal of Computer-Mediated Communication, 13*, 1–9.

Broydrick, S. C. (1998). Seven laws of customer value. *Executive Excellence, 15*, 15.

Buchanan, R., & Gilles, C. (1990). Value managed relationship: The key to customer retention and profitability. *European Management Journal, 8*(4). doi:10.1016/0263-2373(90)90115-M

Buhalis, D. (1996). Information and telecommunication technologies as a strategic tool for tourism enhancement at destination regions. In S. Klein, B. Schmid, A. Tjoa & H. Werthner (Eds.). *Information and communication technologies in tourism, Conference Proceeding, ENTER'96,* (pp. 131-142). 17-19 January, Innsbruck. Spinger-Verlag, Vienna.

Buhalis, D. (1998). Strategic use of information technologies in the tourism industry. *Tourism Management, 19*(5), 409–421. doi:10.1016/S0261-5177(98)00038-7

Buhalis, D. (1999). Marketing the competitive destination for the future. *Tourism Management, 21*(1), 97–116. doi:10.1016/S0261-5177(99)00095-3

Buhalis, D., & Law, R. (2008). Progress in information technology and tourism management: 20 year on and 10 years after the internet, the state of e-tourism research. *Tourism Management, 29*(4), 609–623. doi:10.1016/j.tourman.2008.01.005

Buhalis, D., Tjoa, A. M., & Jafari, J. (Eds.). (1998). *Information and communication technologies in tourism: IT and the dynamic tourism marketplace*. Vienna, Austria: Springer-Verlag.

Burger, F. (Ed.). (1997). TSI@Web-databased supported tourist information on the web. In A. M. Tjoa (Ed.), *Information and communication technologies in tourism,* (1987). New York, NY: Springer-Wien.

Bushman, R. M., & Smith, A. J. (2001). Financial accounting information and corporate governance. *Journal of Accounting and Economics, 32*, 237–333. doi:10.1016/S0165-4101(01)00027-1

Carrasco, J., Hogan, B., Wellman, B., & Miller, E. J. (2006*). Collecting social network data to study social activity-travel behaviour: An egocentric approach.* Paper presented at the 85th Transportation Research Board Meeting, Washington DC, January 22-26.

Chaffey, D. (2009). *E-business and e-commerce management strategy. Implementation and practice*. Financial Times Prentice Hall.

Chatterjee, P. (2001). Online reviews: Do consumer use them? In M. C. Gilly & J. Myers-Levy (Eds.), *Proceeding of the ACR* (pp. 129-134). Provo, UT: Association for Consumer Research

Chen, J., Geyer, W., Dugan, C., Muller, M., & Guy, I. (2009). Make new friends, but keep the old: Recommending people on social networking sites. In *CHI '09: Proceedings of the 27th international conference on Human factors in computing systems*, New York, NY, USA, pp. 201-210, ACM.

Chen, M., & Xie, J. (2008). Online consumer review: Word-of-mouth as a new element of marketing communication mix. *Management Science*, *54*(3), 477–491. doi:10.1287/mnsc.1070.0810

Chen, M., Zhang, D., & Zhou, L. (2005). Providing Web services to mobile users: the architecture design of an m-service portal. *International Journal of Mobile Communications*, *3*(1), 1–18. doi:10.1504/IJMC.2005.005870

Chen, Z., & Dubinsky, A. J. (2003). A conceptual model of perceived customer value in e-commerce: A preliminary investigation. *Psychology and Marketing*, *20*(4), 323–347. doi:10.1002/mar.10076

Compete, Inc. (2006). *Embracing consumer buzz creates measurement challenges for marketers*. Retrieved 05 June, 2010, from: http://www.cymfony.com/

comScore. (2007). *Social networking goes global*. Reston, VA. Retrieved September 9, 2007 from http://www.comscore.com/press/release.asp?press=1555

Constantinide, E., & Fountain, S. J. (2008). Web 2.0: Conceptual foundations and marketing issues. *Journal of Direct. Data and Digital Marketing Practice*, *9*, 231–244. doi:10.1057/palgrave.dddmp.4350098

Cooper, C. (2008). *Tourism: Principles and practice. Pearson Education*. England: FT Prentice Hall.

Corvi, E., & Bonera, M. (2005). *Web orientation and value chain evolution in the tourism industry*. Paper n. 51, Dipartimento di economia aziendale, Università degli studi di Brescia, Italy.

Cova, B., & Pace, S. (2006). Brand community of convenience products: New forms of custodies empowerment. *European Journal of Marketing*, *40*(9-10), 1087–1105. doi:10.1108/03090560610681023

Cristopher, M., Payne, A., & Ballantyne, D. (1991). *Relationship marketing*. Butterworth-Henemann.

Cronin, J. J., & Taylor, S. A. (1992). Measuring service quality: A re-examination and extension. *Journal of Marketing*, *56*, 55–68. doi:10.2307/1252296

Culnan, M. J., & Armstrong, P. K. (1999). Information privacy concerns, procedural fairness and impersonal trust: An empirical investigation. *Organization Science*, *10*, 104–115. doi:10.1287/orsc.10.1.104

Dalgety, D. (2002). *Online travel continues to grow*. Retrieved from Visitearth.com

Davis, F. D. (1989). Perceived usefulness, perceived ease of use, and user acceptance of information technology. *Management Information Systems Quarterly*, *13*, 319–340. doi:10.2307/249008

Davis, F. D., Bagozzi, R. P., & Warshaw, P. R. (1989). User acceptance of computer technology: A comparison of two theoretical models. *Management Science*, *35*, 982–1003. doi:10.1287/mnsc.35.8.982

Dellarocas, C. (2003). The digitalization of word of mouth: Promise and challenges of online feedback mechanisms. *Management Science*, *49*(10), 1407–1424. doi:10.1287/mnsc.49.10.1407.17308

Detlor, B., Sproule, S., & Gupta, C. (2003). Prepurchase online information seeking: Search versus browse. *Journal of Electronic Commerce Research*, *4*(2), 72–84.

Devaraj, S., & Kohli, R. (2003). Performance impacts of Information Technology: Is actual usage the missing link? *Management Science*, *49*(3), 273–289. doi:10.1287/mnsc.49.3.273.12736

Dwyer, C., Hiltz, S. R., & Passerini, K. (2007). Trust and privacy concern within social networking sites: A comparison of Facebook and MySpace. *Proceedings of AMCIS 2007*, Keystone, CO. Retrieved September 21, 2007 from http://csis.pace.edu/~dwyer/research/DwyerAMCIS2007.pdf

Fabris, G. (2008). *Societing*. Milano, Italia: Egea.

Fairnet, A., & Plonchard, E. (2002). *Customer relationship management*. Milano, Italy: Etas.

Ferri, M. A., Basile, G., & Aiello, L. (2010, April). *The relationship between integrated cultural product brand and individual: A conceptual framework*. Paper presented at the 7th CIRCLE Conference for Consumer Behavior and Retailing Research, Estoril Higher Institute for Tourism and Hotel Studies (ESHTE) Estoril, Portugal.

Ferri, M. A., & Celant, A. (2009). I canali della distribuzione. In *L'Italia. Il declino economico e la forza del turismo*. Italy: Marchesi.

Fodness, D., & Murray, B. (1997). Tourist information search. *Annals of Tourism Research, 24*, 503–523. doi:10.1016/S0160-7383(97)00009-1

Ford, D. (Ed.). (1990). *Understeing business markets: Interaction, relationship e network*. London, UK: Academic Press.

Friedman, B., Kahn, P. H., & Howe, D. C. (2000). Trust online. *Communications of the ACM, 43*(12), 34–40. doi:10.1145/355112.355120

Gale, B. T., & Chapman, R. W. (1994). *Managing customer value: Creating quality and service that customers can see*. New York, NY: Free Press.

García-Crespo, A., Chamizo, J., Rivera, J., Mencke, M., Colomo-Palacios, R., & Gómez-Berbís, J. M. (2008). SPERA: Social pervasive e-tourism advisor. *Telematics and Informatics, 26*(3), 306–315. doi:10.1016/j.tele.2008.11.008

Ghosh, A. K. (1998). *E-commerce security: Weak links, best defenses*. John Wiley and Sons.

Giacomazzi, F., & Camisani-Calzolari, M. (2008). *Impresa 4.0. Marketing e comunicazione digitale a 4 direzioni*. Prentice Hall, Financial Times.

Gillin, P. (2007). *The new influencers, a marketer's guide to the new social media*. Quill Driver Books\ Word Dancer Press, Inc, CA, USA.

Go, F. M. (1992). The role of computerized reservation systems in the hospitality industry. *Tourism Management, 13*(1), 22–26. doi:10.1016/0261-5177(92)90026-4

Golinelli, G. M. (2010). *Viable systems approach (VSA), governing business dynamics* (pp. 28–396). Padova, Italy: Cedam.

Gordon, I. (1999). *Relationship marketing: New strategies, techniques and technologies to win the customers you want and keep them forever*. John Wiley and Sons Publishers.

Gratzer, M., & Winiwarter, W. (2003). A framework for competitive advantage in e-tourism. In *Proceedings of the 10th International Conference on Information Technology and Travel & Tourism*, Springer-Verlag.

Gretzel, U. (2006). Consumer generated content: Trends and implication of consumer behaviour. In Buhalis, D., & Costa, C. (Eds.), *Tourism business frontiers* (pp. 9–18). Butterworth-Heinernann. doi:10.1016/B978-0-7506-6377-9.50009-2

Gronroos, C. (1997). From marketing mix to relationship marketing: Towards a paradigm shift in marketing. *Management Decision, 35*(4), 322–339. doi:10.1108/00251749710169729

Guatri, L., & Massari, M. (1992). *La diffusione del valore*. Milano, Italy: Egea.

Gummesson, E. (1994). Marketing relationship marketing operational. *International Journal of Service Industry Management, 5*(5), 5–20. doi:10.1108/09564239410074349

Gummesson, E. (2008). *Total relationship marketing*. Oxford, UK: Butterworth-Heinemann, Elsevier Ltd.

Haubl, G., & Trifts, V. (2000). Consumer decision making in online shopping environments: The effects of interactive decision aids. *Marketing Science, 19*, 5–21. doi:10.1287/mksc.19.1.4.15178

Hawkins, D., & Schuchter, A. (1995, June). *Travel marketing on the Internet*. Paper presented at the biannual Conference of the International Academy for the Study of Tourism, 24-30 June, Cairo, Egypt.

Heiman, A., & Muller, E. (1996). Using demonstrations to increase new product acceptance: Controlling demonstration time. *JMR, Journal of Marketing Research, 33*, 422–430. doi:10.2307/3152213

Hendrix, P. E. (1999). Build it, and they will come. *Marketing Management, 8*, 31–35.

Hitt, L., & Brynjolfsson, E. (1996). Productivity, business profitability and consumer surplus: Three different measures of Information Technology value. *Management Information Systems Quarterly, 20*(2), 121–142. doi:10.2307/249475

Hoffman, D. L., & Novak, T. P. (1996). Marketing in hypermedia computer-mediated environments: Conceptual foundations. *Journal of Marketing, 60*(3). doi:10.2307/1251841

Hoffman, D. L., Novak, T. P., & Peralta, M. (1999). Building consumer trust online. *Communications of the ACM, 42*(4), 80–85. doi:10.1145/299157.299175

Holland, J., & Baker, S. M. (2001). Customer participation in site brand loyalty. *Journal of Interactive Marketing, 15*, 34–45. doi:10.1002/dir.1021

Hopper, L. (1990). Rattling SABRE new ways to compete on information. *Harvard Business Review, 68*(3), 118–125.

Hoque, F. (2000). *E-enterprise: Business models, architecture and components*. Cambridge, UK: Cambridge University Press. doi:10.1017/CBO9780511584923

Howard, T. (2005). Viral advertising spreads through marketing plans. *USA Today*.

Jarvenpaa, S. L., & Todd, P. A. (1997). *Is there a future for retailing on the Internet? Electronic marketing and the consumer*. Thousand Oaks, CA: Sage.

Johansen, J. I., & Monthelie, C. (1996). *Lojalitetsredovisning*. Gothenburg, Sweden: InfoNet Scandinavia.

Kerlinger, F. N., & Lee, H. B. (2000). *Foundations of behavioral research*. New York, NY: Harcourt College Publishers.

Kotler, P., Armstrong, G., Saunders, J., & Wong, V. (1999). *Principles of marketing* (2nd ed.). Prentice Hall Europe.

Kozinets, R. (1999). E-tribalized marketing? The strategic implications of virtual communities of consumption. *European Management Journal, 17*(3), 252–264. doi:10.1016/S0263-2373(99)00004-3

Kriebel, C., & Kauffman, R. (1988). *Measuring and modelling the business value of Information Technologies*. ICIT Press.

Kurtz, D. L. (2008). *Contemporary marketing* (p. 322). Mason, OH: South-Western Cengage Learning.

Lamnek, S. (2005). *Qualitative Sozialforschung Lehrbuch, 4 auflage. Weihnhein, Germany & Basel*. Switzerland: Beltz Verlag.

Law, R., Shansan, Q., & Buhalis, D. (2009). Progress in tourism management: A review of website evaluation in tourism research. *Tourism Management, 31*, 297–313. doi:10.1016/j.tourman.2009.11.007

Li, H., Kuo, C., & Russell, M. G. (1999). The impact of perceived channel utilities, shopping orientation, and demographics on the consumer's online buying behaviour. *Journal of Computer Mediated Communication,* December.

Lohse, G. L., & Spiller, P. (1998). Electronic shopping. *Communications of the ACM, 14*, 81–87. doi:10.1145/278476.278491

Lonardi, G. (May 17, 2010). *Bravofly: è italiano il leader europeo nella vendita di biglietti aerei low cost.* Repubblica, sezione Affari & Finanza, (p. 15).

Lorenzoni, G., & Lipparini, A. (1999). The leveraging of interfirm relationships as a distinctive organizational capability: A longitudinal study. *Strategic Management Journal, 20*, 317–338. doi:10.1002/(SICI)1097-0266(199904)20:4<317::AID-SMJ28>3.0.CO;2-3

Lynch, J. C., & Ariely, D. (2000). Wine online: Search costs and competition on price, quality, and distribution. *Marketing Science, 19*, 83–105. doi:10.1287/mksc.19.1.83.15183

MacKay, K., McVetty, D., & Vogt, C. (2005). Web-based information search and use: Is it new tourism reality? A preliminary examination of visitors to Canada's Four Mountain National Parks. In A. Rice & M. Conlin (Ed.), *Proceedings of the TTRA Canada 2005 conference. Ottawa, Canada, Travel and Tourism Research Association Canada Chapter.*

Marcati, A., Guido, G., & Peluso, A. (2007). *What is marketing for SME entrepreneurs? The need to market the marketing approach.* Paper presented at the 6th International Conference in Marketing Trends, Paris, Venice. 26th-27th of January.

Marcus, B. W., & Wallace, S. L. (1997). *New dimensions in investor relations: Competing for capital in the 21st century. (Wiley Frontiers in Finance)*. John Wiley & Sons.

Marcussen, C. H. (2001). *Trends in European Internet distribution of travel and tourism services.* Research Centre of Bornholm, Denmark, 27 March.

McKinsey. (2007). How business are using Web 2.0: A McKinsey global survey. *The McKinsey Quarterly*. Retrieved August 7, 2010, from http://www.mckinseyquarterly.com/Marketing / How_businesses_are_using_Web_20_A_McKinsey_Global_Survey_1913_abstract

Merten, K. (1997). *Kommunikation: Eine Begriffs- und prozessanalyse*. Opladen, Germany: Westdeutscher Verlag.

Methlie, L. B. (2000). A business model for electronic commerce. *Teletronikk, 96*(2), 8–19.

Miceli, N., Ricotta, F., & Costabile, M. (2007). Customizing customatization: A conceptual framework for interactive personalization. *Journal of Interactive Marketing, 21*(2), 6–25. doi:10.1002/dir.20076

Money, B. R., & Crotts, J. C. (2003). The effect of uncertainty avoidance on information search, planning and purchases of international travel vacations. *Tourism Management, 24*(2), 263–277. doi:10.1016/S0261-5177(02)00057-2

Montoya-Weiss, M. M., Voss, G. B., & Grewal, D. (2003). Determinants of online channel use and overall satisfaction with a relational, multichannel service provider. *Journal of the Academy of Marketing Science, 31*(4), 448–458. doi:10.1177/0092070303254408

Murray, K. B. (1991). A test of service marketing theory: Consumer information acquisition activities. *Journal of Marketing, 55*, 10–15. doi:10.2307/1252200

Nault, B. R., & Dexter, A. S. (1995). Added value and pricing with information technology. *Management Information Systems Quarterly*, 449–464. doi:10.2307/249628

Newman, J. W., & Werbel, R. A. (1973). Multivariate analysis of brand loyalty for major household appliances. *JMR, Journal of Marketing Research, 10*, 404–409. doi:10.2307/3149388

Nielsen, J., & Norman D. A. (2000). Usability on the Web isn't a luxury. *InformationWeek*, 65–73.

Novak, T. P., Hoffman, D. L., & Yung, Y. F. (2000). Measuring the customer experience in online environments: A structural modelling approach. *Marketing Science, 19*, 22–42. doi:10.1287/mksc.19.1.22.15184

Nysveen, H., Leif, B. M., & Pedersen, P. E. (2003). Tourism websites and value-added services: The gap between customer preferences and websites offering. *Information Technology & Tourism, 5*, 165–174. doi:10.3727/109830503108751117

Nysveen, H., & Lexhagen, M. (2002). Reduction of perceived risk through on-line value-added services offered on tourism businesses websites. In *Proceedings of the 9th International Conference on Information and Communication Technologies in Tourism, Innsbruck* (pp. 488–496). Vienna, Austria: Springer-Verlag.

Palmer, J. W., Bailey, J. P., & Faraj, S. (1999). The role of intermediaries in the development of trust on the WWW: The use and prominence of trusted thirds parties and privacy statements. *Journal of Computer-Mediated Communication*, 9.

Pantano, E. (2010). New technologies and retailing: Trends and directions. *Journal of Retailing and Consumer Services, 17*(3), 171–172. doi:10.1016/j.jretconser.2010.03.004

Pantano, E., & Tavernise, A. (2009). Learning cultural heritage through Information and Communication Technologies: A case study. In Chang, M., & Kuo, C.-W. (Eds.), *Learning culture and language through ICTs: Methods for enhanced instruction* (pp. 103–119). Hershey, PA: IGI Global. doi:10.4018/9781605661667.ch007

Park, D. H., Lee, J., & Han, J. (2007). The effect of online consumer reviews on consumer purchasing intention: the moderating role of involvement. *International Journal of Electronic Commerce, 11*(4), 125–148. doi:10.2753/JEC1086-4415110405

Payne, J., Bettman, J. R., & Johnson, E. J. (1991). Consumer decision making. In Kassarjian, H. H., & Robertson, T. S. (Eds.), *Handbook of consumer behavior* (pp. 50–84). Prentice-Hall.

Peppard, J. (Ed.). (1993). *IT strategy, firm business*. London, UK: Pitman.

Peppers, D., & Rogers, M. (1997). *Enterprise one to one. Tools for competing in the interactive age*. New York, NY: Currency Doubleday.

Peterson, R., & William, R. (1992). Measuring customer satisfaction: Fact and artefact. *Journal of the Academy of Marketing Science, 20*(Winter).

Pew Internet & American Life Project. (2006). *Internet evolution, Internet penetration and impact*. Retrieved 21 May, 2010, from http://www.pewinternet.org/PPF/r/182/report_display.asp

Pollock, A. (1995). The impact of information technology on destination marketing. *EIU Travel and Tourism Analyst, 3*, 66–83.

Porter, M. (1985). Technology and competitive advantage. *The Journal of Business Strategy*, (Winter): 60–70. doi:10.1108/eb039075

Porter, M. E. (2001). Strategy and the Internet. *Harvard Business Review*, 63–78.

Pride, W., & Ferrell, O. C. (2005). *Marketing*. Milano, Italia: EGEA.

Prideaux, B. (2002). Airline distribution systems: The challenge and opportunity of the Internet. In Buhalis, D., & Laws, E. (Eds.), *Tourism distribution channels: Practices, issues and transformations*. London, UK: Continuum.

Quelch, J. A., & Kein, L. R. (1996). The Internet and international marketing. *Sloan Management Review*, (Spring): 60–75.

Reicheild, F. F., & Sasser, E. (1990). *Zero defections: Quality comes to services. Harvard Business Review*. September-October.

Richardson, P. S., Dick, A. S., & Jain, A. K. (1994). Extrinsic and intrinsic cue effects on perceptions of store brand quality. *Journal of Marketing, 58*, 28–36. doi:10.2307/1251914

Richins, M. (1983). Negative word-of-mouth by dissatisfied consumers: A pilot study. *Journal of Marketing*, *47*(Winter), 68–78. doi:10.2307/3203428

Robles, P. (2010). *Virgin America: Flying social media into the abyss*. Retrieved June 24, 2010, from http://econsultancy.com/blog/6133-virgin-america-flying-social-media-into-the-abyss

Rogers, E. M., & Kincaid, D. L. (1981). *Communication networks: Towards a paradigm for research*. New York, NY: Free Press.

Ryan, T. M., & Jacobs, C. A. (2005). *Using investor relations to maximize equity valuation*. John Wiley & Sons, Inc.

Schipul, E. (2006). The Web's next generation: Web 2.0. *Tactics,* March, 23.

Schmitt, B. H. (1999). *Experiential marketing: How to get customers to sense, feel, think, act, relate*. Free Press.

Scott, D. (2007). *The new rules of marketing and PR: How to use news releases, blogs, podcasting, viral marketing, & online media to reach buyers directly*. New York, NY: John Wiley & Sons, Inc.

Sharkar, V., & Malthouse, C. (2007). The growth of interactions and dialogs in interactive marketing. *Journal of Interactive Marketing, 21*(2), 2–4. doi:10.1002/dir.20080

Sheldon, P. J. (1997). *Tourism Information Technology*. UK: CAB International.

Siano, A. (2001). *Competenze e comunicazione del sistema d'impresa. Il vantaggio competitivo tra ambiguità e trasparenza*. Milano, Italy: Giuffrè.

Singer, P. (2002). *Economia digitale, dinamica evolutiva e valutazione delle imprese internet. Elementi di teoria ed evidenze empiriche*. Padova, Italy: Cedam.

Singer, P., & Cacia, C. (2009, October). *The role of Web investor relations for mitigating and manage stock exchange liquidity and enterprise risks*. Paper presented at the 9th Global Conference on Business & Economics 2009 Program, Cambridge University, UK.

Singer, P., Ferri, A., Aiello, L., & Cacia, C. (in press). -2010). Internet as a "point of synergy" between communication and distribution: Hypothesis of model applied to tourism. *Journal of Digital Content and Technological Application.*

Singh, J., & Sirdeshmukh, D. (2000). Agency and trust mechanisms in consumer satisfaction and loyalty judgments. *Journal of the Academy of Marketing Science*, *28*, 150–167. doi:10.1177/0092070300281014

Stamboulis, Y., & Skayannis, P. (2003). Innovation strategies and technology for experience based tourism. *Tourism Management*, *24*(1), 35–43. doi:10.1016/S0261-5177(02)00047-X

Stiglitz, J., & Weiss, A. (1981). Credit rationing in markets with imperfect information. *The American Economic Review*, *71*(3), 93–410.

Stipanuk, D. M. (1993). Tourism and technology. Interactions and implications. *Tourism Management,* August.

Strassmann, P. (1990). *The business ~due of computers: An executive guide*. Connecticut: The Information Economics Press.

Swaminathan, V., Lepkowska-White, E., & Rao, B. P. (1999). Browsers or buyers in cyberspace? An investigation of factors influencing electronic exchange. *Journal of Computer-Mediated Communication*, *5*(2), 1–24.

Sward, D. (2006). *Measuring the business value of information technology*. Intel Press.

Teas, R. K., & Agrawal, S. (2000). The effect of extrinsic product cues on consumers' perceptions of quality, sacrifice, and value. *Journal of the Academy of Marketing Science*, *28*, 278–290. doi:10.1177/0092070300282008

Thorbjørnsen, M., Supphellen, M., Nysveen, H., & Pedersen, P. E. (2002). Building brand relationship online: A comparison of two interactive applications. *Journal of Interactive Marketing*, *16*(3), 17–34. doi:10.1002/dir.10034

Timmers, P. (1998). Business models for electronic markets. *Electronic Markets*, *8*, 3–8. doi:10.1080/10196789800000016

Urban, G. (2003). *Customer advocacy: Is it for you?* MIT Sloan School of Management. Center for E-Business, Paper 175.

Varki, S., & Colgate, M. (2001). The role of price perceptions in an integrated model of behavioral intentions. *Journal of Service Research*, *3*(3), 232–240. doi:10.1177/109467050133004

Verity, J., & Hoff, R. (1994, 14 November). The internet: How it will change the way you do business. *Business Week*.

Vesanen, J., & Raulas, M. (2006). Building bridges for personalization: A process model for marketing. *Journal of Interactive Marketing*, *20*(1), 5–20. doi:10.1002/dir.20052

Vickery, G., & Wunsch-Vincent, S. (2007). *Participative Web and user-created content. Web 2.0, wikis and social networking*. OECD.

Wang, Y. (2008). Examining the level of sophistication and success of destination marketing systems: Impacts of organizational factors. *Journal of Travel & Tourism Marketing*, *24*(1), 82–98. doi:10.1300/J073v24n01_06

Watzalawick, P., Beavin, J., & Jackson, D. (1967). *Pragmatics of human communication. A study of interactional patterns, pathologies, and paradoxes*. New York, NY: W.W. Norton e Co., Inc.

Webster, F. E. (1996). The changing role of marketing in the corporation. *Journal of Marketing, 56,* October 1992, 1-17. In G. Cozzi, & G. Ferrero (Eds.) *Le frontiere del marketing*. Torino, Italy: Giappichelli.

Werthener, H. (2001). *E-commerce and travel/tourism issues and challengers. EC3 – E-Commerce Competence Center*. Vienna, Austria: IFIT.

Werthner, H., & Klein, S. (1999). *Information Technology and tourism. A challenging relationship*. Vienna, Austria: Springer Computer Science.

Wirtz, B. W. (2001). *Electronic business*. Wiesbaden,G ermany: Gabler Verlag.

Wolfinbarger, M., & Gilly, M. C. (2001). Shopping online for freedom, control, and fun. *California Management Review*, *43*, 34–55.

Wong, J., & Law, R. (2005). Analysing the intention to purchase on hotel websites: A study of travellers to Hong Kong. *International Journal of Hospitality Management*, *24*(3), 311–329. doi:10.1016/j.ijhm.2004.08.002

Woodruff, R. B. (1997). Customer value: The next source of competitive advantage. *Journal of the Academy of Marketing Science*, *25*, 139–153. doi:10.1007/BF02894350

Woodside, A. G., Frey, L. L., & Daly, R. T. (1989). Linking service quality, customer satisfaction, and behavioral intention. *Journal of Health Care Marketing*, *9*, 5–17.

Yin, R. K. (2008). Applications of case study research. *Applied Social Research Methods Series, 34.*

Zeithaml, V. A., Parasuraman, A., & Malhotra, A. (2002). Service quality delivery through websites: A critical review of extant knowledge. *Journal of the Academy of Marketing Science, 30*, 362–375. doi:10.1177/009207002236911

ADDITIONAL READING

Barlow, A. K. J., & Noreen, Q. S. (2004). Developments in information and communication technologies for retail marketing channels. *International Journal of Retail & Distribution Management, 32*(3). doi:10.1108/09590550410524948

Chang, C. A., & Burke, R. R. (2007). Consumer choice of retail shopping aids. *Journal of Retailing and Consumer Services, 14*, 339–346. doi:10.1016/j.jretconser.2006.12.003

Dean, J. W., & Sharfman, M. P. (1996). Does Decision Process Matter? A Study of Strategic Decision-Making Effectiveness. *Academy of Management Journal, 39*(2), 368–396. doi:10.2307/256784

Deng, L., Turner, D. E., & Gehling, R., Brad Prince. (2010). User experience, satisfaction, and continual usage intention of IT. *European Journal of Information Systems, 19*(1), 60–75. doi:10.1057/ejis.2009.50

Haubl, G., & Trifts, V. (2000). Consumer Decision Making in Online Shopping Environment: The Effects of Interactive Decision Aids. *Marketing Science, 19*(1), 4–21. doi:10.1287/mksc.19.1.4.15178

Henard, D. H., & Dacin, P. A. (2010). Reputation for Product Innovation: Its Impact on Consumers. *Journal of Product Innovation Management, 27*(3), 321–335. doi:10.1111/j.1540-5885.2010.00719.x

Howard, J. A. (1989). *Consumer Behavior in Marketing Strategy*. Englewood Cliffs, NJ: Prentice Hall.

Jarvenpaa, S. L., & Todd, P. A. (1997). Consumer Reactions to Electronic Shopping on the World Wide Web. *International Journal of Electronic Commerce, 1*(2), 59–88.

Kemperman, A. M., Borgers, A. W. J., & Timmermans, H. (2009). Tourist shopping behavior in a historic downtown area. *Tourism Management, 30*(2), 208–218. doi:10.1016/j.tourman.2008.06.002

Krafft, M., & Mantrala, M. K. (2010). *Retailing in the 21st century: current and future trends.* Springer.

Kwon, O., & Wen, Y. (2010). An empirical study of the factors affecting social network service use. *Computers in Human Behavior, 26*, 254–263. doi:10.1016/j.chb.2009.04.011

Mills, J. E., & Law, R. (Eds.). (2004). *Handbook of consumer behaviour, tourism, and the internet.* Haworth Hospitality Press, Binghamton, NY, Co-published as *Journal of Travel and Tourism Marketing, 17 (2/3).*

Pantano, E., & Naccarato, G. (2010). Entertainment in retailing: The influences of advanced technologies. *Journal of Retailing and Consumer Services, 17*(3), 200–204. doi:10.1016/j.jretconser.2010.03.010

Park, I., Bhatnagar, A., & Rao, H. R. (2010). Assurance Seals, On-Line Customer Satisfaction, and Repurchase Intention. *International Journal of Electronic Commerce, 14*(3), 11–34. doi:10.2753/JEC1086-4415140302

Peter, P. J., & Jerry, C. O. (2002). *Consumer Behavior and Marketing Strategy* (6th ed.). New York, NY: McGraw-Hill/Irwin.

Yang, K. (2010). Determinants of US consumer mobile shopping services adoption: implications for designing mobile shopping services. *Journal of Consumer Marketing*, *2*(3), 262–270. doi:10.1108/07363761011038338

KEY TERMS AND DEFINITIONS

Consumer Behavior: Consumer behaviour refers to the dynamic process of understanding the exchanges of opinion and beliefs of consumers and industry, a process that permits forecasting of future trends in purchasing behaviour. It can be considered the essence of three interacting spheres: 1) natural distinctive features of the individual, which are not easily altered; 2) personality, the complex of mental, moral and intellectual qualities that distinguish the actor; 3) the social environment, in which a person lives.

Consumption Experience: Consumption is an experience coming out of the interaction between a subject – the consumer – and an object – a product, an event, an idea, a person, a place, or any other thing within a given context" (Addis M., 2001). This also reflects the multiple benefits that consumers receive from the total consumption experience - experiential benefit (e.g., entertainment) - as well as utilitarian benefit (e.g., convenience). The consumption experiences is a phenomenon directed toward the pursuit of fantasies, feelings, and fun. Value in consumer behaviour does not reside in the object purchased (good or service) but rather pertains directly to the consumption experiences. Value is the principal outcome of a consumption experience. If managers do not recognize these dimensions of value they neglect an important source of competitive advantage.

E-Relationship: Electronic relationship includes all the relationships, network and interaction based on IT. The relationship, in this contest, happen into market-space (from marketplace to market-space).

E-Retailer: the e-retailers are the online retailers that use the main characteristic of the current technologies applied to retailing, such as the interactivity. The advanced technologies applied to retailing are usually based on pervasive environments and mobile and ubiquitous computing. The e-retailers use the interactive technologies in order to enhance consumers shopping experience.

E-Satisfaction: Satisfaction is the summary psychological state resulting when the emotion surrounding disconfirmed expectations is coupled with a consumer's prior feelings about the consumer experience. So, satisfaction may be best understood as an ongoing evaluation of the surprise inherent in a product acquisition and/or consumption experience. In this research, e-satisfaction is defined as the contentment of the customer with respect to his prior purchasing experience with a given e-commerce firm. E-satisfaction can be significantly moderated by variables such as trust and perceived value and can allow firms to create business value.

IT Business Value: The term IT business value is commonly used to refer to the organizational performance impacts of IT, including productivity enhancement, profitability improvement, cost reduction, competitive advantage, inventory reduction, and other measures of performance.

Point of Synergy: Is a new model and a relevant example of knowledge "contamination" where elements of communication are integrated with distribution by means of typical technology tools, particularly internet.

Social Shopping: The social shopping is becoming a purchase modality diffuse between e-user, basing its decision to purchase on suggestions community. The social shopping (i.e. Kaboodle) is a merger between social media and e-commerce, in which users can share contents with their friends.

ENDNOTES

1 Although the views expressed in the chapter belong to all of the authors, the Introduction is attributed to Pierpaolo Singer, paragraphs 1, 3 and 7 are attributed to Lucia Aiello, paragraphs 4, 6 and 8 to Claudia Cacia, the Conclusion is attributed to Antonella Ferri and paragraphs 2 and 5 to all the authors.

2 The authors wish to thank Dr. Maria Teresa Rangheri, Marketing Director, Bravofly, Dr. Rosangela Leone, Media Relations, and Dr. Antonio Antonaci, Product Manager, Bravofly, for their cooperation and willingness to share information with us.

Chapter 9
Customer Intelligence as the Powerful Means for Turning Information into Profit

Sanda Renko
University of Zagreb, Croatia

ABSTRACT

Increasing competition and decreasing customer loyalty forced retailers to obtain accurate information about customers' existing and future needs, their profitability, behaviour, and trends in purchasing. Due to the rapid advancement in technology, retailers have easy access to vast amounts of information about their customers. They can collect and manage customer data and understand their behaviour patterns. The main purpose of the Customer Intelligence is to provide insight into customer's needs, attitudes, and behaviors towards particular retailer, and all elements of its business as well. In such a way, the retailer is able to build deeper and more effective customer relationships and to improve company's strategic decision. This chapter focuses on different aspects of Customer Intelligence and the growing interest and importance for its implementation in the praxis. Moreover, the chapter is trying to clarify some misunderstandings of the concept. The study conducted among retail companies dealing with ICT equipment and services on the Croatian market pointed out that Customer Intelligence provided retailers with a successful decrease of the rate of customer defection and a increase in revenues generated by customers.

INTRODUCTION

Rapidly changing industry dynamics forced companies to be market oriented and to get to know their customers as better as possible. Regardless of the size and the type of the industry, businesses are trying to find a way to discover all the facts about their customers and to improve their knowledge about them. It can help companies to deliver higher added value to the customer, because only those companies that follow well known slogan "customer is the king" can be the successful ones. Today, all functions of a company work together to respond, to serve and to satisfy the customer and the customer is at the centre of the company. In addition, the rapid growth of new technologies

DOI: 10.4018/978-1-60960-738-8.ch009

and Internet has greatly increased market opportunities for companies and has transformed the way how relationships between companies and their customers are managed.

Retailing is recognized as one of the most dynamic industries and the situation on the retail market has dramatically changed over the last thirty years. Retailing companies face many challenges: increasing information availability, more critical, more selective and sophisticated customers, the increase in the number of products and services offered, the number of retail formats, etc.

That increasingly dynamic retailing of the twenty-first century forces managers to monitor, to understand and timely respond to their customers. In such a situation, the most important competitive tools are timely and effective information which give detailed understanding of the experience customers have in interacting with a retailer. The extensive body of work indicates that IT solutions have helped companies to focus on customers and to build relationships with them (Apte, et al., 2005; Davenport, Harris & Kohli, 2001; Novo, 2001; Qiu, Li, & Wu, 2008; Phan & Vogel, 2010; Rubin, & Leigh, 2001; Speier, & Venkatesh, 2002; Trim, & Lee, 2006). This chapter focuses on the Customer Intelligence which is the process of interacting retailing companies with customer services, accounts payable, and all customer activities, in order to build deeper and more effective customer relationship. The chapter begins with the theoretical background where the insight into definitions of Customer Intelligence, motives for implementing it, benefits of its usage, the discussion about the necessary information technology, etc. is given. The review of the available literature showed a lack of consensus regarding the most appropriate way in which Customer Intelligence should be defined because it has been identified with Customer Relationship Management very often. In an attempt to explain the true meaning of Customer Intelligence, the review of the literature from the domain of Business Intelligence and customer relationship management was conducted, too. Aside from published academic papers, articles posted on key CRM portals and the top CRM software manufacturers and providers (e.g. Oracle, SAS) were evaluated. Additionally, relying on the study conducted among retailers in Croatia, the chapter considers the technology which allows to storage and to maintain with information about customers, the impact of the Customer Intelligence implementation on the customer loyalty and on the level of customer retention as well.

THEORETICAL BACKGROUND

In the early twentieth century, retailers had no problem in managing their relationships with customers because they had fewer customers than today and most of them were local. Recordkeeping was done in the retailer's head or in a simple ledger (Seeman & O'Hara, 2006). Selling places looked like friendly establishments where retailers knew who their customers were. Retailers were well-informed about customers' preferences. They listened to what their customers wanted, and continuously changed business in ways that reflected their customers' changing preferences. As last few decades have seen the growth in retailing formats and Internet, companies have accumulated large amount of data from various channels. Retail managers are concerned how to reduce these data into manageable amount of reliable and useful information for decision making. At this point we come to the concept of Customer Intelligence. The author of "Customer Intelligence: From Data to Dialogue", Kelly (2006) identified that the first stage of customer information utilization lasted from the 1990s till 2000, although very many business are still mired in the first stage. The second stage began in the 2000s when retailers stopped with monitoring and recording customer data and started with communicating with their customers.

One more comprehensive explanation of the term is given with following sentences by an expert in the Customer Intelligence – Tony Coretto (2010d): "Customer Intelligence is the combination of art, science and technology providing you insight to your customer's needs, attitudes and behaviors towards your product set. These relationships then allow you to define acquisition, growth and retention strategies using the voice a customer will hear, telling them the key things about the product that they would care about via the media they are most likely to use".

Different Aspects of Customer Intelligence

The literature review suggests that Customer Intelligence can be observed from three different aspects:

- Customer Intelligence as the process
- Customer Intelligence as the set of knowledge
- Customer Intelligence as the strategic tool

There is no doubt that Customer Intelligence is the process, because today's competitive market situation forces companies to obtain information about customers' existing and future needs, their behaviour and trends in decision making processes in order to build deeper and more effective customer relationships and to improve company's strategic decision making. Truly successful Customer Intelligence requires continued monitoring, gathering and analysing information regarding customers. Therefore it is considered to be the process. Typical questions to be answered in Customer Intelligence processes are:

- who are the customers and what needs and preferences do they have
- how often do the customers have transactions with the company
- when was the latest transaction of the customer with the company
- how much money do the customers spend on average in retailer's store
- what activities of the retailer are most effective in generating its customer satisfaction
- how loyal are the customers and who are the most "valuable" customers for the company

In business, Customer Intelligence primarily aims at gaining a comprehensive understanding of customer and their behaviour by means of intelligent tools, which enable a more pointed customer contact and a higher degree of customer loyalty (Decker & Höppner, 2006)

The term intelligence and knowledge are used interchangeably through the chapter of Zablah, Belenger and Johnson (2004) and refer to a "justified belief that increases an entity's capacity for effective action" (Alavi & Leidner, 2001). Coretto (2010a) also pointed out that over the last ten years there was a lot of confusion around Customer Intelligence and customer knowledge. The fact is that Customer Intelligence begins with basic data about the customer such as customer's name, address, demographics and psychographics in some instances. This data is then supplemented with transactional data i.e. reports of customer past activity and behavioural data which involves capturing customer events and purchasing actions over time. Many companies have accumulated large amount of data from various channels. Customer Intelligence consists of data collected from the company's environment. They are gathered from various relationships between company and its current and potential customers. (Panian & Ćurko, 2010, p.175)

Company has to get acquainted with its customers, to find out customers' needs and preferences precisely, and to identify most important characteristics of their purchasing behaviour.

Interactions with customers may be used to develop customer profiles and to predict future actions. For example, understanding the impact of marketing campaign on customer purchasing behaviour may be used to improve the effectiveness of future campaigns.

By hiring analysts skilled in sophisticated statistical methods, companies can mine gathered data and garner fresh insights into neglected customer segments, recent customer trends and other useful information (Kotler & Keller, 2006, p.74)

It has been expected that Customer Intelligence allow retailers to contact their customers with full knowledge of a customers' relationship in advance. Therefore, it consists of large amount of readily available customer data. The main question of the retailer relates how to store and to analyze internal data in conjunction with external data in an easy and profitable way. It is necessary to scrutinize the list of current customers (Van Raaij, 2005) because many databases contain details of customers who no longer have a relationship with the company (Mulhern, 1999). Although companies have searched for way to discover the truth about their customers, they face uncertainties about how many customers they really have and which ones are worth keeping (Joch, 2005)

Customer Intelligence explains how retail companies can structure customer data, gathered through loyalty programs or other vehicles, so as to finally understand where their profits originate – in terms of both products and customers (Hawkins, 2003)

In general, strategy provides a collective direction for the company as a whole and it is frequently called a game plan for getting what a company wants to achieve. According to Newman and Cullen (2002, p. 99), strategy covers corporate policies, the allocation of resources, customer markets and the competitive environment in which a retailer operates. Additionally, a strategy comprises a retailer's basic reason for being in business, such as: to increase profitability, to reach more customers and to extend its market share and consequently to be the market leader. Retail strategy can be understand as a plan of actions, which identifies directions and necessary changes that retailers must make in order to adapt to the dynamic marketplace. Well defined strategy will lead a retailer to a better market position in comparison to its competitors.

In the past, retailing companies were trying to stay ahead of competition by concentrating on the products in their stores. Those were the decades of pre-marketing era when retailers were conviced that consumers favoured products of higher quality and performance and that an aggressive selling and promotion effort could lead to higher profit. In the mid-1950s business shifted from a product-centered, "make-and-sell" philosophy to a customer-centered, "sense-and-respond" philosophy (Kotler & Keller, 2006, p.16) Companies understand that the competitive arena lies not in product-related activities but in understanding their customers and to find the right products for their customers. Usually (Porter, 1980, ch.2), companies are pursuing one of three Porter's generic strategies: overall cost leadership, differentiation and focus. Each of those strategies requires understanding customers, creating and delivering customer value. Unfortunately, many retailers tend to build the advantage on pricing strategies. In other words, they follow overall cost leadership strategy relying on the lowest distribution costs and the monitoring and reduction of all retailers' expenditures. Thus, it can result in lower prices than competitors'. However, companies have realized that a low prices can increase market-share but not market loyalty. Also, the price is the company's value that competitors can easily copy, but it can not help retailer to develop a long-term relationship with customers.

As customers are not equal, different customers' segments require different retailing strategies. Therefore, retailers that want to create and implement the strategy which will deliver high value to their customers have faced with strategically important decision: how much to invest in building and in using customer data. Namely, building and maintaining a customer database require large investments in computer hardware, database software, analytical programs, communication links, and skilled personnel (Kotler & Keller, 2006, p.165). However, with the knowledge about their customers, retailers can classified their customers into profitable and unprofitable ones and to identify their characteristics. Moreover, they are able to estimate the impact of marketing campaign on particular customer segment and to understand what actions could be conducted for making strategic decisions in an order to attract and to keep profitable customers. The goal of the customer intelligence process is to produce information that can be used in the strategic positioning of the company (Gilad, & Gilad, 1983)

The strategy has to be designed to create perceived customer value which is about managing knowledge-based strategies that deploy on organization's intangible assets: customer relationships, innovative products and services, high-quality and responsive operating processes, information technology and databases, and employee capabilities, skills and motivations (Kaplan & Norton, 2001, p.2).

All above mentioned can lead to conclusion that Customer Intelligence can be considered as the most important strategic tool.

The Impact of Customer Intelligence on Profitability

Customer Intelligence has been defined as activities of monitoring, understanding and responding to customers and thus has been recognised as an important element of marketing. The fact is that collecting and analyzing information about customers help retailer to determine and to understand its customers. As previously mentioned Customer Intelligence provides retailers with the large amount of readily available customer data and enables them to classify their customers into profitable and unprofitable ones. However, in order to fully understand what motivates customers (wholesalers, retailers and consumers), it is necessary to establish what underpins customer needs, wants and demands (Trim, & Lee, 2006). One thing is sure: retailers have to build the right type of relationship with each individual customer because it can have a substantial positive impact on its profitability. Customer intelligence can help retailers to reactivate customer purchases, to identify the best prospects and to convert them into their customers as well.

All customers are equally valuable and therefore, maximum profitability can only be achieved when available resources are invested in customer relationships that provide a desired level of return (Ryals, 2003). Combined with insights in customer needs and company capabilities, data about the costs associated with various service levels can be translated into segment-specific service concepts (Van Raaij, 2005). Accordingly, retailers can offer different service levels to their customers. The least profitable customers will receive standardised service, while the degree of customised services is increasing in accordance with the increase of customer profitability level.

Retailers' 'day-to-day' activities should be driven by an understanding of customers' evolving needs. If a retailer can see all interactions with customer service, accounts payable, sales and marketing, reactions to marketing campaigns, its strategy may be successfully adapted to new trends on the market. Consequently, customers will have more choices.

In general, retailers have to evaluate the cost of doing business with each customer, and to understand customer profitability. Customer Intelligence allows retailers to estimate future revenues and costs, thus helping retailers in developing value-driven differentiated customer service strategies (Van Raaij, 2005).

Very interesting observation about Customer Intelligence was given by Coretto (2010b). Namely, he suggested that Customer Intelligence should be considered as a two-way street. In the case that customers feel comfortable with the retailer (concerning its assortment, prices, services, staff personnel, location, atmosphere, etc.) and if customers are persuaded that the retailer is seeking to deliver them superior customer value (relying on their respectful and trusting interactions), then customers will be more engaged with the retailer. Thus, customers are going to satisfy more of their needs in the stores of a particular retailer, understanding that more they contribute to the generated profit, the better the retailer's offer will be. Successful Customer Intelligence requires continued communication between the retailer and its customers. In such a way, retailer is fully aware of customers' questions, desires, complaints, and put more effort to deliver products and services promised. Consequently, this will lead to high level of customer engagement and satisfaction. Finally, customer satisfaction is the main key to customer retention. In other words, satisfied customers stay loyal longer, talk favourably about the company and its products and services, pay less attention to competing stores and are less sensitive to pricing strategies. Myron (2009, p. 6) suggested that "we are entering a new era in Customer Intelligence, which will reveal a wealth of attitudinal data". The same author pointed out that social media enabled customers to express their feelings about particular company, product or service – information that companies have been clamouring to obtain for years.

The literature suggests that effective customer relationship management demands that a firm, at a minimum, is capable of (1) gathering intelligence about its current and prospective customers (Campbell, 2003) and (2) applying that intelligence to shape its subsequent interactions with them (Hirschowitz, 2001)

Misunderstanding of the Concept

Decker and Höppner (2006) considered that there was neither a unanimous understanding of customer intelligence in marketing science nor a well-defined set of Customer Intelligence methods. Customer Intelligence is frequently mentioned in the same situations as customer relationship management. The reason is obvious: every company can be successful if it is managed in customer-oriented way. As companies have to be more customer-centric oriented, they have to capture, monitor and to analyze detailed and up-to-date customer data. Conclusively, Customer Intelligence is a key component of effective Customer Relationship Management and when effectively implemented it is a rich source of insight into the customers' behaviour and experience. As the CRM tool, it enables firms to use the power of database, data mining and interactive technologies (internet) to collect and store amounts of customer data, and to build knowledge from data. It enables firm to continuously deliver customer value.

Customer relationship management involves acquiring and continuously updating knowledge about customer needs, motivations and behaviour over the lifetime of the relationship and applying customer knowledge to continuously improve performance through a process of learning from successes and failures (Özgener, S. & Iraz, R., 2006). Conclusively, customer relationship management is defined as "a set of practices that provide a consolidated, integrated view of customers across all business areas to ensure that each customer receives the highest level of service (Karakostas, Kardaras & Papthanassiou, 2005).

According to Yang (2008), customer relationship management consists of the four perspectives of customer knowledge, customer interaction, customer value, and customer satisfaction. The author explained customer knowledge as the status of the customer and customer data management with the focus on technology learning, understanding customer needs, and customer profiles. Some suggest that customer relationship management is a specialized collection of technological tools, other stress it is a set of business processes that focus on managing the customer experience, and others propose that it is best conceptualized as a comprehensive strategy for customer retention (Zablah, Bellenger & Johnston, 2004).

The literature review (Brohman, Watson, Piccoli & Parasuraman, 2003; Qiu, Li & Wu, 2008;

Phan & Vogel, 2010) reveals some instruments needed for implementing Customer Intelligence that are often associated with the concept: on-line analytical process (OLAP), data mining techniques, data warehouse and Business Intelligence. Coretto (2010c) pointed out that Customer relationship management, Database Marketing, and Business intelligence can be critical components supporting the development and deployment of Customer Intelligence, but alone they are not enough to be a sufficient Customer Intelligence solution. Additionally, the author added Sales-Force Automation and Social Networking to the list of concepts that may be related to customer intelligence.

Data mining, as one of the information technology (IT) services needed by organizations, provides such a technique for the exploration and analysis of these raw data so as to reveal hidden information and knowledge. It has also been recognized as an important way for discovering knowledge from the data and converting "data rich" to "knowledge rich" so as to assist strategic decision making (Qiu, Li & Wu, 2008). DM is the process of analysing marketing-related data from different perspectives, summarising this into useful information for planning and/or decision making (Fleisher, Wright & Allard, 2008) and it can help to determine relationships between marketing mix elements (for example, price, service levels, store atmosphere) and customer demographics. Market research has been in use since the beginning of the twentieth century due to the importance of market information, and particularly customer information, to marketing planning and management decisions (Market Research Society, 2007)

Data mining is the technology that allows searching through large amounts of data for meaningful patterns of consumer behaviour such as switching behaviours, fraud patterns, market basket analysis, and consumer trends (Phan & Vogel, 2010). It is often defined as the – more or less – automatic search and identification of patterns in large amounts of data (Decker, R. & Höppner, M. (2006). Data mining involves the use of sophisticated statistical and mathematical techniques such as cluster analysis, automatic interaction detection, predictive modelling, and neural networking (Kotler & Keller, 2006, p. 164).

Online Analytical Processing - OLAP is a tool that supports multi-dimensional analysis, enabling users to view data in vast data warehouses in different dimensions that normal database queries would not be able to do so quickly (Phan & Vogel, 2010).

Data warehousing technology is the critical technology for drawing data from multiple vendor databases, and reorganizing it for efficient analysis (Brohman, Watson, Piccoli & Parasuraman, 2003). However, customer intelligence isn't a technology. Well-trained and educated company's personnel is needed to use technological infrastructure to capture, query and analyze the data.

Several related terms include competitive intelligence, market intelligence, customer intelligence, competitor intelligence, strategic intelligence and technical intelligence (Lonnqvist & Pirttimaki, 2006)

THE ROLE OF TECHNOLOGY

By among other things, technology does play a substantial role in companies' efforts to interact with their customers and to improve customer service quality. According to Coltman (2007), the contribution of information technology to business performance has been under scrutiny for more than two decades. There has been an evident growth in organizational utilisation of competitive intelligence, customer relationship management, data mining and market research. Except competitive intelligence, all the terms are explained in the text above. Broadly defined, competitive intelligence is the strategic process of focusing the intelligence effort around factors that are critical for the success of the company. It is critical in better understanding and addressing competitors, and to learn about customers and to act upon that knowledge (Fleisher, Wright & Allard, 2008).

Technology is seamlessly linking front and back office function to provide for the efficient and effective management of interactions across different customer touch-points (web sites, a physical store location, sales reports and mail campaigns) (Deck, 2001). Kotler and Keller (2006, p. 163) mentioned a customer purchase, a customer-requested service call, an online query, or a mail-in rebate card as touch points that allow a company to capture information about customers. Regular customer cards can serve as the basis for a customer data base, too. The card system presents a bonus system, by which the customer is rewarded with bonus points for being loyal to the company, or it provides the bearer several kinds of special offers meant only for the card holder (Boedeker, 1997).

Using the data achieved from these customer touch points, analysts can develop a complete view of each customer and pinpoint where additional services are needed.

Decision support systems, data mining tools, and other business analytic software help firms in analyzing business data to forecast market trends, to identify patterns, and to make better business decisions (Cha, Pingry & Thatcher, 2009). Companies may use these business analytic tools to segment customers more effectively and to engage in targeted marketing strategies which lead to more revenues. Business analytic tools generally fall into four categories (http://www.crm2day.com/content/t6_librarynews_1.php?id): statistical analysis, on-line analytical processing (OLAP), data mining, and text mining. Statistical analysis is used to process large amounts of data to uncover key facts, patterns and trends and it helps a retailer to segment and to classify customers. Online Analytical Processing (OLAP) enables easy data analysis and its presentation in different formats. Data mining provides predictions about the future and therefore it leads to an easy decision making. Text mining is searching for patterns and relationships within thousands of information sources.

Gilad and Gilad (1985) pointed out that there was no intelligence without data collection. However, managers are frequently facing the challenge of effective Customer Intelligence management. Namely, they have to consider how to reduce a growing amount of customer data into manageable amount of reliable and useful information for decision making (Smalltree, 2006). Few decades ago, companies needed time for the access to available data and the time for identifying and extracting the data. Also, they managed dozen of records for the same customer and had to re-enter customer name few times in order to have a complete picture about particular customer. Some of them relied mostly on demographic data such as age, gender, geographical location, household income, but this data didn't do much for customer retention efforts and might lead to the wrong conclusion. Nowadays, technology advancement increases the availability of inexpensive and massive data storage and data warehousing techniques. Moreover, new technology can give company directions for reduction and transformation of this large amount of data into manageable amount of valuable information. In other words, managers can look at one screen to see what is going on with their customers.

Open-standards-based technology called Data hubs provides a single, central place to hold accurate data (Joch, 2005). Data hubs let companies manage data quality from a central repository and also feed accurate information to operational systems. Companies can also enrich data by supplementing internal data with information from external sources. Companies must implement IT solutions that enable them to possess resources that work together in order to develop customer knowledge i.e. customer intelligence and to adapt their behaviour towards individual customers or prospects based on that intelligence. Resulting intelligence provides retailers with the accurate selection of right customers and is productively managing interactions with them as well.

On the example of "so-called preference data generated by means of conjoint analysis" and mining technique to media usage data Decker and Höppner (2006) explained how different the data and methods which define the basis of Customer Intelligence could be in the practice.

Electronic point of sale (EPOS) terminals collect a vast amount of data about products and customer buying patterns (Newman & Cullen, 2002, p. 381) and enable retailers to capture the information about their customers. Then the intelligence generation process attempts to convert data that has been collected into actionable intelligence (through analysis techniques, data mining and modelling methods).

Rubin and Leigh (2001) pointed out benefits that small business could obtain with the implementation of the database, such as: to communicate with the clients by service/product, to evaluate revenue of new client, to measure the effectiveness of marketing programs, to reduce client communication costs while increasing effectiveness, etc. The same authors discussed about two major kinds of database systems: 'flat-file' (database system that causes data to be structured as a two-dimensional table of rows and columns, like Excel) and 'relational' (database system that allows retailer to interlink two or more tables automatically, like Oracle)

Database technology is needed to store vast amounts of customer data and to derive and disseminate actionable intelligence to all members of the company who either have direct contact with customers or have an influence over the marketing mix elements of a company's operation (Campbell, 2003). Jones, Clarke-Hill, Hillier and Comfort (2005) argued that the use and RFID generated data (RFID is the generic name for technologies that use radio waves to identify and track objects automatically) to build segmentation models and analyse, and accurately predict customer sales patters will provide a much richer picture of customers than EPOS and scanning data systems. EPOS data (with thousands of transactions on a daily basis) are available quickly and provides accurate information about customer purchases, the sales of individual merchandise lines and other data the retailer wishes to capture, covering, for example, how the customer has paid for the item, a loyalty card reference, the time and date of transaction, and amount spent (Gilbert, 2003, p.346). In praxis, there are data analytics programs such SAS *Enterprise Miner* (http://www.sas.com/solutions/crm/customer-intelligence.html) or SPSS. The SAP Targeted Marketing for Retail package empowers retailers to learn more about customers by collecting information on loyalty-related behaviour. The IBM and Oracle Retail approach relies on a centralized source of information that synthesizes market, trend and customer demographic data on which to base merchandising decisions. This retail solution can help to improve the checkout process and leverage a deeper knowledge about the customer to enhance the in-store or online experience.

SHORTCOMINGS

The main goal of Customer Intelligence is to provide insight into customers needs, attitudes and behaviors towards particular company, but it has difficulties collecting consumer information, especially about their transactions with other businesses. It cannot collect detailed consumer information about their future needs and their past transactions with other vendors (Orman, 2007). O'Mahoney, Hurley, Kushmerick and Silvestre (2004) suggested alleviating the problem by surveying the consumers about their preferences, and by recording their past transactions. However, their recommendations have a number of limitations due to the fact that it does not suggest collection of information about consumers' intentions and tasks. An excellent example of failing to collect adequate information about consumers' needs and preferences is given by Orman (2007) in his case of Amazon's book recommendation system. Namely, this system is unable to distinguish among one-time gift purchases, professional purchases made on behalf of others, sporadic special-occasion purchases and personal purchases. The author (Orman, 2007) concluded that consequently, a gift purchase made for a child would lead to continuing recommendations for children's books irrespective of the consumer's current needs.

According to Stone et al. (2003), successful customer management needs a strong customer information infrastructure to support it. The quality of customer data has to be ensured (Marsh, 2005). Customer relationship management and Business Intelligence are failing because of poor quality of the underlying data (Joch, 2005). However, a company may face the problem of customer attitude towards their privacy and security. Some customers do not want to share their personal information with the company.

Another shortcoming is the result of too much focus on technology application and investing large amounts of money in software. Namely, complex IT solutions and the absence of skilled within the staff to manage information can lead to unexpected costs instead of adequate return on investments. Obviously, it is not enough to buy and to install a piece of software in order to obtain all useful information for strategic decision making. The truth is that there is no Customer Intelligence without the appropriate technological and procedural infrastructure to capture and store customer data. However, this infrastructure does not have to be complex. It can be spreadsheet software that is currently in widespread use such as Excel.

Possible shortcoming can appear in the case that customer feedback results are not gathered immediately after the service interaction. In such situation they are affected by several factors including the time allowed to answer the questions (in exit interviews and customer surveys), and the involvement the customer has with the product/service. Consequently, it may lead to inaccurate assumptions of how customers will behave when applying those results to the decision making and formulating the strategy (Monger & Keen, 2004). Therefore, a program to measure customer experience and satisfaction immediately after the interaction should be set up instead of using a delayed survey methodology.

RESEARCH METHOD

For the purpose of this chapter, a research study examining the adoption of IT-enabled Customer Intelligence, the technology which allows to storage and to maintain with information about customers, the impact of the Customer Intelligence implementation on the customer loyalty and on the level of customer retention as well, etc. on the sample of 50 Croatian retailers was conducted. The methods used in this study were face-to-face interview and an e-mail based questionnaire. The companies chosen were retail companies dealing with ICT equipment and services on the Croatian market and abroad. Similar to Coltman (2007) pre-survey telephone calls were made at each participant to identify whether they would be prepared to participate in the survey or whether they could provide contact details for the most appropriate person in the firm. The research was conducted in the period February – March 2010.

A total of 50 completed questionnaires were received, but one questionnaire was eliminated due to a large number of unanswered questions. The research questionnaire consisted of 28 questions divided in three segments: (1) general questions on company, (2) questions on data storage and warehousing technology, (3) questions regarding monitoring and managing customer satisfaction. The collected data were analyzed using SPSS. Except from descriptive statistics calculations, testing the reliability with Cronbach's Alpha coefficient was conducted. Before using items for further analysis, the reliability testing was

Figure 1. Reducing the rate of customer defection

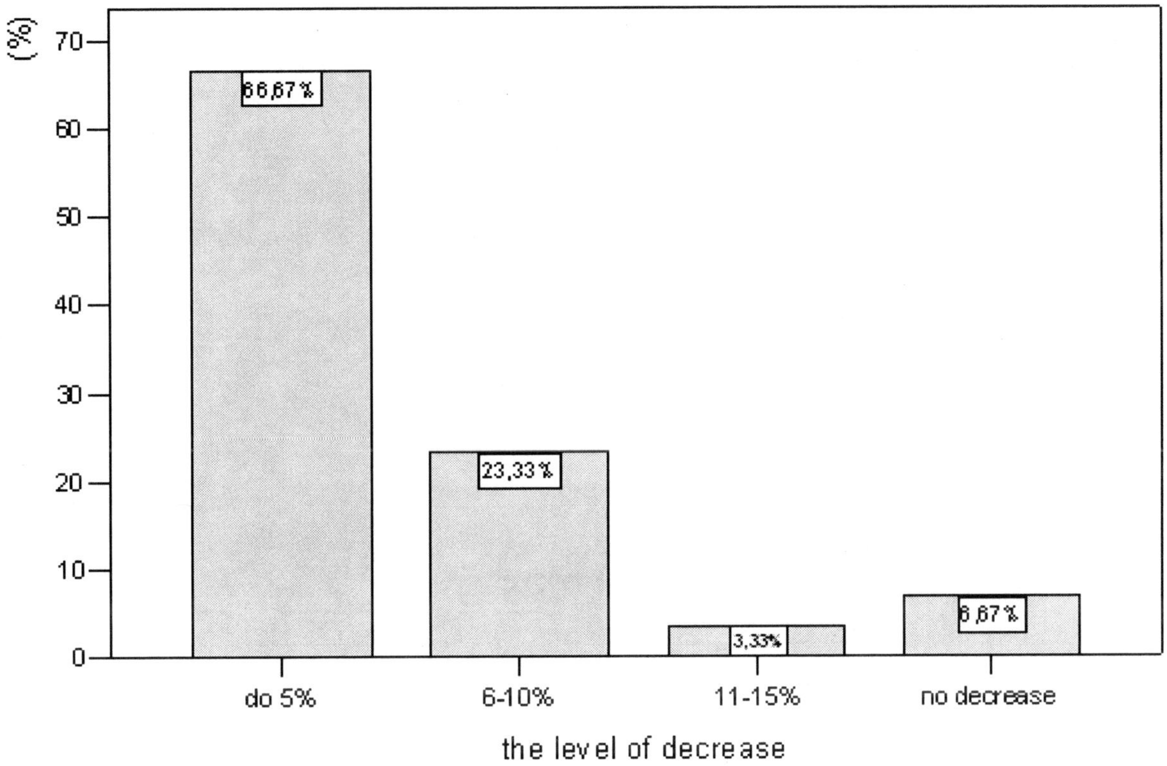

conducted. The value of 0.80 suggested very good internal consistency reliability for the scale used in this research. Since data were not normally distributed, a significance of the findings and the relationship between CI tools implementation was explored using Spearman correlation coefficient.

Additionally, an interview with one retailer who was willing to describe its successfully creating and use of Customer Intelligence throughout the organization in detail, was conducted.

The Findings of the Study among Retail Companies

The sample consists of 49 retail companies dealing with ICT equipment and services and the largest percentage of them operate in the Croatian market (61%). They are mostly large companies with 51 to 250 employees (75% of the sample). The results also show that reporting and analyzing of all business segments on daily basis is the most utilized IT solution in the sampled companies (93% of respondents use this IT solution). Customer relationship management (CRM) is on the second place (used by 79% of respondents), followed by on-line analytical process (OLAP) (75% of respondents use it), data warehouse, and data mining techniques (used by 60% and 48% respectively).

The main purpose of the research was to find out whether Customer Intelligence enabled companies to succeed in reducing the rate of customer defection. Figure 1 shows that due to CI implementation, the largest percentage of companies (66.67%) reduced the rate of customer defection by 5%. It should be pointed out that no matter how effective they use CI, neither company from the sample succeeded in reducing the rate of cus-

Figure 2. Attracting new customers

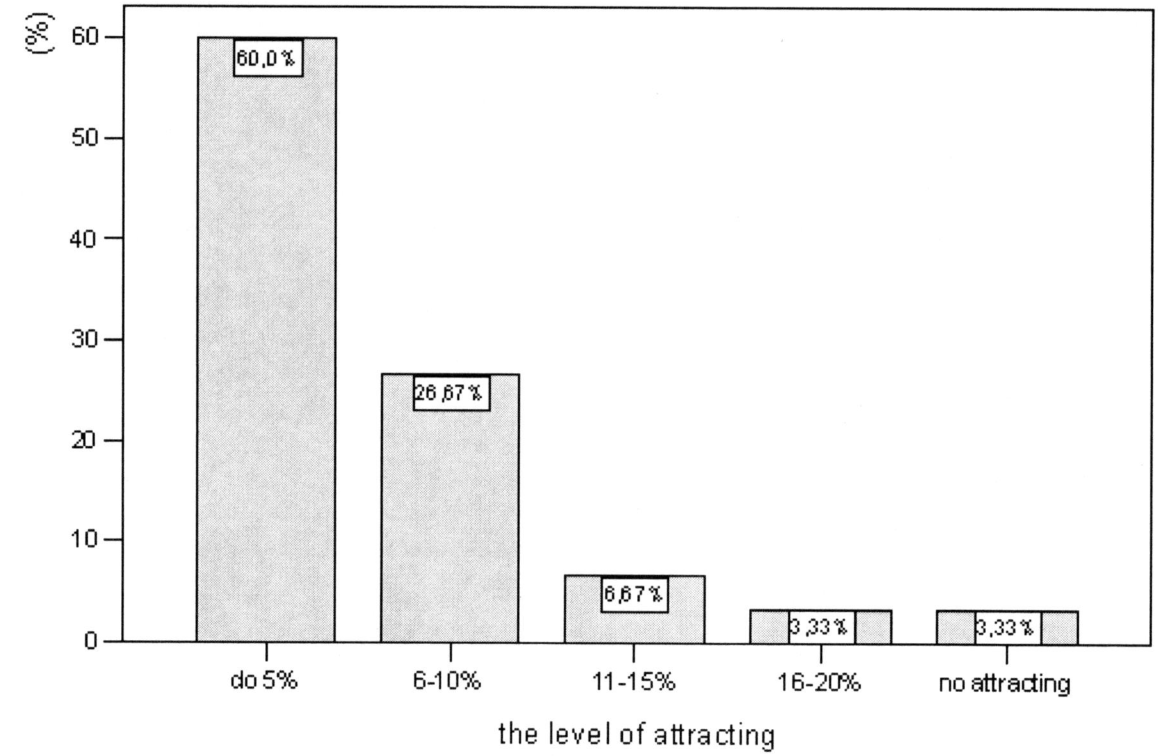

tomer defection more than 15% and that 6.67% of the sample did not reduce the rate of customer defection, at all.

Additionally, respondents were asked whether continued monitoring, gathering and analysing information regarding current customers had a positive impact on attracting new customers. As Figure 2 suggests, the largest percentage of companies (60%) extended their market share by 5%. The smallest percentage of companies did not succeed in attracting new customers.

Since respondents reported successful use of Customer Intelligence, the effect of its implementation on the revenue generated by customer on average was investigated. More than half of the sample reported an increase in revenues generated by customer due to Customer Intelligence by 5%. However, respondents suggested that they did not implement Customer Intelligence for the purpose of revenue increase but for the purpose of customer retention.

A noteworthy result of the research is the significant importance to keep a record on customers' past purchases of ICT equipment and services for all companies in the sample.

In order to find out how respondents measure customer satisfaction, the research instrument asked respondents to rate following ways of monitoring customer satisfaction: a) through direct contacts between salesperson and a customer on a daily basis, b) periodically while offering services, and c) by market research agency. On the scale from 1 to 5, where 1 indicates the lowest grade, respondents highly evaluated daily direct contacts between salesperson and their customers (average score = 4.20) while, surveys conducted by market research agencies obtained average score of 2.00.

The study also suggests that the largest percentage of companies used customer data to set directions for new products and services. Namely, 76.7% of respondents reported that customers' comments, complaints and questions helped them to improve their products and services and to create new ones.

It should be pointed out that correlation analysis found out strong association between data warehousing and the creating of new products (r=0,680**, p=0,000). Namely, companies that continuously monitor their customers, build and maintain a customer database, are more successful in creating new products and services. There is also strong association between building key customers database and products and services database improvements (r=0,824**, p=0,000). In other words, companies that have customers database usually make improvements in their market offer.

An Example of Successful Customer Intelligence in the Croatian Retailing Company

Investigated company is the retailer who is highly specialized in selling ICT equipment and services on the Croatian market and abroad. For confidentiality issues the name of the retailer is not mentioned. The company has got the central transaction system which represents the company's basic data source and covers all of the customer data. Company started with monitoring its customers and collecting basic data about them from the beginning of its business. At first, those data includes personal demographic data (with characteristics such as age, household and personal income, credit card debt, home ownership and net worth, education level, etc.) and geographic data that covers information about the country of origin. This data was then supplemented with transactional data about customers' purchase patterns and services and IT solutions that customers purchased.

As a consequence of aggressive strategy of a regional competitor to extend its business, analyzed company decided to find a way to strengthen current customer loyalty, to attract new customers and to keep them. The company was completely aware that the entry of new competitor could cause some customer defection. Moreover, the company was expecting competitor's larger investments which might result in very low prices and additional benefits for customers. Based on the previously conducted analysis, the company segmented market and conducted the investigation on typical profiles and behaviour patterns of each particular customer segment. It was important to identify loyal customers and who were the most "valuable" customer for the company. Namely, the company needed to figure out how much it would cost to satisfy and to retain current "active" customers and to be sure that costs were allocated to those customers.

Accordingly, the company developed monitoring system enabled to monitor the level of satisfaction of its customers. The system paid special attention to new customers' needs and wants. Since attracting new customers is far more expensive than retaining existing ones (Donio, Massari, & Passiante, 2006; Özgener & Iraz, 2006; Phan & Vogel, 2010) the system had the task to evaluate the profile of the typical customer who was non-loyal and complaining. This task of the system had to be realized before the entry of new competitor in the region. Conclusively, this monitoring system was important strategic tool for the company because it could serve as the support system for top management in making decisions during the period of aggressive competition battle.

Company's monitoring system consisted of two basic components. The first one was designed to loyal customers and it included customer rating model. This rating model helped the retailer to evaluate customers and to have a good insight in the distribution of customers' profitability. The second component of the monitoring system had intention to identify and to analyze the causes of customer defection. The company was considering to reactivate dissatisfied customers, because it is often easier to re-attract non-loyal customers (because the company has already collected their data) than to find new ones (Kotler & Keller, 2006, p. 157) Therefore, the company conducted an e-mail based survey. The results of the survey gave answers to the following questions: what are the main characteristics of the non-loyal customer segments, what are the main motivators for non-loyal customer segments to switch to another retailer, what are the main motivators for non-loyal customer segments to re-activate their purchases and to increase business with the company.

Then the retailer's expert team had to select IT solutions. During the process of brainstorming, team members (sales representatives, marketing managers, and the chief planner of company's information system) had to identify key indicators for customer rating. Key indicators should relate to:

a. customer profitability,
b. customer loyalty, and
c. customer future prospects.

Customer profitability indicators included: last six months purchases, costs of promotional campaigns in the last six months, costs of entertainment and gifts given to customers. A history-oriented customer profitability analysis was made at similar levels as those described by Donio, Massari and Pasiante (2006): 1) gross contribution margin was calculated based on "sales revenue less all product-related expenses for all products sold to an individual customer during a period of time"; 2) all expenses traceable to the individual customer were subtracted, and the result was the operating profit generated by the customer. Customer loyalty indicators gave information about the frequency of customer purchasing, customers' reactions on promotional campaigns, and points of the customer loyalty card. Customer future prospects related to the trends in purchasing in the last two quarters and the preference to buy new products and services during the year. Loyalty card scheme of the company provided majority of necessary information. Inserted in existing CI infrastructure, key indicators created an integrated platform for identifying those offers that were most likely to be attractive to specific group of customers. This solution had the capability to personalise needs of individual customer by differentiating equipment for each unique customer and the ability to identify key factors that influence customer's satisfaction. In such a way, it suggested an appropriate allocation of company's resources to sustain customers' loyalty and to increase the effectiveness of promotional campaigns.

FUTURE RESEARCH DIRECTIONS

The topic is focused on very promising area for future research, because Customer Intelligence is interdisciplinary and has got touch points with customer relationship management, business intelligence, data warehouse, customer loyalty and satisfaction, retail competitive strategy, marketing, IT solutions, and others. Moreover, with growing competition from both traditional and online business (Phan & Vogel, 2010), keeping customers satisfied, increasing potential sales, improving and exploiting customer relationships, and using customer data in order to maintain customer loyalty become strategically important to whole business sector. Therefore, further investigation of how collecting and monitoring customer data affect the customer loyalty and the level of customer retention is necessary. Additionally, to help advance a cohesive body of knowledge on this topic of growing interest and importance, future efforts should attempt to develop deeper conceptualization and description of Customer Intelligence. Future research should capture Customer Intelligence management process and present findings of some case studies in that domain.

CONCLUSION

"Know your customers and give them what they want" (http://www.crm2day.com/content/t6_librarynews_1.php?id) is the best explanation of today's market situation. More than ever, retailers are aware that customers are their most important value. The more the "voice" of the target customer is brought into the company and acted upon, the better the positional advantage and subsequent performance of the business are (Day, 1994). Greater knowledge about what customers want should lead to more effective marketing targeting, product development and positioning (Hunt, & Morgan, 1995). Additionally, rapid technology development enables retailers to communicate with their customers and to interact with them across different customer touch-points. Customers are giving information to retailers about themselves all the time. They are revealing some facts about themselves through every response, transaction, Web site hit. The retailer's challenge is to capture these data, understand it and manage it in order to make decision. Customer Intelligence consists of customer data collected from various relationships between company and its current and potential customer. Its main goal is to produce information that can be used in the strategic positioning of the company. Armed with the information about its segmented market, retailer is able to create typical profiles and behaviour patterns of each particular customer segment. Retailer can identify loyal customers and the most "valuable" customers for the company. Namely, Customer Intelligence can help retailers to figure out how much it would cost to satisfy and to retain current "active" customers and to be sure that costs were allocated to those customers. The results of the study conducted among Croatian retail companies dealing with ICT equipment and services confirmed that Customer Intelligence can reduce the rate of customer defection and succeed in attracting new customers. Finally, it should be pointed out that Customer Intelligence can fail as the consequence of too much focus on complex and expensive technology application and because of poor quality of the underlying data.

REFERENCES

Alavi, M., & Leidner, D. E. (2001). Knowledge management and knowledge management systems: Conceptual foundations and research issues. *Management Information Systems Quarterly*, *25*(1), 107–136. doi:10.2307/3250961

Apte, C., Liu, B., Pednault, E., & Smyth, P. (2005). Business applications of data mining. *Communications of the ACM*, *45*(8), 49–53.

Boedeker, M. (1997). Relationship marketing and regular customer cards: Daily product retailing in Finland. *Marketing Intelligence & Planning*, *15*(6), 249–257. doi:10.1108/02634509710184820

Brohman, M. K., Watson, R. T., Piccoli, G., & Parasuraman, A. (2003). Data completeness: A key to effective Net-based customer service systems. *Communications of the ACM*, *46*(6), 47–51. doi:10.1145/777313.777339

Campbell, A. (2003). Creating customer knowledge competence: Managing customer relationship management programs strategically. *Industrial Marketing Management*, *32*, 375–383. doi:10.1016/S0019-8501(03)00011-7

Cha, H. S., Pingry, D. E., & Thatcher, M. E. (2009). What determines IT spending priorities? *Communications of the ACM*, *52*(8), 105–110. doi:10.1145/1536616.1536644

Coltman, T. (2007). Why build a customer relationship management capability? *The Journal of Strategic Information Systems*, *16*, 301–320. doi:10.1016/j.jsis.2007.05.001

Coretto, T. (2010a). *Customer intelligence = profits*. Retrieved June 15, 2010, from http://nyreport.com/articles/73044/customer_intelligence_profits

Coretto, T. (2010b). *7 characteristics of successful customer intelligence*. Retrieved June 15, 2010, from http://blog.pntmarketingservices.com

Coretto, T. (2010c). Six things customer intelligence is not. Retrieved June 15, 2010, from http://www.nyreport.com/articles/74949/six_things_customer_intelligence_is_not?page=0,1

Coretto, T. (2010d). *Does customer intelligence include prospecting?* Retrieved June 15, 2010, from http://altairdata.wordpress.com/2010/06/08/does-customer-intelligence-include-prospecting/

Davenport, T. H., Harris, J. G., & Kohli, A. K. (2001). How do you know your customers so well? *Sloan Management Review*, *42*, 63–73.

Day, G. S. (1994). The capabilities of market-driven organizations. *Journal of Marketing*, *58*(4), 37–52. doi:10.2307/1251915

Deck, S. (2001). *What is CRM?* Retrieved June 15, 2010, from www.cio.com/research/crm/edit/crmabc.html

Decker, R., & Höppner, M. (2006). Strategic planning and customer intelligence in academic libraries. [from http://www.emeraldinsight.com/Insight/ViewContentServlet?Filename=Publisher]. *Library Hi Tech*, *24*(4), 504–514. Retrieved June 15, 2010. doi:10.1108/07378830610715374

Donio, J., Massari, P., & Passiante, G. (2006). Customer satisfaction and loyalty in a digital environment: An empirical test. *Journal of Consumer Marketing*, *23*(7), 445–457. doi:10.1108/07363760610712993

Fleisher, C. S., Wright, S., & Allard, H. T. (2008). The role of insight teams in integrating diverse marketing information management techniques. *European Journal of Marketing*, *42*(7/8), 836–851. doi:10.1108/03090560810877187

Gilad, B., & Gilad, T. (1985). A system approach to business intelligence. *Business Horizons*, *8*(5), 65–70. doi:10.1016/0007-6813(85)90070-9

Gilbert, D. (2003). *Retail marketing management*. England: Pearson Education Limited.

Hawkins, G. E. (2003). *Customer intelligence*. New York, NY: Breezy Heights Publishing.

Hirschowitz, A. (2001). Closing the CRM loop: The 21st century marketer's challenge: Transforming customer insight into customer value. *Journal of Database Marketing*, *10*(2), 168–178.

Hunt, S. D., & Morgan, R. M. (1995). The comparative advantage theory of competition. *Journal of Marketing*, *59*(2), 1–15. doi:10.2307/1252069

Joch, A. (2005). *Capturing customer intelligence*. Retrieved June 15, 2010, from http://www.oracle.com/oramag/profit/05-may/p25feature.html

Jones, P., Clarke-Hill, C., Hillier, D., & Comfort, D. (2005). The benefits, challenges and impacts of radio frequency identification technology (RFID) for retailers in the UK. *Marketing Intelligence & Planning*, *23*(4), 395–402. doi:10.1108/02634500510603492

Kaplan, R. S., & Norton, D. P. (2001). *The strategy-focused organization: How balanced scorecard companies thrive in the new business environment*. Boston, MA: Harvard Business Review.

Karakostas, B., Kardaras, D., & Papthanassiou, E. (2005). The state of CRM adoption by the financial services in the UK: An empirical investigation. *Information & Management*, *42*(4), 853–863. doi:10.1016/j.im.2004.08.006

Kotler, P., & Keller, K. L. (2006). *Marketing management* (12th ed.). Upper Saddle River, NJ: Pearson Education, Inc.

Lonnqvist, A., & Pirttimaki, V. (2006). The measurement of business intelligence. *Information Systems Management*, (Winter): 32–40. Retrieved from www.ism-journal.com. doi:10.1201/1078.10580530/45769.23.1.20061201/91770.4

Market Research Society. (2007). *Qualifications for market researchers*. Retrieved June 15, 2010, from www.mrs.org.uk/qualifications/index.htm

Marsh, R. (2005). Drowning in dirty data? It's time to sink or swim: A four stage methodology for total data quality management. *Journal of Database Marketing and Customer Strategy Management*, *12*(2), 105–112. doi:10.1057/palgrave.dbm.3240247

Monger, J., & Keen, C. (2004). The research proves it. We cannot wait to measure the customer experience. *DM Review Magazine*, June 2004. Retrieved June 15, 2010, from http://www.crm2day.com/content/t6_librarynews_1.php?id=

Mulhern, F. J. (1999). Customer profitability analysis: Measurement, concentration, and research directions. *Journal of Interactive Marketing*, *13*(1), 25–40. doi:10.1002/(SICI)1520-6653(199924)13:1<25::AID-DIR3>3.0.CO;2-L

Myron, D. (2009). Social media spawns a new era in customer intelligence. *CRM Magazine*, *13*(6), 6–6.

Newman, A. J., & Cullen, P. (2002). *Retailing: Environment & operations*. London, UK: Thomson Learning.

Novo, J. (2001). *Drilling down: Turning customer data into profits with a spreadsheet.* Booklocker.com

O'Mahoney, M., Hurley, N., Kushmerick, N., & Silvestre, G. (2004). Collaborative recommendation: A robustness analysis. *ACM Transactions on Internet Technology*, *4*(4), 344–377.

Orman, L. V. (2007). Consumer support systems. *Communications of the ACM*, *50*(4), 49–54. doi:10.1145/1232743.1232747

Özgener, S., & Iraz, R. (2006). Customer relationship management in small-medium enterprises: The case of Turkish tourism industry. *Tourism Management*, *27*, 1356–1363. doi:10.1016/j.tourman.2005.06.011

Panian, Ž., & Ćurko, K. (2010). *Poslovni informacijski sustavi*. Zagreb, Croatia: Element.

Phan, D. D., & Vogel, D. R. (2010). A model of customer relationship management and business intelligence systems for catalogue and online retailers. [from www.elsevier.com]. *Information & Management*, *47*, 69–77. Retrieved June 15, 2010. doi:10.1016/j.im.2009.09.001

Porter, M. E. (1980). *Competitive strategy: Techniques for analyzing industries and competitors*. New York, NY: The Free Press.

Qiu, L., Li, Y., & Wu, X. (2008). Protecting business intelligence and customer privacy while outsourcing data mining tasks. [from www.springerlink.com]. *Knowledge and Information Systems*, *17*, 99–120. Retrieved June 15, 2010. doi:10.1007/s10115-007-0113-3

Rubin, R., & Leigh, W. E. (2001). Mining customer intelligence: A practical framework for the development of a database marketing system for the small business. *Proceedings of the Second Annual USASBE/SBIDA Joint National Conference Proceedings* 2001: An Entrepreneurial Odyssey, February 7-10, 2001, Orlando, Florida. Retrieved June 15, 2010, from https://usasbe.org/knowledge/proceedings/proceedingsDocs/USASBE2001proceedings-071.pdf

Ryals, L. (2003). Making customers pay: Measuring and managing customer risk and returns. *Journal of Strategic Marketing*, *11*, 165–175. doi:10.1080/0965254032000133476

Seeman, E. D., & O'Hara, M. (2006). Customer relationship management in higher education: using information systems to improve the student-school relationship. *Campus-Wide Information Systems*, *23*(1), 24–34. doi:10.1108/10650740610639714

Smalltree, H. (2006). *The best practices in customer intelligence*. Retrieved June 15, 2010, from www.SearchDataManagement.com

Speier, C., & Venkatesh, V. (2002). The hidden minefields in the adoption of sales force automation technologies. *Journal of Marketing*, *66*(3), 98–111. doi:10.1509/jmkg.66.3.98.18510

Stone, M., Fos, B., & Henderson, I. (2003). The quality of customer information management in customer life cycle management. *Journal of Database Marketing and Customer Strategy Management*, *10*(3), 240–254. doi:10.1057/palgrave.jdm.3240113

Trim, P. R. J., & Lee, Y.-I. (2006). The role of marketing intelligence officers in strategy formulation and implementation. In *Handbook of business strategy*, (pp. 125-130).

Van Raaij, E. M. (2005). The strategic value of customer profitability analysis. *Marketing Intelligence & Planning*, *23*(4), 372–381. doi:10.1108/02634500510603474

Yang, Y.-F. (2008). The roles of human resources, information technology, and marketing knowledge capabilities in performance: An extension of the resource-based theory perspective. *Social Behavior and Personality*, *36*(9), 1269–1282. doi:10.2224/sbp.2008.36.9.1269

Zablah, A. R., Bellenger, D. N., & Johnston, W. J. (2004). Anticipating consumer behavior with analytics. *Industrial Marketing Management*, *33*, 475-489. Retrieved June 15, 2010, from http://www.crm2day.com/content/t6_librarynews_1.php?id

ADDITIONAL READING

Hawkins, G. E. (2003). *Customer Intelligence*. New York: Breezy Heights Publishing.

http://nyreport.com/articles, by Tony Correto

Kelly, S. (2006). *Customer Intelligence: From Data to Dialogue*. Chichester, England: John Wiley & Sons Ltd.

KEY TERMS AND DEFINITIONS

Customer Data: All available information about particular customer (age, gender, education level, number of children, etc.) enriched with information about customer's purchases, customer cards, and other transactions with the company.

Customer Defection: The decrease of customer loyalty caused by some wrong strategic decisions of the company.

Customer Intelligence: The process of collecting customer data during company's interactions with customers, their storing and analyzing.

Customer Profitability: An indicator of the business success of the company; the ability of the company to earn the profit based on increasing purchases and the relationship with customers in comparison to its costs.

Customer Relationship Management: All activities conducted by a company in an effort to keep its current customers and to attract the new ones.

Knowledge: A set of data, information and skills collected through the lifetime.

Retail Strategy: All activities conducted by a retailer in order to increase its market share and to increase customers' loyalty as well.

Chapter 10
Give to Get:
An Experimental Study to Explore Information Giving in New Technology-Based Retail

Katia Premazzi
Bocconi University and SDA Bocconi, Italy

Monica Grosso
Bocconi University and SDA Bocconi, Italy

Sandro Castaldo
Bocconi University and SDA Bocconi, Italy

ABSTRACT

The development of new technologies opens many opportunities for retailers. Nevertheless, the adoption of many such technologies requires retailers to address customers' privacy concerns to encourage them to share information with the firm. Indeed, a large amount of customer information is required before a retailer can exploit the opportunities that new technologies offer. This chapter seeks to help retailers reach this goal by investigating the effects that two variables, which have emerged as relevant in the literature, have on information sharing: trust and compensation (as a form of incentive). The results of two experimental studies focusing on the online setting show the key role that trust plays in increased online information sharing. These results are a starting point for research on the privacy-related issues of new technologies, as the online setting could be perceived as very risky in terms of the invasion of privacy.

INTRODUCTION

The development of information and communication technologies (ICT) has spurned several suggestions for and practical attempts to apply them in a retail setting, both offline (in physical stores) and online. Several goals can be pursued; these afford the possibility to simultaneously increase retail efficiency and effectiveness.

Reducing costs throughout the entire supply chain, as well as at the store level can be achieved

DOI: 10.4018/978-1-60960-738-8.ch010

by optimizing processes. These processes could include rolling out radio frequency identification (RFID) in the logistics and warehouse management systems. RFID uses electronic tags to store data on products (Roussos G Kostakos, 2009; Pantano and Naccarano, 2010). The tag – also known as an electronic label, transponder, or code plate – is made up of a chip attached to an antenna. Like bar codes, RFID tags identify items. However, unlike bar codes, which must be in close proximity of and in the scanner's line of site to be read, RFID tags do not require line of sight and can be embedded within packages. Depending on the type of tag and application, RFID tags can be read at various distances. In addition, the reading of RFID-tagged cartons rolling on a conveyer belt is much swifter than the reading of bar-coded boxes, thus increasing efficiency.

Other goals, related to enhancing the customer shopping experience could be establishing an ongoing dialogue with customers and offering richer and customized proposals to make shopping more informed, convenient, pleasant, and exciting. Examples of such proposals are: user-friendly interfaces that provide detailed, updated, and targeted information on the product assortment and promotions, as well as personal digital assistants.

One of the most advanced and extended experimental applications of the emerging technologies in retail is the well-known Metro Group future store initiative. This initiative involves cooperation with consumer goods manufacturers, IT specialists, and service providers. Among other in-store innovations, the personal shopping assistant (PSA[1]) was tested early on, followed more recently by the mobile shopping assistant (MSA). Thanks to a software platform, the MSA (Pantano and Naccarano, 2010) offers customized services on mobile phones, which have become customers' permanent companions. A shopping list can be created prior to or during the store visit; it can be updated by simply scanning the product bar code on the item with an autofocus camera on the mobile phone. Detailed information about a product and its price can then be accessed at any time. Automatic item scanning and help with finding the store products are also available. Customers can also pay by wireless. Additional services are being developed: for example, a coaching program that helps one adhere to a specific diet.

Experimental applications can also be found in the non-grocery sector. In the fashion industry, for instance, many possibilities are available. These take the form of a personalizing interaction with individual customers when they enter the store (e.g., with a personalized welcome); the prompting of cross-selling by suggesting complements to a primary purchase (e.g., RFID loyalty cards provide sales personnel with customer insights); entertaining and informing customers about the brand philosophy and the in-store product selection (e.g., through large-screen walls or electronic communication); and enhancing the store visit and post-shopping experience by exploiting emotional and social aspects (besides functional ones). Additional examples are digital mirrors that allow customers to view themselves wearing a certain garment or accessory and to compare different total looks without having to actually try on the garments; sending pictures and messages through MMS to relatives and friends asking their opinion about a potential purchase; fingerprint technology that allows automated bill generation for hassle-free payment; and the creation of exclusive communities.

All these innovations do, however, require significant investments. Before deciding if and how to experiment with advanced technology applications, retailers can analyze the technical, economic, and cultural feasibility. The latter is important as many of the proposed or experimental solutions may trigger customers or prospects' privacy concerns. Indeed, to benefit from most of these solutions, customers must first register and disclose personal information, or consent to having their behavior tracked.

This chapter seeks to help retailers understand how to enhance information disclosure from their customers. In particular, we will report on two experimental studies[2] aimed at helping online retailers overcome customers' privacy concern and increase personal information disclosure. This leverages two variables that have emerged in the literature: trust and incentives.[3] These studies' results, although focused on a specific technology-based retail environment, are a starting point for investigating privacy-related issues in the new technology adoption domain and in other retail settings. In some of these settings, however, the development of a similar empirical study would be extremely difficult as it would require researchers to have advanced technology interfaces in order to conduct the experiments.

THEORETICAL BACKGROUND

The main contributions on privacy-related issues linked to new technology adoption focus on the online setting, which developed most during the past decades' technological revolution. It is easier to grasp the relevance of these studies if we consider that an online channel can be, and often is, added to stores to complete and complement the bricks-and-mortar offering (Steinfield, 2004). With this in mind, a growing number of traditional retailers now pursue multi-channel strategies (Müller-Lankenau et al., 2005). Electronic channels are extremely flexible. They allow a firm to optimize its marketing information mix, to automatically suggest complementary products, and to implement relationship-friendly tools such as product comparison aids (Viswanathan, 2005). On the one hand, this implies that customers benefit from multi-channel retail which offers new services and the greater convenience of interacting with it both online and offline (Otto and Chung, 2000) as required. On the other hand, retailers can leverage their physical infrastructure, achieve synergies, and increase their profits (Steinfield, 2004). Furthermore, multi-channel shoppers have a higher purchase volume and are more profitable (Venkatesan et al., 2007).

Secondly, e-tailing is the most complex of the retail sectors in terms of privacy concerns. Customers are required to provide their data to an unknown entity, usually located far from them. Within physical stores, the demand for personal data can be perceived as less risky as customers feel they have greater control of their data usage due to the collecting entity's proximity.

In the online setting, the development of ICT has provided firms with the opportunity to acquire a huge quantity and variety of data from online visitors and shoppers (Cespedes and Smith, 1993; DeCew, 1997). The ease with which data can be acquired and disseminated across the Internet, as well as the electronic setting's peculiarities have led to growing concerns about consumers safeguarding their privacy (e.g., Culnan, 1993; Milne and Gordon, 1993; Milne, 2000; Phelps et al., 1999). The spatial and temporal separation between customers and e-vendors, and the information asymmetry between them (Hee-Woomg et al., 2004), lead to customers not knowing what online firms do with their personal information (lack of control). They are therefore less willing to provide such information to foreign or unknown online retailers.

The literature on privacy concerns and information provision in online settings is useful to gain insights on potential levers to act on. We focus our attention on trust and compensation.

This chapter seeks to analyze the role of trust and compensation in customers' willingness to divulge sensitive information as well as their actual disclosure behavior.

The main contribution of our studies is the use of a controlled experimental setting that allows respondents' actual behavior, attitudes, perceived intentions, and/or past behavior to be measured. Referring to intentions and past behavior (as most survey and experimental studies do) may be insufficient, as they might be not a good proxy for actual customer behaviors. Some authors

(Norberg et al., 2007) have pointed out that there might even be a difference between declared and actual information provision (called the privacy paradox: when actual provision is higher than declared provision).

TRUST AND COMPENSATION: LITERATURE REVIEW AND HYPOTHESIS DEVELOPMENT

In this section, we will review the studies on trust and incentives (e.g., compensation) to develop our hypotheses. Table 1 provides an overview of some relevant and representative studies.

Trust in the Relationship

It is widely agreed that trust is critical in exchanges involving interdependence, uncertainty, and risk (e.g., Milne and Boza, 1999) as, for example, in online relationships and information exchanges. Trust can be a shortcut and act as a mechanism to reduce the complexity of human conduct when people have to deal with uncertainty (Luhmann, 1989). In similar situations, trust can be defined

Table 1. Review of some examples of the main relevant studies on consumers' online privacy concerns, trust, compensation, and information provision

Authors Journal and Year Title	Goals/Object	Methodology	Sample	Main results
Milne, Boza Journal of Interactive Marketing, 1999 Trust and Concern in Consumers' Perceptions of Marketing Information Management Practices	Examines. consumers' trust and concerns (as well as their relationships) regarding information management practices (direct marketing).	US consumer survey in 1997. Three separate studies: 1) to examine the relationship between trust and concern regarding 17 industries and their information-handling practices; 2) to examine the antecedents (perceived control, knowledge, and attitude towards relationship marketing) and consequences of trust, and concern regarding direct marketing and non-traditional direct marketing industries; 3) qualitative data suggests that what consumers feel leads them to trust an organization with their personal information. Multiple regression analysis.	1,508 direct marketing consumers.	Improving trust and reducing concerns are two distinct approaches to managing consumer information. Consumers' perceptions of trust and concern levels vary per industry, and both explain consumers' attitudes to information practices and purchase behavior. On the basis of qualitative insights, experience, reputation, contractual obligations, and regulation are the broad reasons for trusting an organization with their personal information.
Sheehan, Hoy Journal of Advertising, 1999 Flaming, complaining, abstaining: How Online Users Respond to Privacy Concerns	Consumer online privacy concerns and reactions.	E-mail survey. Assessment of respondents' concerns about a series of situations (scenarios describing online marketing practices) which affect privacy online. Correlation of the overall level of concern with the frequency with which respondents adopted seven different online behaviors to safeguard their personal privacy online.	US sample of individuals with personal e-mail account (randomly generated - probability sampling method). Survey sent to 3724 individuals, 889 completed surveys.	The frequency with which five of the seven investigated online behaviors (provide incomplete information, notify Internet Service Providers about unsolicited mails, request removal from mailing lists, send a "flame" to online entities sending unsolicited e-mail) were adopted increased as the respondents' privacy concern increased.

continued on following page

Table 1. Continued

Authors Journal and Year Title	Goals/Object	Methodology	Sample	Main results
Sheehan, Hoy Journal of Public Policy and Marketing, 2000 Dimensions of Privacy Concerns among Online Users	Examines the extent to which knowledge of privacy concerns in traditional direct marketing applies to the online context. The underlying factors of consumer online privacy concerns. Checks whether the five core principles developed by the FTC to guide online development of privacy policy in order to increase online commerce (notice, choice, access, security, redress) address consumers' privacy concerns.	E-mail survey. Assessment of respondents' concerns about a series of situations (scenarios describing online marketing practices) which affect privacy online. The scenarios represented three different levels of a predicted privacy concern (low, moderate, and high) for each of the five dimensions (awareness of data collection, information use, information sensitivity, familiarity with entity, and compensation for information provision).	Same as Sheehan and Hoy, 1999	Online privacy concerns are somewhat more complex than concerns in the traditional marketing communications context. Three factors that influence privacy concern are identified: 1) Control over collection and usage of information; 2) short-term, transactional relationship; 3) established, long-term relationship. People's privacy concerns are influenced by what information is collected in which manner by which entity to be used for what purpose. Many of the study's findings mesh well with the FTC(1998)'s five core principles of fair information practice online. Dimensions not directly incorporated in these five principles, such as relationships and exchange, may influence consumers' privacy concern, too.
Phelps, Nowak, Ferrell Journal of Public Policy and Marketing, 2000 Privacy Concerns and Consumer Willingness to Provide Personal Information	Examines the potential relationships between categories of personal information (demographic and lifestyle information vs. financial, purchase-related, and personal identifier information), beliefs about direct marketing, situational characteristics, specific privacy concerns, and consumers' direct marketing shopping habits. Assesses the trade-offs consumers are willing to make when they exchange personal information for shopping benefits. Aims to assist public policymakers and marketers identify specific information practices and situations that foster consumer privacy concerns.	US cross-sectional mail survey in 1995. Full-profile conjoint scenarios. Input factors hypothesized to affect consumers' overall concerns about the ways companies use personal information and consumer beliefs regarding marketers' information practices: 1) type of personal information requested; 2) amount of information control offered in exchange; 3) potential consequences and benefits; 4) consumer characteristics. The outcomes of the overall concerns and beliefs regarding marketers' information practices are hypothesized to influence consumers' future behavioral and attitudinal responses. Multiple regression analysis.	Two sampling frames: known and recent catalog shoppers and a fairly extensive database containing more than 100 million residential addresses. Surveys were then sent to a subsample of 500 randomly selected names from each sampling frame. A total of 556 completed and usable surveys were received.	Strong relationship between respondents' level of concern about the ways companies use personal information and respondents' information-related beliefs and behaviors. Six factors are found to be important correlates of privacy concern: the type of personal information requested, the ability and desire to control subsequent dissemination of personal information, consumers' perceptions regarding marketers' knowledge about them and their interests, consumers' attitude towards direct mail, consumers' preferences with respect to catalog and advertising mail volume, and previous name removal request behavior.

continued on following page

Table 1. Continued

Authors Journal and Year Title	Goals/Object	Methodology	Sample	Main results
Phelps, D'Souza, Nowak Journal of Interactive Marketing, 2001 Antecedents and Consequences of Consumer Privacy Concerns: An Empirical Investigation	Examines the interrelationships between the potential antecedents and consequences of privacy concerns. Assesses whether and how privacy concerns relate to the purchase decision process and purchase behavior. Aims to guide the development of effective policies and practices to reduce privacy concern.	US cross-sectional mail survey. Series of full-profile conjoint scenarios in the context of shopping for clothing as stimuli. Investigates whether privacy concerns and/or related variables (i.e. willingness to provide information, the amount of control desired, purchase history, and attitude toward direct marketing) play an indirect role in the purchase decision-making process. Potential antecedents of privacy concern: the amount of information control desired (+) and consumers' attitude towards direct marketing (-). Privacy concern is hypothesized to be negatively related to purchase behavior. Simultaneous equation regression model.	Same as Phelps et al. (2000)	Among other things, a consumer's attitude towards direct marketing and his/her desire for information control act as antecedents to privacy concerns. Privacy concern are negatively related to purchase behavior and the purchase decision process.
Andrade, Kaltcheva, Weitz Advances in Consumer Research, 2002 Self-Disclosure on the Web: The Impact of Privacy Policy, Reward and Company Reputation	Explores three approaches to encourage self-disclosure of personal information. Examines the impact of: the privacy policy's comprehensiveness (indicating how the disclosed information will be used), the company's reputation for trustworthiness, as well as the offer of a reward for disclosing information on consumer concerns about disclosing personal information. Hypothesizes a negative relationship between these variables, and that the nature of the information affects concerns about self-disclosure.	Exploratory experimental study. 2 (company reputation: high vs. low) x 2 (privacy policy: extensive vs. brief) x 2 (an offer of a reward: present vs. lacking) between-subjects design. Paper & pencil, mentioning existing retailers' (pretested for high vs low reputation) websites.	114 undergraduate students from a large Southeastern US university.	A privacy policy's comprehensiveness and the company's reputation reduce the level of concern about self-disclosure, while the offer of a reward heightens concern. The type of information which the company requires influences the disclosure concerns. Sensitive information induced stronger concerns than identification information, which, in turn, induced stronger concerns than the disclosure of preferences and habits.

continued on following page

Table 1. Continued

Authors Journal and Year Title	Goals/Object	Methodology	Sample	Main results
En, Hock-Hai, Wen Wan Marketing Letters, 2006 Volunteering Personal Information on The Internet: Effects of Reputation, Privacy Notices, And Rewards On Online Consumer Behavior	Examines the effects of reputation, privacy notices, and rewards on online consumer behavior (declared) in volunteering two types of personal information on the Internet: demographic information and personally identifiable information.	2 (Reward: present vs. absent) × 2 (Reputation: high vs. low) × 2 Privacy Notices (present vs. lacking) factorial design. Dependent variable = online consumers' willingness to provide personal information (list of several types of information including demographic information and personally identifiable information, asking to provide vs. not provide). Vignette technique aimed at eliciting online consumers' willingness to provide accurate information when confronted with the need to provide information before they can use an online store's services. Each treatment is illustrated by a different vignette presented via an Internet-based system developed with Active Server Pages (ASP). Multivariate regression.	E-mail sent to 400 potential subjects recruited from the customer database of a professional special interest web site in Singapore. A total of 147 subjects were chosen.	Rewards (+ for personally identifiable data), privacy notices (+ for both demographic and personally identifiable data), and reputation (+) greatly influence consumers' intention to provide accurate personal information over the Internet. Such effects vary according to the sensitivity of the requested information. Reputation has a moderating influence on rewards and privacy notices' effects on accurate information provision.
Norberg, Horne, and Horne The Journal of Consumer Affairs, 2007 The Privacy Paradox: Personal Information Disclosure Intentions versus Behaviors	Exploratory study to investigate the privacy paradox: whether people say one thing (intend to limit disclosure) and then do another (actually provide personal details) during marketing exchanges.	Two studies (experiment with scenarios) using a repeated-measures design: willingness to disclose specific pieces of information (Phase 1), and actual provision (Phase 2). Regression analysis to examine the impact of perceived trust and perceived risk on intentions and actual behavior in the second study. Hypothesizes that there is a significant difference between one's intention to disclose and actual disclosure due to different frames of reference. When an individual is directly asked about intentions (willingness to provide personal information), risk is expected to significantly influence the response. When an individual is in an actual disclosure situation (asked for information during a marketing exchange), trust as an environmental cue is expected to be relied upon and significantly influence response.	Study 1: Sample of 23 graduate students at a university in the northeastern US. Study 2: total sample of 68 graduate students (28 low trust scenario, 40 high trust scenario).	Existence of the privacy paradox: difference between information actually provided compared to a willingness to provide (in an off-line marketing exchange situation). Individuals' considerations of risk and trust help explain why this occurs. Risk is activated (salient) and significantly influences individuals' intentions to provide, while not influencing their actual behavior. The hypothesis that trust has a greater positive influence on disclosure behavior than it has on intention to disclose is not supported.

as the belief that a party's word or promise is reliable and will fulfill obligations in an exchange relationship (Schurr and Ozanne, 1985).

The lack of trust has been identified as one of the greatest barriers to Internet transactions (e.g., Hoffman et al., 1999). Many studies on online trust focus on trust building in the relationship's initial period, thus referring to a sort of initial trust. McKnight et al. (2004) distinguish between the relationship's introductory stage – when users try to assess a specific unfamiliar website through second-hand information – and the exploratory stage – when consumers have obtained limited first-hand information, or have limited familiarity. Jarvenpaa et al. (2000), for example, found that perceived reputation is positively related to the consumers' initial trust in an Internet store.

As noted by Norberg et al. (2007: 107), "in consumer privacy and in online contexts, trust has been measured directly (Schoenbachler and Gordon 2002) as well as indirectly operationalized as a company's reputation (Andrade, Kaltcheva, and Weitz 2002)".

With respect to database marketing, Milne and Boza (1999) examine consumers' sense of privacy and found that, under certain circumstances, improving trust is more effective than efforts to reduce concern. In the context of a database or relationship marketing, the willingness to disclose information has been found to be dependent upon the level of trust in the requesting organization (Schoenbachler and Gordon, 2002). A similar result with respect to online settings has been found by Cranor et al. (1999), Gefen et al. (2003), and Hoffman et al. (1999). Unfortunately, as Norberg et al. (2007: 108, 118) note, little is known about the effects of trust on willingness versus actual disclosure in any particular marketing exchange context; this remains an open question. By measuring actual behavior, we seek to help fill this gap.

Based on the available research focused on information disclosure intent rather than effective behavior, our first hypothesis is:

H_1: The higher the trust, the higher the behavioral information disclosure.

Compensation

Companies might use incentives to induce consumers to disclose information A certain form of compensation or reward to a customer who provides information can be representative of incentives. The provision of a compensation for information sharing can work as an automatic announcement to consumers that personal data are being collected (Sheehan and Hoy, 2000: 64). At the same time, research has found that providing compensation reduces/offsets some consumer privacy concerns upfront, particularly in situations such as market research (Milne and Gordon, 1993).

The use of compensation as an incentive indicates an exchange of benefits in the act of providing information (Sheehan and Hoy, 2000). Westin-Harris's survey (1997) shows that people often consider the nature of the benefit offered in exchange for information when deciding whether or not an activity violates their personal privacy. This is because interactivity involves exchange, and consumers often weigh the benefits of the exchange when sharing information with firms. Previous studies point out that consumers are aware that not all relationships are mutually beneficial and, consequently, they don't want to enter into long-term relationships or value relationships with organizations (Szmigin and Bourne, 1998; Phelps et al., 2000: 64). Receiving some forms of incentive/compensation also supports Milne and Gordon's (1993) position that some people are willing to give up a degree of privacy to obtain products and services. Consumers may not mind receiving unsolicited marketing communications about products and services in which they are interested, even if bits of their personal information are used to identify them as prospects (Sheehan and Hoy, 2000).

Hence, consumers' willingness to concede a certain amount of privacy could also be increased by providing them with some sort of compensation. This compensation can take several – monetary or non-monetary – forms. Empirical evidence shows that consumers who receive tangible benefits (e.g., discounts, access to websites, future savings, and rewards) may be less concerned with privacy. They feel that an equal exchange has been established (Goodwin, 1991).

Again, little is known about actual disclosure behavior. On the basis of available studies, we expect information sharing to be higher if there is compensation. Our second hypothesis is:

H_2: When compensation is offered, customers will provide more information.

It would be very interesting, from a managerial point of view, to investigate the interaction between trust and compensation, as online retailers might use these variables simultaneously. To the best of our knowledge, there are no experimental studies that simultaneously investigate trust, compensation, and actual information provision in an online setting involving transactions with consumers. However, some useful insights may be drawn from previous studies regarding disclosure intentions.

In an exploratory paper and pencil experimental study, Andrade, Kaltcheva, and Weitz (2002) examined the effects of developing a reputation for trustworthiness, providing a comprehensive privacy policy, and offering a reward to reduce consumers' concerns regarding disclosing personal information. They found that the comprehensiveness of the privacy policy and the company reputation reduce the level of concern. Conversely the offer of a reward heightens the level of concern. They argue that "the subjective assessment of the concern over disclosure in the place of a behavioral measure of actual disclosure presents a weakness" (p. 352).

In their study involving vignettes, Xie et al. (2006) found a highly positive relationship between company reputation (one of the main antecedents of trust) and online consumers' willingness to reveal personal information accurately. They also found that reputation has a moderating influence on the effects of rewards, while privacy notices have one on information provision willingness.

Building on these findings, we suggest that trust could be a moderating variable in the relationship between compensation and information disclosure. In particular, we suggest that by besides strong trust, offering compensation makes individuals more inclined to provide information online, as the two stimuli are reinforced:

H_3: Under a high trust condition, subjects will provide more information when offered compensation than when offered no compensation.

The hypothesized relationships were investigated empirically in two separate laboratory experiments that differ regarding the type of incentive (compensation) offered. In the next sections, we describe the details of the experiments, the measures used, and the results obtained.

STUDY 1

Study 1's objective was to test the impact of trust and immediate/certain compensation of different types (monetary versus non-monetary) on online behavior disclosure (H_1 and H_2). This study also sought to test trust's moderating role in the relationship between compensation and information disclosure (H_3).

Design and Procedure

The study was designed as a 2 (trust: high versus low) x 3 (compensation type: no compensation,

Table 2. Study 1: The number of participants in each cell

		Compensation condition			
		No compensation	**Monetary (certain) Compensation**	**Non monetary (certain) Compensation**	*Total*
Trust condition	Low Trust	29	28	25	*82*
	High Trust	27	29	25	*81*
	Total	*56*	*57*	*50*	*163*

monetary compensation, and non-monetary certain compensation) between-subjects design. Data were collected from 163 undergraduate and graduate students enrolled at a large university in northern Italy. The number of subjects in each of the 6 cells ranged from 25 to 29 (as detailed in Table 2).

The use of student samples is often criticized. However, undergraduate and graduate students represent a significant portion of online customers, as they are among the primary buyers and users of mobile phones and services, especially in Italy.

The experiment was run in a controlled laboratory setting. The experimental materials consisted of a fictitious prototype, a limited-scope "preview" foreign company website. After reading through introductory printed pages, the participants browsed through the website and answered a list of questions on a handout. Similar methods have been used in various studies (e.g., Gefen, 2000; Jarvenpaa et al., 2000; Koufaris and Hampton-Sosa, 2002; Pavlou, 2003). The experimental section lasted approximately 45 minutes.

The participants were solicited from several courses and received a symbolic gift participating. Although this convenience sample was considered adequately representative of the university's student body, we assigned the subjects randomly to the treatment conditions to prevent confounds. The subjects were recruited under the pretext that they would participate in market research. The scenario was that a UK retailer considered expanding its business electronically to different countries, including Italy. The choice of a fictitious retailer was meant to increase the control over trust manipulation. Using a real retailer would have meant influencing the manipulation through existing subject-retailer relationships. This study is therefore focused on initial trust (McKnight et al., 2004).

The subjects were first exposed to the fictitious company by means of a short pamphlet describing its profile. They were asked to read the pamphlet in order to answer questions about the company and its offering. The company supposedly needed various local universities to cooperate to help it fine-tune its international marketing policy. The students' opinions were therefore highly sought as they were a primary target group. To further encourage the careful reading of the pamphlet, the students were also told they might be asked about the firm during the academic term.

The experiment was modeled on the stimulus materials presented by Raman, Brudvig, and Hofacker (2006). Trust was manipulated by preparing two different versions of the pamphlet offering varying descriptions of the fictitious company. We decided on a fictitious company to ensure that the retailer was not familiar to the subjects, as is the case in many internationalization ventures.

The pamphlet description consisted of excerpts from fake articles about a retail company in the UK mobile phone business. These articles supposedly featured on the online version of *The Wall Street Journal* – a well-known and credible source. A fake company rating was also included as a supposedly objective evaluation. The same format and content type was used in the two versions; the main difference lay in the company profile. In the high trust condition, the company was described as having the highest J.D. Power customer satisfaction rating, delivering the "best ever network performance," and having the highest mobile connection success rate. The company was also described as being upgraded by S&P to a "stable outlook." In contrast, in the low trust condition, the company was described as having stagnating sales growth, having the lowest J.D. Power customer satisfaction rating, and "delivering inadequate customer service." The company was also described as having been downgraded by S&P. For plausibility reasons, the pamphlet was presented in English, and all the participants were highly proficient in English.

To further reinforce the marketing research alibi, the first part of the questionnaire required data about the respondents' buying habits. This questionnaire was completed after exposure to the trust manipulation, but prior to browsing the fictitious website. This part also included two questions intended to measure a variable considered a potential covariate: attitude towards online shopping in general. This part ended with the trust manipulation check, consisting of a multi-item question and using a 7-point scale (1 = strongly disagree and 7 = strongly agree).

The subjects were then instructed to visit and thoroughly view the fictitious company's supposed beta-test website. In contrast to Norberg et al. (2007), Xie et al. (2006), and Andrade et al. (2002), we provided a purpose-built website instead of employing vignettes, scenarios, and descriptions.

After the subjects had thoroughly viewed the site's pages, they were instructed to proceed to the registration page and provide information for future transactions with the firm. This registration page requested personal and financial information from the subjects: their name, address, city, state, zip code, e-mail, phone number, credit card type, number, and expiration date, as well as Italy's equivalent of a social security number.

The compensation manipulation occurred during this stage. There were three different versions of the website (randomly assigned to the subjects), each reflecting one of the compensation conditions. In the "no compensation" condition, the subjects were simply required to provide the data indicated above. In the "monetary (certain) compensation" condition, the subjects were informed that if they registered, they would receive a 20 EUR coupon to spend in one of Italy's main electronic goods retail chains. In the "non-monetary (uncertain) compensation" condition, the participants were told that they would receive a specific 20 EUR wireless headphone as a gift. In addition, an individual code providing access to the website registered the personal information that each subject provided and matched this with the corresponding hard copy questionnaire.

Once this task was completed, the participants were asked to complete a questionnaire measuring willingness to provide information online (dependent variable), as well as the control variables: privacy concern, attitude towards online shopping, and involvement with mobile phone services.

Measures

We measured behavioral information sharing in the following steps. First, we measured the quantity of information that the subjects provided on the experimental website (calculated variable = N_provided). This quantity was then matched with the questionnaire data (calculated variable = N_matches). Specifically, the quantity of information provided (N_provided) is the sum of the number of identifying information items – name, address, zip code, phone number, email, SSN, and credit card – provided by each subject. A dummy variable, called information matching, was created for every information item to identify whether the provided information corresponds to the true information. The sum of the amount of matching information is the match variable (N_matches). Information sharing (calculated variable = information provided) is the mean of the two variables. The higher the mean, the higher the sharing.

Previous research suggests that the level of an individual's privacy concerns influences his or her willingness to share information (e.g., Phelps, Nowak and Ferrell 2000; Sheehan and Hoy 2000). Consequently, the level of privacy concern was treated as a within-subject factor and was used as a covariate in this analysis. Privacy concern was measured using an 11-item index, with each item comprising a 7-point scale (1 = strongly disagree and 7 = strongly agree). This index was adapted from the concern for information privacy (CFIP) instrument (Smith, Milberg and Burke, 1996). These 11 items are shown in Table 3. The items were factorized to create a single measure of privacy concern in which high scores indicate a higher level of privacy concern (Cronbach alpha = 0.868).

We introduced two other covariates that we believed could impact the results: attitude towards online shopping behavior and involvement with mobile phone services. Table 4 shows the scales used to measure these two variables. The items were again factor analyzed. This resulted in a single measure for each variable, with an alpha = 0.749 for attitude towards online shopping behavior, and 0.839 for involvement with mobile phone services.

To perform the manipulation check, trust was measured using a 7-point (1 = strongly disagree and 7 = strongly agree) multi-item scale adapted from existing scales in the literature (Table 5). The items were factor analyzed to create a single measure of trust, with high scores indicating a higher level of trust in the website (Cronbach alpha = 0.949).

Results

Tests were conducted to ensure that the statistical assumptions associated with the analysis of variance (ANOVA) and covariance (ANCOVA) were met. Levene's test of the equality of the error variance was not rejected. Tests were also conducted to ensure there was no an interaction effect between the covariate and any of the three other factors. These indicated that the assumption of the homogeneity of the covariance regression coefficients had not been violated. A one-way ANOVA was used to check the trust manipulation. The participants in the high trust condition

Table 3. Privacy concern measure

Dimension	Items
Collection	- When companies ask me for personal information, I sometimes think twice before providing it. - It bothers me to give personal information to so many companies. - I am concerned that companies are collecting too much personal information about me.
Access	- Companies should devote more time and effort to preventing unauthorized access to personal information. - Companies should take more steps to make sure that unauthorized people cannot access personal information in their computers.
Control	- Companies should take more steps to make sure that the personal information in their files is accurate. - Companies should have better procedures to correct errors in personal information. - Companies should devote more time and effort to verifying the accuracy of the personal information in their databases.
Use	- When people give personal information to a company for some reason, the company should never use the information for any other purpose. - Companies should never sell the personal information in their computer databases to other companies. - Companies should never share personal information with other companies unless it has been authorized by the individuals who provided the information.

Table 4. Other covariates measure

Variable	Items
Attitude toward online shopping behavior	Please indicate your agreement with the following statements about online shopping (1=strongly disagree, 7 strongly agree). Online shopping… - …results in lower prices for the consumer - … is convenient for the consumer - … stimulates the development of new products & services - … helps save the consumer time - … allows for comparative shopping - … is a fun way to shop - … is hassle free - … provides wider selection
Involvement with mobile phone services	Please, indicate your feelings about cell phone service: - Important to me 1 2 3 4 5 6 7 not important to me - Of no concern to me 1 2 3 4 5 6 7 of concern to me - Irrelevant 1 2 3 4 5 6 7 relevant - Very meaningful to me 1 2 3 4 5 6 7 means nothing to me - Matters to me 1 2 3 4 5 6 7 doesn't matter - Interesting 1 2 3 4 5 6 7 not interesting - Significant 1 2 3 4 5 6 7 insignificant - Boring 1 2 3 4 5 6 7 am concerned that companies are collecting too much personal information about me.

Table 5. Trust measure

Trust Items
Based on what you have read, how strongly do you agree or disagree with the following? - I feel I can trust the company - This company makes truthful claims - This company is honest - I do not believe what this company tells me - I can rely on this company - This company's brand is safe - I will feel secure when I buy from this company because I know it will never let me down - This company cannot be counted on to do its job

group reported a higher level of trust than those in the low trust condition (M_{HIGH} = 4.6971, M_{LOW} = 3.5449; $F(1, 161) = 85.152$, $p=0.000$).

A factorial analysis of covariance (ANCOVA) was conducted using trust (high or low) and compensation (no compensation, non-monetary certain compensation, or monetary certain compensation) as the independent variables. Information sharing was the dependent variable. Attitude towards online shopping, involvement with mobile phone services, and privacy were used as covariates, while the information provided was the dependent variable. The only significant covariate was attitude towards online shopping. We therefore re-ran the analysis with just this covariate. The beta parameter is 0.423, implying a positive relationship between attitude towards online shopping and information disclosure.

Table 6 shows the generalized linear model (GLM) results when information sharing is the dependent variable.

These results show that the main effects of trust (M_{HIGH} = 3.7284, M_{LOW} = 2.8537) and compensation ($M_{NO\ COMP.}$ = 2.5089, $M_{NON\ MONETARY\ COMP.}$ = 3.2, $M_{MONETARY\ COMP.}$ = 4.1316) are both significant (at 5%), whereas the interaction effect between trust and compensation is significant (at 10%). H_1, H_2, and H_3 are therefore confirmed. The interaction effect between trust and incentive is illustrated in Figure 1.

Table 6. Study 1: GLM result of information sharing

Source	*df*	*Mean Square*	*F-value*	*P-value*
Trust (T)	1	18.318	8.706	0.004
Compensation (C)	2	36.451	17.324	0.000
Attitude towards online shopping (A)*	1	18.276	8.686	0.004
T*C	2	5.185	2.464	0.088
Error	156	2.104		

As noted, there are different types of compensation in practice. Indeed, when combining two types of compensation (money versus a gift) and two probability levels (immediate and certain prizes versus lotteries), at least four different reward systems can be identified. In Study 1, we concentrated on two compensations types: certain monetary compensation and certain non-monetary compensation. Past research (e.g., Deutskens et al., 2004) provides empirical evidence that compensation effectiveness varies with the type of incentive provided. This suggests that Study 1's findings may change with the type of compensation offered in the manipulation. To verify this, we undertook a second study, using the compensations neglected by Study 1 (those based on a lottery) as manipulations. We next present Study 2 and its results.

STUDY 2

This experiment is aimed at directly testing the impact of trust and different types (monetary versus non-monetary) of uncertain compensation (lottery) on actual online behavioral information sharing (H_1 and H_2). In addition it tests the interaction between the two independent variables (H_3).

Figure 1. Study 1: Interaction effect of trust and (certain) incentive on information disclosure

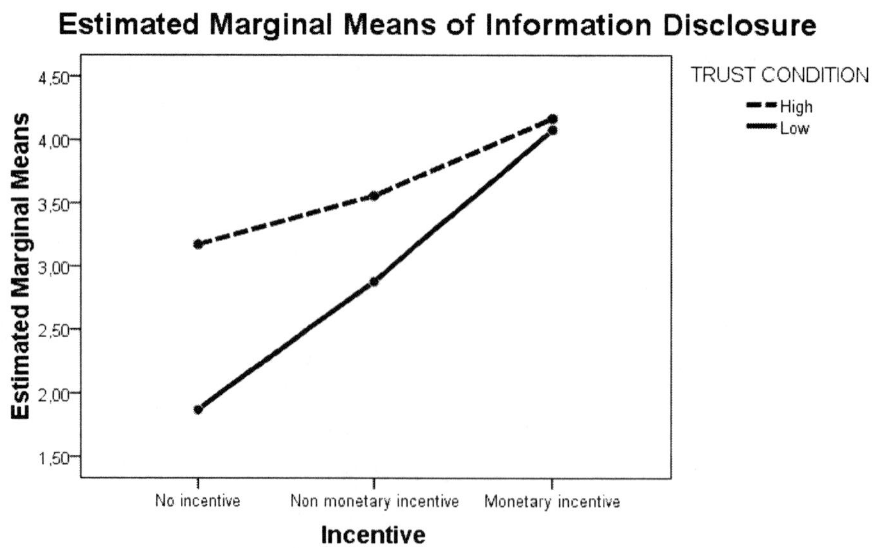

Table 7. Study 2: The number of participants in each cell

		Compensation condition			
		No compensation	Monetary (uncertain) Compensation	Non monetary (uncertain) Compensation	*Total*
Trust condition	Low Trust	20	20	18	58
	High Trust	23	23	21	67
	Total	43	43	39	*125*

Design and Procedure

Study 2's design and procedure were the same as those of Study 1, with the exception that the incentive manipulation was based on a lottery. Hence, the study was a 2 (trust: high versus low) x 3 (uncertain compensation: no compensation, monetary lottery, and non-monetary lottery) between-subjects design. In the monetary, uncertain compensation condition, the participants had the opportunity to participate in a lottery to win 500 EUR in cash. In the non-monetary, uncertain compensation condition, they had the opportunity to participate in a lottery to win a phone to the value of 500 EUR.

Data were collected from 125 undergraduate and graduate students enrolled at a large university in northern Italy. These subjects were randomly assigned to the treatment groups. The number of subjects in each of the 6 cells ranged from 18 to 23 (as indicated in Table 7).

Measures

The dependent variable, covariates, and trust were operationalized exactly as in Study 1 to allow for comparability of findings. Table 8 reports the results of the reliability analysis of the scales.

Manipulation Check and Assumptions

Tests were again conducted to ensure that the statistical assumptions associated with the analysis of variance (ANOVA) and covariance (ANCOVA) were met. Also the manipulation check was successful, as the participants in the high trust condition group reported a higher level of trust than those in the low trust condition ($M_{HIGH} = 4.2945$ $M_{LOW} = 3.4753$; $F(1, 114) = 36.658$, $p=0.000$).

Results

A factorial analysis of covariance (ANCOVA) was conducted using trust (high/low) and compensation (no compensation, non-monetary uncertain compensation, and monetary uncertain compensation) as independent variables. The actual information sharing was the dependent variable. Attitude towards online shopping, involvement with mobile phone services, and privacy were used as covariates. The information provided was

Table 8. Study 2: Scale reliability

Variable	*Cronbach Alpha*
Trust	0.887
Attitude to online shopping behavior	0.736
Commitment to mobile phone services	0.839
Privacy concern	0.859

Table 9. Study 2: GLM result of information disclosure

Source	*df*	*Mean Square*	*F-value*	*P-value*
Trust (T)	1	24.007	8.212	0.005
Compensation (C)	2	2.186	0.748	0,476
Attitude towards online shopping (A)	1	10.736	3.673	0,05
T*C	2	1.286	0.44	0,645
Error	118	2.923		

used as the dependent variable. As in Study 1, the only significant covariate was attitude towards online shopping. We therefore re-ran the analysis with just this covariate. The beta parameter is 0.368, implying a positive relationship between attitude towards online shopping and information disclosure. Table 9 shows the GLM results when information disclosure is used as the dependent variable.

These results demonstrate that only one of the two predicted main effects (trust) was significant ($M_{HIGH} = 3.0517$, $M_{LOW} = 2.0299$). Indeed, only H_1 was confirmed, while H_2 was rejected. H_3 was also rejected; trust emerged as the only variable influencing the level of information disclosure, as can be inferred from Figure 2.

Discussion

The results of our experiments show that, at the earliest relationship stage, trust is a key driver of consumers' information provision in an online context. This is consistent with studies on e-trust that identified trust as the enabling factor of online transactions (e.g., Hoffman et al., 1999) and of increasing consumers' willingness to disclose information (e.g., Grabner-Krauter, 2002).

The experiment results show that compensation's impact depends on the nature of the incentive offered. Only certain compensations have a positive effect on consumers' information disclosure. Uncertain compensation did not show significant effects. Trust's moderating effect was

Figure 2. Study 2: Interaction effect of trust and (uncertain) incentive on information disclosure

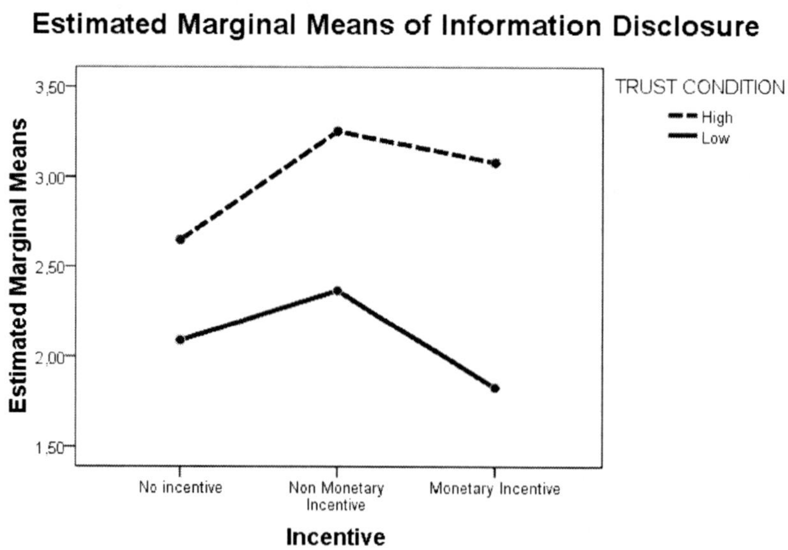

only supported for certain compensations; the effect depicted in Figure 1.

The results of the two studies suggest that, in order to exploit opportunities disclosed by new technologies, retailers should first address trust. To this end, firms' efforts could be directed towards leveraging consumers' trust antecedents, as identified in the literature.

Actions aimed at creating a trustworthy reputation may be the first step. Perceived reputation has been found to be an important antecedent of trust (e.g., Jarvenpaa et al., 2000; Yoon, 2002; Pavlou, 2003; Hee-Woong, Xu and Kho, 2004; Koufaris and Hampton-Sosa, 2004). Hence, retailers should, for example, invest in communication targeted at their customers to build a strong reputation. Online customer trust may also be increased by action in the institutional and/or technologic dimensions. Navigation, user-friendliness (Koufaris and Hampton-Sosa, 2004; Newholm et al., 2004; Bart et al., 2005), error-freeness (Newholm et al., 2004; Bart et al., 2005), website design (Shneiderman, 2000), the presence of a virtual advisor (Urban et al., 2000; Hee-Woong et al., 2004; Bart et al., 2005) and of a community (Smith et al., 2000; Luo 2002; Newholm et al., 2004) have, for example, been shown to be influential. All these aspects could also be relevant for those technology-based tools that support shoppers' in-store decisions (e.g., the magic mirror, the mobile shopping assistant, etc.). Retailers should therefore invest in these features to guarantee that the trust antecedents are conveyed to their potential customers.

The partial confirmation of the hypotheses on compensation (which only has a significant effect when certain compensation is offered) is another important result for practitioners. Many online operators are investing resources in creating more innovative compensation systems to increase their relationships with shoppers, but this strategy is costly and could be useless.

Hence, if retailers – besides leveraging trust antecedents – want to provide potential customers with incentives to increase their information sharing, they should test which forms of compensation they most value. If students are as the main target as in this study, a company should offer certain forms of compensation (coupons, discounts, and/or gifts), as uncertain compensation (based on a lottery) has proven less effective.

FUTURE RESEARCH DIRECTIONS

In our view, many future research directions can lead to an increased understanding of the privacy issues involved in adopting new technology in retail. We describe these issues in terms of two main categories: directions to improve studies on the online setting and to conduct studies in other technology-based retail settings.

Improving Studies on the Online Context

One starting point for future research would be to design further studies to overcome the limitation of the those described above. Caution must therefore be exercised in seeking to generalize our results. Future investigations could also enlarge the scope of our studies. Our study's main limitations and our suggestions for increased generalizability are:

- Our studies were focused on only one e-tailing format and in the context of a specific retail setting (mobile phones and services). The results could vary in other e-tailing contexts involving different types of goods and services.
- We focused on the early stage of the firm-customer relationship and only considered a company seeking to establish a virtual presence (a website) in one country. Investigating the relationship between the same constructs in the context of multi-channel firms – including physical and other direct channels – could be interesting.

- Students were used as a convenience sample. To test our model, we presumed that there were few systematic differences between students and online shoppers in general. If such differences do exist, students in a laboratory might be less cautious than online shoppers. This would indicate that our results are conservative. A future study can ascertain whether students in a laboratory do actually behave like real shoppers.
- The studies were carried out in only one country, Italy. Replication in other countries might be interesting. As noted by Xie et al. (2006: 72), "privacy concerns and attitudes toward providing information online may vary according to the political, economic, legal and cultural contexts." Technology makes physical distance less relevant for a retailer exploiting online settings, but the cultural distance might be very important.
- Although we used a fictitious website, an attempt was made to make it as realistic as possible. A related issue is that the website might have been of little interest to the participants. However, we chose a website that would at least be of moderate interest: the promise of sales of mobile phones and related services. Given Italy's very high penetration of mobile phones and the mobile service expenditure, students were presumed to have at least some interest. As noted, we place a high premium on future research regarding increasing our findings' generalizability by investigating information sharing in different countries, thus controlling for country-related cultural effects.

Given the differing results of the type of compensation offered, future research should investigate different compensation levels' impact on information sharing in greater depth. This could be an interesting research direction, especially regarding the managerial implications of the use of incentives.

These studies have focused on individual trust in an online firm. However, e-trust is a multi-level construct. A peculiarity of the online setting is that it involves trust. This trust is not simply between the customer and the vendor (as investigated here), but also between the buyer and the transaction medium – the technology – and the Internet environment (e.g., Harris and Goode, 2004; Lee and Turban, 2002; McNight et al., 2002; Shankar et al., 2002). Besides considering different antecedents of trust in the vendor, other levels might also be considered (e.g., trust in technology and institutional trust).

Organizational reputation is only one of the factors that might contribute to the establishment of trust in a retailer. Other factors may be the experience with the retailer, or the perception of opportunistic behaviors. Real companies could also be considered, thus allowing brand equity's effects and the dynamic of trust to be investigated. Future research should therefore examine multiple factors' combined contribution. Collaborating with a real retailer may be another possible and desirable direction for future research. This could contribute to closing the potential gap between academic research and managerial practice.

Future Research in Other Retail Settings

Other technology-based retail settings are still uncharted territory for research on privacy concerns. The starting point here could be investigating new retail media that are increasing their market share due to their synergies with computer-based online settings (e.g., the mobile setting). The mobile setting is becoming an increasing relevant marketing and retail tool. It allows the customer to be contacted in her everyday life, offering a greater customization than the traditional online context. Mobile phones are (almost) always with customers, thus allowing firms to follow and intervene in their lives. However, this characteristic – which is a major opportunity for marketers using this medium – is also the main obstacle to its diffusion on the consumer side. Customers might even consider being contacted on their mobiles as a greater invasion of privacy. Similarly, sharing personal data through their mobiles could raise greater privacy concerns than if this were done in a traditional online setting.

We therefore recommend that the differences between the two media should be investigated in terms of customer behavior. We cannot simply assume that e-commerce results would be the same as those of m-commerce. A starting point would therefore be to conduct a study as described; that is, one that includes another independent variable: the media through which the customer and firm interact. This would lead to a further research stream: investigating the privacy-related issues regarding retailers deciding to adopt new technologies within their physical stores, thus enriching the offline offer. As noted, research on this area would require real interaction between the subjects and the new in-store technologies. We believe it is critical that researchers collaborate with retailers to conduct field experiments in real stores.

CONCLUSION

Our study sought to empirically investigate the effect of trust and compensation on information sharing, a prerequisite for many new technologies' adoption. The studies we described focus on the online setting. They are meant to be the starting point for future research into other technology-based retail settings. We specifically considered a retailer in the mobile phone business selling products and services to end customers.

Building on previous studies, and to overcome a frequent limitation, we measured behavioral information sharing – not just attitudes to information sharing or past behavior. What consumers say they do, could differ from what they actually do, also with respect to privacy-related issues (e.g., Norberg et al., 2007). Attitudinal studies (the prevalent approach in this field) are unlikely to give us the whole story. Response style and social desirability are two of the many problems with attitudinal surveys (Aaker et al., 2001, pp. 219-222). In particular, social desirability implies that subjects tend to produce responses that place them in the most desirable light possible. In this case, they might gloss over the possibility that their privacy and trust can somehow be bought.

Our results underline trust's key role in increasing customers' information sharing with e-vendors. We also indicated directions to increase the generalizability of the results in the online and other retail contexts. If future studies support our findings, marketers would have clear evidence of the relevance of managing relationships with customers even if technologies actually reduce the firm's direct contact with them. This paradox could become a key challenge for retailers deciding to invest in new technologies.

REFERENCES

Aaker, D. A., Kumar, V., & Day, G. S. (2001). *Marketing research* (7th ed.). New York, NY: John Wiley and Sons.

Andrade, E. B., Kaltcheva, V., & Weitz, B. (2002). Self-disclosure on the Web: The impact of privacy policy, reward and company reputation. *Advances in Consumer Research. Association for Consumer Research (U. S.)*, *29*(1), 350–353.

Bart, Y., Shankar, V., Sultan, F., & Urban, G. L. (2005). Are the drivers and role of online trust the same for all websites and consumers? A large-scale exploratory empirical study. *Journal of Marketing*, *69*(4), 133–152. doi:10.1509/jmkg.2005.69.4.133

Brown, S. R., & Melamed, L. E. (1990). *Experimental design and analysis (quantitative applications in the social sciences)*. London, UK: Sage Publishers.

Cespedes, F. V., & Smith, H. J. (1993). Database marketing: New rules for policy and practice. *Sloan Management Review*, *34*(Summer), 8–12.

Cranor, L. F., Reagle, J., & Ackerman, M. S. (1999). Beyond concern: Understanding Net users' attitudes about online privacy. *Proceedings of the 1st ACM Conference of Electronic Commerce* (pp.1-8). New York, NY: ACM.

Culnan, M. J. (1993). How did they get my name? An exploratory investigation of consumer attitudes toward secondary information use. *Management Information Systems Quarterly*, *17*(3), 341–363. doi:10.2307/249775

DeCew, J. (1997). *In pursuit of privacy. Law, ethics, and the rise of technology*. Ithaca, NY: Cornell University Press.

Gefen, D. (2000). E-commerce: The role of familiarity and trust. *The International Journal of Management Science*, *28*(6), 725–737.

Gefen, D., Karahanna, E., & Straub, D. W. (2003). Trust and TAM in online shopping: An integrated model. *Management Information Systems Quarterly*, *27*(1), 51–90.

Goodwin, C. (1991). Privacy: Recognition of a consumer right. *Journal of Public Policy & Marketing*, *10*(1), 149–166.

Grabner-Kraeuter, S. (2002). The role of consumers' trust in online-shopping. *Journal of Business Ethics*, *39*(1/2), 43–50. doi:10.1023/A:1016323815802

Harris, L. C., & Goode, M. M. H. (2004). The four levels of loyalty and the pivotal role of trust: A study of online service dynamics. *Journal of Retailing*, *80*(2), 139–158. doi:10.1016/j.jretai.2004.04.002

Hee-Woomg, K., Xu, Y., & Koh, J. (2004). A comparison of online trust building factors between potential customers and repeat customers. *Journal of the Association for Information Systems*, *5*(10), 392–420.

Hoffman, D., Novak, T. P., & Peralta, M. A. (1999). Building consumer trust online environment: The case for information privacy. *Communications of the ACM*, *42*(4), 80–85. doi:10.1145/299157.299175

Jarvenpaa, S. L., Tractinsky, J., & Vitale, M. (2000). Consumer trust in an Internet store. *Information Technology Management*, *1*(1/2), 45–71. doi:10.1023/A:1019104520776

Jobson, J. D. (1991). Applied multivariate data analysis: *Vol. 1. Regression and experimental design*. New York, NY: Springer.

Koufaris, M., & Hampton-Sosa, W. (2002). *Customer trust online: Examining the role of the experience with the website*. New York, NY: Zicklin School of Business, Baruch College Department of Statistics and Computer Information Systems Working Paper Series.

Koufaris, M., & Hampton-Sosa, W. (2004). The development of initial trust in an online company by new customers. *Information & Management*, *41*(3), 377–397. doi:10.1016/j.im.2003.08.004

Lee, M. K. O., & Turban, E. (2002). A trust model for consumer Internet shopping. *International Journal of Electronic Commerce*, *6*(1), 75–91.

Luhmann, N. (1989). Familiarity, confidence and trust: Problems and alternatives. In Gambetta, D. (Ed.), *Trust: Making and breaking cooperative relations* (pp. 123–140). Oxford, UK: Basil Blackwell.

Luo, X. (2002). Trust production and privacy concerns on the Internet: A framework based on relationship marketing and social exchange theory. *Industrial Marketing Management, 31*(2), 111–118. doi:10.1016/S0019-8501(01)00182-1

McKnight, D. H., Choudhury, V., & Kacmar, C. (2002). The impact of initial consumer trust on intention to transact with a website: A trust building model. *The Journal of Strategic Information Systems, 13*(4), 297–323. doi:10.1016/S0963-8687(02)00020-3

McKnight, D. H., Kacmar, C., & Choudhury, V. (2004). Shifting factors and the ineffectiveness of third party assurance seals: A two-stage model of initial trust in a Web business. *Electronic Markets, 14*(3), 252–266. doi:10.1080/10196780420002452 63

Milne, G. R. (2000). Privacy and ethical issues in database/interactive marketing and public policy: A research framework and overview of the special issue. *Journal of Public Policy & Marketing, 19*(1), 1–6. doi:10.1509/jppm.19.1.1.16934

Milne, G. R., & Boza, M. E. (1999). Trust and concern in consumers' perceptions of marketing information management practices. *Journal of Interactive Marketing, 13*(1), 5–24. doi:10.1002/(SICI)1520-6653(199924)13:1<5::AID-DIR2>3.0.CO;2-9

Milne, G. R., & Gordon, M. E. (1993). Direct mail privacy-efficiency trade-offs within an implied social contract framework. *Journal of Public Policy & Marketing, 12*(2), 206–215.

Müller-Lankenau, C., Wehmeyer, K., & Klein, S. (2005). Multi-channel strategies: Capturing and exploring diversity in the European retail grocery industry. *International Journal of Electronic Commerce, 10*(2), 85–122. doi:10.2753/JEC1086-4415100204

Newholm, T., McGoldrick, P., Keeling, K., Macaulay, L., & Doherty, J. (2004). Multi-story trust and online retailer strategies. *International Review of Retail, Distribution and Consumer Research, 14*(4), 437–456. doi:10.1080/0959396042000260889

Norberg, P. A., Horne, D. R., & Horne, D. A. (2007). The privacy paradox: Personal information disclosure intentions versus behaviors. *The Journal of Consumer Affairs, 41*(1), 100–126. doi:10.1111/j.1745-6606.2006.00070.x

Otto, J. R., & Chung, Q. B. (2000). A framework for cyber-enhanced retailing: Integrating e-commerce retailing with brick-and-mortar retailing. *Electronic Markets, 10*(3), 185–191. doi:10.1080/10196780050177099

Pantano, E., & Naccarato, G. (2010). Entertainment in retailing: The role of advanced technologies. *Journal of Retailing and Consumer Services, 17*(3), 200–204. doi:10.1016/j.jretconser.2010.03.010

Pavlou, P. A. (2003). Consumer acceptance of electronic commerce - integrating trust and risk with the technology acceptance model. *International Journal of Electronic Commerce, 7*(3), 69–103.

Phelps, J., Nowak, G. J., & Ferrell, E. (2000). Privacy concerns and consumer willingness to provide personal information. *Journal of Public Policy & Marketing, 19*(1), 27–4. doi:10.1509/jppm.19.1.27.16941

Phelps, J. E., D'Souza, G., & Nowak, G. J. (2001). Antecedents and consequences of consumer privacy concerns: An empirical investigation. *Journal of Interactive Marketing, 15*(4), 2–17. doi:10.1002/dir.1019

Premazzi, K., Castaldo, S., Grosso, M., & Hofacker, C. (in press). Supporting retailers to exploit online settings for internationalization: The different role of trust and compensation. *Journal of Retailing and Consumer Services*.

Raman, P., Brudvig, S., & Hofacker, C. F. (2006, October). *To give or not to give: Providing personal information to online firms.* Paper presented at the Direct Marketing Association Educators' Conference, San Francisco, CA.

Roussos, G., & Kostakos, V. (2009). RFID in pervasive computing: State-of-the-art and outlook. *Pervasive and Mobile Computing, 5*(1), 110–131. doi:10.1016/j.pmcj.2008.11.004

Schoenbachler, D. D., & Gordon, G. L. (2002). Trust and customer willingness to provide information in database-driven relationship marketing. *Journal of Interactive Marketing, 16*(3), 2–16. doi:10.1002/dir.10033

Schurr, P. H., & Ozanne, J. L. (1985). Influences on exchange processes: Buyer preconceptions of a seller's trustworthiness and bargaining toughness. *The Journal of Consumer Research, 11*(4), 939–953. doi:10.1086/209028

Shankar, V., Urban, G. L., & Sultan, F. (2002). Online trust: A stakeholder perspective, concepts, implications and future directions. *The Journal of Strategic Information Systems, 11*(4), 325–344. doi:10.1016/S0963-8687(02)00022-7

Sheehan, K. B. (2005). Public opinions of online privacy: Definitions, assessment and implications for industry and public policy. In Krishnamurthy, S. (Ed.), *Contemporary research in e-marketing* (pp. 186–199). Hershey, PA: Idea Group Publishing.

Sheehan, K. B., & Hoy, M. G. (1999). Flaming, complaining, abstaining: How online users respond to privacy concerns. *Journal of Advertising, 28*(3), 37–51.

Sheehan, K. B., & Hoy, M. G. (2000). Dimensions of privacy concern among online consumers. *Journal of Public Policy & Marketing*, (Spring): 62–73. doi:10.1509/jppm.19.1.62.16949

Shneiderman, B. (2000). Designing trust into online experiences. *Communications of the ACM, 43*(12), 57–59. doi:10.1145/355112.355124

Smith, H. J., Milberg, S. J., & Burke, S. J. (1996). Information privacy: Measuring individuals' concerns about organizational practices. *Management Information Systems Quarterly, 20*(2), 167–196. doi:10.2307/249477

Smith, M., Bailey, J., & Brynjolfsson, E. (2000). Understanding digital markets: Review and assessment. In Bryonjolfsson, E., & Kahim, B. (Eds.), *Understanding the digital economy.* Cambridge, MA: MIT Press.

Steinfield, C. (2004). The missing link: Connecting physical and virtual channels through click & mortar electronic commerce. In Stanoevska-Slabeva, K. (Ed.), *Digital economy* (pp. 141–157). Heidelberg, Germany: Anspruch und Wirklichkeit.

Szmigin, I., & Bourne, H. (1998). Consumer equity in relationship marketing. *Journal of Consumer Marketing, 15*(6), 544–557. doi:10.1108/07363769810240545

Urban, G. L., Sultan, F., & Qualls, W. J. (2000). Placing trust at the center of your internet strategy. *Sloan Management Review, 42*(1), 39–48.

Venkatesan, R., Kumar, V., & Ravishankar, N. (2007). Multichannel shopping: Causes and consequences. *Journal of Marketing, 71*(2), 114–132. doi:10.1509/jmkg.71.2.114

Viswanathan, S. (2005). Competing across technology-differentiated channels: The impact of network externalities and switching costs. *Management Science, 51*(3), 483–496. doi:10.1287/mnsc.1040.0338

Westin-Harris. (1997). *Privacy and American business study, a survey.*

Xie, E., Hock-Hai, T., & Wen, W. (2006). Volunteering personal information on the Internet: Effects of reputation, privacy notices, and rewards on online consumer behavior. *Marketing Letters*, *17*(1), 61–74. doi:10.1007/s11002-006-4147-1

Yoon, S. J. (2002). The antecedents and consequences of trust in online purchase decisions. *Journal of Interactive Marketing*, *16*(2), 47–63. doi:10.1002/dir.10008

ADDITIONAL READING

Castaldo, S. (2007). *Trust in Market Relationships. Camberley*. Edward Elgar Publishing Ltd.

Castaldo, S., Hofacker, C., Grosso, M., & Premazzi, K. (2010). Trust in on line customer-firm interaction: a literature review and directions for research. In Latusek Dominika, D., & Gerbasi Alexandra, A. (Eds.), *Trust and Technology in a Ubiquitous Modern Environment* (pp. 207–304). Hershey, PA: IGI Global. doi:10.4018/978-1-61520-901-9.ch018

Castaldo, S., Hofacker, C., Grosso, M., & Premazzi, K. (2010). Customers' information sharing with e-vendors: The role of initial trust and incentives" (co-authors: S. Castaldo, C. Hofacker, K. Premazzi, P. Raman and S. Brudvig). *International Journal of Electronic Commerce*, *14*(3), 63–91.

Everard, A., & Galletta, D. F. (2005). How Presentation Flaws Affect Perceived Site Quality, Trust, and Intention to Purchase from an Online Store. *Journal of Management Information Systems*, *22*(3), 55–95.

Franco, M., & Mariano, S. (2009). Information Technology Repositories and Knowledge Management Processes: A Qualitative Analysis. *The Journal of Information and Knowledge Management Systems*, *37*(4), 440–451.

Long, G., Hogg, M. K., Hartley, M., & Angold, S. J. (1999). Relationship Marketing And Privacy: Exploring The Thresholds. *Journal of Marketing Practice: Applied Marketing Science*, *5*(1), 4–20. doi:10.1108/EUM0000000004548

Miller, M., & Brunner, C. (2008). Social impact in technologically-mediated communication: An examination of online influence. *Computers in Human Behavior*, *24*(6), 2972–2991. doi:10.1016/j.chb.2008.05.004

Miyazaki, A. D., & Krishnamurthy, S. (2002). Internet Seals of Approval: Effects on Online Privacy Policies and Consumer Perceptions. *The Journal of Consumer Affairs*, *36*(1), 28–49. doi:10.1111/j.1745-6606.2002.tb00419.x

Muller-Seitz, G., Dautzenberg, K., Creusen, U., & Stromereder, C. (2009). Customer acceptance of RFID technology: evidence from the German electronic retail sector. *Journal of Retailing and Consumer Services*, *16*(1), 31–39. doi:10.1016/j.jretconser.2008.08.002

Newsom, M., Collier, D., & Olsen, E. (2009). Using "biztainment" to gain competitive advantage. *Business Horizons*, *52*, 167–176. doi:10.1016/j.bushor.2008.10.005

Nowak, G. J., & Phelps, J. (1997). Direct Marketing And The Use Of Individual-Level Consumer Information: Determining How And When Privacy Matters. *Journal of Direct Marketing*, *11*(4), 94–108. doi:10.1002/(SICI)1522-7138(199723)11:4<94::AID-DIR11>3.0.CO;2-F

Pagani, M. (2007). A vicarious innovativeness scale in the domain of 3G mobile services: integrating the Domain Specific Innovativeness Scale with psychological and rational indicators. *Technology Analysis and Strategic Management*, *19*(6), 709–728. doi:10.1080/09537320701711207

Pagani, M. (2007). Challenges of usability evaluations in the emerging multimedia environment. *Journal of Information Science Technology*, *3*(3), 1–4.

Pagani, M. (2008). A Value-Choice model to forecast market consequences of 3G mobile service design decisions. *International Journal of Mobile Marketing*, *3*(1), 23–31.

Pantano, E. (2010). New Technologies and Retailing: trends and directions. *Journal of Retailing and Consumer Services*, *17*(3), 171–172. doi:10.1016/j.jretconser.2010.03.004

Sayre, S., & Horne, D. A. (2000). Trading Secrets for Savings: How Concerned Are Consumers about a Privacy Threat from Club Cards? *Advances in Consumer Research. Association for Consumer Research (U. S.)*, *27*(1), 151–155.

Sicilia, M., & Ruiz, S. (2010). The Effect of Web-Based Information Availability on Consumers' Processing and Attitudes. *Journal of Interactive Marketing*, *24*(1), 31–41. doi:10.1016/j.intmar.2009.10.001

Söderlund, M., & Julander, C. (2009). Physical attractiveness of the service worker in the moment of truth and its effects on customer satisfaction. *Journal of Retailing and Consumer Services*, *16*(3), 216–226. doi:10.1016/j.jretconser.2008.11.008

Spiekermann, S., Grossklags, J., & Berendt, B. (2001). E-Privacy in 2nd Generation ECommerce: Privacy Preferences vs. Actual Behavior. Paper presented at the ACM Conference on Electronic Commerce, Tampa, FL.

KEY TERMS AND DEFINITIONS

Compensation: Something firms give customers in exchange for their personal data provision; this may take several forms (money, a gift, etc.).

E-Commerce: The buying and selling of products or services over an electronic medium such as the Internet and other computer networks. When it involves final customers it is also known as e-tailing.

Experiment: Quantitative methodology that allows to test the causality between one or more independent variables on a dependent variables, by controlling for others intervenient variables.

Information and Communication Technology (ICT): These are meant to increase the information and communication exchange between parties over time and place. However, they give rise to privacy concerns on the part of many users.

Information Disclosure: Customers' provision of personal information to a seller.

Privacy Concern: Concern about the extent of safety and non-consensual usage of personal data provided to some entity (such as a firm).

Trust: A subject's belief in a risky situation that another subject, over whose behavior she has no control, will act according to her expectations.

ENDNOTES

1. The personal shopping assistant (PSA) is a tool capable of supporting consumers by allowing them to swiftly compare different proposals, besides providing detailed and complete information on products and services (Pantano and Naccarano, 2010, p. 202).
2. The following paragraphs are adapted from Premazzi et al. (2010).
3. Readers not familiar with experimental design and analysis are referred to Jobson (1991) as well as Brown and Melamed (1990).

Chapter 11
You Never Get a Second Chance to Make a First Impression:
Meet Your Users' Expectations Regarding Web Object Placement in Online Shops

Javier A. Bargas-Avila
University of Basel, Switzerland

Sandra P. Roth
University of Basel, Switzerland

Alexandre N. Tuch
University of Basel, Switzerland

Klaus Opwis
University of Basel, Switzerland

ABSTRACT

Knowledge about users' expectations and mental models is a key aspect of interface development. By meeting users' expectations, errors may be prevented and interaction quality enhanced. In the case of online shops, this means that it is crucial to know where users expect to find the most common Web objects such as the search field, shopping cart, or navigation. In this chapter, we show how users' mental models of an online shop can be analyzed and validated empirically. The resulting model shows where users typically expect to find the most common Web objects within an online store. This knowledge can be used to improve the first impression, orientation, and usability of your website.

INTRODUCTION

Imagine yourself visiting different online shops, looking for items that you want to buy. Usually – if you do not know in advance which shops are the most convenient for your needs – you look up different stores by using a search engine. This will lead you to very different venues with varying designs. Imagine yourself arriving on the start page of these shops. Within a very short fraction of time you scan the site with its shapes,

DOI: 10.4018/978-1-60960-738-8.ch011

colors, fonts, pictures, navigation elements, and text components. Within seconds you will know whether you are going to keep on visiting this place, or whether you will hit the back button and return to your search results to visit the next store.

How does this first impression develop? Is it a feeling that is based solely on randomness? On personal preferences, moods, motivations and experiences? Or can we deconstruct and analyze the most important factors that contribute to this impression? Research has shown that this first impression is created within the first 50 milliseconds of perception, and that it remains remarkably stable through time (Lindgaard, Fernandes, Dudek and Brown; 2006). Website owners may therefore want to keep in mind the importance of the first impression and shape it to their advantage.

Since the early stages of the World Wide Web, designers as well as researchers have been confronted with these topics in their pursuit of successful website design. Within this area – as in many interdisciplinary fields – there are varying and sometimes even conflicting viewpoints. Some designers regard the graphic design, composition and structure of a website as an exclusively artistic matter, in which they do not wish or allow themselves to be influenced by others. For them, the website is the final product of a creative process and should not be deconstructed and analyzed. On the other hand, there are researchers who believe we should be able to deconstruct web design into its components and analyze them. In their opinion, by gaining a better understanding of each element, we will be able to create more appealing websites. As is often the case with this kind of almost ideological discussion, the truth is probably to be found somewhere in-between: Although the creation of an appealing website is, of course, also subject to an artistic process, it is usually worthwhile to understand the psychological components that lead users to experience a positive perception and interaction.

In this chapter, we will start with a short introduction into the field of mental models. We will then present research that was conducted in order to understand the expectations that users have formed through time regarding the placement of the most frequent objects within online shops. We will also show that users are able to perceive and find those objects faster when these expectations are met. The results of these studies are not a plea for a standard design, where one website looks like another. Of course, there must always be room for surprising elements in website design, elements that trigger curiosity, demonstrate novelty or are fun. But knowing users' expectations gives website developers the possibility of deciding whether they intentionally want to meet or breach them.

BACKGROUND

What Are Mental Models of Web Pages?

Websites exist in never-ending variations. Try to imagine a prototypical online shop: What does it look like? Probably your shop will have a company logo placed at the top of the page. Certainly you will have a search field, where users are able to enter search queries. Where is this box placed in your imaginary shop? Probably not at the bottom of the website, right? Either way, there will be certain objects that you certainly expect to find in an online shop. This is the case because people form internal mental models of things which they know from their everyday life (Norman, 1983; Rouse and Morris, 1986).

During the 1980s, the term *mental model* was introduced to describe knowledge representation in the field of cognitive psychology (e.g. Gentner and Stevens, 1983; Johnson-Laird, 1983). Johnson-Laird (1983) points out that mental models play a central role in representing objects, situations, orders of events, and the social actions of daily life. He states that mental models allow people to understand phenomena and to make inferences and predictions. The theoretical constructs *mental model*, *conceptual model*, *mental representation*, and *schemata* are also discussed widely in the field of human–computer interaction (HCI) regarding their practical implications (Ben-Ari and Yeshno, 2006; Farris, Jones and Elgin, 2002; Green, 1994; Hsu, 2006; Otter and Johnson, 2000; Payne, 2007). Users' mental models of how a device works may influence their interaction with it and therefore it is crucial to understand these models in order to improve the object's design. Norman (1983) describes some distinctions that must be considered in this context: the *target system*, the *conceptual model of the target system*, and the *user's mental model of the target system*. The target system represents the real object being learned or used. The user's mental model develops during interaction with the system, whereas the conceptual model is a tool for understanding and teaching a system. Users' mental models are neither accurate nor robust. They are built instantaneously as volatile constructs through interactions with the target system and experiences with similar systems in the past, and are also dependent on the users' background knowledge and expertise. These representations of semantic knowledge are then used to match new input with previously known and formed patterns (Bartlett, 1932). As mimics of the physical world, mental representations are also needed to make predictions and to simulate real-world behavior (Craik, 1943). In optimal cases, the conceptual model is very similar to the user's mental model to enable easy comprehension and improve intuitive interactions with the system. The concept of mental models is used in many guidelines for interface and website design (e.g. Apple Inc., 2008; IBM, 2008, International Organization for Standardization, 1998). These guidelines mention the need to reflect and know users' mental models and take into account users' expectations in order to prevent errors and enhance usability.

Because mental models evolve naturally by interacting through experience with, and knowledge of a system or interface, it can be assumed that over the course of time users of the Internet have developed certain expectations of individual web objects or even whole web pages. Spool (2008), for example, found that the login area is expected to consist of two labeled boxes of equal size in close proximity to each other. Users expect that the first box is for the login name and the second is for the password. If this expectation is not met, users may have trouble finding and identifying the login fields and using the login correctly. Positioning navigational regions at expected locations and using specific web objects consistently is also very important for enhancing memorability of web pages (Oulasvirta, 2004). Not only the layout and naming of individual web objects, but also their particular locations lead to certain expectations. Several studies have shown that people expect to find web objects such as the search field, home button, or navigation at specific locations (e.g. Bernhard, 2001; Oulasvirta, 2005). Bernard (2002) investigated user expectations about the location of web objects and found general agreement among the participants, hinting that there seems to be a common mental model of typical websites. Shaik and Lenz (2006) examined whether these expectations had shifted since Bernard's studies and found changes in the expected location of the search field, navigation area, and advertisements. However, it must be noted that they asked participants for expectations of common web objects located on "basic informational websites". This task was more specific than Bernard's, who used the general term "typical web page within a website". Nevertheless, Shaik and Lenz suggest that these expectation shifts occurred as technology changed the appearance of the Internet. In a recent online study, 500 Internet users were asked to construct three different types of web pages according to their expectations: online shops, online newspapers, and companies (Roth, Schmutz, Pauwels, Bargas-Avila and Opwis; 2010). Participants could place a wide selection of web objects by means of drag and drop and adjust their sizes individually.

Results indicate that distinct mental models seem to exist for different web page types, i.e. people agree on many but not all web objects' location. The study provides data about location typicality for web objects. In this chapter, we will present the summarized results of Roth et al.'s study of online shops and parts of a follow-up study that aimed at validating these results.

What Do We Need Mental Models of Websites For?

Fast orientation on web pages is desired by Internet users as well as designers because it enhances efficiency (e.g. Tuch, Bargas-Avila, Opwis and Wilhelm, 2009), reduces cognitive load (e.g. Schmutz, Heinz, Métrailler and Opwis, 2009), and increases users' satisfaction (e.g. Leuthold, Bargas-Avila and Opwis, 2008). The faster the visual recognition of web objects happens, the easier orientation will be. For guidance of attention and visual recognition of objects, two interacting and cooperating mechanisms, top-down and bottom-up processes, are at work (for an overview see Kverga, Ghuman and Bar, 2007). Bottom-up processes operate on raw sensory input and are stimuli driven, so attention is oriented spontaneously towards salient visual features of a scene (Braun and Sagi, 1990; Kowler, Anderson, Dosher and Blaser, 1995; Reinagel and Zador, 1999). Top-down processes require longer-term cognitive strategies, they are goal-directed, and they guide attention through stored knowledge (Pelz and Canosa, 2001; Shinoda, Hayhoe and Shrivastava, 2001).

Bottom-up and top-down processes are constantly at work in the natural environment and also during website navigation. On the one hand, web designers are able to guide attention to desired locations by manipulating object features and triggering bottom-up mechanisms. On the other hand, arranging web objects on web pages according to users' expectations trigger top-down processes. The ability to use stored knowledge, enable quicker interpretations of internal and external events, and faster, more exact, and efficient responses – recognition can be improved by using learned information about temporal, spatial, and semantic circumstances between stimuli in our environment. Several studies have shown that recognition of objects is facilitated when they are placed in their proper context (Biederman, Mezzanotte, and Rabinowitz, 1982; Davenport and Potter, 2004; Palmer, 1975).

Studies in the field of HCI have reported similar results. Oulasvirta, Kärkkäinen and Laarni (2005), Vaughan and Dillon (2006), and Santa-Maria and Dyson (2008) examined the effect of fulfilling or violating mental models of web pages. In Oulasvirta et al.'s (2005) eye-tracking experiments, the link panels were either place only on the left, placed only on the right, or split between both sides of the web pages. When participants were asked to look for a specific link, they fixated first on the left-hand side of the web page even if the links were divided between both sides. First saccades were more often directed to the left-hand side. Additionally, the participants indicated that they would guess the target link to be located on the left-hand side. Besides the increased speed of finding the target, they also found that people were able to remember specific web objects better if they were placed according to their mental models of web pages. Vaughan and Dillon (2006) examined the difference between typically and atypically designed online newspapers. The typical design led to significantly better performance and free recall of news content. Even after a course of five sessions, the participants using the atypical design did not perform as well as the other participants. Santa-Maria and Dyson (2008) constructed conventional and convention-violating web forums, compared them, and showed that participants browsing the convention-violating forum made more page revisits and used the back button more often than users of the conventional web forum. After solving five to six tasks, participants appeared to learn to orient themselves and performance leveled out. The violations of the visual conventions had – at least for the first few visits – a negative influence on orientation and performance.

The studies summarized indicate that it can be beneficial to design certain types of web pages according to expectations, thus triggering top-down processes and increasing orientation. It is a common procedure in user-centered design to include users' expectations and mental models in the design process of web pages and applications. Applying mental models to web page design helps users to orient themselves – at least during the first few visits – and will influence the first impression.

THE MENTAL MODELS OF AN ONLINE SHOP

The work presented in this chapter is based on the publications of Roth et. al. (2010) and Roth, Tuch, Mekler, Bargas-Avila and Opwis (submitted). Although the focus in these works is on mental models for online shops, news portals and company web pages, we will focus here only on expectations regarding online shops.

How to Determine the Mental Model of an Online Shop

To determine a valid mental model of an online shop, we first have to define which web page objects are usually present and therefore expected in a prototypical online shop. To identify these objects, the following procedure was used: The start pages of the 100 most visited web pages of the USA, Switzerland, Germany, and Austria were screened (Alexa - The Web Information Company, 2008) and all objects were listed. Only objects on the first layer (the start page) were collected. In a second step, these objects were used in an online survey to ask n = 136 participants, which of the collected web objects they expected to appear in an online shop. Subsequently, web objects that were

Table 1. Web objects in online shops (adapted from Roth et al., 2010). Bold numbers represent the web objects selected for the study

Web object	Online shop (%)
about us	**70.9**
conditions of use	**97.4**
archive	5.1
shopping cart	**98.3**
privacy notice	**88.9**
FAQ	**92.3**
help	**96.6**
back to homepage	**95.7**
contact	**92.3**
external links	29.9
logo	**91.5**
main area	**93.2**
navigation area	**96.6**
newsletter (link)	**67.5**
search field	**94.0**
sign in/login field	**90.6**
to the top (link)	**51.3**
advertisement area	**50.4**

chosen by more than 40% of the participants were included in the study (see Table 1). The results show that users' expectations about web objects are quite congruent to the website's purpose: In online shops, users do not expect an *archive* or *external links*. This procedure led to a final list of 17 typical web objects for online shops.

To determine the mental model of an online shop, an explorative design was used. Drawings and sketches have shown to be promising tools to visualize people's conceptualizations or mental models (Denham, 1993; Kerr, 1990; Thatcher and Greyling, 1998). At the core of the online study was an application where users were able to construct their prototypical web site with just a few clicks. They were provided with the corresponding web page objects identified in the preliminary study (see Table 1). These web objects were to be distributed and resized in a browser frame. The survey consisted of three parts: (1) the instructions, including a demographic questionnaire and short exercise for learning about the handling of the application, (2) the core application of the study, where people could construct their websites, and (3) the final questions section. For a more detailed description, see Roth et al. (2010). In the main part, participants were instructed to "Build a typical online shop according to your (their) expectations". The instructions also stated that participants could decide which web objects to place and which to leave out.

A total of n = 516 participants (334 women, 179 men, 3 did not indicate their gender) completed this online experiment. The mean age was 28.2 years (SD = 9.1), ranging from 13 to 67 years. Participants rated as fairly high their ability to handle a computer, the Internet, and online shopping. They used computers and the Internet almost daily for business as well as private reasons. Most participants shopped online, ranging from at least several times a year to three times a month. For a detailed descriptive overview see Roth et al. (2010).

The Mental Model of an Online Shop

The experiment delivered for each participant a picture of how he/she expects an online shop to look. To calculate a mean (a common mental model), pixel-based raw data were reduced into a 12x8 grid. If at least 50% of the web object covered a cell, it was counted as a "hit". The higher the number of hits, the darker the color of the cells (see Figure 1). The percentages are calculated from the number of placed objects (see Table 2), e.g. 30-34% means between 30-34% of the people who placed the object chose this location. Cells were not marked if covered by less than 5% of the participants. The respective values and coloring of cells were chosen to facilitate a direct comparison with the results of Bernard's studies (Bernard, 2001, 2002).

An overview of the number of placements in percentages and sizes in pixels of each web object is displayed in Table 2. Each object was placed by at least 47% of the participants. The web objects *main area, search field, sign in/login field, logo,* and *shopping cart* were chosen by almost all participants (at least 90%). Regarding the sizes of the web objects, *main area* is the largest object in both width and height. Its location is always expected to be at the center of the page. The logo is expected as an elongated rectangle extending across almost the whole width of the page. Regarding location, the upper-right corner was used for the *shopping cart*. The *search field* was placed more diffusely, centered across the upper half of the page. The *sign in/login* field is expected in the top-right corner of the web page, whereas the *back to homepage* link was placed in the top-left corner at about the same location as the *logo,* which stretched further across the top width. The *advertisement area* was mostly expected on the right-hand side and across the top width of the web page. The *conditions of use, privacy notice,* and *to the top* links were placed across the bottom width of the page. The links *about us, contact, FAQ,* and *help* share roughly the same area of the web page; that is to say, part of the left-hand side, bottom width and top-right corner. Similar expected locations were apparent for *newsletter,* namely across the left- and right-hand sides of the page. See Figure 1 for an overview of all locations.

To support website developers in the placement of web objects, a consolidated model for a prototypical online store was created. Most locations and sizes of web objects described are not overlapping and can be easily put together. There are some instances where a mix-up can be seen. This is the case for the *login/sign in, search field* and *shopping cart.* In these situations the most obvious compromise was chosen. Figure 2 shows an approximation of a consolidated mental model.

Figure 1. Percentage of placed web objects on online shop web pages (adapted from Roth et al., 2010)

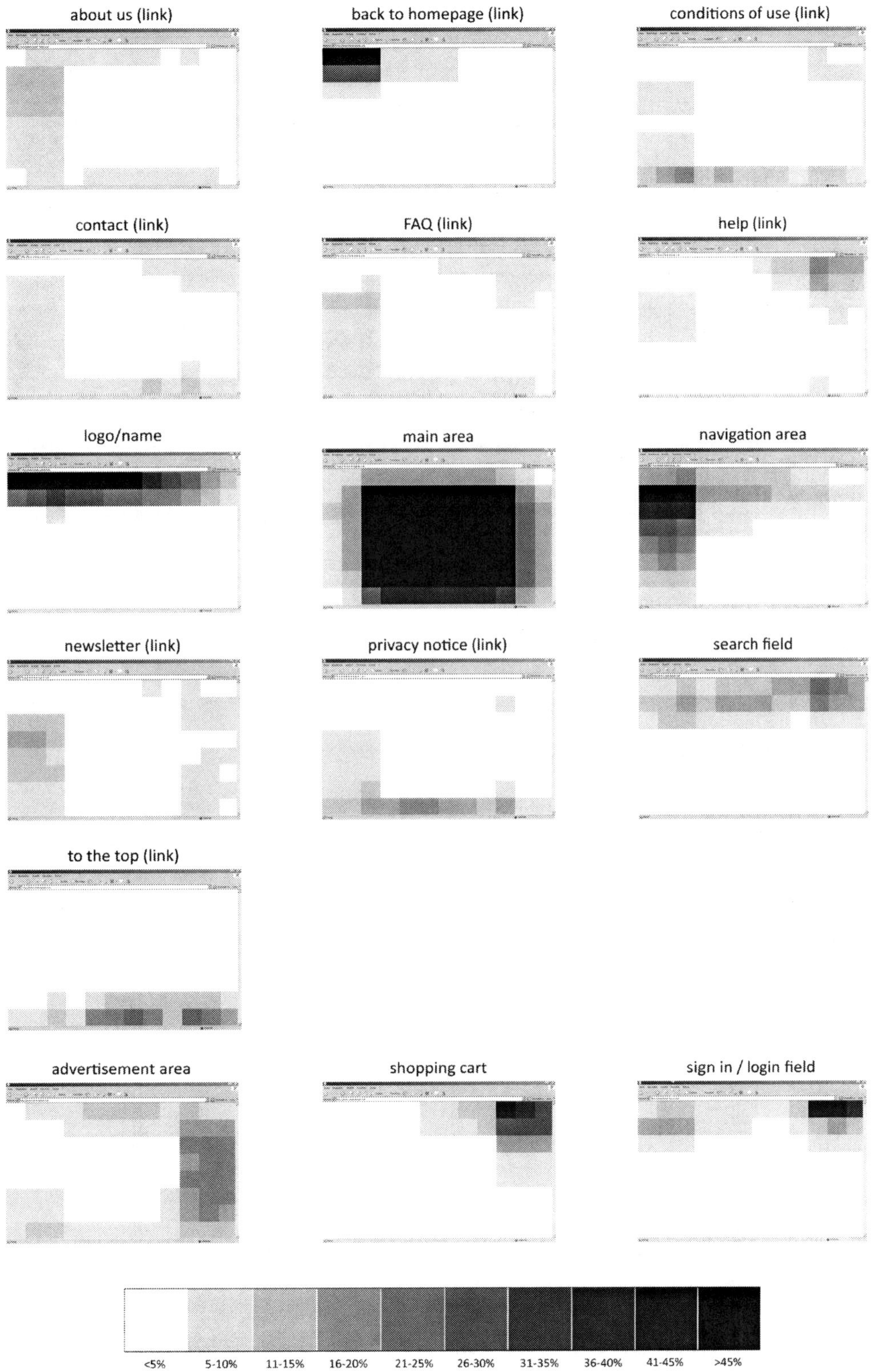

Table 2. Web object properties: number of placements and size (adapted from Roth et al., 2010).

Web object	placements in % [1]	mean width [2]	mean height [2]
about us	75%	140	21
advertisement area	65%	183	82
back to homepage	81%	149	24
conditions of use	83%	139	19
contact	88%	141	20
FAQ	79%	140	19
help	80%	138	20
logo/name	93%	293	47
main area	90%	392	235
navigation area	74%	216	93
newsletter	56%	141	21
privacy notice	76%	142	20
search field	91%	168	24
shopping cart	94%	146	41
sign in/login field	94%	149	25
to the top	54%	146	21

[1] percentage of all participants (n = 516)
[2] in pixels

Figure 2. The image represents a consolidated mental model of an online shop (Roth et al., 2010)

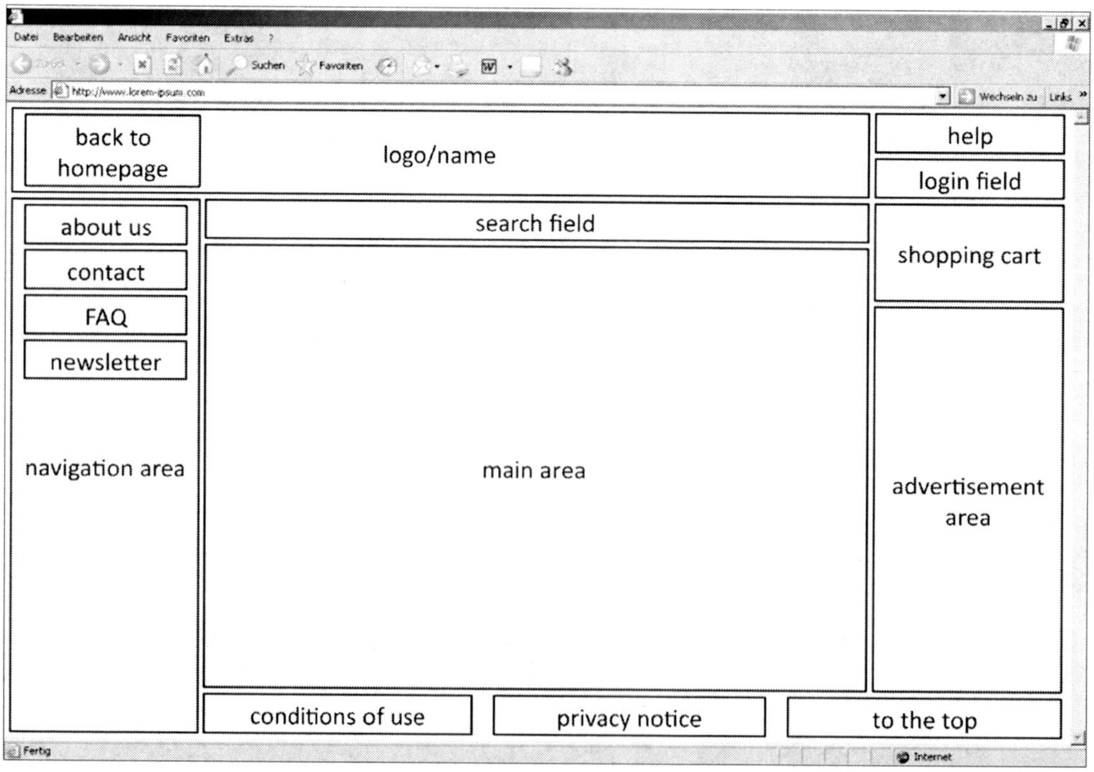

This study helped in determining the mental model of an online shop that Internet users developed over time.

The patterns of the locations of the web objects differ to those found by Bernard (2001). The *navigation area* (called *internal website links* in Bernard's study) is the only web object that was placed similarly. Differences were found e.g. with *advertisement*, *search field*, and a little less markedly with the *back to homepage* link. Taking into account Shaik and Lenz's (2006) similar findings concerning respective web objects, these differences could indeed be due to changes in the Internet's appearance over time. The important point is that people browsing the Internet seem to build up certain expectations of the location of common web objects on web pages. Regardless of the study environment, laboratory setting or online survey, a range of common web objects seems to be expected at certain locations.

Linking the Mental Model to Online Shop Design. How Does this Affect Users' Behavior?

It is one thing to determine the mental model for an online shop – but it is an entirely different to explore the consequences that arise when websites are developed according or contradictory to this model. Are users faster if the model is applied? Can we measure higher user satisfaction or lower error rates? Do people feel more comfortable while using shops that match their expectations? Or do they feel bored, because there are few elements to trigger curiosity, surprise, or fun?

A first step toward exploring the consequences of online shops with high vs. low prototypicality was conducted in a laboratory study using the mental model that we have previously described (Roth et al., 2010). The study applied the mental model and tested its external validity using real website screenshots (Roth et al., submitted). The term "prototypicality" describes how well a website fits the users' mental model. It was assumed that the higher the location typicality of a specific web object on a real web page, the faster it would be spotted on the screen and the fewer fixations would be needed until it was found.

As experimental stimuli, 108 real web page screenshots of three different website types were used: *online shop*, *online newspaper*, and *company web pages*. In this chapter, we will focus only on the results of the online shop stimuli (36 screenshots). Four different web objects were chosen because of their obvious importance for an online shop: *shopping cart*, *navigation*, *login*, and *search*. In order to obtain a wide range of location typicality of web objects, for each object nine different web pages were selected. This resulted in 36 web pages. Real websites were used to ensure a high ecological validity of the experiment. All screenshots were controlled for size, taken at the same resolution (1280x1024 pixels) and pasted into a browser window to increase realism. To obtain these stimuli set, first 100 web pages were randomly selected on the Internet. These screenshots were then examined by three independent raters. They determined how typically the respective web objects were placed for each screenshot according to Roth et al.'s (2010) data, using a matrix approach. Interrater reliability was high, with Cronbach alpha =.95. The exact procedure for the ratings can be found in Roth et al. (submitted).

Forty participants took part in a study using eye-tracking equipment (a device to record and analyze the users' looking behavior). The study had a within-subject design and was conducted in the usability lab of the University of Basel, Switzerland. The independent variable was the location typicality of specific web objects (low vs. high typicality); the dependent variables were *time to first fixation (TFF)* of the target web object and number of *fixations before target (FBT)*. Participants were instructed to spot a target web object as fast as possible on the respective web page stimulus. The target cue was displayed in written form for two seconds before stimulus

Table 3. Statistics for time to first fixation (TFF) in ms on web object targets with low and high typicality (adapted from Roth et al., submitted)

Target	Low typicality M (SD)	High typicality M (SD)	Statistics T	N[1]	p
search	2974 (2030)	949 (492)	18.0	38	<.001
shopping cart	4899 (2138)	2841 (2224)	143.0	39	<.001
navigation	934 (608)	647 (289)	155.0	39	<.001
login	3511 (1358)	902 (1144)	26.0	38	<.001

[1] Different sample sizes due to eye-tracking data loss

Table 4. Statistics for fixations before target spotting (FBT) on web object targets with low and high typicality (adapted from Roth et al., submitted)

Target	Low typicality M (SD)	High typicality M (SD)	Statistics T	N[1]	p
search	10.6 (4.4)	4.6 (1.6)	6.0	39	<.001
shopping cart	19.4 (8.1)	10.0 (5.3)	65.5	39	<.001
navigation	4.4 (1.8)	3.5 (0.8)	198.0	39	<.01
login	13.7 (4.1)	4.0 (1.5)	0.0	38	<.001

[1] Different sample sizes due to eye-tracking data loss

onset in the center of the screen. As soon as they spotted the target object on the web page, participants were supposed to press the space bar. The next target object was presented subsequently. To ensure that the respective mental model was triggered, participants were informed in advance that screenshots of online shops would follow. Stimuli were presented in random order.

In order to investigate the influence of location typicality on specific web objects regarding user performance, analyses were conducted for every single target web object separately. Web pages were median-split in two groups: low vs. high typicality. Performance data were then averaged over web pages, so that finally each participant had a low and high location typicality performance score for each target object. To compare TFF and FBT on target objects with low and high location typicality, Wilcoxon signed-rank tests were applied.

Results show that high location typicality reduced the TFF and decreased the FBT on all web objects significantly (see Tables 3 and 4). Participants were faster and needed fewer fixations before spotting the target object when the location typicality of the object was high. The averaged differences of TFF between low and high typicality ranged between 300ms up to over 2s. Finding atypically placed web objects took between one and nine fixations more than finding typically placed objects. These data show that the empirically accessed mental models match the behavioral data. Real web pages were tested to examine whether web objects that are located according to users' expectations are found faster than uncommonly placed objects. As hypothesized, participants spotted various web objects faster when they were placed at expected locations. Consequently, fewer fixations were needed to find the target web object. On online shop pages, as results show, the *login area* and *search field*, followed by the *shopping cart*, seem to be especially sensitive to location typicality.

As mentioned in the background section, guidance of attention is an unconscious process controlled by a combination of bottom-up (Braun and Sagi, 1990; Kowler et al., 1995; Reinagel and Zador, 1999) and top-down processes (Pelz and Canosa, 2001; Shinoda et al., 2001). Bottom-up processes imply that the eyes are primarily guided by salience. Top-down processes assume that saccade planning and direction are influenced by expectations, knowledge, and/or anticipation of following stimuli and search task. The results presented indicate that mental models are indeed able to trigger top-down processes of website perception and are therefore relevant to website design. Participants' eye movements were apparently steered by expectations of the target location. Irrespective of high or low typicality values, a certain area was fixated after stimulus onset. As expected, attention seems to be influenced by the participants' mental models of web object locations, which match to a great extent the data of Roth et al. (2010).

In summary, we can say that people have a strong idea of where important web objects in online shops are commonly placed and subconsciously use this idea for orientation. To decrease search time for the first visits to web pages, the location of commonly used web objects' should conform to expectations.

FUTURE RESEARCH DIRECTIONS

Although this first empirical study shows that conforming designs lead to faster search times, further studies would be insightful. If web objects placed at typical locations are found faster and remembered better, perception of satisfaction and the appeal of respective web pages may be improved. What is the relation between typicality index and usability measures of web pages? Casaló, Flavian and Guinalíu (2008) showed for instance that familiarity has a moderating role, when it comes to website loyalty and research within the field of the Technology Acceptance Model (TAM) has shown that website design influences consumer acceptance of online shopping (e.g. Ha & Stoel, 2009). Adapting website design to mental models also has implications for the aesthetic perception of a website (e.g. van Schaik and Ling, 2009, Wang, Hernandez and Minor, 2010): What is the relationship between prototypicality and aesthetic perception? There are many more facets of visual aesthetics that influence the first impression and use of a website such as, for example, symmetry, harmony, modernity, color, complexity, variety, grouping, structure, graphics, and simplicity (for an overview see Moshagen and Thielsch, in press). How does prototypicality relate to these factors?

It has to be stressed again that we do not suggest using the mental model, or users' expectations of web pages reported here as strict guidelines for web page design. Further studies should explore when web page design congruent to users' expectations is beneficial and when it should or may be broken. We do not know yet whether people prefer web pages that are consistent with their expectations or whether they prefer to be surprised by unexpected designs. Also, it may make sense to place some specific web objects at expected, and others at unexpected locations. Future studies should measure under what circumstances users have fewer problems solving specific tasks, and are more satisfied with the interface. Future studies should also investigate the influence of bottom-up processes, for example by examining the influence of object design on recognition. Additionally, more holistic tasks, including real interaction, should be studied.

CONCLUSION

In this chapter, we have shown how the mental model for an online shop was determined and validated. People browsing the Internet seem to build up certain expectations for the location of common web objects on web pages. Regardless of the study environment, laboratory setting or online survey, a range of common web objects seems to be expected at certain locations. These expectations are unconsciously used to trigger top-down processes that support the orientation when arriving for the first time at an online shop. This pattern of expectations can be used to support decisions regarding the design of online shops.

REFERENCES

Alexa - The Web Information Company. (2008). *Top sites by country*. Retrieved August 20, 2008, from http://www.alexa.com/site/ds/top%5C_500

Apple Inc. (2007). *Apple human interface guidelines*. Retrieved on September 8, 2008, from developer.apple.com/documentation/UserExperience/Conceptual/AppleHIGuidelines/index.html

Bartlett, F. C. (1932). *Remembering: A study in experimental and social psychology*. Cambridge, UK: Cambridge University Press.

Ben-Ari, M., & Yeshno, T. (2006). Conceptual models of software artifacts. *Interacting with Computers*, *18*(6), 1336–1350. doi:10.1016/j.intcom.2006.03.005

Bernard, M. (2001). *User expectations for the location of web objects*. Paper presented at the CHI - Interactive Posters.

Bernard, M. (2002). *Examining user expectations for the location of common e-commerce web objects*. Retrieved on August 21, 2008, from http://www.surl.org/usabilitynews/41/web_object-ecom.asp

Biederman, I., Mezzanotte, R. J., & Rabinowitz, J. C. (1982). Scene perception: Detecting and judging objects undergoing relational violations. *Cognitive Psychology*, *14*(2), 143–177. doi:10.1016/0010-0285(82)90007-X

Braun, J., & Sagi, D. (1990). Vision outside the focus of attention. *Perception & Psychophysics*, *48*(1), 45–58. doi:10.3758/BF03205010

Casalò, L., Flaviàn, C., & Guinalìu, M. (2008). The role of perceived usability, reputation, satisfaction and consumer familiarity on the website loyalty information process. *Computers in Human Behavior*, *24*(2), 325–345. doi:10.1016/j.chb.2007.01.017

Craik, K. (1943). *The nature of explanation*. Cambridge, UK: Cambridge University Press.

Davenport, J. L., & Potter, M. C. (2004). Scene consistency in object and background perception. *Psychological Science*, *15*(8), 559–564. doi:10.1111/j.0956-7976.2004.00719.x

Denham, P. (1993). Nine- to fourteen-year-old children's conception of computers using drawings. *Behaviour & Information Technology*, *12*(6), 346–358. doi:10.1080/01449299308924399

Farris, J. S., Jones, K. S., & Elgin, P. D. (2002). Users' schemata of hypermedia: What is so 'spatial' about a website? *Interacting with Computers*, *14*(5), 487–502. doi:10.1016/S0953-5438(02)00011-5

Gentner, D., & Stevens, A. L. (1983). *Mental models*. Hillsdale, NJ: Lawrence Erlbaum Associates.

Green, A. J. K. (1994). Interacting cognitive subsystems: A framework for considering the relationship between performance and knowledge representation. *Interacting with Computers*, *6*(1), 61–85. doi:10.1016/0953-5438(94)90005-1

Ha, S., & Stoel, L. (2009). Consumer e-shopping acceptance: Antecedents in a technology acceptance mode. *Journal of Business Research*, *62*(5), 565–571. doi:10.1016/j.jbusres.2008.06.016

Hsu, Y. (2006). The effects of metaphors on novice and expert learners' performance and mental-model development. *Interacting with Computers*, *18*(4), 770–792. doi:10.1016/j.intcom.2005.10.008

IBM. (2008). *Design principles.* Retrieved August 19, 2008, from https://www-01.ibm.com/software/ucd/designconcepts/designbasics.html

International Organization for Standardization. (1998). *9241-11 Ergonomic requirements for office work with visual display terminals (VDT)s - part 11 guidance on usability.* International Standards for Business, Government and Society.

Johnson-Laird, P. N. (1983). *Mental models.* Cambridge, MA: Harvard University Press.

Kerr, S. (1990). Wayfinding in an electronic database: The relative importance of navigational cues vs. mental models. *Information Processing & Management, 26*(4), 511–523. doi:10.1016/0306-4573(90)90071-9

Kowler, E., Anderson, E., Dosher, B., & Blaser, E. (1995). The role of attention in the programming of saccades. *Vision Research, 35*(13), 1897–1916. doi:10.1016/0042-6989(94)00279-U

Kveraga, K., Ghuman, A. S., & Bar, M. (2007). Top-down predictions in the cognitive brain. *Brain and Cognition, 65*(2), 145–168. doi:10.1016/j.bandc.2007.06.007

Leuthold, S., Bargas-Avila, J. A., & Opwis, K. (2008). Beyond Web content accessibility guidelines: Design of enhanced text user interfaces for blind internet users. *International Journal of Human-Computer Studies, 66*(6), 257–270. doi:10.1016/j.ijhcs.2007.10.006

Lindgaard, G., Fernandes, G., Dudek, C., & Brown, J. (2006). Attention Web designers: You have 50 milliseconds to make a good first impression! *Behaviour & Information Technology, 25*(2), 115–126. doi:10.1080/01449290500330448

Moshagen, M., & Thielsch, M. T. (in press). Facets of visual aesthetics. *International Journal of Human-Computer Studies.*

Norman, D. A. (1983). Some observations on mental models. In Gentner, D., & Stevens, A. L. (Eds.), *Mental models* (pp. 7–14). Hillsdale, NJ: Lawrence Erlbaum Associates.

Otter, M., & Johnson, H. (2000). Lost in hyperspace: Metrics and mental models. *Interacting with Computers, 13*(1), 1–40. doi:10.1016/S0953-5438(00)00030-8

Oulasvirta, A. (2004). Task demands and memory in web interaction: A levels of processing approach. *Interacting with Computers, 16*(2), 217–241. doi:10.1016/j.intcom.2003.12.004

Oulasvirta, A., Kärkkäinen, L., & Laarni, J. (2005). Expectations and memory in link search. *Computers in Human Behavior, 21*(5), 773–789.

Palmer, S. E. (1975). The effects of contextual scenes on the identification of objects. *Memory & Cognition, 3*(5), 519–526. doi:10.3758/BF03197524

Payne, S. J. (2007). Mental models in human–computer interaction. In Sears, A., & Jacko, J. A. (Eds.), *The human-computer interaction gandbook: Fundamentals, evolving technologies, and emerging applications* (pp. 63–76). New York, NY: Lawrence Erlbaum Associates. doi:10.1201/9781410615862.ch3

Pelz, J. B., & Canosa, R. (2001). Oculomotor behavior and perceptual strategies in complex tasks. *Vision Research, 41*(25-26), 3587–3596. doi:10.1016/S0042-6989(01)00245-0

Reinagel, P., & Zador, A. M. (1999). Natural scene statistics at the centre of gaze. *Network (Bristol, England), 10*(4), 341–350. doi:10.1088/0954-898X/10/4/304

Roth, S. P., Schmutz, P., Pauwels, S. L., Bargas-Avila, J. A., & Opwis, K. (2010). Mental models for Web objects: Where do users expect to find the most frequent objects in online shops, news portals, and company web pages? *Interacting with Computers, 22*(2), 140–152. doi:10.1016/j.intcom.2009.10.004

Roth, S. P., Tuch, A. N., Mekler, E. D. Z., Bargas-Avila, J. A., & Opwis, K. (in press). *Meet the users' expectations - Web objects are found faster on web pages when they are placed according to users' mental models*.

Rouse, W. B., & Morris, N. M. (1986). On looking into the black box: Prospects and limits in the search for mental models. *Psychological Bulletin, 100*(3), 349–363. doi:10.1037/0033-2909.100.3.349

Santa-Maria, L., & Dyson, M. C. (2008). *The effect of violating visual conventions of a website on user performance and disorientation. How bad can it be?* Paper presented at the SIGDOC.

Schmutz, P., Heinz, S., Métrailler, Y., & Opwis, K. (2009). Cognitive load in e-commerce applications - measurement and effects on user satisfaction. *Advances in Human-Computer Interaction, 2009*, 121494. doi:10.1155/2009/121494

Shaik, A. D., & Lenz, K. (2006). *Where's the search? Re-examining user expectations of web objects*. Retrieved August 21, 2008, from http://www.surl.org/usabilitynews/81/webobjects.asp

Shinoda, H., Hayhoe, M. M., & Shrivastava, A. (2001). What controls attention in natural environments? *Vision Research, 41*(25-26), 3535–3545. doi:10.1016/S0042-6989(01)00199-7

Spool, J. M. (2008). *The wheres and whens of users' expectations*. Retrieved August 19, 2008, from http://www.uie.com/articles/user_expectations/

Thatcher, A., & Greyling, M. (1998). Mental models of the Internet. *International Journal of Industrial Ergonomics, 22*, 299–305. doi:10.1016/S0169-8141(97)00081-4

Tuch, A. N., Bargas-Avila, J. A., Opwis, K., & Wilhelm, F. H. (2009). Visual complexity of websites: Effects on users' experience, physiology, performance, and memory. *International Journal of Human-Computer Studies, 67*(9), 703–715. doi:10.1016/j.ijhcs.2009.04.002

van Schaik, P., & Ling, J. (2009). The role of context in perceptions of the aesthetics of web pages over time. *International Journal of Human-Computer Studies, 67*(1), 79–89. doi:10.1016/j.ijhcs.2008.09.012

Vaughan, M. W., & Dillon, A. (2006). Why structure and genre matter for users of digital information: A longitudinal experiment with readers of a web-based newspaper. *International Journal of Human-Computer Studies, 64*, 502–526. doi:10.1016/j.ijhcs.2005.11.002

Wang, Y., Hernandez, M., & Minor, M. (2010). Web aesthetic effects on perceived online service quality and satisfaction in an e-tail environment: The moderating role of purchase task. *Journal of Business Research, 63*(9-10), 935–942. doi:10.1016/j.jbusres.2009.01.016

ADDITIONAL READING

Ben-Ari, M., & Yeshno, T. (2006). Conceptual models of software artifacts. *Interacting with Computers, 18*(6), 1336–1350. doi:10.1016/j.intcom.2006.03.005

Gentner, D., & Stevens, A. L. (1983). *Mental Models*. Hillsdale, New Jersey: Lawrence Erlbaum Associates.

Goodwin, K. (2009). *Designing for the Digital Age: How to create human-centered products and services*. Wiley Publishing.

International Organization for Standarization. (1998). *9241-11 Ergonomic requirements for office work with visual display terminals (VDT) s - Part 11 Guidance on usability*. International Standards for Business, Government and Society.

Jersey: Lawrence Erlbaum Associates.

Johnson-Laird, P. N. (1983). *Mental Models*. Cambridge: Harvard University Press.

Kveraga, K., Ghuman, A. S., & Bar, M. (2007). Top-down predictions in the cognitive brain. *Brain and Cognition*, *65*(2), 145–168. doi:10.1016/j.bandc.2007.06.007

Lindgaard, G., Fernandes, G., Dudek, C., & Brown, J. (2006). Attention web designers: You have 50 milliseconds to make a good first impression! *Behaviour & Information Technology*, *25*(2), 115–126. doi:10.1080/01449290500330448

Norman, D. A. (1983). Some observations on mental models. In Gentner, D., & Stevens, A. L. (Eds.), *Mental Models* (pp. 7–14). Hillsdale, New Jersey: Lawrence Erlbaum Associates.

Norman, D. A. (2002). *The Design of Everyday Things*. Reading, MA: Perseus Books.

Oulasvirta, A. (2004). Task demands and memory in web interaction: a Levels of Processing approach. *Interacting with Computers*, *16*(2), 217–241. doi:10.1016/j.intcom.2003.12.004

Oulasvirta, A., Kärkkäinen, L., & Laarni, J. (2005). Expectations and memory in link search. *Computers in Human Behavior*, *21*(5), 773–789.

Payne, S. J. (2007). Mental models in human–computer interaction. In Sears, A., & Jacko, J. A. (Eds.), *The Human-Computer Interaction Handbook: Fundamentals, Evolving Technologies, and Emerging Applications* (pp. 63–76). New York: Lawrence Erlbaum Associates. doi:10.1201/9781410615862.ch3

Roth, S. P., Schmutz, P., Pauwels, S. L., Bargas-Avila, J. A., & Opwis, K. (2010). Mental models for web objects: Where do users expect to find the most frequent objects in online shops, news portals, and company web pages? *Interacting with Computers*, *22*(2), 140–152. doi:10.1016/j.intcom.2009.10.004

Roth, S. P., Tuch, A. N., Mekler, E. D. Z., Bargas-Avila, J. A., & Opwis, K. (manuscript submitted for review). Meet the users' expectations - Web objects are found faster on web pages when they are placed according to users' mental models.

Schmutz, P., Heinz, S., Métrailler, Y., & Opwis, K. (2009). Cognitive load in eCommerce applications - measurement and effects on user satisfaction. *Advances in Human-Computer Interaction*, *2009*, 121494. doi:10.1155/2009/121494

Shinoda, H., Hayhoe, M. M., & Shrivastava, A. (2001). What controls attention in natural environments? *Vision Research*, *41*(25-26), 3535–3545. doi:10.1016/S0042-6989(01)00199-7

Tuch, A. N., Bargas-Avila, J. A., Opwis, K., & Wilhelm, F. H. (2009). Visual complexity of websites: Effects on users' experience, physiology, performance, and memory. *International Journal of Human-Computer Studies*, *67*(9), 703–715. doi:10.1016/j.ijhcs.2009.04.002

KEY TERMS AND DEFINITIONS

Bottom-Up Processes: Bottom-up processes operate on raw sensory input and are stimuli driven, so attention is oriented spontaneously towards salient visual features of a scene.

First Impression: The first overall gut feeling that users have when visiting a website for the first time.

Location of Web Objects: The exact placement of a website's key objects chosen by the web designer.

Mental Models: A virtual construct that explains someone's thought process about how something works in the real world.

Screen Design: The chosen way of presenting content via graphical user interfaces.

Top-Down Processes: Top-down processes require longer-term cognitive strategies, are goal-directed and guide attention through stored knowledge.

User Expectations: User's predictions of how a website will look like, where key elements are placed and how the website will behave while it is being used.

Section 3
Impact of Advanced Technologies on Consumer Behaviour

The chapters included in this section aim to investigate the impact of advanced technologies on consumer behavior, in terms of consumer opinions, interaction modalities, and purchasing decision, in order to outline the characteristics capable of major influencing their behavior, as well as the directions for innovative and effective retailing strategies.

Chapter 12
Recommendations to Buy in Online Retailing and Their Acceptance

Daniel Baier
Brandenburgische Technische Universität Cottbus, Germany

Eva Stüber
Brandenburgische Technische Universität Cottbus, Germany

ABSTRACT

Personal welcomings, individual assistance, as well as recommendations to inform and buy are becoming an integral part in online retailing. These new so-called personalization elements are assumed to increase the retailer's share of wallet and the customer's satisfaction. However, up to now only little is known about which external factors influence the customer's acceptance of such personalization elements. This chapter discusses the forms of recommendations to buy and how their acceptance can be measured using the well-known Technology Acceptance Model (TAM) approach. An experiment is used, where volunteers are offered an online shopping experience with individually generated recommendations to buy. The experiment shows how high the acceptance of the generated recommendations is and how close this acceptance is connected to the quality and shopping relevance of the recommendations. Even though the results are limited to the specific recommendation types used, they give important implications for an adequate design of modern online shops.

INTRODUCTION

For some years, online shopping gains growing importance (Kukar-Kinney, Close 2010). So, e.g., for Germany, the yearly large-scale W3B survey of Fittkau & Maaß shows that more than 65% of the German internet users have already bought products in the internet at more than 20 occasions. Additionally, already 10% of these online shoppers buy products in online shops every day (Fittkau & Maaß 2009, 2010). However, at the same time, the number of online shops and the competition between these shops is already high and still increases. Online shops that only rely on

DOI: 10.4018/978-1-60960-738-8.ch012

an anonymous appearance and a static presentation of the products restrict their success in the customer communication. Instead, the development of an own brand identity seems to be necessary (Hausmann, Siepke 2009).

Here, the personalization of the customer communication using personal welcomings, offerings of individualized recommendations to buy as well as virtual shopping advisers seems to be a promising possibility to differentiate from the competitors (Barlow et al. 2004; Holzwarth et al. 2006). As the technology to implement such an interactivity is available, the key to system acceptance is an appropriate mode of information presentation (Hausmann, Siepke 2009).

However, up to now only few studies have analyzed, which external factors influence the customer's acceptance of such an assistance. So, as one of the few studies, Fittkau & Maaß (2010) found that 27.6% of the German internet users rate recommendations to buy theoretically useful. However, only 15.4% of them have a positive attitude towards them, 60.2% have a neutral, whereas 24.3% have a negative one. The main reasons: 53.7% expect data security problems, 51.4% feel observed in their privacy, 16.5% ignore them (Fittkau & Maaß 2010). This chapter closes the gap in acceptance analysis by an experiment, where a modified version of the Technology Acceptance Model (TAM), originally developed by Davis (1989) and Davis et al. (1989) and extended by Venkatesh and Davis (2000) is used for measuring the customer's acceptance of recommendations to buy. For online retailing, O'Cass and Fenech (2003) as well as Ha and Stoel (2009) have already used this model in a similar setting.

After a short background section on recommendations to buy, the measurement approach is shortly presented, followed by a discussion of the experiment and the results. The chapter closes with future research directions and a short conclusion. The chapter here is an extended, updated, and for this issue modified version of Baier, Stüber (2010).

BACKGROUND: RECOMMENDATIONS TO BUY IN ONLINE RETAILING

Online Shopping

Online shopping – from the sellers point of view – is usually defined as "an example of a direct electronic channel linking sellers and customers" that "allows customers to select and purchase product items over an interactive electronic medium, typically through interactive television or the internet" (Chau et al. 2000). Since the electronic channel linking reduces the possibilities for the seller to demonstrate the solidity and the capabilities of the products and the sales organization, trust (Gefen et al. 2003) and content (Chau et al. 2000) play an important role in the shoppers selection and buying process.

Here, with respect to content and trust, gender is assumed to have an influence. Women are assumed to be emotionally less gratified and have less satisfaction from online shopping than men (Comegys et al. 2006, Hansen, Møller-Jensen 2009). Men would also be more trusting than women and think that online shopping is more convenient (Comegys et al. 2006). Additionally, women perceive higher risks than men in online purchasing in both probabilities and consequences (Garbarino, Strahilevitz 2004).

However, as online shopping has increased both male and female portions, these differences seem to get less and less important (Hansen, Møller-Jensen 2009). Now, more and more websites and online shops are more frequently used by women than men. So, e.g., in online clothes shopping, women are far more active than men. Consequently, for all online shops, an adequate placement of trustworthy electronic recommendations can play an essential role in the competition between on- and offline shops.

Recommendations to Buy

Recommendations to buy are the offers of alternatives or additional products in online retailing (or by sellers in traditional retailing). The main target group are the general online purchaser for whom these recommendations can provide advice, personality, and convenience. A well-known firm, that makes heavy use of such recommendations is amazon.com. Well-known online clothes shops in Germany, that use such recommendations, are esprit.de (see Figure 1 for an example during a session) or mexx.de: Depending on already selected products, such online shops make suitable offers. So, e.g., if the online purchaser has selected specific trousers, the shop proposes – depending on a database of suitable as well as often selected combinations – matching sneakers, shirts, waistbelts, or handbags.

Recommender systems use the consumer's saved and aggregated historical data to provide recommendations; they register the latest navigation and consumer behavior and consider additional information, or rather, they generate various combinations from suitable data sources. In so doing, the recommender systems in use have, amongst other things, a potential to generate attention and register consumer behavior as well as examine customer satisfaction and customer loyalty (Ochi et al. 2010). They work like "a salesperson who is highly knowledgeable about both the alternatives and the consumer's tastes" (Ariely et al. 2004), and so the offering of recommendations to buy is a possibility for recommendation systems' output (see Figure 2).

Such recommendations to buy are the internet's successful answer to the old-fashioned upselling or prompting at the point of sale (Ebster et al. 2006). So, e.g., Forrester Research estimates an amount from 10 to 30% of an online retailer's sales coming directly from recommendations (Mulpuru 2007). However, in the past, only the implementation and the functioning of the automatic generation of such recommendations – the so-called recommender systems and their algorithms – have been in the focus of research (e.g., Iacobucci et al. 2000, Wang, Benbasat 2007; Bodapati 2008). Similar to conjoint analysis, recommendation systems screen attractive alternatives. But they are models for individual-level prediction that can be useful even if there are a few alternatives (in contrast to conjoint analysis, which is typically used to screen many products to find a few attractive market options) (Ansari et al. 2000). As a function of the context, different recommendation methods can be used (see Table 1).

Figure 1. Recommendations to buy during an online session at a German textile shop

Figure 2. Classification of recommendation systems (Hansen et al. 2007)

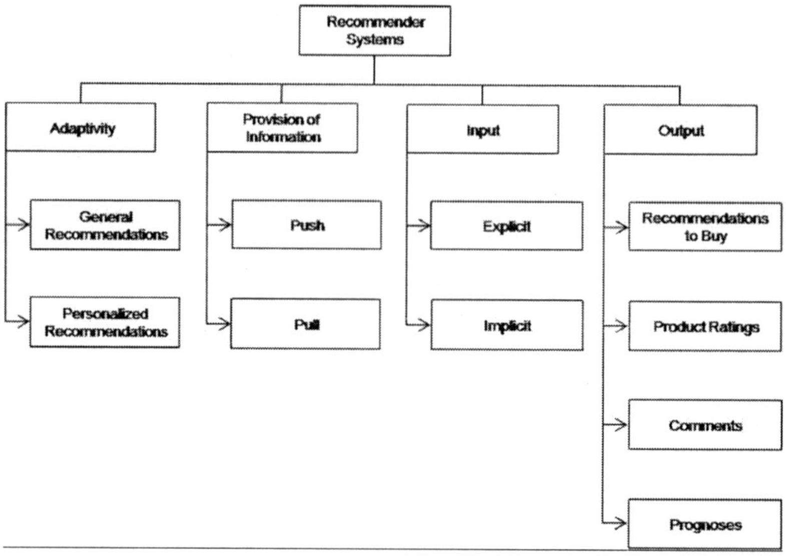

Up to now, only limited research on the customer's acceptance of such recommendations to buy in online shops and their satisfaction with these recommendations is available. As such, computer based recommendations have a significant positive impact on product selection, customer satisfaction and customer loyalty (Senecal, Nantel 2004; Aksoy et al. 2006), even though it is not clear, whether the automatically generated recommendations really help consumers discover new products or whether they only reinforce the popularity of already-popular products (Fleder, Hosanagar 2009). At least, it is agreed that the effort required to search for products can be reduced and the quality of buying decisions can be increased (Häubl, Trifts 2000). In addition

Table 1. Recommendation methods

	Collaborative filtering methods	**Content-based methods**	**Hybrid methods**
Approach	Usage of buyers' ratings of products rather than content information and finding of people-to-people similarities between buyers	Usage of content informa-tion to build profiles of pro-ducts and buyers that are then used to calculate the match between a specific buyer and product	Combination of different rudiments; diverse types, e.g., weighted, switching, mixed, feature combination, cascade, feature augment-tation, and meta-level
Example	amazon.com	Internet search engines	
Limitations	Poor recommendation qua-lity under data sparsity Limited ability to recom-mend new products Usage of ad hoc prediction algorithms, which are not based on a statistical model (uncertainty)	Bad recommendations since it only considers the prespe-cified contents for products Bias to restrict the scope of the recommendation to items similar to those the consumer already rated No recommendations for people who provide no pre-ference information	Depending on the method's combination
References	Iacobucci et al. 2000; Herlocker et al. 2002; Mild, Reutterer 2003; Takács et al. 2009	Ansari et al. 2000; Ansari, Mela 2003; Fleder, Hosanagar 2009	Kim, Kim 2001; Burke 2002; Li et al. 2005; Fleder, Hosanagar 2009

to the positive impact on customer satisfaction, it is also possible to determine an increase in confidence (Fitzsimons, Lehmann 2004).

However, open research questions still are which consumers actively use these recommendations, how, why and whether they are satisfied by these recommendations to buy.

MAIN FOCUS: HOW TO MEASURE ACCEPTANCE FOR RECOMMENDATIONS

Issues, Controversies, Problems: The Technology Acceptance Model

The Technology Acceptance Model (TAM) is one of the most influential measurement approaches for relating design alternatives to user acceptance of new information technologies or systems. TAM was proposed by Davis (1989) and Davis et al. (1989) and bases essentially on Fishbein and Aijzen (1975)'s Theory of Reasoned Action (TRA), a theory that relates the general construct behavioral intention to two other constructs, namely attitude and subjective norm. „The goal of TAM is to provide an explanation of the determinants of computer acceptance that is general, capable of explaining user behavior across a broad range of end-user-computing technologies and user populations, while at the same time being both parsimonious and theoretically justified." (Davis et al. 1989). The great importance can be seen from the study by Lee et al. 2003 who found out that Davis (1989) and Davis et al. (1989) were cited in 698 papers between 1989 and 2003.

On the other side, as a major weak point of the TAM approach, often its theoretical foundation is named. So, e.g., Bagozzi (2007) questioned the general approach by doubting that usage behavior could be predicted by salient beliefs, especially by "adding up measures for perceived usefulness and perceived ease of use". Also, the deterministic modelling approach was questioned by him allowing no subjective evaluations and reflections in the course of action. Chuttur (2009) has summed up the scepticism of many critical authors w.r.t. to "TAM as a theory" by its lack of falsifiability, its questionable heuristic value, its limited explanatory and predictive power, its triviality, and its lack of any practical value.

However, in many applications TAM has shown its practical value as a flexible modelling instrument. So, e.g., when analyzing the acceptance of a new information system, it helps to essentially relate the intention to use the new system to two constructs: the perceived usefulness ("the degree to which a person believes that using a particular system would enhance his or her job performance", Davis 1989) and the perceived ease-of-use of the system ("the degree to which a person believes that using a particular system would be free from effort", Davis 1989). Figure 3 visualizes the relations between the main constructs of TAM in the right box.

Figure 3. Modified Technology Acceptance Model (TAM) approach

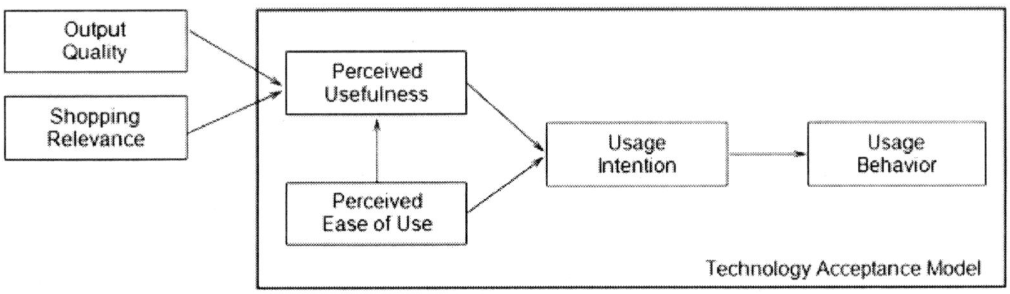

For these constructs, robust and valid questionnaire instruments have been developed. In a huge number of follow-up studies, these measurement scales have been refined and extended to different settings, nowadays TAM is also very popular in online retailing. Here, first of all the focus of the investigations is trust as a key factor (e.g., Gefen et al. 2003; Pavlou 2003; Pavlou, Fygenson 2006) or rather the adoption of the internet as a distribution channel and their influencing factors (e.g., O`Cass, Fenech 2003; Lee et al. 2006; Kamaruizaman 2007; Ha, Stoel 2009). The studies investigate personality traits (O`Cass, Fenech 2003), shopping experiences on the internet, involvement with product categories (McKechnie et al. 2006) as well as risk propensity and affinity to computers (Stern et al. 2008).

Applied to recommendations to buy as the information system under study, questionnaire items for the construct "Perceived ease of use" could be "Additional information is easy to get.", "The handling of the system is easy." For the construct "Perceived usefulness" items like "Recommendations are useful for me." or "Recommendations improve my performance." could be derived from the literature. On the other side, for the behavioral constructs of the TAM approach, for the construct "Usage intention" items like "I will use the recommendations for searching.", "I will recommend the usage to my friends.", "I will use the recommendations for shopping.", or "I will rethink my recommendation usage positively.", for "Usage behavior" items like "I use the recommendations very often.", "I use the recommendations for additional products." or "I use the recommendations to improve my results." could be useful.

Venkatesh and Davis (2000) extended the original TAM model to explain perceived usefulness and usage intention in terms of social influence and cognitive instrumental processes. In our modified TAM approach, we use similar extensions, in Figure 1 these constructs are already added: Two constructs for measuring the cognitive instrumental process, output quality and shopping relevance, are added to the traditional TAM approach. In the recommendation to buy example, for "Output quality" the items "The quality of recommendations is high.", "Recommended products fit to my wants." or "Recommended products fit to my desire." could be adequate, for "Shopping relevance" the items "Recommendations make shopping more simple.", "Recommendations make shopping more convenient.", "Recommendations lead to desired products." or "Recommendations lead to new relevant products." could be used. Constructs for measuring the social influence, e.g., "Subjective norm" with items like "People who influence my behavior think that I should use the recommendations." or "Image" with items like "People who use the recommendations have more prestige than others that do not." (cf. Venkatesh, Davis 2000) were not included in the approach, since for online shopping a more private usage situation is expected.

Solutions and Recommendations: An Experimental Application

In order to test the above TAM approach for recommendations to buy, a laboratory experiment with 100 students was conducted at a German university. The gender distribution in the sample was 65% male and 35% female, they were 19 to 32 years old with an average of 24.1 years. 82% of them were Germans. So the sample matches the gender and age distribution of online shoppers in Germany with more male and younger people in the population (Fittkau & Maaß 2009, 2010). The data collection form was a standardized one-to-one interview. The students were introduced to the topic of recommendations to buy during online-shopping and – then – received a first questionnaire. Standardized items on online purchasing behavior, advantages and disadvantages of online shopping, importance of brands, of advice, and of recommendations to buy in general were used to introduce the topic.

Then, the students were asked to visit a preselected online shop for buying clothes and to "purchase" a leisure outfit consisting of 2-4 parts. "Purchase" means that they only had to select their parts during the session and to put them in the online market basket. A clothes online shop was selected in order to analyze whether gender (or topic) differences play a role in technology acceptance.

The chosen online shop (slightly modified for the experiment) uses hybrid methods (cf. Table 1) for the generation of its recommendations to buy: Users can register via a user profile and declare their preferences w.r.t. the products offered in the shop. So, collaborative filtering methods (using e.g. clustering), can be applied to recommend similar additional products to similar people. Also, the shop uses content-based methods to generate recommendations to buy: For each clothing (e.g. trousers) exists a list of suitable accessoires (e.g. suitable shirts, waistbelts, shoes w.r.t. to style and color) which is used for recommendations to buy. During the session, the shop derived recommendations to buy. Using a web proxy, the students' activities on the website and the systems derived recommendations to buy were tracked for a later analysis.

Afterwards, a second questionnaire was presented, now containing the items of the TAM approach as discussed in section 2 w.r.t. the constructs "shopping relevance", "output quality", "perceived ease of use", "perceived usefulness", "usage intention", and "usage behavior". All items of the approach were measured on a 7 point Likert scale (with "1"="strongly disagree", ... "7"= "strongly agree"). Additionally, socio-demografic data were collected.

Nearly all participants of the study were active online shoppers and aware that recommendations to buy appear quite regularly during an online shop session. Nevertheless, the first questionnaire was also used to focus their attention on the appearing recommendations to buy during the shop visit so that in the second interview the questions w.r.t. "output quality", "perceived ease of use", "usage intention" and so on could be answered sufficiently.

For measuring the construct "usage behavior" also the tracking files of the shop visits could be used where the "purchasing" activities and the system's recommendations to buy were stored. However, as usual with TAM approaches, also standard items for measuring "observable" "usage behavior" were used in the second questionnaire (see Porter, Donthu 2006).

From the first questionnaire, various descriptive results could be obtained. So, e.g., the students supported with their answers the traditional advantages of online shopping: As important advantages were named the possibilities to shop 24 hours, to save time and to compare products and prices easily. In contrast the missing possibility to touch products and to evaluate their quality were named as disadvantages. As product categories where they expect that recommendations to buy are most useful, PC hard- and software and consumer electronics were named.

From the second questionnaire the results with respect to the TAM approach could be derived. Table 2 shows the construct results with the literature source. For each construct, the items, their mean values and standard deviations as well as their principal component loadings, the corresponding variance explained and Cronbach's alpha as a reliability measure are displayed. One can see that the reliability measures and factor loadings are quite good, supporting the reliability of the TAM measurement constructs.

As a gender influence was expected in the experiment, also the constructs were analyzed with respect to significant difference (see Table 3). Here, one can see, that with the general adequacy of the topic, also usage intentions and usage behavior of the recommendations is rated more positively.

The results with respect to the modified TAM approach are reflected in Table 4. The dependencies of the constructs were analyzed using OLS

Table 2. Summary of construct results

Construct	Mean	Standard deviation	Component loadings	Variance explained	Relia-bility	Literature Source
Shopping relevance				*79.5%*	*0.913*	*Königsdorfer, Gröppel-Klein (2007)*
Recommendations …						
… make shopping more simple.	4.61	1.36	0.905			
… make shopping more convenient.	4.45	1.34	0.892			
… lead to desired products.	5.44	1.43	0.902			
… lead to new relevant products.	5.14	1.48	0.867			
Output quality				*89.9%*	*0.841*	*Venkatesh, Davis (2000)*
The quality of recommendations is high.	4.41	1.28	0.893			
Recommended products fit to my wants.	4.15	1.49	0.851			
Recommended products fit to my desire.	4.03	1.48	0.877			
Perceived ease of use				*85.2%*	*0.826*	*Davis (1989) O'Cass, Fenech (2003)*
Additional information is easy to get.	5.10	1.34	0.923			
The handling of the system is easy.	4.95	1.40	0.923			
Perceived usefulness				*82.5%*	*0.787*	*Venkatesh, Davis (2000) O'Cass, Fenech (2003)*
Recommendations …						
… are useful for me.	4.69	1.64	0.908			
… improve my performance.	4.57	1.60	0.908			
Usage intention				*66.0%*	*0.826*	*Venkatesh, Davis (2000) Moon, Kim (2001)*
I will …						
… use the recommendations for searching.	5.12	1.43	0.716			
… recommend the usage to my friends.	4.69	1.61	0.828			
… use the recommendations for shopping.	4.36	1.85	0.840			
… rethink my recommendation usage positively.	4.42	1.53	0.859			
Usage behavior				*66.58%*	*0.747*	*Porter, Donthu (2006)*
I use the recommendations …						
… very often.	3.76	1.94	0.812			
… for additional products.	4.34	1.73	0.836			
… to improve my results.	3.95	1.72	0.799			

mean values w.r.t. "1"="strongly disagree", … "7"= "strongly agree"

(ordinary least squares). Table 4 shows the respective results for the total sample and also for the gender-specific sub-groups. One can see, that expected TAM dependencies are confirmed by the model. The constructs perceived usefulness and perceived ease of use have an influence on usage intention and usage behavior.

Additionally to the analysis via OLS, a PLS (partial least squares) analysis was used. PLS is a group of alternating least squares algorithms that extend principal components analysis and canonical correlation analysis, which have been originally designed by Wold (1974, 1982, 1985) for the analysis of high dimensional data. Over the years the methods have been improved and extended (see e.g. Tenenhaus et al. 2005 for an overview). Nowadays, PLS is a standard tool for developing and analyzing path models. Standard software, e.g. SmartPLS (Ringle et al. 2005, Henseler et al. 2009), is available for the analysis.

The main idea behind PLS is a combination of two sets of linear equations, the so-called inner and outer model. Whereas the inner model relates unobserved or latent variables, the outer model relates latent variables and its observed or manifest variables. Figure 5 already shows the modified TAM approach from Figure 4 in a PLS setting.

Linear relationships between the latent variables (here: output quality, perceived usefulness, …) form the inner model, linear relationships between the latent variables and indicators (here, e.g., OQ1, OQ2, …) form the outer model. The

Table 3. Gender differences with respect to the constructs

Construct	female		male		T-value
	Mean	Standard deviation	Mean	Standard deviation	
Shopping relevance					
Recommendations …					
… make shopping more simple.	4.61	1.42	4.61	1.32	0.71
… make shopping more convenient.	4.71	1.41	4.30	1.29	1.46
… lead to desired products.	5.60	1.37	5.35	1.46	0.82
… lead to new relevant products.	5.22	1.42	5.09	1.51	0.44
Output quality					
The quality of recommendations is high.	4.60	1.26	4.31	1.29	1.09
Recommended products fit to my wants.	4.20	1.45	4.12	1.52	0.25
Recommended products fit to my desire.	4.11	1.53	3.98	1.47	0.41
Perceived ease of use					
Additional information is easy to get.	5.31	1.32	4.98	1.35	1.17
The handling of the system is easy.	5.17	1.27	4.83	1.46	1.16
Perceived usefulness					
Recommendations …					
… are useful for me.	4.74	1.72	4.66	1.51	0.24
… improve my performance.	4.86	1.61	4.42	1.62	1.32
Usage intention					
I will …					
… use the recommendations for searching.	5.83	1.15	4.74	1.43	3.89**
… recommend the usage to my friends.	5.20	1.47	4.41	1.63	2.37**
… use the recommendations for shopping.	4.66	1.78	4.20	1.87	1.18
… rethink my recommendation usage positively.	4.80	1.59	4.21	1.47	1.84**
Usage behavior					
I use the recommendations …					
… very often.	3.74	1.99	3.77	1.93	-0.64
… for additional products.	4.80	1.75	4.09	1.68	1.98**
… to improve my results.	4.40	1.96	3.70	1.55	1.94**

mean values w.r.t. "1"="strongly disagree", … "7"= "strongly agree", **: significant differences with $p<0.1$

Table 4. Summary of regressions w.r.t. the technology acceptance model

Predicted construct	Predictor	All (n=100)		Female (n=35)		Male (n=65)	
		Path	R^2	Path	R^2	Path	R^2
Perceived usefulness			.62		.73		.58
	Shopping relevance	.418**		.411**		.368**	
	Output quality	.280**		.573**		.127**	
	Perceived ease of use	.206**		-.106		.309**	
Usage intention			.27		.32		.23
	Perceived ease of use	.308**		.302**		.292**	
	Perceived usefulness	.277**		.347**		.246**	
Usage behavior			.32		.35		.28
	Usage intention	.564**		.590**		.531**	

outer model is specified in the reflective mode which means that causal relationships from the latent variables to the indicators are assumed. The PLS algorithm for estimating the relationship parameters (weights) is now a sequence of regressions for approximating the latent variable scores

Figure 4. Modified TAM approach with results using OLS (standardized regression coefficients and R2 measures; all coefficients are significant with p<0.1)

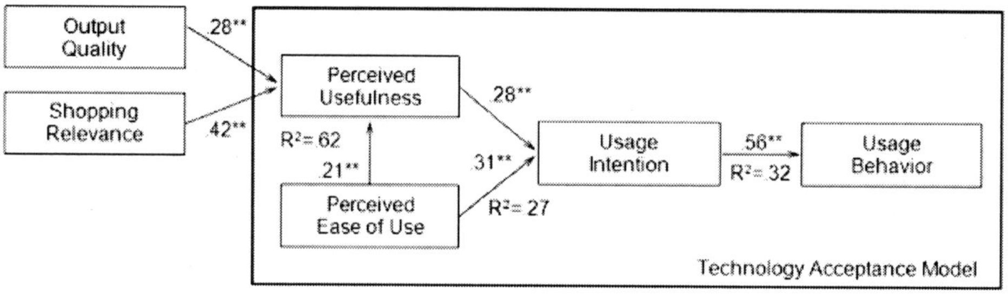

Figure 5. Modified TAM approach with results using PLS (n=100)

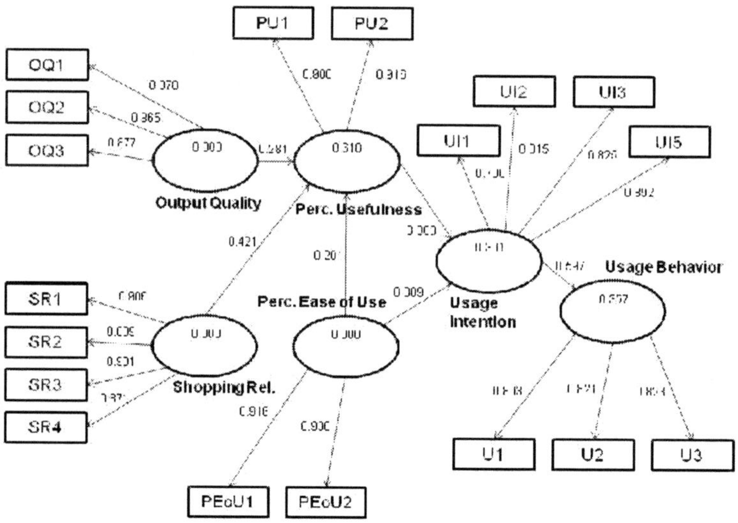

followed by an estimation of the path coefficients. Figure 5 shows quite similar path coefficients as in the OLS approach in Figure 4 which supports the above findings.

Additionally, for analyzing gender specific path differences, a multi-group analysis was performed. Following Keil et al. (2000) as well as Henseler et al. (2009), the male and female (sub)groups were exposed to bootstrap analyses, each with 200 bootstrap samples. Mean paths and standard errors as well as t-statistics following Keil et al. (2000) were calculated (see Table 5).

Again, as with the OLS results, main gender specific differences could be found w.r.t. the effect of perceived ease of use and perceived usefulness. The gender specific difference for these path coefficients shows to be highly significant.

FUTURE RESEARCH DIRECTIONS

Further research needs to be done, e.g., concerning group-specific differences between users and non-users of recommendations and the effects of recommendations on consumer's satisfaction. One of the main problems is the subjectivity of the user's responses during the survey. Stüber (2011) has extended the above approach by incorporating a more objective component by using an eye tracking approach for measuring the time the user

Table 5. PLS-bootstrapping-based multi-group analysis w.r.t. the technology acceptance model

Predicted construct	Predictor	Female (n=35)		Male (n=65)		
		Path	Std.err.	Path	Std.err.	t-Value
Perceived usefulness	Shopping relevance	.438	.071	.342	.098	-.279
	Output quality	.565	.098	.253	.091	-.733
	Perceived ease of use	-.129	.073	.315	.086	2.984**
Usage intention	Perceived ease of use	.341	.089	.291	.122	-.169
	Perceived usefulness	.340	.103	.295	.115	.153
Usage behavior	Usage intention	.600	.057	.600	.056	-.001

**: significant differences with $p<0.1$; 200 samples

looks at the recommendations and whether this duration is correlated to the TAM latent constructs and the concrete selling success. Moreover, the laboratory experiment should be replaced by a more convenient data collection where the users are interviewed and observed in their normal online buying situation at home.

CONCLUSION

The modified TAM approach seems to be viable approach to measure user's acceptance for recommendations to buy. The two key beliefs (perceived usefulness and perceived ease-of-use) especially account for technology usage. Additionally, shopping relevance and output quality have an influence. However, if the topic is adequate to the interested user (as in the case of recommendations in clothes online shopping), the effect of perceived ease of use could not be confirmed.

REFERENCES

Aksoy, L., Bloom, P. N., Lurie, N. H., & Cooil, B. (2006). Should recommendation agents think like people? *Journal of Service Research*, *8*(4), 297–315. doi:10.1177/1094670506286326

Ansari, A., Essegaier, S., & Kohli, R. (2000). Internet recommendation systems. *JMR, Journal of Marketing Research*, *37*(3), 363–375. doi:10.1509/jmkr.37.3.363.18779

Ansari, A., & Mela, C. F. (2003). E-customization. *JMR, Journal of Marketing Research*, *40*(2), 131–145. doi:10.1509/jmkr.40.2.131.19224

Ariely, D., Lynch, J. G. Jr, & Aparicio, M. IV. (2004). Learning by collaborative and individual-based recommendation agents. *Journal of Consumer Psychology*, *14*(1/2), 81–95. doi:10.1207/s15327663jcp1401&2_10

Bagozzi, R. P. (2007). The legacy of the technology acceptance model and a proposal for a paradigm shift. *Journal of the Association for Information Systems*, *8*(4), 244–254.

Baier, D., & Stüber, E. (2010). Acceptance of recommendations to buy in online retailing. *Journal of Retailing and Consumer Services*, *17*(3), 173–180. doi:10.1016/j.jretconser.2010.03.005

Barlow, A. K. J., Siddiqui, N. Q., & Mannion, M. (2004). Development in Information and Communication Technologies for retail marketing channels. *International Journal of Retail and Distribution Management*, *32*(3), 157–163. doi:10.1108/09590550410524948

Bodapati, A. V. (2008). Recommendation systems with purchase data. *JMR, Journal of Marketing Research*, *45*(1), 77–93. doi:10.1509/jmkr.45.1.77

Burke, R. (2002). Hybrid recommender systems: Survey and experiments. *User Modeling and User-Adapted Interaction, 12*(4), 331–370. doi:10.1023/A:1021240730564

Chau, P. Y. K., Au, G., & Tam, K. Y. (2000). Impact of information presentation modes on online shopping: An empirical evaluation of a broadband interactive shopping service. *Journal of Organizational Computing and Electronic Commerce, 10*(1), 1–22.

Chuttur, M. Y. (2009). *Overview of the technology acceptance model: Origins, developments and future directions*. Indiana University, USA. Sprouts: Working Papers on Information Systems, 9 (37).

Comegys, C., Hannula, M., & Väisänen, J. (2006). Longitudinal comparison of Finnish and US online shopping behaviour among university students: The five-stage buying decision process. *Journal of Targeting. Measurement and Analysis for Marketing, 14*(4), 336–356. doi:10.1057/palgrave.jt.5740193

Davis, F. D. (1989). Perceived usefulness, perceived ease of use, and user acceptance of information technology. *Management Information Systems Quarterly, 13*(3), 319–340. doi:10.2307/249008

Davis, F. D., Bagozzi, R. P., & Warshaw, P. R. (1989). User acceptance of computer technology: A comparison of two theoretical models. *Management Science, 35*(8), 982–1003. doi:10.1287/mnsc.35.8.982

Ebster, C., Wagner, U., & Valis, S. (2006). The effectiveness of verbal prompts on sales. *Journal of Retailing and Consumer Services, 13*(3), 169–176. doi:10.1016/j.jretconser.2005.08.003

Fishbein, M., & Ajzen, I. (1975). *Belief, attitude, intention, and behavior: An introduction to theory and research*. Reading, MA: Addison-Wesley.

Fittkau and Maaß. (2009). *World Wide Web-Benutzer-Analyse (W3B), 27*. Fittkau and Maaß. Retrieved from www.w3b.org

Fittkau and Maaß. (2010). *World Wide Web-Benutzer-Analyse (W3B), 29*. Fittkau and Maaß. Retrieved from www.w3b.org

Fitzsimons, G. J., & Lehmann, D. R. (2004). Reactance to recommendations: When unsolicited advice yields contrary responses. *Marketing Science, 23*(1), 82–94. doi:10.1287/mksc.1030.0033

Fleder, D., & Hosanagar, K. (2009). Blockbuster culture's next rise or fall: The impact of recommender systems on sales diversity. *Management Science, 55*(5), 697–712. doi:10.1287/mnsc.1080.0974

Garbarino, E., & Strahilevitz, M. (2004). Gender difference in the perceived risk of buying online and the effects of receiving a site recommendation. *Journal of Business Research, 57*(7), 768–775. doi:10.1016/S0148-2963(02)00363-6

Gefen, D., Karahanna, E., & Straub, D. W. (2003). Trust and TAM in online shopping: An integrated model. *Management Information Systems Quarterly, 27*(1), 51–90.

Ha, S., & Stoel, L. (2009). Consumer e-shopping acceptance: Antecendents in a technology acceptance model. *Journal of Business Research, 62*(5), 565–571. doi:10.1016/j.jbusres.2008.06.016

Hansen, H. R., Knotzer, N., & Madlberger, M. (2007). Empfehlungssysteme zur Verkaufsberatung im Internet. State-of-the-Art und Konsumentenakzeptanz. *Wirtschaftsinformatik, 49*(Sonderheft), 50-61.

Hansen, T., & Møller-Jensen, J. (2009). Shopping orientation and online clothing purchases: The role of gender and purchase situation. *European Journal of Marketing, 43*(9/10), 1154–1170. doi:10.1108/03090560910976410

Häubl, G., & Trifts, V. (2000). Consumer decision making in online shopping environments: The effects of interactive decision aids. *Marketing Science, 19*(1), 4–21. doi:10.1287/mksc.19.1.4.15178

Hausmann, A. V., & Siekpe, J. S. (2009). The effect of Web interface features on consumer online purchase intentions. *Journal of Business Research*, *62*(1), 5–13. doi:10.1016/j.jbusres.2008.01.018

Henseler, J., Ringle, C. M., & Sinkovics, R. R. (2009). The use of partial least squares path modeling in international marketing. *Advances in International Marketing*, *20*, 277–319.

Herlocker, J. L., Konstan, J. A., & Riedl, J. T. (2002). An empirical analysis of design choices in neighborhood-based collaborative filtering algorithms. *Information Retrieval*, *5*(4), 287–310. doi:10.1023/A:1020443909834

Holzwarth, M., Janiszewski, C., & Neumann, M. M. (2006). The influence of avatars on consumer shopping behavior. *Journal of Marketing*, *70*(4), 19–36. doi:10.1509/jmkg.70.4.19

Iacobucci, D., Arabie, P., & Bodapati, A. (2000). Recommendation agents on the internet. *Journal of Interactive Marketing*, *14*(3), 2–11. doi:10.1002/1520-6653(200022)14:3<2::AID-DIR1>3.0.CO;2-J

Kamaruizaman, Y. (2007). Adoption of travel e-shopping in the UK. *International Journal of Retail and Distribution Management*, *35*(9), 703–719. doi:10.1108/09590550710773255

Keil, M., Tan, B. C., Wei, K.-K., Saarinen, T., Tuunainen, V., & Wassenaar, A. (2000). A cross-cultural study on escalation of commitment behaviour in software projects. *Management Information Systems Quarterly*, *24*(2), 299–325. doi:10.2307/3250940

Kim, B.-D., & Kim, S.-O. (2001). A new recommender system to combine content-based and collaborative filtering systems. *Journal of Database Marketing*, *8*(3), 244–252. doi:10.1057/palgrave.jdm.3240040

Königstorfer, J., & Gröppel-Klein, A. (2007). Experiences of failure and anger when using the mobile and wired Internet: The interference of acceptance-and resistance-driving factors. *Marketing Journal of Resource Management*, *29*(1e), 34–47.

Kukar-Kinney, M., & Close, A. G. (2010). The determinants of consumers' online shopping cart abandonment. *Journal of the Academy of Marketing Science*, *38*(2), 240–250. doi:10.1007/s11747-009-0141-5

Lee, H.-H., Fiore, A. M., & Kim, J. (2006). The role of the technology acceptance model in explaining effects of image interactivity technology on consumer responses. *International Journal of Retail and Distribution Management*, *34*(8), 621–644. doi:10.1108/09590550610675949

Lee, Y., Kozar, K. A., & Larsen, K. R. (2003). The technology acceptance model: Past, present, and future. *Communications of AIS*, *12*, 752–780.

Li, Y., Lu, L., & Li, X. (2005). A hybrid collaborative filtering method for multiple-interests and multiple-content recommendation in e-commerce. *Expert Systems with Applications*, *28*(1), 67–77. doi:10.1016/j.eswa.2004.08.013

McKechnie, S., Winkihofer, H., & Ennew, C. (2006). Applying the technology acceptance model to the online retailing of financial services. *International Journal of Retail and Distribution Management*, *34*(4/5), 388–410. doi:10.1108/09590550610660297

Mild, A., & Reutterer, T. (2003). An improved collaborative filtering approach for predicting cross-category purchases based on binary market basket data. *Journal of Retailing and Consumer Services*, *10*(3), 123–133. doi:10.1016/S0969-6989(03)00003-1

Moon, J., & Kim, Y. (2001). Extending the TAM for a World-Wide-Web context. *Information & Management*, *38*(4), 217–230. doi:10.1016/S0378-7206(00)00061-6

Mulpuru, S. (2007). *Which personalization tools work for e-commerce - and why*. Forrester Research.

O'Cass, A., & Fenech, T. (2003). Web retailing adoption: Exploring the nature of Internet users Web retailing behaviour. *Journal of Retailing and Consumer Services*, *10*(2), 81–94. doi:10.1016/S0969-6989(02)00004-8

Ochi, P., Rao, S., Takayama, L., & Nass, C. (2010). Predictors of user perceptions of Web recommender systems: How the basis for generating experience and search product recommendations affects user responses. *International Journal of Human-Computer Studies*, *68*(8), 472–482. doi:10.1016/j.ijhcs.2009.10.005

Pavlou, P. A. (2003). Consumer acceptance of electronic commerce: Integrating trust and risk with the technology acceptance model. *International Journal of Electronic Commerce*, *7*(3), 101–134.

Pavlou, P. A., & Fygension, M. (2006). Understanding and predicting electronic commerce adoption: An extension of the theory of planned behavior. *Management Information Systems Quarterly*, *30*(1), 115–143.

Porter, C. E., & Donthu, N. (2006). Using the technology acceptancem to explain how attitudes determine Internet usage: The role of perceived access barriers and demographics. *Journal of Business Research*, *59*(9), 999–1007. doi:10.1016/j.jbusres.2006.06.003

Ringle, C. M., Wende, S., & Will, A. (2005). *SmartPLS2.0 M3*. Retrieved from www.smartpls.de

Senecal, S., & Nantel, J. (2004). The influence of online product recommendations on consumers' online choices. *Journal of Retailing*, *80*(2), 159–169. doi:10.1016/j.jretai.2004.04.001

Stern, B. B., Royne, M. B., Stafford, T. F., & Bienstock, C. C. (2008). Consumer acceptance of online auctions: An extension and revision of the TAM. *Psychology and Marketing*, *25*(7), 619–636. doi:10.1002/mar.20228

Stüber, E. (2011). *Kaufempfehlungen als Personalisierungsansatz im Internethandel – Eine empirische Untersuchung der Akzeptanz am Beispiel der Bekleidungsbranche*. Gabler.

Takács, G., Pilászy, I., Németh, B., & Tikk, D. (2009). Scalable collaborative filtering approaches for large recommender systems. *Journal of Machine Learning Research*, *10*, 623–656.

Tenenhaus, M., Esposito Vinzi, V., Chatelin, Y.-M., & Lauro, C. (2005). PLS path modeling. *Computational Statistics & Data Analysis*, *48*(1), 159–205. doi:10.1016/j.csda.2004.03.005

Venkatesh, V., & Davis, F. D. (2000). A theoretical extension of the technology acceptance model: Four longitudinal field studies. *Management Science*, *46*(2), 186–204. doi:10.1287/mnsc.46.2.186.11926

Wang, W., & Benbasat, I. (2007). Recommendation agents for electronic commerce: Effects of explanation facilities on trusting beliefs. *Journal of Management Information Systems*, *23*(4), 217–246. doi:10.2753/MIS0742-1222230410

Wold, H. O. (1974). Causal flows with latent variables: Partings of the ways in the light of NIPALS modeling. *European Economic Review*, *5*(1), 67–86. doi:10.1016/0014-2921(74)90008-7

Wold, H. O. (1982). Soft modeling: The basic design and some extensions. In Jöreskog, K. G., & Wold, H. O. (Eds.), *Systems under indirect observations, part II* (pp. 1–54). Amsterdam.

Wold, H. O. (1985). Partial least squares. In Kotz, S., & Johnson, N. L. (Eds.), *Encyclopedia of statistical sciences, 6* (pp. 581–591). New York.

ADDITIONAL READING

Ahn, H. J. (2006-2007). Utilizing Popularity Characteristics for Product Recommendation. *International Journal of Electronic Commerce*, *11*(2), 59–80. doi:10.2753/JEC1086-4415110203

Ahn, T., Ryu, S., & Han, I. (2004). The Impact of the Online and Offline Features on the User Acceptance of Internet Shopping Malls. *Electronic Commerce Research and Applications*, *3*(4), 405–420. doi:10.1016/j.elerap.2004.05.001

Ajzen, I. (1985): From Intentions to Actions: A Theory of Planned Behavior. Kuhl, J.; Beckmann, J. (Hrsg.): *Action Control: From Cognition to Behavior*, Berlin, 11–39.

Ajzen, I. (1991). The Theory of Planned Behavior. *Organizational Behavior and Human Decision Processes*, *50*(2), 179–211. doi:10.1016/0749-5978(91)90020-T

Ajzen, I., & Fishbein, M. (1980): *Understanding Attitudes and Predicting Social Behavior*, Englewood Cliffs.

Benbasat, I., & Barki, H. (2007). Quo vadis, TAM? *Journal of the Association for Information Systems*, *8*(4), 211–218.

Chen, L.-D., & Tan, J. (2004). Technology Adaptation in E-Commerce: Key Determinants of Virtual Stores Acceptance. *European Management Journal*, *22*(1), 74–86. doi:10.1016/j.emj.2003.11.014

Dijkstra, T. (1983). Some Comments on Maximum Likelihood and Partial Least Squares Methods. *Journal of Econometrics*, *22*(1-2), 67–90. doi:10.1016/0304-4076(83)90094-5

Fornell, C., & Larcker, D. F. (1981). Evaluating Structural Equation Models with Unobservable Variables and Measurement Error. *JMR, Journal of Marketing Research*, *18*(1), 39–50. doi:10.2307/3151312

Franke, G. R., & Park, J.-E. (2006). Salesperson Adaptive Selling Behavior and Customer Orientation: A Meta-Analysis. *JMR, Journal of Marketing Research*, *43*(4), 693–702. doi:10.1509/jmkr.43.4.693

Gretzel, U., & Fesenmaier, D. R. (2006). Persuasion in Recommender Systems. *International Journal of Electronic Commerce*, *11*(2), 81–100. doi:10.2753/JEC1086-4415110204

Jarvenpaa, S. L., Tractinsky, N., & Vitale, M. (2000). Consumer trust in an Internet Store. *Information Technology Management*, *1*(1-2), 45–71. doi:10.1023/A:1019104520776

Kramer, T. (2007). The Effect of Measurement Task Transparency on Preference Construction and Evaluations of Personalized Recommendations. *JMR, Journal of Marketing Research*, *44*(2), 224–233. doi:10.1509/jmkr.44.2.224

Pavlou, P. A., & Gefen, D. (2004). Building Effective Online Marketplaces with Institution-Based Trust. *Information Systems Research*, *51*(1), 37–59. doi:10.1287/isre.1040.0015

Punj, G. N., & Moore, R. (2007). Smart Versus Knowledgeable Online Recommendation Agents. *Journal of Interactive Marketing*, *21*(4), 46–60. doi:10.1002/dir.20089

Song, J. H., & Zinkhan, G. M. (2008). Determinants of Perceived Web Site Interactivity. *Journal of Marketing*, *72*(2), 99–113. doi:10.1509/jmkg.72.2.99

Song, M., Parry, M. E., & Kawakami, T. (2009). Incorporating Network Externalities into the Technology Acceptance Model. *Journal of Product Innovation Management*, *26*(3), 291–307. doi:10.1111/j.1540-5885.2009.00659.x

Stüber, E., & Brusch, M. (2009): Personalisation in Retailing: Differences and Commonalities in Selected Branches. Stauss, B.; Brown, S. W.; Edvardsson, B.; Johnston, R. (Eds.): *QUIS 11 - Moving Forward with Service Quality*, Wolfsburg, 654–663.

van der Heijden, H., Verhagen, T., & Creemers, M. (2003). Understanding Online Purchase Intentions: Contributions from Technology and Trust Perspectives. *European Journal of Information Systems*, *12*(1), 41–48. doi:10.1057/palgrave.ejis.3000445

Vesanen, J. (2007). What is Personalization? A Conceptual Framework. *European Journal of Marketing*, *41*(5/6), 409–418. doi:10.1108/03090560710737534

KEY TERMS AND DEFINITIONS

Online Shopping: Usage of an electronic channel that allows customers to select and purchase product items over an interactive electronic medium.

Partial Least Squares (PLS) Analysis: PLS is a group of alternating least squares algorithms that extend principal components analysis and canonical correlation analysis for the analysis of high dimensional data. The main idea behind PLS is a combination of two sets of linear equations, the so-called inner and outer model. Whereas the inner model relates unobserved or latent variables, the outer model relates latent variables and its observed or manifest variables. PLS can be used to calibrate TAM.

Perceived Ease-Of-Use (PEOU): PEOU is defined as the degree to which a person believes that using a particular new information technology or system would be free from effort. According to TAM PEOU is a major factor that influences the decision of a person about how and when to use the new information technology or system.

Perceived Usefulness (PU): PU is defined as the degree to which a person believes that a particular new information technology or system would enhance his or her performance in a specific application setting. According to TAM PU is a major factor that influences the decision of a person about how and when to use the new information technology or system.

Personalization: An attempt to improve the success of online shopping or other customer services through a more individual interaction between the customer and the producer or retailer. Personal welcomings, individual assistance as well as recommendations to inform and buy are examples of personalization in online shopping. In online shopping personalization is assumed to increase the retailer's share of wallet and – ideally at the same time – the customer's satisfaction.

Recommendations to Buy: Offering of alternative or additional products or services during an (online) shop visit.

Recommender System: An information system that attempts to recommend items to a person that are likely to be of interest for her or him. During an (online) shop visit where a recommender systems is implemented the recommended items are typically alternative or additional products or services. Recommender systems use the consumer's saved and aggregated historical data to provide recommendations; they register the latest navigation and consumer behavior and consider additional information, or rather, they generate various combinations from suitable data sources.

Technology Acceptance Model (TAM): A theory that relates design alternatives for information technologies or systems to user acceptance. TAM was proposed by Davis (1989) and Davis et al. (1989) and bases essentially on Fishbein and Aijzen (1975)'s theory of reasoned action (TRA), a theory that relates the general construct behavioral intention to two other constructs, namely attitude and subjective norm.

Chapter 13
From User Cognition to User Interaction Modalities in Consumer Behaviour

Barry Davies
University of Gloucestershire, UK

Eleonora Bilotta
University of Calabria, Italy

Kevin Hapeshi
University of Gloucestershire, UK

Emanuela Salvia
University of Calabria, Italy

Rocco Servidio
University of Calabria, Italy

ABSTRACT

Information and Communication Technology (ICT) has revolutionized science and commerce and has increased the innovation and the spread of a variety of virtual environments applications. These innovations are the result of the both technological development and cognitive studies. The chapter aims to underline the relationships between Human-Computer Interaction (HCI) and consumer behaviour, focusing the attention on the 3-D virtual environments dedicated to electronic and Internet e-commerce (e-retail) services. We introduce how the 3-D interfaces can contribute to the successful impact of online retail. The importance of the relationship between customer and system concerns the effective potentiality of the user interface. If a user interface is ineffective, the system's functionalities and usefulness will be limited and the users will be confused, frustrated, and annoyed, and therefore less likely to use the system again. Finally, we aim to outline the cognitive and technological aspects involved in the communication process between user and virtual e-retail system interface and directions for possible future research.

DOI: 10.4018/978-1-60960-738-8.ch013

INTRODUCTION

The application of direct e-retail services through Information and Communication Technology (ICT) to consumer behaviour represents an important challenge for the modern economy. In order to take full advantage of the range and volume of potential customers, a thorough understanding of both the user interface and the customer experience is required. Understanding provides the platform to help developers design and maintain high quality and efficient interactions between the system and final user (Benyon, Turner, & Turner, 2005). Many of the most recent applications of computer technology concern e-commerce systems. To take advantage of the new generation of e-commerce web sites, the goal of these systems is to include more interactive Web 2.0 services (Tatnall, 2010). These technologies adopt new interfaces that improve the communication between user and system. For example, users can personalize the information on their favourite portals to activate specific services like alert messages that inform them that a new product is available.

Effective Internet marketing requires an understanding of the customer's needs, but also of the user's behaviour and habits, such as how the user finds the necessary information. It is therefore important to understand how different models of presenting information can modify user behaviour; how users explore and utilize information about products and the impact that it may have on them. For example, interactive 3-D models of products can make more information available and transform the online experience by greatly enhancing the realism of the product presentation and making it more intuitive and compelling. This suggests that business success can be positively influenced by the use of advanced digital technology designed and applied within a human context. Technological advances should be implemented taking into account the human factors (Bainbridge, 2004; Carroll, 2000; Sears & Jacko, 2009). These include cognitive processes, user behaviour and preferences whilst interacting with products presented in a digital or virtual environment. When done well, businesses can reap rewards afforded by high sales and satisfied returning customers.

Advances in user interface technology, driven by mobile phone and games technology, have led to a wide range of Human-Computer Interfaces (Sears & Jacko, 2009). These interfaces include different interaction modalities such as touch screens, voice control, kinaesthetic interfaces, multimodal interactions, and so on. The main idea of these systems is to augment the communication between user and digital information, worlds or objects. However, despite the support from Cognitive Psychology in the design and development of such interfaces, it is still unclear how these can be fully utilized in a commercial setting. New information technologies are constantly emerging altering the traditional business models, and in particular the relationships between commercial organizations and their customers. Technological innovations open new challenges by developing different devices employing new communication modalities (Tatnall, 2010). Therefore, for successful application, it is essential to understand the impact that different interfaces and interactions modalities can have on customers and on consumer behaviour through different digital products and platforms. We need also to understand how mixed-modality interface integrated with the new generation of technologies offered by mobile systems, ubiquitous and pervasive communications can be combined with the existing marketing channels and practices.

The focus of this chapter is to emphasize the relationships between user cognitive aspects and system interfaces. In particular, we will describe the cognitive functions to be taken into account in the design process, and how it can affect the interaction between user and electronic system. We will also show the potentialities of multi-modal interfaces and virtual environments to business and marketing by: (1) providing an overview of the relationships between Human-Computer

Interaction (HCI) and consumer behaviour; (2) showing how different interaction modalities can enhance the communication process between user and consumer system; and (3) showing how digital and interactive technologies can offer to the consumer many advantages and unique opportunities in exploring information and products.

THEORETICAL BACKGROUND

In this section we describe the conceptual aspects of both *Human-Computer Interaction* and *Consumer Behaviour* by discussing how these two research fields can improve the design process concerning the services quality in electronic retail. The main idea is that navigation in complex virtual environments is often difficult. In particular, when the information is not transparent, users find it difficult to achieve specific objectives, such as buying products or searching information before to complete the purchase process. In our view, the best way to improve the quality of the e-retail web sites is to understand the user's cognitive limitations, characteristics and requirements and then to apply these results to develop efficient environments. The designer also needs to exploit the potentiality of the new environments based on the Virtual Reality applications.

Human-Computer Interaction

Human-Computer Interaction (HCI) is a multidisciplinary research field enabling new technology designs that are easy and pleasant to use for the final users. It requires a creative approach to the design of new products and future vision to think how new technological solutions can improve the lifestyle of the new generations (Bainbridge, 2004; Bilotta, 1996; Carroll, 2000; Sears & Jacko, 2009). HCI has its roots in the main areas of industrial engineering, human factors and cognitive psychology; the aim is to design and develop user-friendly interfaces to computing technology. Traditionally, the research in the HCI area has emphasized the technological aspect of this relationship focus of computer interfaces. More recently, other aspects concerning the organizational, social and cognitive factors have also begun to be considered. With the development of more complex information technology, HCI research aims to facilitate the communication between user and machine. To achieve these aims, this relationship must consider not only the technological viewoint, such as the usability of the system, but also how people use technological tools and how it influences the social and cultural context.

The advent of the Internet, combined with the fast proliferation of personal computers and mobile devices, has stimulated researchers to examine new ways to improve the quality of the user interfaces. These aspects have many implications on HCI research, where the final users expect to use the technological tools everyday, improve their own efficiency and experience an improved quality of the services (Tatnall, 2010). Technological systems are used to solve a variety of problems and, but will only be effective whenr users are motivated to adopt them. An understanding of the motivation and modes of interaction among final users and system interfaces will help in designing solutions that will be used. In many cases, technological tools are designed to investigate potential applications and overcome technological problems; then the results are applied to implement the final product. So that the designer's mental model (Johnson-Laird, 1983) of the system's function is different to that of the final user, who will need to interact with the system to achieve specific goal, whether it is booking a room or buying a product online. The HCI field aims to develop principles to help designers to improve the experience of users.

Research in the HCI field is very extensive and varied, and there is no single common definition of HCI. Here. we use the definition given by the ACM Special Interest Group in Computer-Human Interaction Curriculum Development, which is also cited in most HCI literature. It is considered as an acceptable standard definition (Hewett, Baecker, Card, Carey, Gasen, Mantei, Perlman, Strong, & Verplank, 2009, p. 5): "*Human-computer interaction is a discipline concerned with the design, evaluation and implementation of interactive computing systems for human use in a social context, and with the study of major phenomena surrounding them.*" The focus here is consumer activities in general, including searching and browsing products, making decision, carrying out purchases and sharing the experience with others. All of which can now be done through online technology.

In the last 10 to 20 years, enterprises and academic institutions have taken the study of HCI more seriously at an earlier stage of development process, to improve the design of artificial systems and enhance the communications process among final users and the interface. HCI is fundamentally an information-processing task. Although the user interaction with a computer interface needs to have in mind the planning of goals and sub-goals, today most of the actual system interfaces produced are still considered difficult to use. The Internet is littered with users posting public messages of complaint about the design of interfaces to web pages and electronic devices such as watches, remote controls, Personal Digital Assistants (PDAs) and other mobile products. A Google search on the Internet through using the term "poor design" highlights the extent to which there is user frustration with poorly designed products; most of this frustration is directed towards interfaces of electronic products, and increasingly it focuses on the online consumer experience (Scoresby, 2004).

All users of a computer-based system aim to use it to complete tasks or extract information, but the interface itself sets users a problem, which they would prefer not to have to deal with. Therefore the easier the interface is to use, the better. Krug (2005) expressed the user's position in the Phrase: "don't make me think", which he used in the title of his book. The user would rather concentrate on the task at hand, whether producing a document or purchasing products and services, rather than 'solving' an interface puzzle. An interface that is poorly designed confronts the user with a 'guessing game', which requires they learn how to control or navigate before they can achieve the required task.

Designers must recognize that the user's task is not simple. On any computer-based system, the user starts the interaction by giving to the computer commands that are directed to accomplish the initial goal. The commands may activate software applications designed to execute specific tasks such as word processing, graphical applications, chat program, online retail, and so on. The result of this interaction is a computer output, usually displayed on a screen, providing adequate information for the final user to accomplish the task, to start a new command to obtain the desired output or to complete the next step. Typically, the sequence of interaction is simple and rapid, but in some cases it may be long and complex, and several other actions may be requested before the user can achieve the final goal. Therefore the interaction with the system interface must be designed in accordance with the users' needs and bearing in mind their cognitive resources and abilities. In this scenario system design, usability and interaction are considered the core of the HCI discipline.

Today, deep changes are taking place that touch all aspects of the society: changes in work, home, business, communication, science, technology, and engineering. These changes involving all people and cannot but influence the future of HCI since they relate to how people interact with technology in an increasingly dynamic and complex world (Peachey, Gillen, Livingstone, & Smith-Robbins, 2010). This makes it even more essential for HCI to play a vital role to design new and more complex system interfaces.

The main task for HCI research is to understand the mental model (Johnson-Laird, 1983) of the final user. The user brings extensive prior experience with the real world and therefore already has a mental model that describes the tasks implemented in the system interface. This model arises from a combination of real experiences, which may include the use of previous software interfaces and with the ability of the user to interact with the system. For example, users have stored in their mind how to select an object from a list of products and then to complete all the operations to execute an online shopping. Based on this, each user has a mental model of specific task that includes certain expectations, such as the ability to select a product and then to purchase it. When (for example) a web site ignores the user's mental model and does not meet at least some of the user's expectations, then for the user it will be both complex and unpleasant to use. If an online retail system forces the user to utilize an unfamiliar conceptual model, the system will be seen as difficult to use and the active task abandoned. Understanding the user's mental model and designing the interface to reflect the user's expectation of the tasks (and in particular the logical organization of the visual functions of the interface) will help to improve the quality of the interaction.

Consumer Behaviour

Consumer behaviour is the foundation of the marketing and understanding consumer behaviour is essential for the success of the marketing. Given these aspects, consumer behaviour is an interdisciplinary research field, which aims to investigate how, why and where people consume products and services (Chaudhuri, 2006). This scientific field include research from different disciplines such as psychology, sociology, economics and social aspects of the consumer behaviour. All human behaviours can broadly be attributed to three wide factors: *internal* that represents the characteristics of the individual; *external* that includes environmental stimulus; and finally the *cognitive* factors that constitute the mental process. The result of this process is the subject's behaviour response.

Figure 1. A simple consumer behaviour model

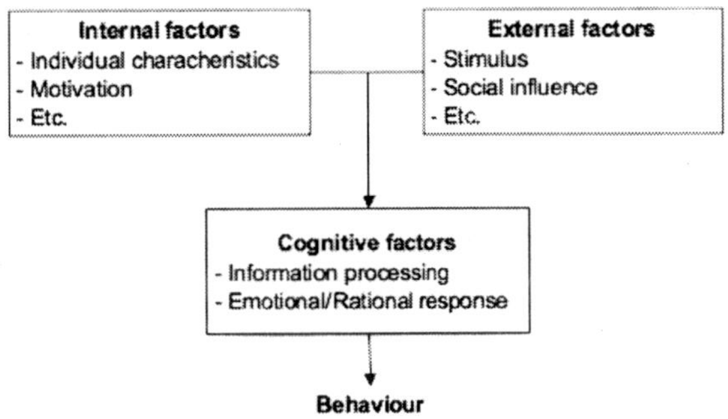

Figure 1 provides a graphical representation of this mechanism.

Consumer behaviour studies attempt to identify the principal variables that influence the decision-making process in individuals and in particular, how people execute their final choice. The understanding of these processes is very complex because many aspects can influence the human decision-making strategies. For example, decisions are influenced from groups, family, friends, and other reference people as well as society in general that reflect the social trends.

Another important question is why people would prefer to use the Internet to buy a product. These questions concern the relationships between personal motivations and Internet services, which offer new alternatives in comparison to the traditional economical opportunity. According to Anolli (2005), motivation is a mental state or a state of needs that push the individual towards certain types of actions that are considered as likely to bring satisfaction. Essentially, in relation to decision-making process, motivation is a part of the consumption process and it is stimulated by internal and external mechanisms under the control of the cognitive functions. Internal mechanisms concern personal factors such as psychological and social events; while external mechanisms refer to environmental stimulation. Taking into account these limits, Internet technology has created new generation of online services dedicated to support both company and customer. Recently, online consumers have come to represent a new category of subjects that use the Information and Communication Technology (ICT) services to purchase everything they can. For this reason, online consumer behaviour represents a new field, the aom of which is the study the customer's motivation and behaviour while searching, selecting, deciding and completing purchases online.

Consumer behaviour and business strategy has undergone a rapid transformation with the spread of ICT and particularly Internet services like recent Web 2.0 applications (Blanca, Julio, & Martin, 2009; Tatnall, 2010). These innovations make it possible for users to access a large amount of information through many different tools, technologies and personalized services. Today consumers are able to shop from anywhere and at any time by 'surfing' commercial web sites and by using few clicks of the mouse or equivalent. Although many users report they prefer to buy by Internet, only a relatively few users complete the purchase process. In fact, much evidence suggests that many online consumers abandon their 'shopping carts' prior to 'checkout' and of those that progress another majority of users abandon them at the point of sale. These results suggest that while Internet has become a significant source of product information, several barriers exist that inhibits the transition in consumer behaviour from information search to final product purchase. Among the reasons commonly cited are: (a) difficulty in locating products, (b) technical problems in interacting with the web sites, (c) tedious checkout processes, (d) a reluctance to supply personal and credit card information, although the commercial web sites usually assure complete privacy and security, (e) hidden charges. These results underline that today many users are still hesitant in purchasing products online. Consumers are reluctant in relation to to final online purchase, because they think that the system security level is low and therefore it is easy for hackers to extract personal details and data. as well as personal banking information.

Cognitive Processing Approach and Virtual Economics System

The aim of this section is to identify possible relationships between cognitive processing and online consumer behaviour, which uses advanced tools such as Virtual Reality applications. These aims will be addressed from both an economical and psychological standpoint. Although online environments represent a new way for consumers to access products, in the previous section we have discussed how today many customers have some difficulties in completing the online purchase process. Online retail *should* increase product purchases, because the Internet offers to the users both practical services and advantages (e.g., time, product costs, etc.). It could be argued that the first generation of commercial web sites was influenced by behaviourism, which considered the human brain a "black-box". These initial virtual environments were developed taking into account the idea that we (the system designers) can understand and control the subject's behaviour through the way it is affected by controlled external stimuli (e.g., product price) or uncontrolled internal stimuli (e.g., subject's motivation or needs). This vision emphasized the idea that it was possible to shape consumer behaviour by manipulating only the external stimulus. Behaviourists set aside cognitive processes, because they appeared to be impossible to objectively measure as they were private events (Stenberg, 2009).

More recently, commercial web sites appear to have adopted the cognitive science paradigm, which takes the view that the subject or customer is an information processing system, able to elaborate and form decisions in their own mind regarding external data. With this view, new approaches to study consumer behaviour have emerged. An important effect of this new approach is the re-evaluation of the cognitive processes. Cognitivism emphasized the role of the mental processes to interact with the environment. Consequently, commercial web sites have begun to include personalization in order to reduce the subject's cognitive effort while they interact with the electronic pages, and then to furnish them new services able to support the user's needs (Tatnall, 2010). In fact, new user interfaces (i.e., user profile) have been designed which aim to facilitate the interaction between customer and the artificial system.

Recently consumer neuroscience studies, taking into account cognitive science research, have investigated economic problems, integrating neuroscientific findings into economic science (Fugate, 2007; Lee, Broderick, & Chamberlain, 2007). Consumer neuroscience is a sub-area of neuroeconomics (Balconi & Antonietti, 2009; Sanfey, Loewenstein, McClure, & Cohen, 2006) that addresses marketing relevant problems with methods and approaches from brain research. Although consumer neuroscience is still in its infancy, it has the potential to be an innovative research field for further investigation concerning decision-making behaviour. The results of these studies attempt to understand and predict human behaviour, in order to improve the personalization of the e-retail environment. A deeper comprehension of the user as consumer will make it possible to derive new theories for marketing research. In addition, these results will help the designer to develop more suitable web sites, which will improve the user experience during the purchase process. The determination of the cerebral areas that are involved during consumer decision-making process is important for various reasons. In particular, the neuroscience approaches will enable the researchers to reassess the existing theories that explain the consumer behaviour, during interaction with complex virtual digital objects and environments.

Another concurrent challenge is the use of the Virtual Reality to design more attractive online economic systems. Virtual Reality technologies refer to "a computer simulation that creates an image of a world that appears to our senses in much the same way we perceive the real world, or 'physical' reality" (Craig, Sherman, & Will, 2009, p. 1). It can be used to create online shopping centre where consumer can interact with the objects that they desire to buy and then to interrogate virtual agents or avatars that will furnish information about the product. The future scenario is to realize virtual environments, like Second Life, that include social services and then the consumer can see the authenticity of the products to acquire (Brunetti & Servidio, 2009; Peachey, Gillen, Livingstone, & Smith-Robbins, 2010). According to Sharma, Anantaram, and Hiranmay (2009), online web-based shopping environments lack of the social interaction. Usually, users buy online products individually, but in real context, interpersonal interactions and stimulation of multiple modalities are an important aspect of the shopping activities because they support the decision-making process. Market research has shown that the Internet environment is not typically a place where consumers make always purchases. Consumers utilize the web because it is convenient, user-friendly and in particular consider it as a new shopping option that offers real opportunity (Schiffman, Kanuk, & Hansen, 2008).

There are several example of the benefit of providing opportunities to share experiences online. Microsoft has made available a free online service called Photosynth (http://photosynth.net/) that allows users to share places and things creating navigable 3-D virtual environments. The provision of the technologies has facilitated the transparency and improves the user confidence because they can see the products and receive specific information from other consumers. Virtual Reality can be used as marketing strategy demonstrating that a particular enterprise or product is innovative from a technological point of view. In this case, getting people to remember and talk about the product is the goal. For example, as a new and exciting technology, Virtual Reality has been used several times as part of advertising strategy emphasizing particular properties of the product promoted.

To show the close alliance between consumer behaviour and cognitive processes, we introduce some studies dedicated to analysing these relationships. This analysis considers cognitive processes as an important aspect of online retail. The first aspect of this relationships concern the web site usability (Gwizdka & Spence, 2007; Hornbæk, 2006; Nielsen, 1993, 2000). The usability of a web site is important because it can influence the volume of sales., Users are often unwilling to interact with web site pages with low usability. For example, users avoid web sites that include pages that are too hard to read, or too difficult to understand, or pages that differ from user expectations. Usability evaluations enable the designer to discover potential problems and then, by applying specific corrections, re-design the web site to more closely match the users' needs (Servidio, Feraco, & Pantano, 2009).

Effective e-retail must therefore take into account the consumer's cognitive and decision-making processes and their behaviour, in order to develop systems that enable consumers to explore products and services in a manner that is comfortable for them. These e-retail systems could be highly complex and include multimodal interfaces. The challenge for designers and developers is to make the task as easy as possible to match positive real-world shopping experiences, and offer [articular e-retail potential benefits and convenience. The rest of this chapter highlights some of the issues that we need to understand to achieve these aims.

COGNITIVE AND AFFECTIVE RESPONSES IN VIRTUAL SYSTEMS

This section aims to underline the relationships between cognitive and affective responses as a design strategy to improve the quality of the e-retail systems. It is evident that many online retails environments attract attention when customers are able to interact with the product delivered. The system should be able to show to the user the properties of the product so to increase the quality of the visual information. This information enables the user to create a mental model of the product reducing the distance between real and virtual object (Sharma, Anantaram, & Hiranmay, 2009).

Consumer Behaviour and System Interaction

In recent years, with advances in Information and Communications Technology (ICT), other behavioural variables have influenced the consumer behaviour. For example, when a consumer interacts with commercial web sites he/she needs to find and quickly process any visual, textual and (in some circumstances) audio information related to the product to buy. Recent research shows that the first impression is a very important aspect from a marketing point of view (Lindgaard, Fernandes, Dudek, & Browñ, 2006). Thus, in the presence of a positive first impression, a person can choose to continue the interaction with the system and try to evaluate the attributes of the products. On the other hand, a negative effect has a strong impact from a cognitive point of view and the subject's response may be to leave the web site altogether. A limit of most online retail is that the consumer typically only has visual or textual information from which to judge the properties of the product. In many cases, consumers find it difficult to create a mental model of the product. In order to go beyond these restrictions. A possible solution is to propose new visualization technique, which can improve the communication process between consumer and product information stored within the electronic system.

Within current commercial web sites the product information are generally presented using two-dimensional (2-D) digital images and accompanying it by text description (Park, Stoel, & Lennon, 2008). In an online environment consumer have an indirect experience of the product characteristics. The aim of the new technologies, such as Virtual Reality applications, is to stress the importance of creating virtual environments that help consumers to simulate a direct experience with the online products. Technological augmentation such as rotation, zoom, and object manipulation generate a sense of a three-dimensional (3-D) presentation of the product (Kuo & Chu, 2005; Li, Daugherty, & Biocca, 2001) improving the creation of the user mental model (Johnson-Laird, 1983). For example, to measure the object dimensions allows consumers to experience 'virtually' the functionalities of the product by using appropriate visualization tools. Researchers have shown that the virtual product presentation using online environments stimulates consumers to realize behaviours that are analogous to the in-store experience (Li, Daugherty, & Biocca, 2002). Therefore, virtual environments improve not only the quality of the interaction with the system but also support consumers to realize in his or her own mind a mental model of the product. The mental model is affected also by others components such as "cognitive, affective and conative" (Lavidge & Steiner, 1961). When all these dimensions characterize the mental model, its internal organization is more structured because it includes more details about the properties of the product visualized. Each of these dimensions includes the traditional information-processing approaches and each is influenced by personal factors.

Recent research by Park, Stoel, & Lennon (2008) shows that rotation, one type of visual simulation technique used to create 3-D experiences, can create positive consumer responses. The rotation function, used for product presentation within online e-retail, it is useful to help the consumer during online shopping phase. The aim of this strategy is to create pleasurable online shopping experiences emphasizing the emotional mechanism. However, it is clear that emotions play a central role in leading and governing decision-making behaviour, by virtue of their capacity to regulate numerous cognitive and physiological processes (Muramatsu & Hanoch, 2005). In addition, rotation allows consumers to perceive the properties of the product affecting the three types of responses (Park, Stoel, & Lennon, 2008): (1) *cognitive* (how the consumer perceives the information related to the product), (2) *affective* (mood concerning the emotional responses), and (3) *conative* (personal attitude influence the consumer behavioural intention). The current development of the 3-D virtual environments or product model provides the diffusion of new marketing strategies. Differentiation strategies aim to create new market rules and build up business with new and innovative strategies. Designing new marketing strategies with the support of the ICT increase the potentiality of the e-retail elevating the quality of service provided by their online business. In addition, these new environments can also serve as the source of competitive strength among companies. These technological environments promote the sharing of information changing the business process of many companies. The use of 3-D data can therefore significantly improve the communication process of the product. In fact, when a consumer interacts with a 3-D product, he/she will develop a positive mood in relation to the product (Guttentag, 2010). Another advantage to the adoption of virtual environments concerns the production of the 3-D documentation, which shows the properties of the product and thereby stimulate positive behavioural outcomes such as purchase.

Design Usable Virtual Environments Interfaces

In the previous section, we have discussed the potential value in using 3-D environments as a new marketing strategy. A major challenge for Human-Computer Interaction (HCI) researchers is to develop new user interfaces able to engage users and to support their interaction with the system. Clearly, e-retail environments should take into account different ways to support user interactions, aiming to make them as natural and intuitive as possible. For example, one of the general research themes in the field of HCI is the consistency of design across platforms (Ivory & Megraw, 2005; Shneiderman, 1997). When an interface has good consistency, it helps the user to focus on the products and contents delivered, reducing the number of errors made in navigating to the relevant information. Another cognitive benefit of consistency is to allow the user to share the information learned from one application and seamlessly apply it in another situation or application (Mach, Hunter, & Grewal, 2010; Maekawa, Itoh, Kawai, Kitamura, & Kishino, 2009). From a cognitive standpoint, this ability it important, as it supports the users whilst they are interacting with the system, thus avoiding the need to learn fresh commands for a new interface. It is common to assume that when users first learn an interface they develop a mental model (Johnson-Laird, 1983) of the interface commands; the designer should avoid forcing the user to develop new mental models as they navigate through the web site.

Over the years tools used to evaluate the system interface usability and the criteria against which judgements are made have been developed (for a review, see Hornbæk, 2006). However, despite the recent advances in Information and Communication Technology (ICT) in many areas, a consolidated evaluation methodology for e-retail applications is still not available. With the increase of contemporary web technologies dedicated to covering user needs and also more general commercial applications, usability is increasingly recognized as an important facet in evaluating the quality of virtual environments, including traditional Graphical User Interface (GUI) applications, web sites, and mobile devices, etc. However, usability has still not been defined in a consistent way across the standards; it is just mentioned, or represented in the manner of the models described below. Most of the usability definitions or models do not include all major aspects of the user interface.

There are several different standards for usability (e.g., ISO 9241, ISO/IEC 9126, IEEE Std.610.12) as well as a number of conceptual models (e.g., Metrics for Usability Standards in Computing [MUSiC]). However, not all of these standards or models describe the same operational definitions and measures (Seffah, Loewenstein, McClure, & Cohen, 2006). So, to reduce the complexity of the existing usability standards and models, a first tentative effort to unify these into a single consolidated conceptual model has been proposed. The result of this research is the Quality in Use Integrated Measurement (QUIM) that includes 10 factors, each of which corresponds to a specific facet of usability that is identified in an existing standard or model (Seffah, Loewenstein, McClure, & Cohen, 2006). These 10 factors are decomposed into a total of 26 sub-factors, or measurable criteria, that themselves are further decomposed into 127 specific metrics.

Although many current web sites include advanced tools, such as 3-D objects or environments, many final users consider these web interface difficult to use and therefore do not make engage with them. In our opinion, the reasons of these weaknesses can be ascribe to the lack of a conceptual framework that supports both developers during the design phase of the system, and usability evaluators in measuring subsequent system performance. Thevarious standards and metrics are not well integrated into current e-retail development practices in the design of web application dedicated to shopping. A reason for these weaknesses is that perhaps most software developers and web sites designers do not apply correctly any particular model in the evaluation of the system usability.

This is not surprising because currently there are many different criteria available to measure a web site's usability. In fact, in most cases, the practice of evaluating web site usability employs methods with which the developers are familiar, but fails to take into account the user's point-of-view. Also, the developer's task is to ensure a working system, but often at the expense of the user interface. Most of the system functions are not visible to the users, so specific functions need to be tested and evaluated carefully. When we considered most of the current web applications, it was evident that each required specific tools and guidelines to perform the testing evaluation process. That is, there may be many approaches and tools to evaluate the web sites contents, but in this field each evaluator adopts only those methods and standards that they know well. These well-knoq=wn tools of course derive from previous experience, so may not be best suited to developing implementations for e-reail. For example, from this perspective, it would be important to develop tools and criteria for the design and implementation of commercial web sites that include 3-D virtual environments. The aim of these criteria would be to improve the ease and quality of the user interaction with system interface.

Solutions and Recommendations

User interface design and testing is an interdisciplinary activity. In particular, understanding how humans perceive and interact with the system interface can contribute to the implementation of more intuitive Graphical User Interfaces. Therefore, in this section we provide some recommendations concerning navigation and interaction tasks, which can be applied to the design of more user-friendly interfaces for e-retail applications (Peachey, Gillen, Livingstone, & Smith-Robbins, 2010).

When a user interacts with a virtual system, navigation is the cognitive process by which users have previously controlled their route through *real* environments. They will use environmental cues and visual aids such as maps and other information, so that they can manage their behaviour and then to achieve goals, without getting lost or without significant mental effort. A major problem for users when they navigate within a *virtual* environment is how to maintain in short-term memory the position in the location and the orientation while they interact with the system.

An important aspect of navigation is the 'cognitive map' the user develops of the route taken through the navigation structure. Usually, when a user navigates within a virtual environment they will apply the same physical world behaviour, whether it is a strategy or skill, to improve interaction with objects. Consequently, spatial abilities help users operate more easily within the system. These abilities allow users to mentally manipulate an object, or to represent it, by using visualization tools. The new generation of virtual environments should combine users mental models and manual operations, in order to improve user interactions and outcomes.

The development of a cognitive map of a system's navigation structure helps the user operate within a virtual environment, simulating what they do when moving around in the real world. From a cognitive point of view, a cognitive map is a mental representation of the real or virtual environment. To create a cognitive map, each subject needs to operate directly within the real context, or to use technological tools to traverse the virtual context. After this interaction, the user it is able to represent in his or her own mind how the system works and affective 'routes' to be followed. To facilitate this mental representation, system functionality needs to be realized to take into account the cognitive functionalities of the final user. According to Norman (2004), before designing a system interface, the designer must know the mental model of the final user. In particular, designers should know how a final user interacts in the real context and how this might translate to interacting with the virtual objects, in order to help the user.

The interaction between user and interface is an important aspect of these problems and challenges. It concerns the ability of the user to perceive the properties of the object's interface, because these characteristics will determine how it can be used or manipulated. In other words, the affordance

(Gibson, 1986; Norman, 2004) is the cues that an object gives to explain what can be done with it. We think that within an e-retail environment the affordance has an important role because it not only stimulates users to action, but also allows the user to manage behaviour according to their personals needs. The user interface is one of the most important parts of any artificial system, because it is the only way to connect system and user. The designer should be sure to include sufficient cues and information to enable the user to develop a mental map and usable interfaces will facilitate this process.

In recent years, a new generation of Graphical User Interfaces has been proposed. These interfaces try to build virtual environments exploiting the potentiality of the Virtual Reality applications. In general, these interfaces are called 3-D GUI and represent the future challenge to create more interactive virtual environments. The 3-D GUI will facilitate the design of new environments that will make it possible for users to engage using natural skills, not only for navigation and for orientation, but also for improving the quality of the interaction between user and system. In a recent work, Pantano and Servidio (2009) have proposed a 3-D virtual environment dedicated to promote tourist sites. The authors found that this interface helped users create a mental model of the place, improving the decision-making process.

FUTURE RESEARCH DIRECTIONS

It is clear that much work still needs to be done in terms of analysis of user behaviour and user interaction with 3-D virtual environments. These studies should compare different typologies of users, in order to examine the behavioural differences concerning product presentation and perceived information, and the effects on attitude and purchase intention. Future work should continue to explore how purchase motivations between virtual environments (e.g., like Second Life) and the real world relate among different users needs. The current study examined theoretical aspects concerning the relationships between users' cognitive processes and system interfaces.

Future research directions should therefore focus on two major issues. Firstly, research should investigate the relationships between consumer behaviour and Human-Computer Interaction. In particular, researchers should analyze how these interfaces can be applied to create new 3-D virtual e-retail environments. Secondly, research should analyze technological impacts from a cognitive standpoint. In our opinion, it is very important to explore the business applications of the new digital technologies. One common aspect of these two main research fields is the usability problem involved when subjects interact with the artificial environments. Usability studies will investigate whether users are able to comprehend the spatial information of the visual interface, and generate a mental model of the virtual environment and then perform behaviours directed towards pproduct purchase.

The new virtual shopping environments will have to exploit the potentiality offered by the virtual tools. These tools support specific cognitive dimensions, which enable consumers to explore product attributes, making for a more interactive and appropriate shopping experiences. Consumers need to interact with virtual environments that are able to reproduce the principal functions that characterize real shopping behaviour.

Due the limited nature of most current virtual environments, users interact with them mainly through the visual modality. With emerging technologies we can investigate implementations, which could include other interaction modalities, such as sound, touch, proprioception, and smell, all of which would enable the person to process the shopping information in a more direct and realistic way. The new virtual environments could enhance the shopping experience by more closely matching the user's needs and, by linking users, also create an online social context. It would be interesting to identify and categorize the principles of online shopping behaviour, in order for the system to respond appropriately to the user's behaviour, while they interact with virtual shopping environments.

The results of these studies can be applied to investigate the relationship between virtual e-retail and user interaction, taking into account the design of the virtual system. For example, future study should explore the level of reality of a virtual interface used to present and deliver products and identify how the user perceives and responds to the interface's characteristics. Neuropsychological evidence suggests that users often form a mental model of the system interface used to interact with the virtual environment. These mental models create an interconnected neural pathway in the brain, the activation of which depends on different parameters, such as user familiarity with the interface, previous learning with user interfaces, and mapping of the interface commands. When subjects are exposed to the novel stimuli, the previous mental activation was inhibited or abolished. Subjects became new users again.

These results provide evidence to show the role of the cognitive process in interacting with virtual systems. In particular in 3-D environments, where the visual stimulation level it is stronger, ther is a high attention level required to interact with commands of the interface. This interaction process involves different cognitive aspects such as attention, perception, working memory and long long-term memory and in some cases also emotional aspects connected with not only with the product that the virtual store delivers, but also with the visual organization of the interface.

Thus, another important future research investigation should include the identification of some guidelines on the design of 3-D virtual environments dedicated to e-retail applications. These guidelines should evaluate both visual layout and command interface intended to ensure system designers understand how the user interacts with the virtual system. These studies also could enhance the knowledge within the HCI field and help researchers understand user cognitive functions involved during the interaction with 3-D virtual systems for e-retail applications. To test these behavioural aspects, it is important to design controlled experimental settings by using different typologies of 3-D interfaces that include different levels of complexity. The main idea of these studies would be to identify the visual characteristics able to reduce the user's cognitive load, so that they can create mental schemas of the 3-D interface, updating it more easily when subjects see new stimuli.

CONCLUSION

Based on this preliminary empirical analysis, the Information and Communication Technologies (ICT) offer to the consumer a new challenge. In this chapter, we have tried to outline the relationships between consumer behaviour and Human-Computer Interaction (HCI) studies focusing the attention on the design of new generation of Graphical User Interface (GUI). In our opinion with the advent of the pervasive technology dedicated to the online shopping it is useful to study the navigation behaviour between user and system's interface functionalities.

The aim of this review was to contribute to the discussion concerning the design of new 3-D online shopping environments. So, by using advanced visualization tools will be possible to manage the amount of information that consumers perceived when surf in online shopping environments. This new system will provide opportunities to enhance the customer positive attitude while they execute online purchases. The current potentialities of the new generation of virtual environments represent a new way to formulate positive attitude and then to support the purchase intention from a cognitive standpoint.

Future research should examine the effects of increasing the complexity of these environments and interfaces, in order to evaluate how different user motivational mechanisms can influence both cognitive and affective responses. In our opinion, the integration of visualization techniques and new manipulative tools will represent an innovative challenge to present product descriptions in online e-retail because they are associated with positive consumer responses. In addition, the recent technological innovations direct us to improve the web services to increase the development of new behavioural modalities improving the user attitude to purchase intention and purchasing behaviour.

REFERENCES

Anolli, L. (2005). *Psicologia generale*. Bologna, Italia: il Mulino.

Bainbridge, W. S. (2004). *Berkshire encyclopedia of human-computer interaction*. USA: Berkshire Publishing Group.

Balconi, M., & Antonietti, A. (Eds.). (2009). *Scegliere, comprare. Dinamiche di acquisto in psicologia e neuroscienze*. Milano, Italia: Springer-Verlag.

Benyon, D., Turner, P., & Turner, S. (2005). *Designing interactive systems: People, activities, contexts, technologies*. Boston, MA: Addison-Wesley.

Bilotta, E. (1996). *Interfacce multimodali e aspetti psicologici dell'interazione uomo computer*. Cosenza, Italy: Edizioni Bios.

Blanca, H., Julio, J., & Martin, J. M. (2009). The impact of self-efficacy, ease of use and usefulness on e-purchasing: An analysis of experienced e-shoppers. *Interacting with Computers*, *1*(2), 146–156.

Brunetti, G., & Servidio, R. (2009). Conceptual design scheme for virtual characters. In F. Lehmann-Grube, & J. Sablatnig (Eds.), *Proceedings of the First International Conference, FaVE* (pp. 135-150). Germany: Springer.

Carroll, J. (2000). *Making use: Scenario-based design of Human-Computer Interactions*. Cambridge, MA: MIT Press.

Chaudhuri, A. (2006). *Emotion and reason in consumer behavior*. Oxford, UK: Elsevier.

Craig, A. B., Sherman, W. R., & Will, J. D. (2009). *Developing virtual reality applications. Foundations of effective design*. Oxford, UK: Elsevier.

Fugate, D. L. (2007). Neuromarketing: A layman's look at neuroscience and its potential application to marketing practice. *Journal of Consumer Marketing*, *24*(7), 385–394. doi:10.1108/07363760710834807

Gibson, J. J. (1986). *The ecological approach to visual perception*. Hillsdale, NJ: Lawrence Erlbaum Associates.

Guttentag, D. A. (2010). Virtual reality: Applications and implications for tourism. *Tourism Management*, *31*(5), 637–651. doi:10.1016/j.tourman.2009.07.003

Gwizdka, J., & Spence, I. (2007). Implicit measures of lostness and success in Web navigation. *Interacting with Computers*, *19*(3), 357–369. doi:10.1016/j.intcom.2007.01.001

Hewett, T. T., Baecker, R., Card, S., Carey, T., Gasen, J., Mantei, M., et al. Verplank, W. (2009). *ACM SIGCHI curricula for Human-Computer Interaction*. Retrieved June 18, 2010, from http://old.sigchi.org/cdg/index.html

Hornbæk, K. (2006). Current practice in measuring usability: Challenges to usability studies and research. *International Journal of Human-Computer Studies*, *64*(2), 79–102. doi:10.1016/j.ijhcs.2005.06.002

Ivory, M. Y., & Megraw, R. (2005). Evolution of web site design patterns. *ACM Transactions on Office Information Systems*, *23*(4), 463–497. doi:10.1145/1095872.1095876

Johnson-Laird, P. (1983). *Mental models: Toward a cognitive science of language, inference and consciousness*. Cambridge, MA: Harvard University Press.

Krug, S. (2005). *Don't make me thin: A common sense approach to Web usability* (2nd ed.). Berkeley, CA: New Riders Publishing.

Kuo, C.-F., & Chu, C.-H. (2005). An online ergonomic evaluator for 3D product design. *Computers in Industry*, *56*(5), 91–112. doi:10.1016/j.compind.2005.02.002

Lavidge, R. J., & Steiner, G. A. (1961). A model for predictive measurements of advertising effectiveness. *Journal of Marketing*, *25*(6), 59–62. doi:10.2307/1248516

Lee, N., Broderick, A. J., & Chamberlain, L. (2007). What is neuromarketing? A discussion and agenda for future research. *International Journal of Psychophysiology*, *63*, 199–204. doi:10.1016/j.ijpsycho.2006.03.007

Li, H., Daugherty, T., & Biocca, F. (2001). Characteristics of virtual experience in electronic commerce: A protocol analysis. *Journal of Interactive Marketing*, *15*(3), 13–29. doi:10.1002/dir.1013

Li, H., Daugherty, T., & Biocca, F. (2002). Impact of 3-D advertising on product knowledge, brand attitude, and purchase intention: the mediating role of presence. *Journal of Advertising*, *31*(3), 43–57.

Li, P., Huynh, T., Reformat, M., & Miller, J. (2007). A practical approach to testing GUI systems. *Empirical Software Engineering*, *12*(4), 331–357. doi:10.1007/s10664-006-9031-3

Lindgaard, G., Fernandes, G., Dudek, C., & Browñ, J. (2006). Attention web designers: You have 50 milliseconds to make a good first impression! *Behaviour & Information Technology*, *25*(2), 115–126. doi:10.1080/01449290500330448

Mach, Q. H., Hunter, M. D., & Grewal, R. S. (2010). Neurophysiological correlates in interface design: An HCI perspective. *Computers in Human Behavior*, *26*(3), 371–376. doi:10.1016/j.chb.2009.11.008

Maekawa, T., Itoh, Y., Kawai, N., Kitamura, Y., & Kishino, F. (2009). MADO interface: A window like a tangible user interface to look into the virtual world. In N. Villar, I. Shahram, M. Fraser, & S. Benford (Eds.), *Proceedings of the 3rd International Conference on Tangible and Embedded Interaction* (pp. 175-180). New York, NY: ACM.

Muramatsu, R., & Hanoch, Y. (2005). Emotions as a mechanism for boundedly rational agents: The fast and frugal way. *Journal of Economic Psychology*, *26*(2), 201–221. doi:10.1016/j.joep.2004.03.001

Nielsen, J. (1993). *Usability engineering*. London, UK: Academic Press.

Nielsen, J. (2000). *Designing Web usability: The practice of simplicity*. Indianapolis, IN: New Riders Press.

Norman, D. A. (2004). *Emotional design*. New York, NY: Basic Books.

Pantano, E., & Servidio, R. (2009). Pervasive environments for promotion of tourist destinations. In Vrontis, D., Weber, Y., Kaufmann, R., & Tarba, S. (Eds.), *Managerial and entrepreneurial developments in the Mediterranean area* (pp. 1213–1335). Cyprus: EuroMed Press.

Park, J., Stoel, L., & Lennon, J. S. (2008). Cognitive, affective and conative responses to visual simulation: The effects of rotation in online product presentation. *Journal of Consumer Behaviour*, *7*(1), 72–87. doi:10.1002/cb.237

Peachey, A., Gillen, J., Livingstone, D., & Smith-Robbins, S. (2010). *Researching learning in virtual worlds*. London, UK: Springer.

Sanfety, A. G., Loewenstein, G., McClure, S. M., & Cohen, J. D. (2006). Neuroeconomics: Cross-currents in research on decision-making. *Trends in Cognitive Sciences*, *10*, 108–116. doi:10.1016/j.tics.2006.01.009

Schiffman, L. G., Kanuk, L., & Hansen, H. (2008). *Consumer behaviour: A European outlook*. Financial Times/ Prentice Hall.

Scoresby, K. (2004). *Seven deadly excuses for poor design*. Retrieved October 2010, from http://www.zdnet.com/news/seven-deadly-excuses-for-poor-design/138101

Sears, A., & Jacko, A. J. (Eds.). (2009). *Human-Computer Interaction. Development process*. London, UK: Taylor & Francis.

Seffah, A., Donyaee, D., Kline, R. B., & Padda, H. R. (2006). Usability measurement and metrics: A consolidated model. *Software Quality Journal*, *14*(2), 159–178. doi:10.1007/s11219-006-7600-8

Servidio, R., Feraco, A., & Pantano, E. (2009). Usability evaluation of a Calabrian tourist visitor information center website. *Information Technology & Tourism*, *11*(2), 95–110. doi:10.3727/109830509789994766

Sharma, G., Anantaram, C., & Hiranmay, G. (2009). The role of semantics in next-generation online virtual world-based retail store. In F. Lehmann-Grube, & J. Sablatnig (Eds.), *Proceedings of the First International Conference, FaVE* (pp. 91-105). Germany: Springer.

Shneiderman, B. (1997). *Designing the user interface: Strategies for effective human-computer interactions*. Addison Wesley Longman.

Tatnall, A. (Ed.). (2010). *Web technologies: Concepts, methodologies, tools, and applications*. Hershey, PA: IGI Global.

ADDITIONAL READING

Arthur, E. J., & Hancock, P. A. (2001). Navigation Training in Virtual Environments. *International Journal of Cognitive Ergonomics*, *5*(4), 387–400. doi:10.1207/S15327566IJCE0504_2

Barlow, T. (2001). Acquisition of Route Knowledge Through Navigation in a Virtual Environment. *International Journal of Cognitive Ergonomics*, *5*(3), 279–295. doi:10.1207/S15327566IJCE0503_9

Braun, K. A. (1999). Postexperience advertising effects on consumer memory. *The Journal of Consumer Research*, *25*, 319–334. doi:10.1086/209542

Gibson, J. J. (1979). *The Ecological Approach to Visual Perception*. Boston: Houghton-Mifflin.

Hausman, A. V., & Siekpe, J. S. (2009). The effect of web interface features on consumer online purchase intentions. *Journal of Business Research*, *62*(1), 5–13. doi:10.1016/j.jbusres.2008.01.018

Hendaoui, A., Limayem, M., & Thompson, C. W. (2008). 3D Social Virtual Worlds: Research Issues and Challenges. *Internet Computing, IEEE*, *12*(1), 88–92. doi:10.1109/MIC.2008.1

Jiyeon, K., & Sandra, F. (2007). Hedonic usage of product virtualization technologies in online apparel shopping. *International Journal of Retail & Distribution Management*, *35*(6), 502–514. doi:10.1108/09590550710750368

Kanellopoulos, D. N. (2008). An ontology-based system for intelligent matching of travellers' needs for Group Package Tours. *International Journal of Digital Culture and Electronic Tourism, 1*(1), 76–99. doi:10.1504/IJDCET.2008.020136

Kjeldskov, J., & Paay, J. (2005). Understanding and modelling built environments for mobile guide interface design. *Behaviour & Information Technology, 24*(1), 21–35. doi:10.1080/01449290512331319012

Nielsen, J. (1999). *Designing Web Usability: The Practice of Simplicity*. Indianapolis: New Riders.

Norman, D. A. (2004). *Emotional design: why we love (or hate) everyday things*. New York: Basic Books.

Oviatt, S. L. (2008), Multimodal Interface. In Jacko, J. A., & Sears, A. (eds.), *The Human-Computer Interaction Handbook: Fundamentals, Evolving Technologies, and Emerging Applications*. CRC Press (413-433).

Pantano, E. (2010). New Technologies and Retailing: trends and directions. *Journal of Retailing and Consumer Services, 17*(3), 171–172. doi:10.1016/j.jretconser.2010.03.004

Steven, M. E., & Harshavardhan, G. (2010). The novelty of 3D product presentations online. *Journal of Interactive Advertising, 2*(1), 10–18.

Takatalo, Jari, Nyman, G, & Laaksonen, L. (2009). Components of human experience in virtual environments. *Computers in Human Behavior, 24*(1), 5–13.

KEY TERMS AND DEFINITIONS

Consumer Behaviour: Is the study of individuals, groups and organizations and the processes they use to select, use and dispose of products, services, experiences and ideas to satisfy needs and how these processes influence both consumers and society.

Graphical User Interface (GUI): It allows people to interact with computer programs by using visual modalities.

Human-Computer Interaction (HCI): It concerns the study of design, evaluation and implementation of interactive systems for human use, by exploiting the advances in the vision, speech recognition, 3-D graphics fields.

Neuroeconomics: It is a research field that merges methods from neuroscience and economics to understand how the human brain elaborates information and then generate decisions in economic and social contexts.

Virtual Environments: Refer to an organization of sensory to show visual information that leads users to perceptions of an artificial environment as natural.

Virtual Reality: It is a computer technique which allows reproducing and displaying objects and scenarios, by exploiting computer graphic tools.

Web 2.0: It is based on the concept of collaborative and cooperative work. Web 2.0 applications enable the communications in a flat way rather than through a centralized approach. So, these new technologies facilitate the user participation, interactivity and social networking increasing the communication between people and groups.

Chapter 14
Mobile Purchase Decision Support Systems for In-Store Shopping Environments

Tobias Kowatsch
University of St. Gallen, Switzerland

Wolfgang Maass
University of St. Gallen, Switzerland & Hochschule Furtwangen University, Germany

ABSTRACT

Purchase decision-making is influenced by product information available in online or in-store shopping environments. In online shopping environments, the use of decision support systems increases the value of product information as information becomes adaptive and thus more relevant to consumers' information needs. Correspondingly, mobile purchase decision support systems (MP-DSSs) may also increase the value of product information in in-store shopping environments. In this chapter, we investigate the use of a MP-DSS that is bound to a physical product. Based on Theory of Planned Behaviour, Innovation Diffusion Theory, and Technology Acceptance Model, we propose and evaluate a model to better understand MP-DSSs. Results indicate that perceived usefulness influences product purchases and predicts usage intentions and store preferences of consumers. We therefore discuss new business models for retail stores in which MP-DSSs satisfy both the information needs of consumers and the communication needs of retailers.

INTRODUCTION

Consumers depend on precise and comprehensible product information at the point of sale. For example, consumers with food allergies need to know about the ingredients of groceries and consumers that buy a memory card for a digital camera need to know if both products are compatible with each other. Product information therefore strongly influences purchase behaviour as found by consumer research for in-store shopping situations (Tellis & Gaeth, 1990). In the case of online purchase situations, the value of product information can be increased further with the use of purchase DSS,

DOI: 10.4018/978-1-60960-738-8.ch014

also known as recommendation agents, as they "elicit the interests or preferences of individual users for products either explicitly or implicitly, and make recommendations accordingly" (Bo & Benbasat, 2007, p. 137). In this sense, product information provided by purchase DSS becomes adaptive and therefore more relevant to individual consumers' information needs, whereas product information on printed product labels is static by definition. Correspondingly, several studies revealed that online purchase DSS help to reduce search complexity and consumers' information overload (Häubl & Trifts, 2000; Todd & Benbasat, 1999), improve decision quality (Pereira, 2001), increase trust in decisions (Gregor & Benbasat, 1999), and finally, influence consumer behaviour and purchase intentions (Bo & Benbasat, 2007; Häubl & Trifts, 2000; Kamis, Koufaris, & Stern, 2008). In practice, they are restricted to online applications and are adopted by providers of product information, e.g., for car configurations, such as offered by Toyota (carconfig.toyota-europe.com), collaborative product recommendations (e.g., Amazon.com) or recommendations of recipes based on particular ingredients a consumer has available (e.g., allrecipes.com).

As purchase DSS are used via websites at home, they may also be used on mobile devices for in-store product information acquisition. Correspondingly, mobile applications are currently being developed for consumers to communicate with physical products (Maass & Varshney, 2008). Thus, mobile shopping assistants such as Impulse (Youll, Morris, Krikorian, & Maes, 2000), MyGrocer (Kourouthanassis & Roussos, 2003), MASSI (Metro AG), the Tip'n Tell client (Maass & Filler, 2006), the Mobile Prosumer (Resatsch, Sandner, Leimeister, & Krcmar, 2008), EasiShop (Keegan, O'Hare, & O'Grady, 2008), or APriori (von Reischach, Guinard, Michahelles, & Fleisch, 2009) allow to request product information directly at the point of sale. For example, a garment is identified by a mobile barcode or radio-frequency identification (RFID) reader device and then provides its information such as the recommended sales price, its producer or other products that fit with it. In that case, physical products can be enriched with new digital product information services relevant to the consumer. This would not only change the way retail stores are perceived by consumers, e.g., they might request product information directly at the point of sale instead at home, but it would also have managerial implications for retailers and providers of product information. In particular, the use and impact of mobile purchase DSSs (MP-DSSs) in in-store shopping environments are a main concern from both consumer's and retailers' perspective.

Up until now, little research has been conducted on the utility of purchase DSSs for in-store purchase decision-making. For example, Westerman et al. (2007) found that a desktop-based purchase DSS improves the quality of purchase decisions by providing recommendations based on a weighted adding model. In another lab experiment, product information provided by a MP-DSS was perceived as being better than static product information (e.g., information printed on product labels) particularly for product bundle purchases in in-store situations (Kowatsch, Maass, Filler, & Janzen, 2008; Maass & Kowatsch, 2008a). Further, the existence of personalized cues provided by MP-DSSs and indicating the attractiveness of a product improves the quality of product consideration sets (van der Heijden, 2006). But with regard to the literature review in the next section it is still open, (1) whether MP-DSSs are adopted for product information acquisition in retail stores, (2) by which factors they are adopted, (3) if they influence purchase behaviour, and finally, (4) whether MP-DSSs influence the consumers' preferences in selecting retail stores that provide access to them.

This chapter provides first answers to these questions with the help of a lab experiment. It is organized as follows. Next, we will discuss related literature on product information and purchase DSSs before we develop our research model that is based on Theory of Planned Behaviour (Ajzen,

1991), Innovation Diffusion Theory (Rogers, 2003), and Technology Acceptance Model (Davis, 1989). Then, the model is tested empirically as described in the methodology section. The results and limitations of the study are discussed subsequently before we conclude this chapter by a summary and an outlook on future work.

BACKGROUND

Online and mobile E-Commerce applications are specializations of economic markets and span a continuum of institutionalized environments, in which customer-oriented transactions on products and services are performed (Bakos, 1998). In E-Commerce applications, physical products are represented by digital product descriptions, whereas in-store shopping situations support users with direct perceptions of products by touch, smell, and other sensual cues (Citrin, Stem Jr, Spangenberg, & Clark, 2003; McCabe & Nowlis, 2003). Recently, several studies evaluated the impact of digital replications with different sensual experiences by virtual reality simulations, which facilitate further transfer of physical shopping experiences into digital shopping environments (Lee & Chung, 2008; Zhenhui & Benbasat, 2004). By contrast and consistent with our approach, the opposite direction investigates how E-Commerce services can be embedded into physical shopping environments by desktop-based (Westerman, et al., 2007) or mobile and ubiquitous computing technologies (Junglas & Watson, 2006; Kleijnen, de Ruyter, & Wetzels, 2007; van der Heijden, 2006; Watson, Pitt, Berthon, & Zinkhan, 2002). In these scenarios, rich product information becomes available at the point of sale and can be dynamically adapted to consumers' needs by using a product interface that embeds value added services. In this chapter, we focus on a particular service in the form of a purchase DSS that provides product information. As the relevance of DSSs for purchase situations was discussed in the introduction, we describe their interrelation with product information subsequently.

Purchase decision-making depends on product information that can be imperfect for a number of reasons, "such as the proliferation of competing brands, the difficulties of exhaustive search or sampling, biases in product evaluation, constant product innovation or consumer mobility" (Stahl & Freudenschuss, 2006, p. 1932). This information asymmetry between producers and consumers results in emphasizing price and quality attributes during purchase decisions at the point of sale (Tellis & Gaeth, 1990). If a customer knows little about the product's quality, he will optimize his or her choice according to price considerations. With increased product information about expected quality, consumers tend towards rational decisions on the expected utility over both attributes (Tellis & Gaeth, 1990). Correspondingly, MP-DSS use may reduce product information asymmetry between producers and consumers in front of the product shelf. For instance, product reviews provided by professional magazines or user-communities via the MP-DSS may reveal information on the quality of a product, thus may change the purchase behaviour (Kowatsch, Maass, & Fleisch, 2011; von Reischach, et al., 2009).

Furthermore, influence on shopping experience is distinguished into emotional impressions that affect customers' moods and product information that affects rational decision-making (Groeppel & Bloch, 1990). MP-DSSs are intrinsically focused on product information (van der Heijden, 2006). Additionally, product information can be classified into singular product information or relational product information. Singular product information describes a particular product, i.e. product features. By contrast, relational product information describes product sets that can be classified into product complementary sets (e.g., product bundles) and product similarity sets (alternative products). Several techniques have been used for the automatic derivation of product similarity sets (Kurkovsky & Harihar, 2006). Product similarity sets are further classified whether they are solely derived from product features such as content-based recommendations (Maidel, Shoval, Shapira, & Taieb-Maimon, 2008) or indirectly derived via preferences and decisions of other users such as social or collaborative recommendations (Esslimani, Brun, & Bayer, 2008; Yang, Li, & Zhang, 2009; Zhang & Li, 2008). In the current chapter, both singular and relational product information is covered but only product similarity sets based on product features are provided to the consumer through the MP-DSS.

Finally, we investigate a conversational MP-DSS that uses a question-and-answer-based dialog system (Gurevych & Mühlhäuser, 2007; Maass & Janzen, 2007). In this sense, product information is requested by asking questions concerning (1) singular product information, e.g., "What is the price of this product?", or (2) relational product information, e.g., "Are there accessories available?"

RESEARCH MODEL

In the following, we develop our research model that investigates the use of a MP-DSS for product information acquisition in in-store purchase situations. The MP-DSS is implemented on an RFID-enabled mobile device that allows the identification of products and therefore extends traditional product information capabilities on printed product labels by providing relevant product information on demand. We study the adoption of the MP-DSS by applying Innovation Diffusion Theory (IDT, Rogers, 2003), Technology Acceptance Model (TAM, Davis, 1989), and Theory of Planned Behaviour (TPB, Ajzen, 1991) to be consistent with prior work on Information Systems adoption research (Davis, 1989; Kamis, et al., 2008; Venkatesh, Morris, Davis, & Davis, 2003). Further, we apply these theories because we do neither focus on the outcomes such as net benefits of a successful Information Systems implementation, for which the DeLone and McLean Model of Information Systems Success would be appropriate (DeLone & McLean, 1992, 2003), nor do we investigate how information is processed by individual users of MP-DSS, for which information processing choice theory (Bettman, Luce, & Payne, 1998) would be more appropriate.

According to IDT, the MP-DSS represents an innovation that a consumer can adopt for product information acquisition in purchase situations. Another research stream applies intention-based models to understand the adoption of IT. Accordingly, corresponding models such as TPB are grounded in social psychology to identify attitudes, social influences and facilitating conditions that predict the behavioural intention of usage. The behavioural intention to use a MP-DSS for product information acquisition predicts its adoption, respectively. For instance, TAM is based upon this line of research.

Two constructs from prior research are adequate to be utilized in our model. The first is perceived usefulness, which is defined as "the degree to which a person believes that using a particular system would enhance his or her job performance" (Davis, 1989, p. 320). In our context, perceived usefulness refers to the degree to which a customer believes that using a MP-DSS would enhance his or her performance to acquire relevant information for product purchases. Regarding TPB, perceived usefulness as believe towards a future behaviour predicts the intention to use a MP-DSS for product information acquisition. Thus, we hypothesize the following relationship with regard to the first two research questions:

H1 *Perceived usefulness of a MP-DSS has a positive relation with the intention to use that MP-DSS for product information acquisition.*

In our context, a MP-DSS can only be used with products that can be identified, e.g., through barcode or radio frequency identification. Therefore, customers are likely to select only those retail stores that enable them to use the MP-DSS.

We call them MP-DSS-enabled retail stores. In line with TPB, the behavioural intention to prefer MP-DSS-enabled retail stores is predicted by the usefulness of the MP-DSS as perceived by consumers. In addition, Kamis et al. (2008) found that perceived usefulness of an online purchase DSS strongly predicts the intention to return to it. This is also related to our research as return reflects store preference. We postulate therefore the following relationship that addresses the second and fourth research question:

H2 *Perceived usefulness of a MP-DSS has a positive relation with the intention to prefer MP-DSS-enabled retail stores for product information acquisition.*

Next, the MP-DSS supports consumers in buying situations as it helps them to find relevant product information. With higher degrees of perceived usefulness of the MP-DSS's dialog system, buying intentions are increased as relevant information for purchase decisions is provided. This relation is also supported by marketing and information system research (Kamis, et al., 2008; Tellis & Gaeth, 1990). We therefore hypothesize the following by addressing research questions two and three:

H3 *Perceived usefulness of the MP-DSS has a positive relation with the intention to purchase a product after using it for product information acquisition.*

The second TAM construct used in our model as predictor variable is perceived ease of use, which refers to the degree "to which a person believes that using a particular system would be free of effort" (Davis, 1989, p. 320). Here, the MP-DSS supports the consumer to acquire product information in in-store purchase situations. In order to cover the first two research questions and consistent with TPB, TAM and recent work (Hasan & Ahmed, 2007; Maass & Kowatsch, 2008a), we hypothesize the following relationship:

H4 *Perceived ease of use of the MP-DSS has a positive relation with the intention to use it for product information acquisition.*

And finally, in line with Davis (1989) and Kamis et al. (2008), perceived ease of use predicts perceived usefulness for desktop applications, human computer interfaces and online recommendation systems. Additionally, it was found that ease of use predicts relative advantage of dynamic product information provided by MP-DSSs (Maass & Kowatsch, 2008a). Here, relative advantage is similar to the usefulness construct as discussed by Moore and Benbasat (1991). Hence, perceived ease of use of the MP-DSS is also suggested to influence intention to use indirectly through perceived usefulness, which also addresses the first two research questions:

H5 *Perceived ease of use of a MP-DSS has a positive relation with perceived usefulness of that MP-DSS for product information acquisition.*

METHOD

In order to test the research model, we first developed a mobile recommendation agent (MP-DSS) before we conducted a lab experiment. In the following, the implementation of the MP-DSS is described. Then, a detailed overview of the lab experiment is presented.

Mobile Recommendation Agent

In contrast to the barcode-based MP-DSS of van der Heijden (2006), our MP-DSS is implemented on an RFID-enabled PDA (HP iPAQ Pocket PC with an Socket 6E RFID reader) by using the. Net Compact Framework. It uses a dialogue web service and a linguistic knowledge base. Both components are part of the Tip 'n Tell middleware for smart products (Maass & Filler, 2006), which are implemented by a web service architecture on the basis of the Jena 2.0 system (jena.sourceforge.net) that allows the integration of reasoning mechanisms, such as Fact++. Reasoning is used to process questions directed at products by users of the MP-DSS. Correspondingly, the graphical user interface of the MP-DSS decodes user questions into requests, which are sent to the dialogue web service. The PDA is connected to the Tip 'n Tell middleware via wireless LAN technologies.

In our implementation, users are in the role of consumers in in-store shopping situations and initiate a dialog with a product by identifying them with the RFID reader attached to the PDA. Therefore, products need to be annotated with RFID tags (ISO15693, HF range with 13,56 MHz), which carry URL references to the location where their product information is stored. With the help of rules and reasoning mechanisms of the Tip 'n Tell middleware, not only information of a particular product such as its price or description can be requested but also information that allows to ask for the compatibility or complementary of two different products (e.g., for the recommendation of product bundles).

Lab Experiment

A lab experiment was conducted to test the usability of the MP-DSS and to evaluate our research model. The sampling procedure was as follows. We e-mailed an invitation to all bachelor students studying media and computer science at our University. In this invitation, the experiment was promoted by claiming that each participant will receive 8 Dollars. Further, it contained a link to a website, on which the students were able to register for the experiment by choosing one out of 50 time slots – each slot intended for one student. After all slots were booked out, no one could register anymore. Therefore, only the first 50 subjects were allowed to participate and to receive the 8 Dollars after the experiment.

A pre-test with several bachelor students was carried out to get preliminary feedback in advance of the experiment, which was then used to optimize the MP-DSS and the instructions of the questionnaire. The MP-DSS was developed for consumers to acquire relevant product information in retail stores. This is accomplished in two consecutive tasks. First, the user starts the dialog with the MP-DSS by a touching gesture at a product by which the product's ID is read via the RFID reader of the PDA as shown in Figure 1. Second,

Figure 1. The consumer starts the dialog function of the mobile purchase decision support system with a touching gesture on the RFID-tagged product

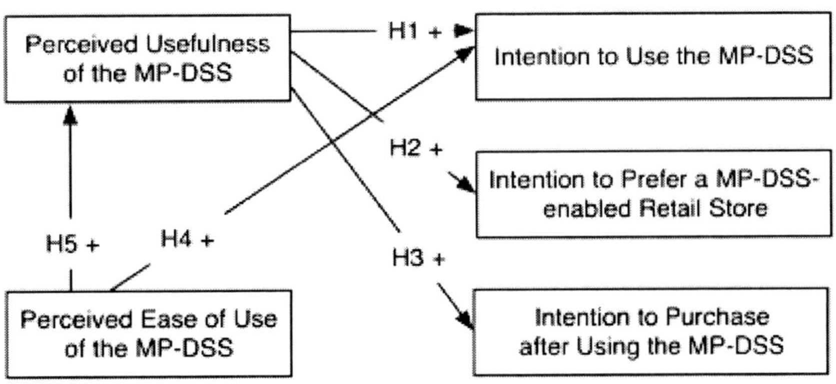

Figure 2. Screenshot of the interface of the purchase decision support system. Here, the consumer asks for compatible accessories with the help of the dialog function.

the user asks for product information by using the question-and-answer-based dialog function of the MP-DSS, which can be seen in Figure 2. In order to test the usability of the MP-DSS, both tasks had to be considered. For this purpose, we evaluated the usability of both the touching gesture and the dialog function with the system usability scale (SUS) which was developed by Brooke (1996). The usability test was required to indicate the maturity of our MP-DSS implementation based upon the research model is evaluated. In this sense, over-average usability scores are required to discuss profound implications of the results.

In the first part of the experiment, the subjects were told that they are customers of a retail store that sells mobile navigation units and accessories as shown in Figure 1. Seven mobile navigation units were equipped with an RFID-tag that was fixed below a tally labelled touch me. The tally was used to inform the subjects on which spot of the product they could start the MP-DSS's dialogue function by the touching gesture. Subjects were instructed to buy one of the available mobile navigation units and one accessory. Product information could only be obtained by using the MP-DSS as no paper-based product labels were shown. This MP-DSS-only treatment implies a possible future retail scenario that is supported by recent empirical findings of a similar lab experiment, in which product information provided by a MP-DSS was perceived as being better than paper-based product information (Kowatsch, et al., 2008; Maass & Kowatsch, 2008a). In that experiment, the implementation of the MP-DSS is highly comparable with the MP-DSS of our work as the implementation differs only in the way the questions towards a product are selected to acquire the corresponding product information, i.e. a pull-down menu with several questions rather than two buttons that represent two questions.

The subjects had to consider the following instructions for the purchase task: the mobile navigation unit had to be low-priced, had to support the USB-standard and had to include a 1GB SD-Card. In addition, the accessory had to be compatible with the mobile navigation unit they chose. These constraints were formulated such that subjects had to ask several questions, thus getting trained with both touching gesture and dialog function. Name, producer, price and a small image of the product were shown immediately at the top of the screen after the pointing gesture was finished as can be seen in Figure 2. The following four questions were available by the MP-DSS's drop down list:

1. Are there product details available?
2. Which standards are supported?
3. Which accessories are available?
4. Are there alternative products available?

Subjects had 10 minutes to finish this part of the experiment. Then, in the second part, subjects were given a questionnaire with the SUS items (Brooke, 1996). The questionnaire was also used to ask for the constructs of our research model. Items regarding the perceived characteristics usefulness and ease of use were adapted from existing scales (Kamis, et al., 2008). Furthermore, three items were used to measure the behavioural intention to use the MP-DSS (Davis, 1989), product purchase intention and the intention to prefer a retail store that allows the usage of such kind of MP-DSS (Kamis, et al., 2008). According to Ajzen and Fishbein (1980), all three statements cover the three behavioural elements action (usage / purchase / prefer), target (MP-DSS) and context (product information / product / retail store) as can be seen in Table 2. All items were based on seven-point Likert scales, ranging from extremely disagree (1) to extremely agree (7). At last, the questionnaire was used to collect demographic data and to ask for the length of the experiment and the comprehensibility of the instructions.

Results

Thirty-eight male and nine female students participated in the lab experiment. Their age ranged from 20 to 24 (n = 31) and from 25 to 29 (n = 16). The instructions of the experiment and the questionnaire were perceived as being reasonable (Mean = 6.64; SD =.53) and acceptable on its length (Mean = 6.49; SD =.98). Reliability of the SUS items was tested with Cronbach's Alpha, which resulted in viable.69 and.83 for the pointing gesture and the dialog function. Because the sample is obviously biased by gender, we conducted an analysis of variance for all theoretical constructs with gender being the independent variable. As a result, we found no significant differences for all constructs, i.e. the level of significance was above.05 and therefore, gender did not influence the following results significantly.

In general, the SUS score ranges from 0 to 100 with 100 representing the best usability value (Brooke, 1996). Our SUS score of the pointing gesture was 78.8 on average and below the dialog function's score that yielded remarkable 85.8. This result can be explained by the free-text feedback of the subjects that predominantly addressed the slow speed of starting the dialog function with the touching gesture. This issue is based upon technical restrictions regarding the RFID reader's capabilities. In addition, some subjects requested the ability of the dialog system not only to ask for product information of one product but also to compare different products by their properties. This feature will be considered in the next stage of the MP-DSS's development. Nevertheless, both SUS scores were highly significant above

Table 1. Results of the system usability scale (SUS) (see Brooke 1996) and one sample t-tests for both tasks and 47 subjects)

	Task 1: Starting the dialog by a touching gesture	**Task 2: Usage of the dialog to obtain product information**
Number of Items	10	10
Cronbach's Alpha	.69	.83
SUS score	78.8	85.8
Standard Deviation	12.6	12.1
t-value	16.3	19.6
p-value	<.001	<.001

Note: the average test value of 50 was used for the t-tests as the SUS score ranges between zero and 100

Table 2. Survey instrument and descriptive statistics (N = 47)

	Construct and Items	SFL
	Perceived usefulness of the MP-DSS *Cronbach's Alpha:.861, Mean: 5.73, SD:.81*	
PU1	Using this MP-DSS can improve my performance to acquire product information.	.720
PU2	Using this MP-DSS can improve my productivity to acquire product information.	.887
PU3	Using this MP-DSS can improve my effectiveness to acquire product information.	.897
PU4	I find using this MP-DSS useful to acquire product information.	.847
	Perceived ease of use of the MP-DSS *Cronbach's Alpha:.716, Mean: 6.27, SD:.81*	
PEU1	Learning to use the MP-DSS to acquire product information would be easy for me.	.713
PEU2	My interaction with the MP-DSS is clear and understandable to acquire product information.	.832
PEU3	It would be easy for me to become skilful at using the MP-DSS to acquire product information.	.515
PEU4	I find the MP-DSS easy to use to acquire product information.	.858
	Intention to use the MP-DSS *Mean: 5.77, SD: 1.29*	
IU	I would use the MP-DSS to acquire product information in retail stores.	N/A
	Intention to purchase after using the MP-DSS *Mean: 4.72, SD: 1.61*	
IP	I would purchase a product after I was using the MP-DSS for product information acquisition.	N/A
	Intention to prefer a MP-DSS-enabled retail store *Mean: 4.23, SD: 1.59*	
IPS	I would prefer a retail store that allows me to use the MP-DSS to acquire product information.	N/A

Note: SFL (Standardized Factor Loadings), SD (Standard Deviation)

the neutral test value of 50 by applying t-tests for one sample. As a result, the overall usability of the MP-DSS is promising with regard to its early development stage. Thus, the following test of the research model is based upon a usable MP-DSS implementation. A summary of the descriptive statistics and the results of the usability test are shown in Table 1.

Consistent with prior research (Kamis, et al., 2008; Komiak & Benbasat, 2006), partial least squares (PLS) analysis was used to test our research model. PLS belonging to structural equation modelling (SEM) was chosen over regression analysis, because SEM can analyze all of the paths in one analysis (Barclay, Thompson, & Higgins, 1995; Gefen, Straub, & Boudreau, 2000). PLS allows analyzing (1) the structural model for assessing the relationships among our theoretical constructs and (2) the measurement model for assessing the validity and reliability of our questionnaire items. In our research, all theoretical constructs were modelled as reflective, because their questionnaire items are manifestations of them (Barclay, et al., 1995) and are expected to correlate with each other (Chin, 1998). By using G*Power3 (Faul, Erdfelder, Lang, & Buchner, 2007), a sample size of 31 was calculated for two predictors (Method: F-test, Multiple Regression – Omnibus) which would be good enough to detect PLS path coefficients with large effect sizes (f^2 =.35). A statistical power of.80 was used, which is common in MIS research (Baroudi & Orlikowski, 1989; Cohen, 1977). Thus, the sample size of 47 subjects was sufficient for the lab experiment.

In order to test the validity of our constructs, we performed a confirmatory factor analysis using SEM with SmartPLS 2.0 and the bootstrapping resample procedure (Ringle, Wende, & Will, 2005). Although one item (PEU3) from the perceived ease of use scale had a factor loading below the recommended value of.70, we retained it in order to maintain continuity with prior research using the same scales. All of the items loaded significantly on their assigned latent variables. Thus, our scales show good convergent validity. The Cronbach's alpha values for all constructs are above the recommended.70 value, indicating good reliability (Nunnally, 1967). The factor loadings, Cronbach's alpha values, and descriptive statistics of all constructs are shown in Table 2.

The results of the PLS analysis can be seen in Figure 3. First, the coefficient between perceived usefulness and the intention to use the MP-DSS for product information acquisition is positive and significant (b=.373, $p<.01$). Second, the coefficients between perceived usefulness and the intention to prefer a retail store that allows the usage of the MP-DSS for product information acquisition (β =.324, $p<.05$) and the intention to purchase a product after using the MP-DSS (β=.547, $p<.001$) are both positive and significant. This supports H1, H2 and H3. In addition, the coefficient between perceived ease of use and perceived usefulness is positive and significant (β=.521, $p<.001$) supporting H5. By contrast, perceived ease of use does not significantly predict the intention to use the MP-DSS for product information acquisition (β = -.050, $p>.05$). Therefore, H4 is not supported by our empirical data and possible explanations for this non-significant relationship are provided in the following section.

DISCUSSION

Theoretical Implications

Our study provides evidence for the usability of one particular mobile recommendation agent (MP-DSS) for product information acquisition in in-store purchase situations. This result is valid for both tasks relevant for consumer-product communication. First, consumers start the dialog with the product by pointing at it with an RFID-enabled PDA on which the MP-DSS is implemented. Participants may be already familiar with the semantics of this gesture representing *I mean you* or *I need something from you* or *I am interested in you* when they want to call someone's attention. Second, consumers use the question-and-answer-based dialog system of the MP-DSS

*Figure 3. Results of PLS Analysis. Note: * = p >.05, ** = p <.01, *** = p <.001*

to request product information relevant to them. This dialog-based interaction has similarities with the communication of consumers and sales personnel (Maass & Kowatsch, 2008b). Thus, consumers are able to reuse knowledge learned from prior purchase situations, which may increase the MP-DSS's usability. Although successfully tested for one particular instance of a MP-DSS, the proposed communication design can therefore be reused and tested in other situations within a product's life cycle. For example, purchase transactions or product support activities can be started by touching gestures and supported by a dialog-based MP-DSS.

Beside the over-average usability scores of the MP-DSS, we found that the behavioural intention to use the MP-DSS is strongly predicted by its perceived usefulness, which extends findings of prior research on the adoption of Information Systems (Davis, 1989; Kamis, et al., 2008; Venkatesh, et al., 2003) for MP-DSSs in in-store shopping situations. This indicates also, that the MP-DSS is perceived as being useful and thus will be adopted by at least some of the subjects. Moreover, perceived usefulness of the MP-DSS strongly predicts the intention to prefer a retail store that allows consumers to use it. And second, perceived usefulness also predicts the intention to buy a product after using the MP-DSS. Both findings not only extend the work of Kamis et al. (2008) to in-store shopping situations, but are also relevant to store managers as discussed in the next section.

In contrast to the findings of Davis (1989), perceived ease of use does not significantly predict usage intentions of our MP-DSS. Beside the sample size only indicating big effects in the PLS path model, other predictors from IDT such as compatibility or complexity (Moore & Benbasat, 1991; Rogers, 2003) or other constructs such as trust (Komiak & Benbasat, 2006), risk and convenience (Ponder, Lueg, & Williams, 2006) or self-consciousness when using MP-DSSs in public areas (Serif & Ghinea, 2008) might be more relevant and therefore should be investigated further. Moreover, the interface design of MP-DSSs has been of little importance in our research yet, which might be another reason for this non-significant relationship. In this sense, MP-DSSs are primarily problem-solving intermediaries that are valued by their contribution of solutions to a problem and thus, perceived usefulness is much more important than ease of use. However, ease of use enables the adoption of MP-DSSs as it strongly predicts the usefulness of MP-DSSs, which supports prior research (Davis, 1989; Kamis, et al., 2008).

Managerial Implications

As a result of the empirical findings, the use of MP-DSSs and corresponding infrastructures might increase the sales volume of retail stores through an increase of consumer frequency – predicted by consumers' intention to prefer MP-DSS-enabled retail stores – and the potential of product purchases – predicted by the consumers' intention to purchase a product after using the MP-DSS. Therefore, we recommend store managers to provide access (1) to electronic product information via a free wireless network infrastructure (e.g., WLAN) and (2) to products, which can be easily identified by RFID, barcodes or QR codes. This would not only enhance the retailer's competitive advantage as he or she attracts more consumers, which would otherwise find and purchase adequate products online at home, but this may also increase consumers' shopping experience in retail stores (Groeppel & Bloch, 1990). Hereby, retail stores add a unique selling proposition. The exemplary use of those technologies with a focus on increased customer experience is evaluated and presented at the Future Retail Center in Regensdorf by SAP Research in Switzerland.

We further assume that there is the potential of a new market for companies to develop and sell MP-DSSs for in-store purchase situations. The extension of physical products with product information services from several providers might change the way of retailing. New business models may consider extended product recommendation services such as digital product reviews or dynamic pricing services.

Product review services may be relevant for MP-DSS, because word-of-mouth has long been recognized as a major driver of product sales as quality uncertainty and high search costs for identifying relevant product information may prevent consumers from making purchases (P. Y. Chen, Wu, & Yoon, 2004). Particularly in online purchase situations and even in the late nineties, almost the half of the 5500 participants of a BizRate survey said that they have consulted opinion sites before they made a purchase (Piller, 1999). In addition, the Internet has not only significantly reduced the consumers' information-retrieval cost (Moon, 2004) but also the reviewers' information-delivery cost (Y. Chen & Xie, 2005). Thus, a lot of free digital product reviews are available on Amazon.com, DooYoo.co.uk, Ciao.de, eOpinions.com and Ask.com among other Web 2.0 review platforms. In addition, several expert magazines provide digital product reviews for free, e.g., PC Praxis, Car and Driver or Runner's World. But there are also paid digital product reviews available from experts. For instance, consumer and test magazines such as Consumer Reports, AudioVision, or Stiftung Warentest regularly publish paid reviews on their websites. The adoption of digital product reviews on MP-DSS is promising, as there already exist various services such as Wired Product Reviews, Beauty Product Reviews and ProductWiki among many others. In addition, a recent empirical study has successfully tested adoption behaviour in the context of free and paid digital product reviews for consumer goods (Kowatsch, et al., 2011).

Second, dynamic pricing services may also be relevant for MP-DSS as marketing research has identified several benefits of dynamic pricing in the retail industry (Kowatsch & Maass, 2011). For instance, pricing strategies that address inventory considerations and time horizons (Elmaghraby & Keskinocak, 2003; Gallego & van Ryzin, 1994; Su, 2007), bundling (Bitran & Ferrer, 2007; Gaeth, Levin, Chakraborty, & Levin, 1991) or personalized offerings (Choudhary, Ghose, Mukhopadhyay, & Rajan, 2005; Liu & Zhang, 2006) have been found to increase sales volume, customer satisfaction and to skim reservation prices. Dynamic pricing is therefore highly relevant to retailers but time and costs limit frequent updates of printed price tags in in-store shopping environments. Especially for low-cost products, personalized pricing is not feasible at all and can be only indirectly applied through the use of loyalty cards that promise discounts after or at the purchase. In this sense, retailers lack the capability of applying sales strategies that rely on an instant pricing of products and thus, pricing in retail stores is rather static today. Thus, retailers' operational agility, i.e. the ability to accomplish speed, accuracy, and cost economy in the exploitation of pricing strategies is strongly restricted (Sambamurthy, Bharadwaj, & Grover, 2003). As one solution, pricing services and flexible price delivery infrastructures may address this challenge. Prior research suggests that smart products could support the presentation of dynamic prices as they incorporate information technology for business purposes (Konana & Ray, 2007; Maass & Varshney, 2008). In contrast to Electronic Shelf Labelling Systems (Southwell, 2002), the concept of smart products is more flexible because products can be directly identified through the attached barcode or RFID tag in order to request price information, for instance, on a mobile device. Accordingly, smart product infrastructures together with MP-DSSs can be used to present price information instantly to customers in retail stores (Maass & Filler, 2006). The adoption of these technologies by consumers is also promising as shown in the current chapter and other studies (Kowatsch, et al., 2008; Kowatsch, et al., 2011; Maass & Kowatsch, 2008a).

And finally, for providers of MP-DSS technology and corresponding product services, another implication can be drawn from our subjects' feedback: it is crucial to provide fast product identification technologies and fast mobile applications, such that product information is presented (almost) without any delay to the end user. For instance, this can be done with an efficient interface to the camera of a mobile phone to identify products with barcodes or QR codes, e.g., by using Google's Android development kit or Apple's iPhone platform. But also interfaces to radio frequency devices as discussed in the current chapter and which are also available in Nokia's NFC devices must be used with efficiency optimization in mind assuming the existence of products tagged with RFID antennas. Although some of these applications do already exist (Keegan, et al., 2008; Maass & Filler, 2006; Resatsch, et al., 2008; von Reischach, et al., 2009), they have two major shortcomings: they show a slow performance and lack mature graphical user interfaces that allow access to value added services such as product recommendation services or purchase transaction services that would further increase both perceived ease of use and perceived usefulness of a MP-DSS.

Limitations

This study is limited due to the preliminary character of the experiment and the use of a small sample that is not representative for consumers of retail stores in general. Thus, we found astonishingly clear results that are moderated by the fact that subjects were mostly technically savvy individuals. Additionally, consumers' self-consciousness of using a MP-DSS in public retail stores was not tested in our experiment, but it may play a major role (Serif & Ghinea, 2008). Hence, it is expected that individuals with less technical knowledge and little self-consciousness show smaller effects. Nevertheless, we want to make sure that at least technically savvy individuals adopt our MP-DSS in purchase situations such that we are able to optimize the MP-DSS and then conduct a more representative field study. Another shortcoming of our experiment lies in the subjects' motivation to perform the decision-making task, which was externally motivated rather than intrinsic. Thus, the utility of our MP-DSS may be moderated by the degree of the task motivation of the participants (Chan, 2009). In addition, our findings are based on the consumer electronics domain, in particular on mobile navigation units and accessories. Therefore, the research model needs to be tested for other products to add external validity. And finally, the perceived relevance of product information provided by a MP-DSS when being compared to paper-based information in terms of alternative or complementary information channels was not covered by this study but will guide future work.

FUTURE RESEARCH DIRECTIONS

This initial experiment deliberately focused at one particular instance of a MP-DSS with a limited sample size. Thus, we will conduct a more representative field study, which also investigates the relation between paper-based product information and information provided by a MP-DSS in terms of alternative or complementary information channels. Additionally, in another study, we compare different graphical user interfaces for one platform as well as from different platforms (e.g., Apple's iPhone vs. Nokia N97 vs. a Google Android mobile phone) that implement a MP-DSS with various communication designs across different usage situations within the product life cycle. Here, the ease of use construct will be investigated further regarding its role to predict the behavioural intention to use MP-DSSs. Another issue of future work is the determination and classification of relevant services that are required by consumers in purchase situations and have the potential to be provided through products. For instance, personalized grocery recommendations based upon consumer's allergies or dynamic pricing services for product bundles are relevant for both consumers' information needs and retailers' communication needs and can be provided on demand by products via MP-DSSs.

CONCLUSION

Product information influences product purchases and especially for online purchase situations, the value of product information is increased by the use of recommendation agents as information adapts dynamically to the interests and preferences of consumers. Because unprecedented, we investigated the use of mobile recommendation agents (MP-DSSs) by means of an innovative product interface for product information acquisition in in-store shopping situations. Based on TPB, IDT and TAM we developed a theoretical model and tested it by conducting a lab experiment. Results demonstrate the usability and utility of our MP-DSS implementation. We found that perceived usefulness of the MP-DSS predicts its adoption, purchase intentions and store preferences, which is relevant for store managers in considering new kinds of business models that allow products to communicate with their consumers. Accordingly, the communication between consumers and products generate a new type of commercial situations in which the being of both consumer and product in a physical situation becomes an integral part of the commercial task environment.

REFERENCES

Ajzen, I. (1991). The theory of planned behavior. *Organizational Behavior and Human Decision Processes*, *50*, 179–211. doi:10.1016/0749-5978(91)90020-T

Ajzen, I., & Fishbein, M. (1980). *Understanding attitudes and predicting social behaviour*. Englewood Cliffs, NJ: Prentice Hall.

Bakos, J. Y. (1998). The emerging role of electronic marketplaces on the Internet. *Communications of the ACM*, *41*(8), 35–42. doi:10.1145/280324.280330

Barclay, D., Thompson, R., & Higgins, C. (1995). The Partial Least Squares (PLS) approach to causal modeling: Personal computer adoption and use as an illustration. *Technology Studies*, *2*(2), 285–309.

Baroudi, J. J., & Orlikowski, W. J. (1989). The problem of statistical power in MIS research. *Management Information Systems Quarterly*, *13*(1), 7–106. doi:10.2307/248704

Bettman, J. R., Luce, M. F., & Payne, J. (1998). Constructive consumer choice processes. *The Journal of Consumer Research*, *25*(3), 187–217. doi:10.1086/209535

Bitran, G. R., & Ferrer, J.-C. (2007). On pricing and composition of bundles. *Production and Operations Management*, *16*(1), 93–108. doi:10.1111/j.1937-5956.2007.tb00168.x

Bo, X., & Benbasat, I. (2007). E-commerce product recommendation agents: Use, characteristics, and impact. *Management Information Systems Quarterly*, *31*(1), 137–209.

Brooke, J. (1996). SUS: a quick and dirty usability scale. In Jordan, P., Thomas, B., Weerdmeester, B., & McClelland, A. (Eds.), *Usability evaluation in industry*. London, UK: Taylor and Francis.

Chan, S. H. (2009). The roles of user motivation to perform a task and decision support system (DSS) effectiveness and efficiency in DSS use. *Computers in Human Behavior*, *25*(1), 217–228. doi:10.1016/j.chb.2008.09.002

Chen, P. Y., Wu, S. Y., & Yoon, J. (2004). *The impact of online recommendations and consumer feedback on sales*. Paper presented at the 25th International Conference on Information Systems (ICIS 2004).

Chen, Y., & Xie, J. (2005). Third-party product review and firm marketing strategy. *Marketing Science*, *24*(2), 218–240. doi:10.1287/mksc.1040.0089

Chin, W. W. (1998). Issues and opinion on structural equation modeling. *Management Information Systems Quarterly, 22*(1), vii–xvi.

Choudhary, V., Ghose, A., Mukhopadhyay, T., & Rajan, U. (2005). Personalized pricing and quality differentiation. *Management Science, 51*(7), 1120–1130. doi:10.1287/mnsc.1050.0383

Citrin, A. V., Stem, D. E. Jr, Spangenberg, E. R., & Clark, M. J. (2003). Consumer need for tactile input: An internet retailing challenge. *Journal of Business Research, 56*(11), 915. doi:10.1016/S0148-2963(01)00278-8

Cohen, J. (1977). *Statistical power analysis for the behavioral sciences* (revised ed.). New York, NY: Academic Press.

Davis, F. D. (1989). Perceived usefulness, perceived ease of use, and user acceptance of Information Technology. *MIS Ouarterly, 13*(3), 319–339. doi:10.2307/249008

DeLone, W. H., & McLean, E. R. (1992). Information systems success: The quest for the dependent variable. *Information Systems Research, 3*(1), 60–95. doi:10.1287/isre.3.1.60

DeLone, W. H., & McLean, E. R. (2003). The DeLone and Mc Lean model of Information Systems success: A ten-year update. *Journal of Management Information Systems, 19*(4), 9–30.

Elmaghraby, W., & Keskinocak, P. (2003). Dynamic pricing in the presence of inventory considerations: Research overview, current practices, and future directions. *Management Science, 49*(10), 1287–1309. doi:10.1287/mnsc.49.10.1287.17315

Esslimani, I., Brun, A., & Bayer, A. (2008). Behavioral similarities for collaborative recommendations. *Journal of Digital Information Management, 6*(6), 442–448.

Faul, F., Erdfelder, E., Lang, A.-G., & Buchner, A. (2007). G*Power 3: A flexible statistical power analysis program for the social, behavioral, and biomedical sciences. *Behavior Research Methods, 39*(2), 175–191. doi:10.3758/BF03193146

Gaeth, G. J., Levin, I. P., Chakraborty, G., & Levin, A. M. (1991). Consumer evaluation of multi-product bundles: An information integration analysis. *Marketing Letters, 2*(1), 47–57. doi:10.1007/BF00435195

Gallego, G., & van Ryzin, G. J. (1994). Optimal dynamic pricing of inventories with stochastic demand over finite horizon. *Management Science, 40*(8), 999–1020. doi:10.1287/mnsc.40.8.999

Gefen, D., Straub, D., & Boudreau, M.-C. (2000). Structural equation modeling techniques and regression: Guidelines for research practice. *Communications of AIS, 7*(7), 1–78.

Gregor, S., & Benbasat, I. (1999). Explanations from intelligent systems: Theoretical foundations and implications for practice. *Management Information Systems Quarterly, 23*(4), 497–530. doi:10.2307/249487

Groeppel, A., & Bloch, B. (1990). An investigation of experience-oriented consumers in retailing. *International Review of Retail, Distribution and Consumer Research, 1*(1), 101. doi:10.1080/09593969000000008

Gurevych, I., & Mühlhäuser, M. (2007). Natural language processing for ambient intelligence. *Künstliche Intelligenz, 2*, 10–16.

Hasan, B., & Ahmed, M. U. (2007). Effects of interface style on user perceptions and behavioral intention to use computer systems. *Computers in Human Behavior, 23*(6), 3025–3037. doi:10.1016/j.chb.2006.08.016

Häubl, G., & Trifts, V. (2000). Consumer decision making in online shopping environments: The effects of interactive decision aids. *Marketing Science, 19*(1), 4–21. doi:10.1287/mksc.19.1.4.15178

Junglas, I. A., & Watson, R. T. (2006). The U-Constructs: Four information drives. *Communications of AIS*, *17*, 2–43.

Kamis, A., Koufaris, M., & Stern, T. (2008). Using an attribute-based decision support system for user-customized products online: An experimental investigation. *Management Information Systems Quarterly*, *32*(1), 159–177.

Keegan, S., O'Hare, G. M. P., & O'Grady, M. J. (2008). EasiShop: Ambient intelligence assists everyday shopping. *Journal of Information Science*, *178*(3), 588–611. doi:10.1016/j.ins.2007.08.027

Kleijnen, M., de Ruyter, K., & Wetzels, M. (2007). An assessment of value creation in mobile service delivery and the moderating role of time consciousness. *Journal of Retailing*, *83*(1), 33–46. doi:10.1016/j.jretai.2006.10.004

Komiak, S. Y. X., & Benbasat, I. (2006). The effects of personalization and familiarity on trust and adoption of recommendation agents. *Management Information Systems Quarterly*, *30*(4), 941–960.

Konana, P., & Ray, G. (2007). Physical product reengineering with embedded information technology. *Communications of the ACM*, *50*(10), 72–78. doi:10.1145/1290958.1290965

Kourouthanassis, P., & Roussos, G. (2003). Developing consumer-friendly pervasive retail systems. *Pervasive Computing*, *2*(2), 32–39. doi:10.1109/MPRV.2003.1203751

Kowatsch, T., & Maass, W. (2011. (accepted for publication)). A Design Model for Knowledge-based Pricing Services in the Retail Industry. *International Journal of Web Engineering and Technology.*

Kowatsch, T., Maass, W., Filler, A., & Janzen, S. (2008). *Knowledge-based bundling of smart products on a mobile recommendation agent*. Paper presented at the IEEE 7th International Conference on Mobile Business (ICMB).

Kowatsch, T., Maass, W., & Fleisch, E. (2011). The role of product reviews on mobile devices for in-store purchases: consumers' usage intentions, costs and store preferences. *International Journal of Internet Marketing and Advertising: Special Issue on Information Systems and Marketing*, *6*(3), 226–243. doi:10.1504/IJIMA.2011.038237

Kurkovsky, S., & Harihar, K. (2006). Using ubiquitous computing in interactive mobile marketing. *Personal and Ubiquitous Computing*, *10*(4), 227–240. doi:10.1007/s00779-005-0044-5

Lee, K. C., & Chung, N. (2008). Empirical analysis of consumer reaction to the virtual reality shopping mall. *Computers in Human Behavior*, *24*(1), 88–104. doi:10.1016/j.chb.2007.01.018

Liu, Y., & Zhang, Z. J. (2006). The benefits of personalized pricing in a channel. *Marketing Science*, *25*(1), 97–105. doi:10.1287/mksc.1050.0120

Maass, W., & Filler, A. (2006). Towards an infrastructure for semantically annotated physical products. In Hochberger, C., & Liskowsky, R. (Eds.), *Informatik 2006* (pp. 544–549). Berlin, Germany: Springer.

Maass, W., & Janzen, S. (2007). *Dynamic product interfaces: A key element for ambient shopping environments*. Paper presented at the 20th Bled eConference. Retrieved from http://domino.fov.uni-mb.si/proceedings.nsf/2007

Maass, W., & Kowatsch, T. (2008a). *Adoption of dynamic product information: An empirical investigation of supporting purchase decisions on product bundles*. Paper presented at the 16th European Conference on Information Systems (ECIS).

Maass, W., & Kowatsch, T. (2008b). *Mobile decision support vs. interpersonal sales communication: predictors of buying intentions for price bundles*. Paper presented at the 21st Bled eConference.

Maass, W., & Varshney, U. (2008). Preface to the focus theme section: Smart products. *Electronic Markets*, *18*(3), 211–215. doi:10.1080/10196780802265645

Maidel, V., Shoval, P., Shapira, B., & Taieb-Maimon, M. (2008). *Evaluation of an ontology-content based filtering method for a personalized newspaper*. Paper presented at the ACM Conference On Recommender Systems.

McCabe, D. B., & Nowlis, S. M. (2003). The effect of examining actual products or product descriptions on consumer preference. *Journal of Consumer Psychology*, *13*(4), 431–439. doi:10.1207/S15327663JCP1304_10

Moon, B.-J. (2004). Consumer adoption of the internet as an information search and product purchase channel: some research hypotheses. *International Journal of Internet Marketing and Advertising*, *1*(1), 104–118. doi:10.1504/IJIMA.2004.003692

Moore, G. C., & Benbasat, I. (1991). Development of an instrument to measure the perceptions of adopting an information technology innovation. *Information Systems Research*, *2*, 173–191. doi:10.1287/isre.2.3.192

Nunnally, J. C. (1967). *Psychometric theory*. New York, NY: McGraw-Hill.

Pereira, R. E. (2001). Influence of query-based decision aids on consumer decision making in electronic commerce. *Information Resources Management Journal*, *14*(1), 31. doi:10.4018/irmj.2001010104

Ponder, N., Lueg, J. E., & Williams, Z. (2006). Salesperson assistance versus self-service in retailing: Are they both a matter of convenience? *Marketing Management Journal*, *16*(1), 72–87.

Resatsch, F., Sandner, U., Leimeister, J. M., & Krcmar, H. (2008). Do point of sale RFID-based information services make a difference? Analyzing consumer perceptions for designing smart product information services in retail business. *Electronic Markets*, *18*(3), 216–231. doi:10.1080/10196780802265728

Ringle, C. M., Wende, S., & Will, A. (2005). *SmartPLS* (release 2.0 M3). Hamburg, Germany: University of Hamburg. Retrieved from http://www.smartpls.de

Rogers, E. M. (2003). *Diffusion of innovations* (5th ed.). New York, NY: Free Press.

Sambamurthy, V., Bharadwaj, A., & Grover, V. (2003). Shaping agility through digital options: Reconceptualizing the role of Information Technology in contemporary firms. *Management Information Systems Quarterly*, *27*(2), 237–263.

Serif, T., & Ghinea, G. (2008). Mobile information access in the real world: A story of three wireless devices. *Computers in Human Behavior*, *24*(4), 1385–1403. doi:10.1016/j.chb.2007.07.012

Southwell, M. (2002, July 2). Beyond the POS. *Business Trends*.

Stahl, F., & Freudenschuss, M. (2006). *Mobile business through RFID technology: About the potential of RFID technology for distributing free and paid content on mobile devices in retail business*. Paper presented at the Americas Conference on Information Systems.

Su, X. (2007). Intertemporal pricing with strategic customer behavior. *Management Science*, *53*(5), 726–741. doi:10.1287/mnsc.1060.0667

Tellis, G. J., & Gaeth, G. J. (1990). Best value, price-seeking, and price aversion: The impact of information and learning on consumer choices. *Journal of Marketing*, *54*(2), 34–45. doi:10.2307/1251868

Todd, P., & Benbasat, I. (1999). Evaluating the impact of DSS, cognitive effort, and incentives on strategy selection. *Information Systems Research*, *10*(4), 356–374. doi:10.1287/isre.10.4.356

van der Heijden, H. (2006). Mobile decision support for in-store purchase decisions. *Decision Support Systems*, *42*(2), 656–663. doi:10.1016/j.dss.2005.03.006

Venkatesh, V., Morris, M. G., Davis, G. B., & Davis, F. D. (2003). User acceptance of information technology: Toward a unified view. *Management Information Systems Quarterly*, *27*(3), 425–478.

von Reischach, F., Guinard, D., Michahelles, F., & Fleisch, E. (2009). *A mobile product recommendation system interacting with tagged products*. Paper presented at the Seventh Annual IEEE International Conference on Pervasive Computing and Communications (PerCom 2009).

Watson, R. T., Pitt, L. F., Berthon, P., & Zinkhan, G. M. (2002). U-commerce: Expanding the universe of marketing. *Journal of the Academy of Marketing Science*, *30*(4), 333–347. doi:10.1177/009207002236909

Westerman, S. J., Tuck, G. C., Booth, S. A., & Khakzar, K. (2007). Consumer decision support systems: Internet versus in-store application. *Computers in Human Behavior*, *23*(6), 2928–2944. doi:10.1016/j.chb.2006.06.006

Yang, J.-M., Li, K. F., & Zhang, D.-F. (2009). Recommendation based on rational inferences in collaborative filtering. *Knowledge-Based Systems*, *22*(1), 105–114. doi:10.1016/j.knosys.2008.07.004

Youll, J., Morris, J., Krikorian, R., & Maes, P. (2000). *Impulse: Location-based agent assistance*. Paper presented at the Fourth International Conference on Autonomous Agents.

Zhang, X., & Li, Y. (2008). Use of collaborative recommendations for web search: An exploratory user study. *Journal of Information Science*, *34*(2), 145–161. doi:10.1177/0165551507080413

Zhenhui, J., & Benbasat, I. (2004). Virtual product experience: Effects of visual and functional control of products on perceived diagnosticity and flow in electronic shopping. *Journal of Management Information Systems*, *21*(3), 111–147.

ADDITIONAL READING

Fleisch, E., & Mattern, F. (2005). *Das Internet der Dinge*. Berlin: Springer. doi:10.1007/3-540-28299-8

Floerkemeier, C., Roduner, C., & Lampe, M. (2007). RFID Application Development With the Accada Middleware Platform. *IEEE Systems Journal*, *1*(2), 82–94. doi:10.1109/JSYST.2007.909778

Kowatsch, T., & Maass, W. (2009). Towards a Framework for Knowledge-based Pricing Services Improving Operational Agility in the Retail Industry. Paper presented at the 2nd International Workshop on Enabling Service Business Ecosystems (ESBE'09), co-located with the Mediterranean Conference on Information Systems (MCIS 2009), CEUR-Workshop Proceedings Vol-530.

Maass, W., & Lampe, M. (2007). Integration of Standardized and Non-Standardized Product Data. In *R. Koschke & K.-H. Rödiger. Informatik 2007, Lecture Notes in Informatics* (pp. 141–146). Berlin: Springer.

Mudambi, S. M., & Schuff, D. (2010). What Makes a Helpful Online Review? A Study of Customer Reviews on Amazon.com. *Management Information Systems Quarterly*, *34*(1), 185–200.

Swaminathan, V. (2003). The Impact of Recommendation Agents on Consumer Evaluation and Choice: The Moderating Role of Category Risk, Product Complexity, and Consumer Knowledge. *Journal of Consumer Psychology*, *13*(1/2), 93.

Todd, P., & Benbasat, I. (1992). The Use of Knowledge in Decision Making: An Experimental Investigation of the Impact of Computer-Based Decision Aids. *Management Information Systems Quarterly*, *16*, 373–393. doi:10.2307/249534

von Reischach, F., Michahelles, F., Guinard, D., Adelmann, R., Fleisch, E., & Schmidt, A. (2009). An Evaluation of Product Identification Techniques for Mobile Phones. Paper presented at the 2nd IFIP TC13 Conference in Human-Computer Interaction (Interact 2009), Upsala, Sweden.

Young, E., & Benbasat, I. (2004). A Framework for the Study of Customer Interface Design for Mobile Commerce. *International Journal of Electronic Commerce*, *8*(3), 79–102.

KEY TERMS AND DEFINITIONS

Consumer Behaviour: Behaviour of human beings that is related to buying situations.

In-Store Shopping Environment: A bricks-and-mortar retail store where consumers are able to buy products.

Online Shopping Environment: A store that is available over the WWW and where consumers are able to buy products.

Purchase Decision Support System: An information system that recognizes the interests or preferences of an individual buyer and makes product recommendations accordingly; also known as product recommendation agent.

Chapter 15
Customer Acceptance of a New Interactive Information Terminal in Grocery Retailing:
Antecedents and Moderators

Stephan Zielke
Georg-August-Universität Göttingen, Germanyw

Waldemar Toporowski
Georg-August-Universität Göttingen, Germany

Björn Kniza
Georg-August-Universität Göttingen, Germany

ABSTRACT

This chapter uses an extended version of the Technology Acceptance Model to analyze the customer acceptance of a new interactive information terminal for cooking recipes aimed at grocery shoppers. The results show that perceived usefulness, perceived ease of use, and perceived enjoyment influence the customer acceptance of the terminal via direct and indirect effects. Furthermore, the impact of these variables depends on individual differences in experience of Information Technology and the relevance of the information content. These findings carry several management and research implications.

INTRODUCTION

Information terminals receive an increasing amount of attention in retailing and new applications for such terminals are constantly being developed and introduced (de Moerloose, Antioco, Lindgreen, & Palmer 2005; Rowley 1995; Rowley & Slack 2003). Retailers can use information terminals to stimulate purchases, increase customer loyalty and support or replace sales personnel. Hence, information terminals can have a substantial impact on a retailer's profitability. Furthermore, information terminals enhance the retailers' customer insight by provid-

DOI: 10.4018/978-1-60960-738-8.ch015

ing information about the customers' search and decision processes.

However, these positive effects depend on the customers' acceptance and usage of an information terminal. Therefore, it is important for retailers to understand how their customers accept this technology. Whether the ease or enjoyment of using an information terminal influences the customers' acceptance more strongly has important strategic implications, for example. As retailers often serve different customer segments, it is also important to know, whether the impact of these variables differs between customer groups. The effects of the ease of use and enjoyment might differ between customers with strong and weak information technology experience. Also the relevance of the information content might influence the relative impact of the ease of use and enjoyment on customer acceptance. When retailers know which segments they serve and how these segments respond to variations in the ease of use and enjoyment, they can offer tailored information terminals to their customers. Therefore, retailers should investigate which variables influence the acceptance of information terminals and how the impact of these variables differs between customer segments.

As the literature on information terminals does not discuss these issues thoroughly, the present chapter aims to close this research gap. It analyzes the acceptance of a new interactive information terminal for cooking recipes based on an extended version of the technology acceptance model (Davis 1986, 1989; Davis, Bagozzi, & Warshaw 1989). The model assumes that usefulness, ease of use and enjoyment influence the acceptance of a technology (van der Heijden 2004). Furthermore, the chapter sheds light on how experience of information technology and the relevance of information content moderate the relationships in the proposed model.

The investigated terminal offers to search for cooking recipes for grocery customers and print them out at the point of sale. They can choose the number of people they want to cook for and print out a shopping list with the respective quantities. Retailers can use this terminal to generate sales, promote brands and provide added value to their customers. The information terminal for cooking recipes was selected, as this type of terminal is used in grocery stores where customers with different levels of information technology experience and interest in the information content shop.

The results show that the perceived usefulness, ease of use and enjoyment influence the acceptance of the terminal directly and indirectly. However, the effects of these variables are moderated by the experience of information technology and the relevance of the information content (interest in cooking). While enjoyment is more important for customers with little information technology experience and little interest in cooking, ease of use is more important for those with a lot of information technology experience and a high interest in cooking.

The remainder of this chapter is organized as follows: Section 2 presents the theoretical background on information terminals and the customer acceptance of new technologies. Section 3 develops a framework of hypotheses based on the technology acceptance model. Section 4 describes the method used to test the research model and section 5 presents the results. A discussion follows in section 6. The chapter closes in section 7 with some limitations and future research directions.

THEORETICAL BACKGROUND

Research on Information Terminals

An information terminal or information kiosk is a self-service technology station with an interactive information processing capacity located in public concourse (Rowley & Slack 2007). The literature reports numerous applications for such terminals in different industries (de Moerloose et al. 2005; Slack & Rowley 2002a), particularly in retailing (Rowley 1995; Rowley & Slack 2003).

Information terminals in retailing offer applications customers can use to:

- search for product information
- compare the technical features of products
- find new ideas for product usage
- visualize products and product usage supported by images and videos
- test products, such as CDs, DVDs or games
- find out whether products are in stock and where they are located in the store
- find additional products which are not displayed in the store due to limited space
- receive recommendations for substitutes or complementary products
- receive individualized recommendations based on their search and purchase history
- design their own product
- print out coupons and vouchers
- reserve and order products
- manage their loyalty program membership

More generally, information terminals can be described by certain dimensions. Tung and Tan (1998) suggested characterizing information terminals by the dimensions information dissemination and transaction power. Information dissemination refers to the provision of information on the products and services offered, while transaction power refers to the possibility to perform transactions, especially ordering and buying products. Rowley and Slack (2003) develop this typology by suggesting two additional characteristics: interaction and relationship-orientation. Interaction requires some information input by the customer. Interactive terminals, for example, provide information and recommendations based on personal preferences. Relationship-orientation aims to enhance customer-retailer relationships. Information terminals with a strong relationship-orientation are usually linked to a loyalty scheme. They offer specials, service enhancements and other added value to loyalty card holders (Rowley 2000). In a later paper, Rowley and Slack (2007) extend their initial typology. In addition to the task (e. g. information provision or relationship building), they also suggest describing information terminals based on location (e. g. retail outlets), the actual or intended user group (e. g. shoppers) and technology (e. g. screen with or without card slots or keyboard).

An information terminal for cooking recipes can be set up in retail stores to target their shoppers. Regarding the task, such a terminal can be characterized by a high amount of information dissemination and interaction, as customers search for recipes based on their personal preferences. The transaction power is small as customers do not use the terminal to buy products. However, they can at least print out coupons and vouchers for products from the recipes. The relationship-orientation depends on the involvement in a loyalty card program or at least the possibility to create a personal profile for using the terminal. The necessary technology components (card slots, keyboard, and printer) depend on the respective task of the terminal.

In addition to defining information terminals, it is also important to clarify differences to and similarities with related technologies. Often, information terminals and mobile technologies offer similar features. An iPhone application, for instance, might also recommend cooking recipes based on the product assortment of a specific retailer. However, the most important difference between information terminals and mobile technologies is that information terminals are stationary and accessible to the public. Furthermore, they do not have the disadvantages of mobile technologies, such as small screen sizes and keyboards. Therefore, they are an alternative to mobile technologies for some customer segments (Slack & Rowley 2002b).

Information terminals are also related to product recommendation agents. According to Xiao and Benbasat (2007), these are software agents which make product recommendations based on interests and preferences of individual users. Although, recommendation agents are mostly used in e-commerce, they may also be a feature of an information terminal. With regard to information terminals for cooking recipes, recommendation agents could suggest tailored recipes for individual users.

Although various researchers have published a number of papers about the applications and typologies of information terminals (de Moerloose et al. 2005; Slack & Rowley 2002a; Slack & Rowley 2002b; Tung & Tan 1998; Rowley 1995; Rowley & Slack 2003; Rowley & Slack 2007), empirical studies on the acceptance of such terminals in retailing are scarce. Swoboda (1998) found that users and non-users of an interactive information terminal for music CDs differ in their experience with computers and their involvement. Another study by Lee, Fairhurst, and Lee (2009) demonstrated the positive impact of the service quality delivered by a kiosk system on retail patronage intentions. Chiu, Fang, and Tseng (2010) recently analyzed differences between early and potential adopters of a kiosk technology. As the specific literature on the acceptance of retail information terminals is limited, the more general literature on the acceptance of new technologies might provide a deeper insight.

Research on the Acceptance of New Technologies

Different research streams in the literature analyze the acceptance of new technologies (Moore & Benbasat 1991; Venkatesh, Morris, Davis, & Davis 2003). The most prominent research streams build on the innovation diffusion theory (Rogers 1962) and the technology acceptance model (Davis 1986, 1989; Davis, Bagozzi, & Warshaw 1989). The technology acceptance model in particular was used to explain customer acceptance of retail and service innovations, such as online shopping (Featherman & Pavlou 2003; Gefen, Karahanna, & Straub 2003) or internet banking (Curran & Meuter 2005; Kleijnen, Wetzels, & de Ruyter 2004; Wang, Wang, Lin, & Tang 2003).

The technology acceptance model is an adaptation of the theory of reasoned action (Ajzen & Fishbein 1980; Fishbein & Ajzen 1975). This theory assumes that the attitude towards behavior and social norms jointly determine behavioral intentions, which are then linked to actual behavior. The attitude is a function of salient beliefs regarding the consequences of a behavior multiplied by the evaluation of these consequences. Later, Ajzen (1985, 1991) introduced the perceived behavioral control as a further influencing factor of intentions and behavior (theory of planned behavior).

Similar to the theory of reasoned action, the technology acceptance model assumes that the attitude towards using a technology influences the intention to use, which has an impact on usage behavior. However, the technology acceptance model differs from the theory of reasoned action in the determinants of the attitude towards using the technology, as the perceived usefulness and the perceived ease of use influence the attitude. Furthermore, the perceived ease of use and usefulness are related as the ease of use influences the perceived usefulness (Davis, Bagozzi, & Warshaw 1989).

Perceived usefulness refers to how the usage of a technology can improve the outcome of a certain behavior (e. g. shopping or consumption) while ease of use refers to how easily people can learn and memorize the usage of a technology. Both constructs are influenced by external variables. For online shopping, examples of external variables are flexibility in navigation, perceptions of convenience or the substitutability of personal examination (Childers, Carr, Peck, & Carson 2001). The original model proposed and empirically tested by Davis, Bagozzi, and Warshaw (1989) is presented in Figure 1.

Social norms are not included in the original model, as Davis, Bagozzi, and Warshaw had serious theoretic and psychometric concerns about that construct. Furthermore, they found no em-

Figure 1. The technology acceptance model proposed by Davis, Bagozzi, and Warshaw (1989)

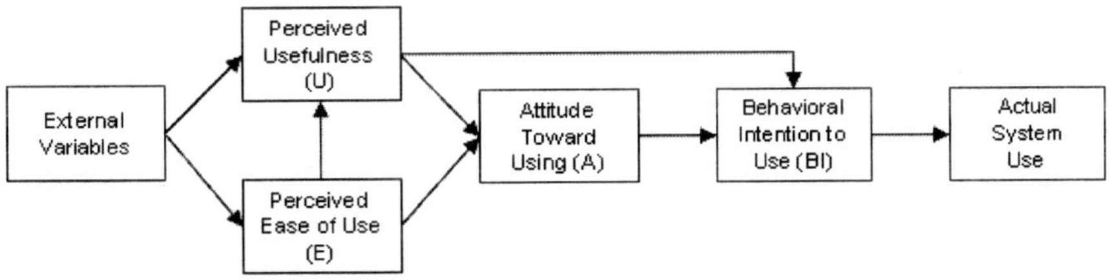

pirical evidence that social norms can have an impact on the attitude towards using a technology (however, in their study, the technology was a word processing program; using that technology is usually not observed by others, which might limit the relevance of social norms).

In previous applications, on average the model explained about 40 percent of the variance in usage intentions and behavior (Venkatesh & Davis 2000). Furthermore, moderators, such as shopping motivation (Childers et al. 2001), the type of respondents and the type of technology or culture, were identified (Schepers & Wetzels 2007). Socio-demographic variables, experience with the technology and voluntariness of use also played a moderating role in the formation of customer acceptance (Venkatesh et al. 2003).

A number of researchers extended and modified the technology acceptance model. Several authors integrated enjoyment or other emotional variables as predictors of customer acceptance (Childers et al. 2001; Dabholkar & Bagozzi 2002; Kulviwat, Bruner, Kumar, Nasco, & Clark 2007; van der Heijden 2004). Other authors extended the model by integrating variables, such as perceived risk or trust in the owner of the technology (e. g. Curran & Meuter 2005; Featherman & Pavlou 2003; Gefen, Karahanna, & Straub 2003; Xiao & Benbasat 2007). However, these variables seem to be more relevant for technologies customers use for monetary transactions, such as online shopping or internet banking. For interactive information terminals in grocery retailing that are characterized by a lower level of transaction power, the perceived enjoyment might especially be relevant.

Further modifications of the model are related to the dependent variables attitude towards use, intention to use and usage behavior. Some studies concentrate on the attitude and the intention (e. g. Wixom & Todd 2005; Kulviwat et al. 2007), while others use only the attitude (Childers et al. 2001) or intention to use as the dependent variable (e. g. Gefen, Karahanna, & Straub 2003; van der Heijden 2004). According to the theory of planned behavior, the perceived behavioral control especially explains any discrepancies between the attitudes, intentions and behavior. As information terminals are usually located in public concourse, they are easily accessible. Therefore, the perceived behavioral control should be large, resulting in strong correlations between the dependent variables. Given these circumstances, it seems reasonable to summarize the attitudes, intentions and behavior under the umbrella of customer acceptance as a single dependent variable.

HYPOTHESES

Based on previous research, an extended version of the technology acceptance model explains the acceptance of new interactive information terminals in grocery retailing (e. g. a new terminal for cooking recipes). The extended version integrates enjoyment as a third predictor after the ease of use and usefulness. Furthermore, it concentrates on acceptance as a single dependent variable. Acceptance reflects a positive attitude towards adopting a technology, the intention to adopt and/or the actual adoption behavior. The model also

Figure 2. Research model

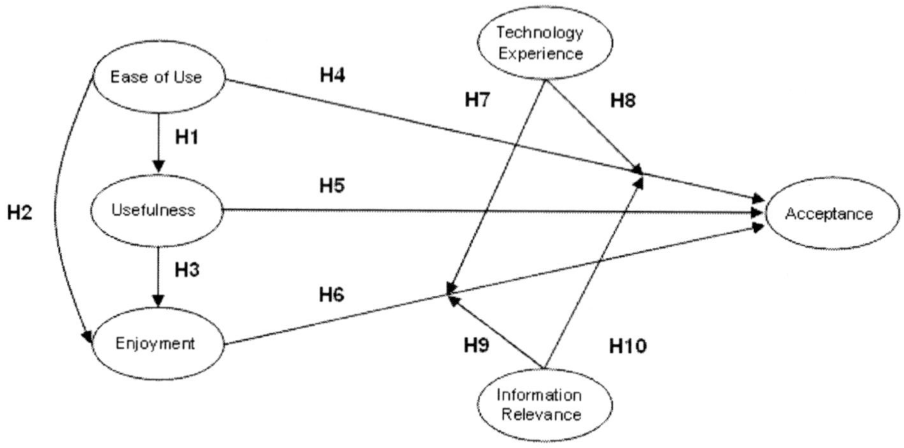

assumes that some relationships are moderated by the individual information technology experience and interest in the information content (Figure 2).

Building on van der Heijden (2004), the model assumes that the perceived ease of use influences the perceived usefulness (**H1**) and the perceived enjoyment positively (**H2**). In addition, a positive effect of usefulness on enjoyment is also hypothesized (**H3**) as a number of studies found that emotions mediated the behavioral impact of perceptions (e. g. Chebat & Michon 2003; Chebat & Slusarczyk 2005). Furthermore, the perceived ease of use (**H4**), perceived usefulness (**H5**) and enjoyment (**H6**) have direct effects on the acceptance (similar to van der Heijden's model, where these variables influence the intention to use directly).

While the proposed model finds a lot of support in previous research, less is known about the moderators of the proposed relationships. Results from Venkatesh et al. (2003) indicate that prior experience of a technology moderates the formation of technology acceptance. Wixom and Todd (2005) explain the perception and acceptance of a technology with variables relating to the system and the information content. Hence, moderators should also relate to both conceptual levels. In the context of information terminals, Swoboda (1998) found that individual predispositions, such as experience with computers and involvement, explain the usage of an information terminal. Similar constructs might moderate the relationships in the technology acceptance model. Accordingly, the present study investigates experience with information technology and the relevance of information content as moderators. The first moderator is related to the information system, the second to the information content.

Regarding experience, the previous research on self-service technologies in retailing has shown that the enjoyment of using a technology mostly influences customer satisfaction for first-time users (Zielke, Lietke, Toporowski, & Boslau) or when the technology was recently introduced (Marzocchi & Zammit 2006). Therefore, it seems plausible to assume that for customers with little information technology experience, enjoyment has a strong impact on the acceptance of a new information terminal, while this effect might be insignificant for experienced users. Furthermore, customers with little technology experience might have more difficulties with using the technology, resulting in a greater impact of the ease of use.

H7: The impact of enjoyment on acceptance is stronger for customers with little informa-

Figure 3. Screenshots from an information terminal for cooking recipes (xplace, Germany)

tion technology experience compared to experienced customers.

H8: The impact of the ease of use on acceptance is stronger for customers with little information technology experience compared to experienced customers.

The relevance of information content relates to the involvement construct. Previous research has argued that information is elaborated more deeply under high involvement conditions, while emotions play a greater role under low involvement conditions (Petty & Cacioppo 1981, 1986). Therefore, it seems plausible that the enjoyment of using the technology has a stronger impact on the acceptance for customers with little interest in information content, while the ease of use has a stronger impact for customers with high interest.

H9: The impact of enjoyment on acceptance is stronger for customers with little interest in the information content compared to customers with strong interest.

H10: The impact of the ease of use on acceptance is stronger for customers with a high level of interest in the information content compared to customers with little interest.

RESEARCH METHOD AND SAMPLE

The hypotheses were tested for a new interactive information terminal for cooking recipes. The terminal offers to search for cooking recipes for grocery customers and print them out. The customers can choose the number of people they want to cook for and print out a shopping list with the respective quantities. Figure 3 presents some screenshots from the terminal.

The data was collected with a paper-and-pencil questionnaire in a real shopping environment. Measures for the ease of use, usefulness, enjoyment and acceptance were adapted from previous research (e. g. Curran & Meuter 2005; Davis 1989; Gefen, Karahanna, & Straub 2003; Kulviwat et al. 2007; Lin & Hsieh 2006; Venkatesh & Davis 2000; Venkatesh et al. 2003; van der Heijden 2004; Weijters, Rangarajan, Falk, & Schillewaert 2007). The acceptance was measured by questions on the attitude towards using the terminal, the re-use intention and the intention to recommend to others. Recommendation to others was included as several previous studies treated word-of-mouth communication as an indicator of attitudinal loyalty or behavioral intentions (e. g. Bitner 1990; Cronin, Brady, & Hult 2000; Rundle-Thiele 2005; Bodet 2008). The measures for the experience of information technology and the relevance of information content were developed specifically for this study. As the study analyses the acceptance of a terminal for cooking recipes, the relevance of the information content was operationalized as the interest in cooking. The complete measures are presented in the appendix. All items were measured on seven-point scales. The questionnaire was pre-tested before conducting the main study.

The main study was conducted in Germany. Interviewers collected the data in three supermarkets immediately after a new information terminal for cooking recipes had been installed. The terminals were placed near the fruit and vegetables just after the entrance to the supermarkets. The interviewers asked the store customers if they were willing to try out the terminal and complete a questionnaire afterwards. If customers agreed to participate in the survey, they received a short introduction from the interviewer before using the terminal on their own. No specific task was given to the customers; the interviewers just asked them to try the terminal without any time restriction. The interviews were then conducted immediately after using the terminal, yielding 216 completed questionnaires in total. After completing their shopping trip, the respondents were approached a second time, and asked about their expenditures and if they had bought any products from the recipes they had printed out.

The sample reasonably represents the customer structure of the grocery stores where the study was conducted. About 70 percent of the respondents were female, 7.9 percent of the respondents were younger than 20, 34.7 percent were aged between 20 and 39, 40.4 percent were between 40 and 59, and 16.7 percent of the respondents were over 60. As regards the household size, 15.4 percent of the respondents lived in single households, 41.1 percent in households with two people and 43.5 percent in households with three or more people. The monthly expenditures for groceries were € 240 on average, the respondents spent 24.98 € on the shopping trip and 20.5 percent indicated that they bought at least one article from the recipes.

RESULTS

Before testing the hypotheses, the convergent and discriminant validity was checked according to the recommendations in the literature (Churchill 1979; Nunnally & Bernstein 1994). Firstly, alpha coefficients were calculated for each construct (Cronbach 1951). All the coefficients are clearly above.70 according to the recommendation of Nunnally and Bernstein (1994), indicating sufficient convergent validity. A confirmatory factor analysis was conducted with Mplus 4.1 (Muthén & Muthén 2007) in order to test the discriminant validity. After eliminating two items with larger cross-loadings, the fit indexes of the model were satisfactory (CFI:.948; TLI:.939; RMSEA:.054; SRMR:.060; χ^2/df: 3416/300). Furthermore, the discriminant validity was sufficient according to the Fornell-Larcker-Criterion, as all the squared correlations between the constructs were smaller than the average variance extracted by each construct (Fornell & Larcker 1981). The construct means, standard deviations, alpha coefficients, correlations and extracted variances are presented in Table 1. The respective factor loadings are presented in the appendix.

As no serious measurement problems were identified, the hypotheses were tested with PLS path models, using smartPLS 2.0 (Ringle, Wende, & Will 2005). PLS was used instead of a covariance structure model as the sample was split for the moderator analysis and PLS is especially recommended for small sample sizes (Chin & Newsted 1999). Table 2 presents the results from the hypotheses tests for the basic model without considering the moderating effects.

The results show that the ease of use influences the usefulness (.31), which has an impact on the enjoyment (.40), thus supporting H1 and H3. However, contrary to H2, the ease of use and

Table 1. Construct means (M), standard deviations (SD), alpha coefficients, correlations (lower triangle) and average variance extracted (diagonal)

	Construct	M	SD	Alpha	1	2	3	4	5	6
1	Technology experience	3.95	1.83	.83	.70					
2	Information content relevance	4.73	1.61	.86	-.14	.62				
3	Ease of use	6.30	1.13	.93	.12	.02	.76			
4	Usefulness	5.42	1.35	.91	-.07	.22	.29	.73		
5	Enjoyment	5.47	1.44	.87	-.21	.17	.11	.40	.67	
6	Acceptance	5.82	1.29	.91	-.05	.33	.41	.72	.42	.65

Table 2. Results without moderating effects

		Direct		Total	
	Hypotheses	Beta	(Sig.)	Beta	(Sig.)
H1	Ease of use → Usefulness	.31	.000	.31	.000
H2	Ease of use → Enjoyment	.01	.951	.13	.121
H3	Usefulness → Enjoyment	.40	.000	.40	.000
H4	Ease of use → Acceptance	.22	.014	.40	.000
H5	Usefulness → Acceptance	.52	.000	.60	.000
H6	Enjoyment → Acceptance	.21	.002	.21	.002

enjoyment are completely unrelated. All three variables influence acceptance, supporting H4, H5 and H6. Usefulness has the strongest direct impact (.52), followed by the ease of use (.22) and enjoyment (.21). However, as the ease of use also influences the acceptance indirectly via the usefulness, the total effect is much stronger than the direct effect (.40). Overall, the three predictors explain 52.3 percent of the variance in the acceptance.

The moderating effects were analyzed using median splits and comparisons of the separate PLS-results for both groups. Table 3 presents the respective results for the experience with information technology. Supporting H7, the effect of enjoyment on the acceptance is significant for customers with little experience (.37), while it is insignificant for customers with strong experience (.15). Surprisingly, and contrary to H8, the ease of use has no direct impact on the acceptance (.12) when the experience is minor, while the impact is strong for highly experienced customers (.30). However, the indirect effects are stronger for customers with little experience, resulting in similar total effects (.43 vs..40). The explained variance in the acceptance is 68.8 percent for customers with little experience and 40.7 percent for customers with a lot of experience.

Table 4 presents the results for moderating effects of information content relevance. Supporting H9, enjoyment has a significant impact on the acceptance for customers with little interest in cooking (.29), while the effect is insignificant for customers with a strong interest (.14). The results also support H10 as the direct effect of the ease of use is only significant for customers with a strong interest in cooking (.37). The total effect is also substantially stronger compared to customers with little interest (.51 vs..34). The explained variance in the acceptance is 59.9 percent for

Table 3. Results for the moderating effects of the experience with information technology

		Little Experience (n=104)				High Level of Experience (n=112)			
		Direct		Total		Direct		Total	
	Hypotheses	Beta	(Sig.)	Beta	(Sig.)	Beta	(Sig.)	Beta	(Sig.)
H1	Ease of use → Usefulness	.40	.000	.40	.000	.20	.083	.20	.083
H2	Ease of use → Enjoyment	.06	.565	.30	.019	.01	.947	.06	.541
H3	Usefulness → Enjoyment	.58	.000	.58	.000	.27	.006	.27	.006
H4	Ease of use → Acceptance	.12	.111	.43	.000	.30	.038	.40	.005
H5	Usefulness → Acceptance	.49	.000	.70	.000	.45	.000	.49	.000
H6	Enjoyment → Acceptance	.37	.000	.37	.000	.15	.107	.15	.107

Table 4. Results for the moderating effects of the information content relevance

		Little Relevance (n=105)				Strong Relevance (n=111)			
		Direct		Total		Direct		Total	
	Hypotheses	Beta	(Sig.)	Beta	(Sig.)	Beta	(Sig.)	Beta	(Sig.)
H1	Ease of use → Usefulness	.36	0.000	.36	0.000	.29	0.016	.29	0.016
H2	Ease of use → Enjoyment	-.02	0.829	.11	0.216	.03	0.815	.15	0.249
H3	Usefulness → Enjoyment	.38	0.000	.38	0.000	.39	0.000	.39	0.000
H4	Ease of use → Acceptance	.10	0.196	.34	0.000	.37	0.010	.51	0.000
H5	Usefulness → Acceptance	.58	0.000	.69	0.000	.41	0.000	.46	0.000
H6	Enjoyment → Acceptance	.29	0.001	.29	0.001	.14	0.156	.14	0.156

customers with little interest in cooking and 46.7 for customers with a strong interest.

DISCUSSION

The results support most of the hypothesized effects and the explained variance in the acceptance is considerably stronger than the average of 40 percent reported by Venkatesh and Davis (2000). Furthermore, the results extend the research of Swoboda (1998), who found that users and non-users of an interactive information terminal for music CDs differed in their experience of computers and their involvement. The present study sheds some additional light on the technology experience and interest in the information content (which is related to involvement) as moderators in the technology acceptance model.

However, two of the hypothesized effects are not supported. Firstly, in none of the models did the ease of use have a significant direct effect on the enjoyment as observed by van der Heijden (2004). This result supports studies which have not modeled a relationship between both constructs (e. g. Childers et al. 2001; Dabholkar & Bagozzi 2002; Kulviwat et al. 2007). Hence, the ease of use and enjoyment are relatively unrelated constructs, which have separate effects on the customer acceptance of the investigated information terminal. Only for customers with little information technology experience does a significant indirect effect via usefulness exist. Therefore, the ease of use can at least be a precondition of usefulness and enjoyment when customers only have little experience of the technology.

Secondly, the results for the moderating effects of technology experience on the impact of the ease of use were unexpected as the ease of use only influenced the acceptance directly for experienced users. One explanation might be that experienced users focus more on the usability of the technology to find and receive cooking recipes in an efficient manner. Furthermore, according to the elaboration likelihood model (Petty & Cacioppo 1981, 1986), emotional aspects might be more relevant than the benefits when the customer's ability to use the terminal is limited. Therefore, inexperienced users should focus less on the ease of use. Another explanation refers to the indirect effects. As discussed before, the ease of use influences usability, resulting in indirect effects on both the enjoyment and customer acceptance. Therefore, the total effects of the ease of use on acceptance are similar for inexperienced and experienced users. For inexperienced users, ease of use is a precondition for usefulness and enjoyment, while it has a direct impact for customers with high level of technology experience.

The main management implication is that information terminals must be easy to use and enjoyable to cover all customer groups. Furthermore, retailers can develop specific strategies for different customer segments. The results show, for example, that customers with little information technology experience and/or little interest in the information content can become loyal users of the terminal if they enjoy using it. Therefore, retailers should actively encourage these customers to try the terminal, and they should communicate the fun customers will have when they use it. Furthermore, developers of such terminals should present the information content in a way that is enjoyable for customers who are only moderately interested in the information provided. When retailers serve customers with a high level of information technology experience and a strong interest in the information content, enjoyment becomes less important than the ease of use (one example might be a terminal for the configuration of computer equipment in a computer store). In this situation, retailers should focus on the ease of using the terminal (perhaps by reducing the number of enjoyable features that might derogate the ease and efficiency of using the terminal, e. g. a stylish but inefficient layout).

The results also support the conceptualization of acceptance, which summarizes attitudes, intentions and behavior under the umbrella of a single dependent variable. As the coefficient alpha (.91) and the average variance explained (.65) were quite large for the acceptance, it seems reasonable to conceptualize the acceptance by a single multi-item construct. However, this might be different for technologies that are not accessible to the public or that are not used voluntarily (e. g. when employees have to use new computer software). In the first case, the positive attitude might be stronger than intentions and behavior, while in the second case, the attitude might be weaker. However, for retail technologies, which are easily accessible without any cost for the customer and which are used voluntarily, the attitude, intention and behavior should correlate strongly.

LIMITATIONS AND RESEARCH DIRECTIONS

The chapter closes with some limitations and implications for future research. Future studies might work with larger sample sizes to develop more differentiated customer segmentations. It will be especially interesting to compare customer groups with different combinations of technology experience and interest in information content. Preliminary analyses of the current data, for example, indicated that in the segment of customers with a high level of experience and strong interest in the information content, the ease of use has an even larger impact on the acceptance than the usefulness, while the effect of enjoyment is close to zero.

Future studies might also integrate additional predictors and analyze their impact on different dependent variables. Possible sources for additional predictors are the innovation diffusion theory (Rogers 1962), the service quality approach (Parasuraman, Berry, & Zeithaml 1988), the perceptual dimensions suggested by Dabholkar (1996) or the integrated models from Moore and Benbasat (1991) and Venkatesh et al. (2003). However, as the three predictors from the extended technology acceptance model explained a substantial amount of variance in the acceptance, the contribution of integrating additional variables is questionable. Nevertheless, it seems important to analyze how retailers can influence these predictors by the design of the terminal, the information content, the presentation of information and additional features (Childers et al. 1991).

It will be also interesting to analyze synergies between stationary terminals and mobile information technologies, such as iPhone applications. The importance of the predictors might differ as different customer segments use both types of information technologies. It would be also of interest to analyze how information terminals can influence the perception of other marketing instruments, such as the assortment or prices. The processibility of price information, for example, is an important dimension of a retailer's price image (Zielke 2006, 2010). Information terminals can support the processing of price information by providing the prices per person for the recipes or by offering price comparisons between alternative products.

Finally, it should be mentioned that the empirical study analyzed an information terminal for cooking recipes, which is a very specific application. Therefore, future studies should analyze the research model presented in this chapter for a broader range of retail technologies.

REFERENCES

Ajzen, I. (1985). From intentions to actions: A theory of planned behavior. In Kuhl, J., & Beckmann, J. (Eds.), *Action control: From cognition to behaviour* (pp. 11–39). Berlin, Germany: Springer.

Ajzen, I. (1991). The theory of planned behavior. *Organizational Behavior and Human Decision Processes*, *50*(2), 179–211. doi:10.1016/0749-5978(91)90020-T

Ajzen, I., & Fishbein, M. (1980). *Understanding attitudes and predicting social behavior*. Englewood Cliffs, NJ: Prentice-Hall.

Bitner, M. J. (1990). Evaluating service encounters: The effects of physical surroundings and employee responses. *Journal of Marketing*, *54*(2), 69–82. doi:10.2307/1251871

Bodet, G. (2008). Customer satisfaction and loyalty in service: Two concepts, four constructs, several relationships. *Journal of Retailing and Consumer Services*, *15*(3), 156–162. doi:10.1016/j.jretconser.2007.11.004

Chebat, J.-C., & Michon, R. (2003). Impact of ambient odors on mall shoppers' emotions, cognition, and spending. A test of competitive causal theories. *Journal of Business Research*, *56*(7), 529–539. doi:10.1016/S0148-2963(01)00247-8

Chebat, J.-C., & Slusarczyk, W. (2005). How emotions mediate the effects of perceived justice on acceptance in service recovery situations: An empirical study. *Journal of Business Research*, *58*(5), 664–673. doi:10.1016/j.jbusres.2003.09.005

Childers, T. L., Carr, C. L., Peck, J., & Carson, S. (2001). Hedonic and utilitarian motivations for online retail shopping behavior. *Journal of Retailing*, *77*(4), 511–535. doi:10.1016/S0022-4359(01)00056-2

Chin, W. W., & Newsted, P. R. (1999). Structural equation modeling analysis with small samples using partial least squares. In Hoyle, R. H. (Ed.), *Statistical strategies for small sample research* (pp. 307–341). Thousand Oaks, CA: Sage.

Chiu, Y.-T. H., Fang, S.-C., & Tseng, C.-C. (2010). Early versus potential adopters: Exploring the antecedents of use intention in the context of retail service innovations. *International Journal of Retail & Distribution Management*, *38*(6), 443–459. doi:10.1108/09590551011045357

Churchill, G. A. (1979). A paradigm for developing better measures of marketing constructs. *JMR, Journal of Marketing Research*, *16*(1), 64–73. doi:10.2307/3150876

Cronbach, L. J. (1951). Coefficient alpha and the internal structure of tests. *Psychometrica*, *16*(3), 297–334. doi:10.1007/BF02310555

Cronin, J. J., Brady, M. K., & Hult, G. T. M. (2000). Assessing the effects of quality, value, and customer satisfaction on consumer behavioral intentions in service environments. *Journal of Retailing*, *76*(2), 193–218. doi:10.1016/S0022-4359(00)00028-2

Curran, J. M., & Meuter, M. L. (2005). Self-service technology adoption: Comparing three technologies. *Journal of Services Marketing*, *19*(2), 103–113. doi:10.1108/08876040510591411

Dabholkar, P. A. (1996). Consumer evaluations of new technology-based self-service options: An investigation of alternative models of service quality. *International Journal of Research in Marketing*, *13*(1), 29–51. doi:10.1016/0167-8116(95)00027-5

Dabholkar, P. A., & Bagozzi, R. P. (2002). An attitudinal model of technology-based self-service: Moderating effects of consumer traits and situational factors. *Journal of the Academy of Marketing Science*, *30*(3), 184–201.

Davis, F. D. (1986). *A technology acceptance model for empirically testing new end-user systems: Theory and results*. Cambridge, MA: MIT.

Davis, F. D. (1989). Perceived usefulness, perceived ease of use, and user acceptance of information technology. *Management Information Systems Quarterly*, *13*(3), 319–340. doi:10.2307/249008

Davis, F. D., Bagozzi, R. P., & Warshaw, P. R. (1989). User acceptance of computer technology: A Comparison of two theoretical models. *Management Science*, *35*(8), 982–1003. doi:10.1287/mnsc.35.8.982

De Moerloose, C., Antioco, M., Lindgreen, A., & Palmer, R. (2005). Information kiosks: The case of the Belgian retail sector. *International Journal of Retail & Distribution Management*, *33*(6), 472–490. doi:10.1108/09590550510603651

Featherman, M. S., & Pavlou, P. A. (2003). Predicting e-services adoption: A perceived risk facets perspective. *Journal of Human-Computer Studies*, *59*(4), 451–474. doi:10.1016/S1071-5819(03)00111-3

Fishbein, M., & Ajzen, I. (1975). *Belief, attitude, intention and behavior: An introduction to theory and research*. Reading, MA: Addison-Wesley.

Fornell, C., & Larcker, D. F. (1981). Evaluating structural equation models with unobservable variables and measurement error. *JMR, Journal of Marketing Research*, *18*(1), 39–50. doi:10.2307/3151312

Gefen, D., Karahanna, E., & Straub, D. W. (2003). Trust and TAM in online shopping: An integrated model. *Management Information Systems Quarterly*, *27*(1), 51–90.

Kleijnen, M., Wetzels, M. G. M., & de Ruyter, K. (2004). Consumer acceptance of wireless finance. *Journal of Financial Services Marketing*, *8*(3), 206–217. doi:10.1057/palgrave.fsm.4770120

Kulviwat, S., Bruner, G. C. II, Kumar, A., Nasco, S. A., & Clark, T. (2007). Toward a unified theory of customer acceptance technology. *Psychology and Marketing, 24*(12), 1059–1084. doi:10.1002/mar.20196

Lee, H.-J., Fairhurst, A. E., & Lee, M.-Y. (2009). The importance of self-service kiosks in developing consumers' retail patronage intentions. *Managing Service Quality, 19*(6), 687–701. doi:10.1108/09604520911005071

Lin, J.-S. C., & Hsieh, P.-L. (2006). The role of technology readiness in customers' perception and adoption of self-service technologies. *International Journal of Service Industry Management, 17*(5), 497–517. doi:10.1108/09564230610689795

Marzocchi, G. L., & Zammit, A. (2006). Self-scanning technologies in retail: Determinants of adoption. *The Service Industries Journal, 26*(6), 651–669. doi:10.1080/02642060600850790

Moore, G. C., & Benbasat, I. (1991). Development of an instrument to measure perceptions of adopting an information technology innovation. *Information Systems Research, 2*(3), 192–222. doi:10.1287/isre.2.3.192

Muthén, L. K., & Muthén, B. O. (2007). *Mplus user's guide* (4th ed.). Los Angeles, CA: Muthén & Muthén.

Nunnally, J. C., & Bernstein, I. H. (1994). *Psychometric theory* (3rd ed.). New York, NY: McGraw-Hill Education.

Parasuraman, A., Berry, L. L., & Zeithaml, V. (1988). SERVQUAL: A multiple-item scale for measuring consumer perceptions of service quality. *Journal of Retailing, 64*(1), 12–40.

Petty, R., & Cacioppo, J. (1981). *Attitudes and persuasion: Classic and contemporary approaches*. Dubuque, IA: Brown.

Petty, R., & Cacioppo, J. (1986). The elaboration likelihood model of persuasion. *Advances in Experimental Social Psychology, 19*, 123–205. doi:10.1016/S0065-2601(08)60214-2

Ringle, C. M., Wende, S., & Will, A. (2005). *SmartPLS 2.0 (M3)*. Hamburg, Germany: University of Hamburg.

Rogers, E. M. (1962). *The diffusion of innovations*. New York, NY: The Free Press.

Rowley, J. (1995). Multimedia kiosks in retailing. *International Journal of Retail & Distribution Management, 23*(5), 32–40. doi:10.1108/09590559510089212

Rowley, J. (2000). Loyalty kiosks: Making loyalty cards work. *British Food Journal, 102*(5/6), 390–397. doi:10.1108/00070700010329236

Rowley, J., & Slack, F. (2003). Kiosks in retailing: The quiet revolution. *International Journal of Retail & Distribution Management, 31*(6), 329–339. doi:10.1108/09590550310476060

Rowley, J., & Slack, F. (2007). Information kiosks: A taxonomy. *The Journal of Documentation, 63*(6), 879–897. doi:10.1108/00220410710836402

Rundle-Thiele, S. (2005). Elaborating customer loyalty: Exploring loyalty to wine retailers. *Journal of Retailing and Consumer Services, 12*(5), 333–344. doi:10.1016/j.jretconser.2004.11.002

Schepers, J., & Wetzels, M. (2007). A meta-analysis of the technology acceptance model: Investigating subjective norm and moderation effects. *Information & Management, 44*(1), 90–103. doi:10.1016/j.im.2006.10.007

Slack, F., & Rowley, J. (2002a). Kiosks 21: A new role for information kiosks? *International Journal of Information Management, 22*(1), 67–83. doi:10.1016/S0268-4012(01)00041-X

Slack, F., & Rowley, J. (2002b). Online kiosks: The alternative to mobile technologies for mobile users. *Internet Research: Electronic Networking Applications and Policy*, *12*(3), 248–257. doi:10.1108/10662240210430928

Swoboda, B. S. (1998). Conditions of consumer information seeking: Theoretical foundations and empirical results of using interactive multimedia systems. *International Review of Retail, Distribution and Consumer Research*, *8*(4), 361–381. doi:10.1080/095939698342715

Tung, L. L., & Tan, J. H. (1998). A model for the classification of information kiosks in Singapore. *International Journal of Information Management*, *18*(4), 255–264. doi:10.1016/S0268-4012(98)00014-0

Van der Heijden, H. (2004). User acceptance of hedonic information systems. *Management Information Systems Quarterly*, *28*(4), 695–704.

Venkatesh, V., & Davis, A. (2000). A theoretical extension of the technology acceptance model: Four longitudinal field studies. *Management Science*, *46*(2), 186–204. doi:10.1287/mnsc.46.2.186.11926

Venkatesh, V., Morris, M. G., Davis, G. B., & Davis, F. D. (2003). User acceptance of Information Technology: Toward a unified view. *Management Information Systems Quarterly*, *27*(3), 425–478.

Wang, Y., Wang, Y., Lin, H., & Tang, T. (2003). Determinants of user acceptance of internet banking: An empirical study. *International Journal of Service Industry Management*, *14*(5), 501–519. doi:10.1108/09564230310500192

Weijters, B., Rangarajan, D., Falk, T., & Schillewaert, N. (2007). Determinants and outcomes of customers' use of self-service technology in a retail setting. *Journal of Service Research*, *10*(1), 3–21. doi:10.1177/1094670507302990

Wixom, B. H., & Todd, P. A. (2005). A theoretical integration of user satisfaction and technology acceptance. *Information Systems Research*, *16*(1), 85–102. doi:10.1287/isre.1050.0042

Xiao, B., & Benbasat, I. (2007). E-commerce product recommendation agents: Use, characteristics, and impact. *Management Information Systems Quarterly*, *31*(1), 137–209.

Zielke, S. (2006). Measurement of retailers' price images with a multiple-item scale. *International Review of Retail, Distribution and Consumer Research*, *16*(3), 297–316. doi:10.1080/09593960600696990

Zielke, S. (2010). How price image dimensions influence shopping intentions for different store formats. *European Journal of Marketing*, *44*(6), 748–770.

Zielke, S., Lietke, B., Toporowski, W., & Boslau, M. (2008). Self-service and traditional check-outs in a retail setting: An analysis of perceptual differences and satisfaction drivers. *Marketing–Journal of Research and Management, 30*(1), 24-37.

KEY TERMS AND DEFINITIONS

Acceptance of a Technology: Customer acceptance is related to the process of adopting a technology. Acceptance reflects a positive attitude towards adopting a technology, the intention to adopt and/or actual adoption behavior.

Ease of Use (Related to a Technology): Ease of use refers to how easily people can learn and memorize the usage of a technology.

Enjoyment (Related to a Technology): Enjoyment refers to the fun experienced while using a technology.

Information Kiosk: The term information kiosk is used synonymously with the term information terminal.

Information Terminal: An information terminal is an interactive self-service technology, which allows users to search for information and/or execute transactions. Information terminals are located in public concourse, such as train stations, airports, retail stores and banks.

Technology Acceptance Model: The technology acceptance model is a general framework that explains the acceptance of technologies. According to the model, the perceived usefulness and ease of use are the key antecedents of customer acceptance.

Usefulness (Related to a Technology): Usefulness refers to how the usage of a technology can improve the outcome of a certain behavior (e. g. shopping or consumption).

APPENDIX

Table 5. Measures and loadings from the confirmatory factor analysis

Construct	Items	Factor Loadings
Usefulness	The information provided by the terminal is helpful	0.83
	The terminal is useful for my shopping	0.93
	The terminal makes my shopping easier	0.86
	The terminal makes it easy to buy all ingredients for a dish	0.74
	The terminal provides inspiration for new dishes*	-
	All in all, I think of the information terminal as useful*	-
Ease of use	It is easy to memorize how to use the terminal	0.75
	The usage of the terminal is easy	0.88
	It was easy to learn how to use the software	0.95
	All in all, I think the information terminal is easy to use	0.97
Enjoyment	unenjoyable – enjoyable	0.66
	boring – exciting	0.74
	unpleasant – pleasant	0.88
	uninteresting – interesting	0.90
Acceptance	I like the information terminal	0.86
	Using the terminal is something positive	0.84
	There is a high probability that I will use the terminal again	0.92
	I intend to use the terminal during my next shopping trip	0.84
	I will not use the terminal again**	0.64
	I will recommend the terminal to others	0.77
Technology experience	I often buy online	0.56
	I have good computer knowledge	0.84
	I often use the internet	0.97
Information relevance	Cooking is my hobby	0.60
	I often cook by myself	0.77
	I often try out new recipes	0.89
	I search actively for new recipes	0.85
* eliminated because of cross-loadings with acceptance; ** reverse coded		

Chapter 16
Factors Affecting WiFi Use Intention:
The Context of Cyprus

Despo Ktoridou
University of Nicosia, Cyprus

Hans-Ruediger Kaufmann
University of Nicosia, Cyprus

Christos Liassides
Columbia Management, Cyprus

ABSTRACT

Wireless communications are here to stay; nevertheless, a number of individuals are still reluctant to use them for accessing the Web. This urges providers to better understand the concerns of consumers in order to better position the products and services in the market and to reduce the barriers that consumers may have in using WiFi – Wireless Fidelity Internet technologies. The present quantitative study was carried out for the purpose of assessing the effect of a number of constructs, identified in the past to affect the use of IT and WiFi use intention, as well as identifying the underlying factor structure of these constructs. It further aimed to assessing the overall attitudes and behavior of consumers towards WiFi use, as well as identifying and comparing WiFi users and non-users' overall behavior towards WiFi use and their perceptions of factors determining WiFi adoption.

INTRODUCTION

The ability to communicate from virtually anywhere and the convergence of web and wireless technologies offer an unprecedented level of flexibility and convenience, particularly, for ubiquitous information access. WiFi internet technology changed the way that IT technology is conventionally used. Today, it is used by most PCs, operating systems, video game consoles, laptops, smart phones, printers, and other peripherals. All of a sudden, the use of technology became much easier and much more virtual than before. People can access and share information from any point, at any time, at their own convenience and pace.

DOI: 10.4018/978-1-60960-738-8.ch016

Although the WiFi technology is a major achievement and provides more convenience to internet users than before, a number of individuals are still reluctant to use it for accessing the World Wide Web. Researchers in the field argue that final purchasing decisions are often the result of perceptions and attitudes that consumer hold towards products and services. It is therefore paramount for providers to understand the concerns of consumers in order to better deliver and position products and services into the market minimizing any hesitation that consumers have in using WiFi internet technologies.

Furthermore, although a number of studies have researched the impact of consumer attitudes on the use of new technologies (e.g. Davis, 1989; Moore and Benbasat, 1991; Compeau and Higgins, 1995; Thompson et al, 1991; Salisbury et al., 2001; Venkatesh et al, 2003), no research up to present has focused on the impact of the perceptions and attitudes of consumers towards the use of wire-free technologies, such as WiFi, Bluetooth, and infrared. Such technologies are incorporated into existing technological devices to make their use much more virtual and mobile, offering higher efficiency, ease of access and convenience. The present research was a first attempt to assess consumer perceptions towards wire-free technologies on the propensity and desire to use such technologies.

Finally, there is total absence in the literature of similarities and/or differences that exist between users and non-users of particular technology in terms of their overall attitude and perceptions towards this innovation and the factors determining IT adoption.

Consequently, this research aims to assess the effect of a comprehensive number of factors, identified in previous studies, which influence the intention to use information technologies on the willingness of consumers to use WiFi Internet technology. This study marries and synthesizes the evidence from previous studies in the field and attempts, for the first time, to assess the concentrated impact of all forces, proven to affect IT use, on the intention to use WiFi. Moreover, the study will pursue to grouping these variables into a smaller more manageable set, with explanatory power relating to the intention to use WiFi.

Specifically, the study aims to:

- Identify the nature of factors that tend to affect WiFi use for accessing the Internet. There is a total absence, at presence, of empirical evidence that examines the factors that tend to affect the use of WiFi for Internet access. Thus, the present research contributes to the literature in the sense that it reveals that set of factors with the highest impact on WiFi use for Internet access. Previous evidence relevant to the forces that affect the use of information technologies is integrated.
- Examine a full range of factors, at once, that have been identified in the past to affect use of IT. No research at present attempted to test the effect of all these factors at once, thus identifying the set of forces with the highest explanatory power. Hence, the research attempts to identify the individual, as well as the combined impact and explanatory power of a range of factors (revealed by previous studied) on the use of WiFi for Internet access purposes.
- Reduce the number of factors that have been revealed to affect the use of IT, into a smaller, more manageable set. Previous studies revealed a large number of forces with effect on propensity to use new technologies. The present study will capitalize on this evidence, attempting to use fewer dimensions or factors that can be used in an effort to explain the factors that impact on IT use.
- Examine the overall attitudes and behavior of consumers towards WiFi use for Internet access.

- Compare WiFi users and non-users' overall behavior towards WiFi use and their perceptions towards factors determining WiFi adoption.

BACKGROUND: BENEFITS AND THREATS

Benefits for the Users

WiFi – wireless fidelity, or otherwise known as the protocol 802.11 has become a portable mechanism for businesspeople, students, and everyone else to get connected (Potter 2006).

In 1997 the 802.11 protocol has been approved as a standard with a sky rock usage of the resulting WiFi technology over the last years. WiFi networks are found almost everywhere: at home, at work, at university campuses for an ubiquitous access incorporating data and services (Hole et al. 2005). According to Stuart (2003), the use of WiFi technology has a number of benefits compared to the conventionally wired technology: Freedom of Movement; Broad device compatibility; Faster connections; Nonstop network connection from a range of sites; Wireless connections= significantly low installation costs.

Drawbacks and Potential Threats

Among the drawbacks of the use of WiFi is the fact that the use of wireless technology cannot guarantee a truly secure connection, especially at public hotspots. WiFi security options are still limited, despite the efforts that have been made over time to make the wireless protocol more accessible (Potter, 2006).

Wireless networks had and will continue to have security issues mostly with unauthorized accesses from hackers. Hackers with wired LANS found inside a building are protected unlike wireless LANS, which are more sensitive as to interceptions from hackers, due to data traveling over radio waves. Hackers are mostly either software or hardware fans that are enthusiastic in abusing the limits of programming code or computer hardware. By uploading a malicious code they aim to disrupt computer systems or networks to steal data or create havoc (Hole et al. 2005).

WiFi networks do still not enjoy adequate protection from a potential attack from hackers. Today, a variety of hacking tools are available to be downloaded from the internet, which anyone can obtain and use to intrude into private and corporate wireless networks (Hole et al. 2005). Outsiders, therefore, can sometimes get into wireless networks as fast and easily as the authorized users can (Stuart, 2003).

According to MiSme cooperation report (MiSme, 2007) other drawbacks of the use of WiFi include:

- **Limited range:** "Although you lose the wires, you're still limited to the base station's range, typically 75 to 150 feet indoors and a few hundred feet outdoors, depending on equipment, radio frequency, and obstructions".
- **Power drain:** "Networks using early versions of Wi-Fi technology tend to quickly gobble power -- a disadvantage for battery-dependent laptop users".
- **Interference:** "Nearby microwave ovens and cordless phones, particularly older models, can slow down Wi-Fi transmissions".

Setting up a Wi-Fi network is very easy and cost effective; nevertheless, ensuring its security requires a lot of effort. Encryption methods are not deployed in Access points unless the network has been enabled. Hackers can easily attack a secure network and steal private information. Moreover, guests utilizing the network's resources can minimize its performance. Notwithstanding, Wifi technology is here to stay and is growing and, based on ongoing technological advancements and

Figure 1. Factors influencing consumer behavior Source: Kotler and Armstrong, 1999; p. 135

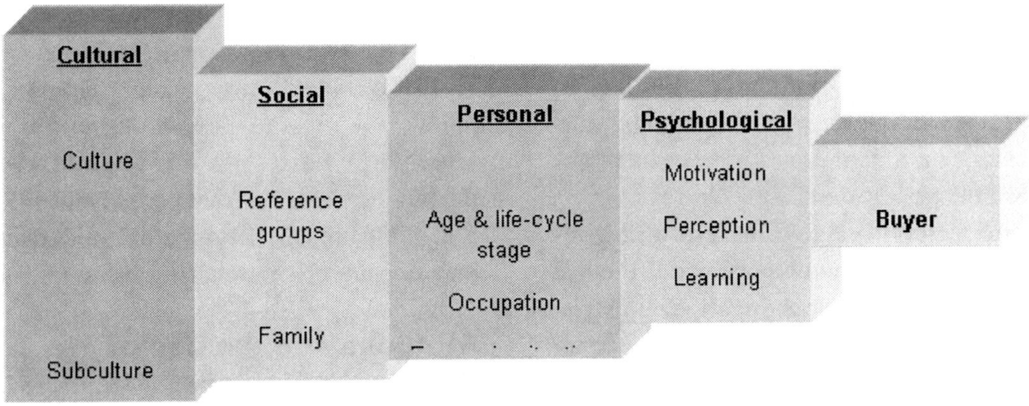

efforts to increase security, it will have a positive influence on our everyday lives and businesses.

CONSUMER BEHAVIOR

Definition and Importance of Consumer Behavior

A definition of Consumer Buyer Behavior is provided by Kotler et al (2005, p. 255): "Consumer behavior refers to the buying behavior of final consumers- individuals and households who buy goods and services for personal consumption".

Researching consumer behavior is very important for businesses since they become aware of the various factors which influence consumers' purchase decisions in terms of what, where, how and how much, when and why they buy. This research also results in understanding of how consumers respond to different product features, prices, and advertising appeals (Kotler et al., 2005).

Factors Affecting Purchasing Decisions

To obtain a better understanding of consumer buying behavior and why consumers buy as they do, many specialists have turned to the behavioral sciences for help (Perreault et al, 2000). From this perspective, consumer purchases are said to be influenced strongly by cultural, social, social, personal, and psychological characteristics which are described in the following paragraph, which is based mainly on Kotler and Armstrong (1999) as well as Kotler et al. (2005).

Figure 1 portrays the factors that have an influence over the buying decisions of final consumers.

Cultural factors often exert the deepest influence on consumer behavior. Culture refers to a common set of basic values, perceptions, wants, and behaviors learned from the family and other important cultural institutions. Certain normative influences might come from subcultures or social classes representing groups of people with shared value systems derived from shared life experiences and situations.

Social factors, such as the quality of relations with other people and the role and status that one has in the society can affect his/ her purchasing behavior as well. Family members, for example, share similar attitudes and values and tend to affect many aspects of consumer behavior. At the same time, reference groups, which include people that an individual wants to become like or imitate, tend to affect purchasing decisions. Individuals tend to adopt values from reference groups and make buying decisions in a way that they are accepted by the respective reference group (Perreault et al, 2000). Regarding the factor of 'social influence', a definition is provided by Venkatesh et al. (2003, p. 451) pointing to normative influences on behavior: It is the "degree to which an individual perceives that important others believe he or she should use the new system".

Personal factors that may affect a buyer's decisions may include the buyer's age and life-cycle stage, occupation, economic situation, lifestyle, and personality and, to a large extent, his/her self-concept or identity (Kaufmann, 2004). The relevance and interdependence of the latter two factors is confirmed by Kitsios and Kaufmann (2009) and Siano, Kaufmann, Basile (2009) who state that social normative influences occur in brand consumption in Cyprus.

Finally, a number of psychological factors, such as the level of motivation, individual perceptions, different approaches of learning and ingrained attitudes and beliefs may also affect consumer behavior (Kotler and Armstrong, 1999; Perreault et al, 2000; Aaker et al. 2004; Churchill and Iacobucci (2005); Kotler et al., 2005; Perreault et al, 2000).

Perceptions, Attitudes and Buying Behavior

As mentioned above consumers' buying behavior is greatly affected by perceptions and attitudes.

Perceptions refer to the way consumers collect information from their surroundings and interpret it. Consumers being constantly exposed to stimuli provided by advertising, products' packages or shop displays apply the following selective processes (Perreault et al, 2000):

- **Selective exposure:** Search for information that meets their preferences.
- **Selective perception:** Monitoring and modification of information, messages and thoughts that conflict with formerly learned approaches and attitudes.
- **Selective retention:** Memorization of preferred information only.

In this context, attitudes representing a person's point of view towards something tend to affect the selective processes, the learning and, eventually, the buying decisions that people make (Perreault et al, 2000) and, hence, have a dominant influence on a consumer's purchasing behavior.

Attitudes and the Use of New Technologies

A number of studies have been conducted for the purpose of identifying the major factors that influence the adoption and use of information technology advances by consumers. Studies performed in the field confirmed that consumer perceptions and attitudes towards new technological advances, such as the Internet and e-commerce, greatly affect the adoption and use of these innovations (e.g. Labuschagne & Eloff, 2000; Nabi, 2005; Porter and Donthu, 2006).

Among the factors that have been identified to have an impact on the use of information technologies were the perceived ease of use and the perceived usefulness from the use of these innovations.

USER ACCEPTANCE RESEARCH

The Technology Acceptance Model (TAM)

The technology acceptance model or TAM is tailored to IS contexts and was designed to predict IT–Information Technology acceptance and usage (see Figure 2). According to this model perceived usefulness and perceived ease of use tend to affect the intention to use information technologies (Venkatesh et al, 2003). The model is based on the theory of reasoned action (TRA) (Adams et

Figure 2. The Technology Acceptance Model

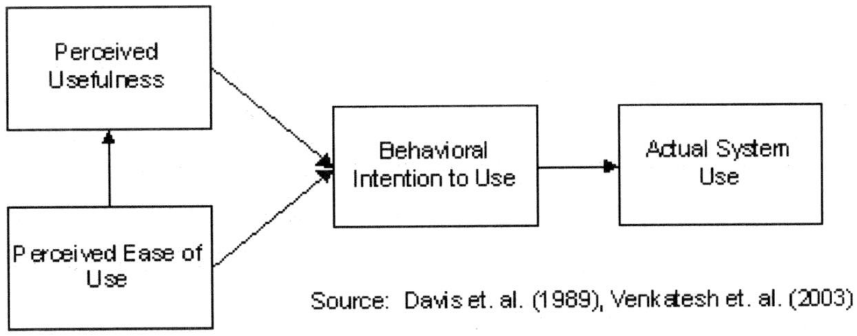

al., 1992, Dishaw and Strong, 1999, Chau, 1996 in Liao, 2007).

Other models developed so far are: the motivational model, the theory of planned behavior, PC utilization theory, the innovation diffusion theory and the social cognitive theory (Saade et al., 2007).

In an effort to identify the factors that tend to affect the usage of WiFi, the present study will formulate a number of hypotheses based on previous theoretical models and evidence in the field of IT adoption.

The study will test for the effect of 16 constructs on the intention of consumers to use the WiFi. (The researchers acknowledged the fact that 16 variables is a very large number to manage. Thus, when developing the Likert scale of the survey instrument, two items were used per variable. In other words, a scale was developed that included 32 items, which, in turn, represented 16 variables.

These 16 variables comprise the following:

1. **Self–efficacy:** watching others performing a behaviour, in this case interacting with a computer system, influences the observers' perceptions of their own ability to perform the behaviour (Compeau and Higgins, 1995; Venkatesh et al.,2003).
2. **Perceived usefulness:** Refers to the degree to which a person believes that using a particular system would enhance his or her job performance. This follows from the definition of the word useful: "capable of being used advantageously. A system high in perceived usefulness is one for which a user believes in the existence of a positive use-performance relationship (Davis, 1989; O'Cass and Fenech, 2003; Vijayasarathy, 2004; Baron et al., 2006; Bruner and Kumar, 2005).
3. **Perceived ease of use:** Refers to the degree to which a person believes that using a particular system would be free of effort. This follows from the definition of "ease": "freedom from difficulty or great effort." All else being equal, an application perceived to be easier to use than another is more likely to be accepted by users (Davis, 1989; Burton-Jones and Hubona, 2005; Baron et al., 2006).
4. **Affect:** Affect refers to the feelings of joy, elation, or pleasure, or depression, disgust, displeasure, or hate associated by an individual with a particular act (Thompson et al. 1991; Dabholkar, 1996; Bruner and Kumar, 2005; Baron et al., 2006).
5. **Perceived long-term consequences:** Another important factor influencing behaviour is the expected consequences of the behaviour. Each act is perceived as having potential consequences that have value, together with a probability that the consequence will occur. *Consequences of use.* These are outcomes that have a payoff in the future, such as increasing the flexibility to change jobs or increasing the

opportunities for more meaningful work. For some individuals, the motivation to adopt and use PCs may relate more to building or planning for the future than to addressing current needs (Thompson et al. 1991).

6. **Complexity:** Complexity is defined as "the degree to which an innovation is perceived as relatively difficult to understand and use. The more complex the innovation, the lower tends to be its rate of adoption (Thompson et al. 1991; Moore and Benbasat, 1991).
7. **Job Fit:** The second near-term component relates to the capabilities of an information technology to enhance an individual's job performance. It measures the extent to which an individual believes that using an IT can enhance the performance of his or her job (e.g., obtaining better information for decision making or reducing the time required for completing important job tasks) (Thompson et al. 1991).
8. **Facilitating conditions:** Behaviour cannot occur if objective conditions in the environment prevent it. In the context of PC use, the provision of support for users of PCs may be one type of facilitating condition that can influence system utilisation. By training users and assisting them when they encounter difficulties, some of the potential barriers to use are reduced or eliminated (Thompson et al. 1991; Baron et al., 2006).
9. **Relative Advantage:** the degree to which an innovation is perceived as being better than its precursor (Moore and Benbasat, 1991).
10. **Compatibility:** the degree to which an innovation is perceived as being consistent with the existing values, needs, and past experiences of potential adopters (Moore and Benbasat, 1991).
11. **Trialability:** the degree to which an innovation may be experimented with before adoption (Moore and Benbasat, 1991).
12. **Image:** perceiving the adoption of innovation as an enhancement of one's image or status in a social group (Moore and Benbasat, 1991; Baron et al., 2006; Venkatesh et al., 2003; Kitsios and Kaufmann, 2009; Siano, Kaufmann and Basile, 2009).
13. **Voluntariness of use** (Moore and Benbasat, 1991).
14. **Result Demonstrability**: Tangibility and visibility of the results of using the innovation leads to higher adoption rates (Moore and Benbasat, 1991).
15. **Visibility.** It appears that the more a potential adopter can see an innovation, the more likely he is to adopt it (Moore and Benbasat, 1991).
16. **Perceived security risks:** Perceived IT security risk is the extent to which one believes that a particular information – related technology is secure to use. In the case of the use of the internet, perceive security risks relate to the perceived security in terms of transmitting sensitive information (Salisbury et al. 2001; Schiffman and Kanuk, 2004).

METHODOLOGY

Model and Hypotheses

Although a number of studies have researched the impact of consumer attitudes on the use of new technologies (e.g. Labuschagne & Eloff, 2000; Venkatesh et al. 2003; Nabi, 2005; Porter and Donthu, 2006), no research up to present has focused on impact of the perceptions and attitudes of consumers towards the use of wire-free technologies, such as WiFi, bluetooth, and infrared. Such technologies are incorporated into existing technological devices to make their use much more virtual and mobile, offering higher efficiency, ease of access and convenience. The present research was a first attempt to assess consumer perceptions towards wire-free technologies on the propensity and desire to use such technologies.

This research represents a first attempt, in the field, in assessing the impact of all forces, proven to affect IT use, on intention to use WiFi. Further, the study pursues grouping these variables into a smaller more manageable set, with explanatory power over intention to use WiFi.

In an effort to identify the factors that tend to affect the usage of WiFi, the present study will formulate a number of hypotheses based on previous theoretical models and evidence in the field of IT adoption.

The study will test for the effect of 16 constructs on the intention of consumers to use the WiFi (see Figure 3). The researcher acknowledged the fact that 16 variables is a very large number to manage. Thus, when developing the likert scale of the survey instrument, he used two items per variable. In other words, a scale was developed that included 32 items, which in turn represented 16 variables (see below for details).

The aforementioned listed constructs represent the independent variables of the study, while the dependent variable will be the intention/ willingness to use WiFi for Internet access.

The hypotheses that will drive the present research are the following:

H1a: The higher the perceived usefulness of WiFi, the higher the intention to use WiFi internet technology is expected to be.

H1b: The higher the perceived ease of use of WiFi, the higher the intention to use WiFi internet technology is expected to be.

H1c: The lower the perceived security risks associated with the use of WiFi, the higher the intention to use WiFi internet technology is expected to be.

H1d: There is a positive relationship between affect towards PC use and the utilization of PCs.

H1e: There is a negative relationship between the perceived complexity of a PC and the utilization of PCs.

H1f: There is a positive relationship between perceived Job fit and the utilization of PCs.

H1g: There is a positive relationship between perceived long-term consequences of use and the utilization of PCs.

H1h: There is a positive relationship between facilitating conditions for PC use and the utilization of PCs.

H1i: The higher the perceived self efficacy relevant to the use of WiFi, the higher the intention to use WiFi internet technology is expected to be.

Figure 3. Research Model

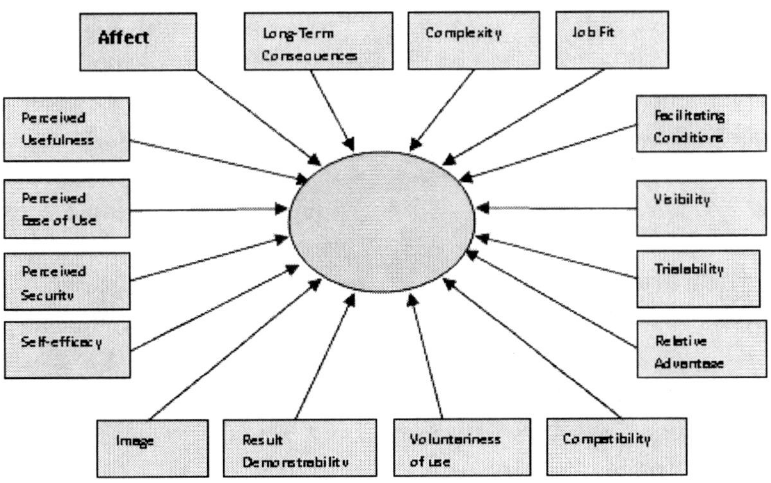

H1j: The more the WiFi is perceived as being better than its precursor, the higher the higher the intention to use WiFi internet technology is expected to be.

H1k: The more the WiFi is perceived as being consistent with the existing values, needs, and past experiences of potential adopters, the higher the intention to use WiFi internet technology is expected to be.

H1l: The more the WiFi is experimented with before adoption, the higher the intention to use WiFi internet technology is expected to be.

H1m: The more the WiFi is perceived to enhance one's image or status, the higher the intention to use WiFi internet technology is expected to be.

H1n: The more the WiFi is perceived as being voluntary, the higher the intention to use WiFi internet technology is expected to be.

H1o: The more amenable to demonstration the WiFi is, the higher the intention to use WiFi internet technology is expected to be.

H1p: The more a potential adopter can see WiFi, the higher the intention to use WiFi internet technology is expected to be.

As mentioned earlier, the study also aims in comparing WiFi users' and non-users' overall behavior towards WiFi use and their perceptions towards factors determining WiFi Adoption. Thus a relevant hypothesis raised at this stage is the following:

H2: Users and non-users of WiFi differ in their overall behavior towards WiFi use and their perceptions towards factors determining WiFi adoption.

Nature of Research and Research Method

The research was descriptive and explanatory and collected data that were quantitative in nature. Since the study aimed in testing the effect of a number of variables (e.g.. the perceived security risk, ease of use, usefulness etc.) on another variable (intentions to adopt WiFi) with the aim to compare also the results with a rich source of evidence that is already available, a quantitative study seemed to be the most appropriate route to follow.

Quantitative research tends to be descriptive in nature and leads in the collection of quantitative data that are data in numerical format (Malhotra and Birks, 2000). Tools usually employed for the collection of quantitative data include structured personal interviews, survey questionnaires, tests/measures and observation (Easterby-Smith et al, 2006).

In the case of the present research, the method of a survey was employed, which involved the use of a structured questionnaire that had to be filled in by a number of individuals within the targeted population (see sections below for details).

The target population in this case was people that were users of the Internet. People outside this group may find the WiFi technology as of no use since they are not using the Internet. Therefore they will not be in a position to answer the questionnaire.

DATA COLLECTION METHOD

Originally, the research targeted to collect data from a sample of 150 Internet users (see Table 1 for details). Eventually, the research yielded 114 completed questionnaires, from a number of individuals across Cyprus.

Stratified sampling was utilized for yielding a representative sample across the country. Respondents were categorized into five major groups according to the district of tenancy, i.e. target population was segmented into the following districts: Nicosia, Limassol, Larnaca, Paphos Famagusta (the area controlled by the Republic of Cyprus).

Table 1. Stratified sample and final distribution of completed questionnaires (Source: Cyprus Statistical Service, 2008)

District	Actual Population	Percentage Distribution	Initial Sample	Final Completed Questionnaires	Distribution of Completed Questionnaires
Famagusta	43,700	5.6%	9	6	5.3%
Larnaca	131,900	16.7%	25	18	15.8%
Limassol	226,700	28.7%	43	33	28.9%
Nicosia	310,900	39.4%	59	47	41.2%
Paphos	76,100	9.6%	14	10	8.7%
Total	**789,300**	**100%**	**150**	**114**	**100.0**

This questionnaire involved closed questions which mainly included dichotomous, multiple choice questions and Likert scale questions.

The first part of the questionnaire included a screening question that left out all individuals that were not internet users. The second part included a number of demographic questions such as district of tenancy, gender, and education. The third and last part assessed the attitudes and perceptions of users towards the WiFi Internet technology. Perceptions, attitudes and intentions were measured using Likert scale questions, which asked from the responses to rate a number of variables/ statements using a scale from 1 to 7 (1=low, 7 = high).

For the purpose of collecting data that would be as valid as possible, thus providing good chances for generalization towards the whole targeted population, the researcher used a systematic random sampling procedure in the course of data collection. Using this technique, the participants in the study were chosen on a purely random basis. In order to make this possible, the researcher chose key-locations in each district, where a large number of people would be found. Since the research was targeting final consumers, key-shopping centers would offer a good location for reaching these individuals. The researcher approached potential participants in the following locations in each city:

Findings

The attitude of people against new technologies is a major issue that deserves attention since, most of the time, the use of new technologies depends on the perceptions and attitudes of consumers towards these inventions.

High tech businesses, like any other businesses, secure their long term existence through sales of their products and services to final consumers. New technologies, however, as products, differ on the fact that many times consumers feel more reluctant to use for a number of reasons. Therefore, high tech firms need to ensure that their products are positioned into the market and that consumers have become familiar with them any kind of consumer concerns about the use of these products have been eliminated.

The research deals with a contemporary theme which is the use of high tech inventions such as the wireless internet. As mentioned above the perceptions of consumers about high tech products determine their willingness to use such products. Specifically, the study will examine the impact of perceived usefulness and ease of use, and perceived security risks on consumers' intention to access wireless internet. Based on evidence in the field the study attempted to provide an analysis of the aforementioned 16 variables for the purpose of assessing their importance in relation to the use of WiFi technology.

Table 2. Multiple regression: Model Summary

Model	R	R Square	Adjusted R Square	Std. Error of the Estimate
1	.933[a]	.870	.846	.35580

Table 3. Multiple Regressions: ANOVA[b]

Model		Sum of Squares	df	Mean Square	F	Sig.
1	Regression	74.598	16	4.662	36.830	.000[a]
	Residual	11.140	88	.127		
	Total	85.738	104			

For the first time, a research was attempting to test for the effect of all the above variables simultaneously on the willingness to use a particular technology. The dependent variable in our analysis was the willingness or intention to use WiFi while the independent variables, included all 16 variables mentioned above.

Standard multiple regression was employed for the analysis. Using this technique, all independent variables are entered into the equation simultaneously. Each independent variable is evaluated in terms of its predictive power, over and above that offered by all the other independent variables. This approach also tells how much unique variance in the dependent variable each of the independent variables explained (Pallant, 2006).

The results from the analysis are presented in Tables 2 and 3. The R Square values in Table 2 (Model Summary) show the percentage of the variance explained by the model (i.e. the percentage of the variance in the dependent variable explained by the 16 independent variables). Overall, the variables under examination are explaining 84.6% of the variance. This is a very high value and shows that the predictive effect of the 16 independent variables is very high and that intention to use WiFi is at a great extent determined by these factors.

The ANOVA table (see Table 3) shows that the model as a whole is statistically significant [p> 0.05]. This means that the 16 measures can predict WiFi use intention very well.

The Coefficients table (see Table 4) is a useful tool for identifying how well each of the variables contributes to the equation (Pallant, 2006). In our case we want to see how well each of the 16 independent variables contributes to the prediction of the dependent variable (i.e. WiFi use intention). Table 3 shows that most of the variables make a statistically significant contribution (i.e. less than 0.05) (see column Sig.).

Amongst the independent variables being tested, perceived usefulness (beta = 2.764), trialability (beta=-2.688), facilitating conditions (beta = -1.559), ease of use (beta = 1.302), and result demonstrability (beta=- 1.109), have the strongest unique contribution in explaining the dependent variable (i.e. WiFi use intention), when the variance explained by the other variables is controlled for.

The findings from the study were very illuminating, revealing a higher contribution of the aforementioned variables in explaining WiFi usage intention. The study reveals that the particular group of variables has very high predictive power, thus the researchers felt confirmed in their choice in assessing the simultaneous effect of all 16 variables on intention to use WiFi.

The hierarchical regression analysis revealed that either aggregately (i.e. all factors combined)

Table 4. Multiple regression: Coefficients

Model	Unstandardized Coefficients		Standard. Coeffic.			95% C.I. for B		Correlations			Collinearity Statistics	
	B	Std. Error	Beta	t	Sig.	Lower	Upper	Zero-order	Partial	Part	Tole-rance	VIF
(Constant)	8.353	1.342		6.223	.000	5.685	11.020					
Self Efficacy	-.151	.177	-.125	-.853	.396	-.503	.201	.762	-.091	-.033	.069	14.520
Perceived usefulness	2.222	.394	2.764	5.647	.000	1.440	3.004	.680	.516	.217	.006	162.291
Ease of use	1.485	.242	1.302	6.134	.000	1.004	1.966	.713	.547	.236	.033	30.496
Affect	-.653	.148	-.403	-4.403	.000	-.948	-.358	.579	-.425	-.169	.176	5.667
long term consequences	-.512	.093	-.387	-5.484	.000	-.697	-.326	-.003	-.505	-.211	.297	3.370
Complexity	-.612	.207	-.776	-2.952	.004	-1.025	-.200	.696	-.300	-.113	.021	46.791
Job fit	.045	.090	.049	.493	.623	-.135	.224	.409	.053	.019	.149	6.717
Facilitating conditions	-1.051	.158	-1.559	-6.654	.000	-1.365	-.737	.420	-.579	-.256	.027	37.174
Relative Advantage	-.843	.201	-.639	-4.193	.000	-1.242	-.443	.604	-.408	-.161	.064	15.727
Compatibility	.416	.118	.393	3.520	.001	.181	.651	.578	.351	.135	.119	8.435
Trialability	-5.784	.608	-2.688	-9.511	.000	-6.993	-4.576	.611	-.712	-.365	.018	54.100
Image	.665	.098	.699	6.807	.000	.471	.859	.320	.587	.262	.140	7.142
Voluntariness	.414	.118	.318	3.520	.001	.180	.648	.344	.351	.135	.181	5.517
Result Demonstrability	1.345	.190	1.109	7.075	.000	.968	1.723	.712	.602	.272	.060	16.631
Visibility	.658	.156	.407	4.217	.000	.348	.969	.503	.410	.162	.159	6.300
Perceived security risks	.084	.085	.076	.991	.324	-.085	.254	.461	.105	.038	.253	3.955

Notes

a) Predictors: (Constant), Perceived security risks, Facilitating conditions, long term consequences, Voluntariness, Image, Affect, Job fit, Ease of use, Compatibility, Visibility, Result Demonstrability,

Self Efficacy, Relative Advantage, Complexity, Trialability, Perceived usefulness

b) Dependent Variable: WiFi use intention

or individually, the variables being tested make a significantly statistical contribution in the explanation of the dependent variable (intention to use WiFi). That means that they succeed in predicting WiFi use intention. This leads us to accept most of the hypotheses raised in the previous section. With the exception of **H1i** (The higher the perceived self efficacy relevant to the use of WiFi, the higher is expected to be the intention to use WiFi internet technology), **H1c** (The lower the perceived security risks associated with the use of WiFi, the higher is expected to be the intention to use WiFi internet technology) and **H1f** (There will be a positive relationship between perceived Job fit and the utilization of PC); the rest of the hypotheses (H1a; H1b; H1d; H1e; H1g; H1h; H1j; H1k; H1l; H1m; H1n; H1o: H1p) is accepted.

Table 5 provides details regarding the percentage of variance explained by each of the sixteen variables being examined. The analysis of the data shows that only three components have eigenvalues higher than 1, which together explain 86.1% of the variance. The same information is portrayed diagrammatically on the scree plot in

Figure 4. Scree plot

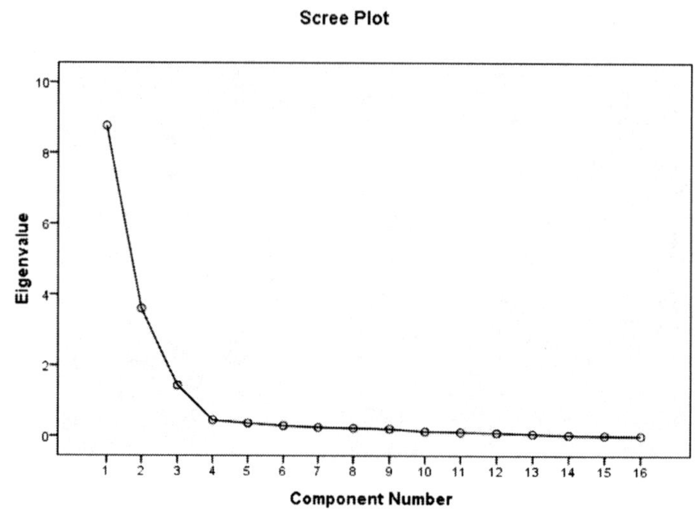

Table 5. Factor Analysis: Total Variance Explained

Component	Initial Eigenvalues			Extraction Sums of Squared Loadings		
	Total	% of Variance	Cumulative %	Total	% of Variance	Cumulative %
1	8.754	54.712	54.712	8.754	54.712	54.712
2	3.603	22.517	77.229	3.603	22.517	77.229
3	1.427	8.916	86.145	1.427	8.916	86.145
4	.449	2.808	88.953			
5	.365	2.281	91.234			
6	.294	1.835	93.069			
7	.245	1.533	94.602			
8	.225	1.404	96.006			
9	.205	1.279	97.285			
10	.130	.814	98.099			
11	.113	.708	98.807			
12	.091	.570	99.377			
13	.057	.354	99.731			
14	.025	.159	99.890			
15	.014	.085	99.974			
16	.004	.026	100.000			

Notes: Extraction Method: Principal Component Analysis.

Figure 4. Considering the results in Table 5, the shape of the scree plot (Table 6), as well as the fact that most items cluster in the first two dimensions (see Table 7), it seems to be most appropriate to retain two factors. Looking at Table 5, we notice that the first two factors explain 77.2% of the total variance meaning that only two of the

Table 6. Factor Analysis: Component Matrix

	Component		
	1	2	3
Result Demonstrability	.929		
Complexity	.916		
Ease of use	.908		
Self Efficacy	.900		
Perceived usefulness	.894	-.351	
Compatibility	.887		
Relative Advantage	.871	-.349	
Trialability	.840	.424	
Affect	.785	.410	
Perceived security risks	.694	.357	.374
Facilitating conditions	.603	-.752	
Voluntariness	.411	.743	
Visibility	.569	.719	
Job fit	.483	-.690	
Image	.369	.615	.591
long term consequences		-.459	.774

Notes: Extraction Method: Principal Component Analysis./ 3 components extracted.

Table 7. Factor Analysis: Rotated Component Matrix

	Component	
	1	2
Facilitating conditions	.947	
Perceived usefulness	.908	.311
Relative Advantage	.890	
Complexity	.836	.430
Compatibility	.824	.400
Job fit	.814	
Ease of use	.809	.450
Result Demonstrability	.714	.593
long term consequences	.458	
Visibility		.916
Trialability	.366	.867
Voluntariness		.833
Affect	.333	.821
Self Efficacy	.538	.756
Perceived security risks		.722
Image		.707

Notes:
Extraction Method: Principal Component Analysis/ Rotation Method: Varimax with Kaiser Normalization/ Rotation converged in 3 iterations.

sixteen variables can explain the majority of the variance in the model.

In an effort to provide an interpretation of the data yielded through the factor analysis and to identify the nature of the two underlying factors, Varimax factor rotation took place (see Table 8). The table presents details regarding the loadings of each item on the two selected factors. The items that load higher on factor 1 include:

- Facilitating conditions (.947)
- Perceived usefulness (.908)
- Relative Advantage (.890)
- Complexity (.836)
- Compatibility (.824)
- On the other hand, the items that load higher on factor 2 include:
- Visibility (.916)
- Trialability (.867)
- Voluntariness (.833)
- Affect (.821)
- Self efficacy (.756)

The items that load highly in each of the two factors are the ones that will eventually assist in the interpretation of the two components. The first dimension is characterized by the high loadings of variables such as facilitating conditions, perceived usefulness, relative advantage, reduced complex-

Table 8. Attitude Towards WiFi Use: Group Statistics

	Use of WiFi for Internet Access?	N	Mean	Std. Deviation	Std. Error Mean
I am positive towards WiFi Internet access	Yes	54	6.00	.000	.000
	No	60	5.15	1.696	.219
It makes sense to use WiFi internet	Yes	54	5.39	.492	.067
	No	60	4.50	.748	.097
People should adopt WiFi internet technology	Yes	54	5.39	.492	.067
	No	60	4.20	.819	.106

ity, etc. One can interpret this dimension as 'the level of support for and advantages of using WiFi' (e.g. support, increased performance, reduced task complexity, more effectiveness than previous technologies, etc.). The second dimension in turn, is characterized by the high loading of components such as visibility, trialability, voluntariness, affect, and self efficacy. These variables give a picture of the theme of the particular dimension, which seems to refer to 'the period one needs for becoming familiar and feeling secure with WiFi use'.

Finally, the study provided all evidence needed for meeting the last objective referring to the assessment of the attitudes of users and non-users of WiFi against WiFi use as well as their differentiated perceptions towards factors determining WiFi Adoption. To investigate these topics a number of independent-samples t-tests were carried out. Firstly, an independent-samples t-test was conducted to compare the importance scores assigned to three statements measuring attitudes towards WiFi use, for users and non-users of WiFi technology (see Tables 8 and 9). Significant differences were identified in all three items (see Sig. 2-tailed value, $p < 0.05$). The study reveals that existing users of WiFi have a more positive attitude towards WiFi use than non-users (the mean response rate of WiFi users is higher in all cases), although both groups exhibit an overall positive stance towards WiFi usage.

Finally, an independent-samples t-test was conducted to compare users and non-users perceptions of a number of factors with potential impact on WiFi usage (see Tables 9 and 10). Significant differences were identified in most of the variables under assessment (see Sig. 2-tailed value, $p < 0.05$). Only in the case of affect, long-term consequences, voluntariness, and perceived security risks no differences have been identified between the two groups.

The study reveals that existing users of WiFi have higher perceived self efficacy regarding the use of Wifi, attach higher perceived usefulness and ease of use to WiFi, and view WiFi as one having more positive long-term consequences, than non-users. At the same time, users attach higher perceived job fit, favorable conditions, relative advantage, compatibility, Trialability, and result demonstrability to WiFi than non- users. On the other hand, non-users attach higher perceived image and visibility to WiFi than users of WiFi (see Table 11).

Discussion

Consistent with Davis' (1989) findings, the research revealed that the higher the perceived usefulness and ease of use of WiFi, the higher will be the intention to use WiFi internet technology. Other findings were also consistent with previous studies in the field mentioned before such as the positive relationship between affect and WiFi usage intention, the negative relationship between perceived complexity of WiFi and the utilization

Table 9. Attitude Towards WiFi Use: Independent Samples Test

		Levene's Test for Equality of Variances		t-test for Equality of Means						
		F	Sig.	t	df	Sig. (2-tailed)	Mean Diff.	Std. Error Diff.	95% C.I. of the Diff. Lower	95% C.I. of the Diff. Upper
I am positive towards WiFi Internet access	Equal variances assumed	182.399	.000	3.682	112	.000	.850	.231	.393	1.307
	Equal variances not assumed			3.883	59.000	.000	.850	.219	.412	1.288
It makes sense to use WiFi internet	Equal variances assumed	11.727	.001	7.408	112	.000	.889	.120	.651	1.127
	Equal variances not assumed			7.565	102.906	.000	.889	.117	.656	1.122
People should adopt WiFi internet technology	Equal variances assumed	5.514	.021	9.263	112	.000	1.189	.128	.935	1.443
	Equal variances not assumed			9.497	98.213	.000	1.189	.125	.940	1.437

of WiFi, the positive link between perceived long-term consequences of use and the utilization of WiFi, the positive relationship between facilitating conditions and WiFi use intention, the positive relationship between perceived advantages of WiFi and the willingness to use WiFi, the positive link between compatibility and WiFi usage intention, positive relationships between trialability, result demonstrability, visibility and intentions to use WiFi, and the positive influence of voluntariness of use and image from the use of WiFi on willingness to use WiFi.

On the other hand, the study provides evidence that contrasts with current findings in the field. Although evidence from previous studies point to a positive link between perceived job fit and intentions to use a particular technology, no such observation was obvious in the current study.

Although findings from previous studies view perceived self-efficacy as an important determinant of intentions to use an innovation, no support for this statement is obvious in the context of the present research. Finally, although a number of scholars view perceived security risks as an important determinant of IT usage, no such relationship was evident in the current study.

Another major contribution of the present study refers to the finding of the researchers regarding the underlying factor structure of the 16 constructs that can predict WiFi or any IT-related usage. Specifically, the study revealed that the factors that can affect usage of a particular innovation such as WiFi, can be divided and measured across two major dimensions, the one being the 'level of support for and advantages of using WiFi' (e.g. support, increased performance, reduced task complexity, more effectiveness than previous technologies, etc.) and the second being the 'period one needs for becoming familiar and feeling secure with WiFi use' (e.g. trying first, observing demonstration and tangible results, feeling confident regarding the use of the technology, etc.). Compared to previous literature, the study innovatively revealed that users and non-users of Wifi have to be differentiated as to their attitudes and perceptions.

Table 10. Perceptions towards Factors Determining WiFi Adoption: Group Statistics

	Use of WiFi for Internet Access?	N	Mean	Std. Deviation	Std. Error Mean
Self Efficacy	Yes	54	5.306	.5861	.0798
	No	60	4.700	.7713	.0996
Perceived usefulness	Yes	54	6.389	.4921	.0670
	No	60	4.400	.5660	.0731
Ease of use	Yes	54	5.3889	.49208	.06696
	No	60	4.2000	.56148	.07249
Affect	Yes	54	4.3889	.49208	.06696
	No	60	4.2000	.60506	.07811
long term consequences	Yes	54	4.6111	.49208	.06696
	No	60	4.4250	.81732	.10552
Complexity	Yes	54	6.0833	.73811	.10044
	No	60	4.2750	.68535	.08848
Job fit	Yes	54	5.8889	.72468	.09862
	No	60	4.5250	.73890	.09539
Facilitating conditions	Yes	54	5.4722	.70320	.09569
	No	60	3.2000	.73761	.09522
Relative Advantage	Yes	54	5.1944	.24604	.03348
	No	60	4.1250	.54946	.07094
Compatibility	Yes	54	5.1944	.24604	.03348
	No	60	4.1000	.88202	.11387
Trialability	Yes	54	3.8611	.47001	.06396
	No	51	3.5882	.31154	.04362
Image	Yes	54	2.4444	.60397	.08219
	No	60	3.2500	1.05163	.13577
Voluntariness	Yes	54	4.6944	.89240	.12144
	No	60	4.9000	.43957	.05675
Result Demonstrability	Yes	54	5.3333	.44510	.06057
	No	60	4.5750	.78560	.10142
Visibility	Yes	54	5.6111	.66351	.09029
	No	60	5.8250	.43007	.05552
Perceived security risks	Yes	54	4.5000	.71375	.09713
	No	60	4.4500	.90056	.11626

CONCLUSION

The major objectives of the present study were to identify the nature of factors that tend to affect WiFi intention usage, to assess the predictive ability of a full set of factors (identifying their individual as well as the combined impact and explanatory power), to identify the underlying factor structure of a number of constructs with potential effect on WiFi usage, the overall attitudes and behavior of

Table 11. Perceptions towards Factors Determining WiFi Adoption: Independent Samples Test

		Levene's Test for Equality of Var.		t-test for Equality of Means						
		F	Sig.	t	df	Sig. (2-tailed)	Mean Diff.	Std. Error Diff.	95% C.I. of the Diff. Lower	Upper
Self Efficacy	Equal variances assumed	2.447	.121	4.679	112	.000	.6056	.1294	.3491	.8620
	Equal variances not assumed			4.746	109.033	.000	.6056	.1276	.3527	.8584
Perceived usefulness	Eq. var. assumed	3.553	.062	19.920	112	.000	1.9889	.0998	1.7911	2.1867
	Eq. var not assumed			20.067	111.874	.000	1.9889	.0991	1.7925	2.1853
Ease of use	Eq. var. assumed	.114	.736	11.964	112	.000	1.18889	.09937	.99200	1.38578
	Eq. var not assumed			12.048	111.927	.000	1.18889	.09868	.99336	1.38442
Affect	Eq. var. assumed	.008	.927	1.816	112	.072	.18889	.10401	-.01719	.39496
	Eq. var not assumed			1.836	110.905	.069	.18889	.10289	-.01499	.39277
long term consequ-ences	Eq. var. assumed	1.389	.241	1.453	112	.149	.18611	.12811	-.06773	.43995
	Eq. var not assumed			1.489	98.338	.140	.18611	.12497	-.06188	.43410
Complexity	Eq. var. assumed	6.285	.014	13.563	112	.000	1.80833	.13333	1.54416	2.07251
	Eq. var not assumed			13.510	108.485	.000	1.80833	.13386	1.54302	2.07365
Job fit	Eq. var. assumed	2.806	.097	9.930	112	.000	1.36389	.13734	1.09176	1.63602
	Eq. var not assumed			9.941	111.160	.000	1.36389	.13720	1.09202	1.63576
Facilitating conditions	Eq. var. assumed	.542	.463	16.789	112	.000	2.27222	.13534	2.00406	2.54039
	Eq. var not assumed			16.831	111.617	.000	2.27222	.13500	2.00473	2.53972
Relative Advantage	Eq. var. assumed	12.538	.001	13.160	112	.000	1.06944	.08126	.90843	1.23046
	Eq. var not assumed			13.634	83.598	.000	1.06944	.07844	.91345	1.22544
Compat-ibility	Eq. var. assumed	49.523	.000	8.811	112	.000	1.09444	.12421	.84834	1.34055
	Eq. var not assumed			9.221	69.068	.000	1.09444	.11869	.85767	1.33122
Trialability	Eq. var. assumed	35.125	.000	3.485	103	.001	.27288	.07830	.11759	.42816
	Eq. var not assumed			3.525	92.550	.001	.27288	.07742	.11912	.42663
Image	Eq. var. assumed	42.783	.000	-4.942	112	.000	-.80556	.16301	-1.12854	-.48257
	Eq. var not assumed			-5.076	95.840	.000	-.80556	.15871	-1.12059	-.49052
Voluntari-ness	Eq. var. assumed	50.385	.000	-1.584	112	.116	-.20556	.12977	-.46269	.05157
	Eq. var not assumed			-1.533	75.442	.129	-.20556	.13405	-.47256	.06145
Result Demonstra-bility	Eq. var. assumed	29.728	.000	6.247	112	.000	.75833	.12140	.51780	.99887
	Eq. var not assumed			6.419	95.122	.000	.75833	.11813	.52382	.99285
Visibility	Eq. var. assumed	42.006	.000	-2.062	112	.042	-.21389	.10372	-.41940	-.00838
	Eq. var not assumed			-2.018	89.201	.047	-.21389	.10600	-.42450	-.00328
Perceived security risks	Eq. var. assumed	26.606	.000	.326	112	.745	.05000	.15334	-.25383	.35383
	Eq. var not assumed			.330	110.290	.742	.05000	.15150	-.25022	.35022

consumers towards WiFi use for Internet access as well as identification and comparison of WiFi users and non-users overall behavior towards iFi use and their perceptions against factors determining WiFi adoption.

The findings from the study provided all information needed for reaching the aforementioned objectives. Firstly, assessing the impact of a number of 16 constructs on intention to use WiFi, the research revealed that most of these variables can affect WIFi usage intention. Further, the study revealed that these constructs have a high explanatory power, both as a group, and stand-alone. Specifically, the 16 variables being tested were identified to have a very high combined ability in terms of predicting WiFi usage, being in a position to explain up to 85% of the variance in the model.

Further, the study succeeded in reducing the number of factors that have been revealed to affect the use of IT, into a smaller, more manageable set. The factor analysis resulted in the identification of two major dimensions that can be used instead of the 16 variables, in terms of explaining the impact of a set of IT-related characteristics on WiFi usage. These dimensions are: "the level of support for and advantages of using WiFi" (e.g. support, increased performance, reduced task complexity, more effectiveness than previous technologies, etc.) and "the period one needs for becoming familiar and feeling secure with WiFi use" (e.g. trying first, observing demonstration and tangible results, feeling confident regarding the use of the technology, etc.).

Moreover, the study succeeded in its efforts to reach the last research objective that was the identification of overall attitudes and behavior of consumers towards WiFi use. The findings showed that users and non-users have to be differentiated as to their attitudes and perceptions towards Wifi and an overall very positive attitude towards the use of WiFi for Internet access and a very high intention/ willingness in terms of utilizing WiFi for Internet access.

The findings from the study can be useful for many stakeholder groups such as managers, policy-makers and researchers. Managers benefit in that they can acquire useful information about the forces that are more likely to affect WiFi usage. Although numerous studies identified evidence relevant to the forces that affect IT usage, for the first time, a study reveals evidence relevant to WiFi use. Thus, managers of organizations that offer WiFi internet services can have a better understanding of the IT characteristics that determine usage of WiFi. Such information can be useful in the sense that businesses can invest in building more positive perceptions among consumers regarding usage of WiFi and in developing effective strategies and tactics to translate willingness to use into actual usage.

Policy-makers also benefit in that they obtain a better understanding of the forces that drive adoption of new technologies. Advancement of economies presupposes adoption of new technologies by people. Thus governments need to capitalize on these IT characteristics that can act as motives of technology usage.

Furthermore, researchers in the field can capitalize on the study's findings. The present study contributed to the field to the extent that it successfully attempted for the first time to assess a large holistic set of factors influencing on the usage of Wifi whilst subdividing this set into two major dimensions. Thus, future research in the field can assess IT usage across these two dimensions. Thus, future research in the field can assess IT usage across these two dimensions. Further, researchers can use the evidence from the study in their efforts to develop more robust IT usage measurement instruments. For example, scales can measure actual or intended use across these two dimensions.

Finally, the study was conducted in the presence of a number of limitations. Research related limitations included the limited data validity, which stemmed from the fact that the study was conducted in the presence of seriously limited evidence in Cyprus regarding the factors that tend to affect adoption of new technologies. Other research-related drawbacks included the difficulties in collecting the necessary evidence, mainly due to the high percentage of people's denial to participate. Further, since the research was conducted in Cyprus, using a relatively small sample (114 individuals), there is a potential problem of generalizing too widely on the basis of a narrow evidential base.

REFERENCES

Aaker, D., Kumar, V., & Day, G. (2004). *Marketing research* (8th ed.). New Jersey: John Wiley & Son, Inc.

Adams, D. A., Nelson, R. R., & Todd, P. A. (1992). Perceived usefulness, ease of use, and usage of information technology: A replication. *Management Information Systems Quarterly*, *16*(2), 227–247. doi:10.2307/249577

Baron, S., Patterson, A., & Harris, K. (2006). Beyond technology acceptance: Understanding consumer practice. *International Journal of Service Industry Management*, *17*(2), 111–135. doi:10.1108/09564230610656962

Bruner, G. C., & Kumar, A. (2005). Explaining consumer acceptance of handheld internet devices. *Journal of Business Research*, *58*(5), 553–558. doi:10.1016/j.jbusres.2003.08.002

Burton-Jones, A., & Hubona, G. S. (2005). Individual differences and usage behavior: Revisiting a technology acceptance model assumption. *The Data Base for Advances in Information Systems*, *36*(2), 58–77.

Compeau, D., & Higgins, C. (1995). Application of social cognitive theory to training for computer skills. *Information Systems Research*, *6*(2), 118. doi:10.1287/isre.6.2.118

Dabholkar, P. A. (1996). Consumer evaluations of new technology-based self-service options: An investigation of alternative models of service quality. *International Journal of Research in Marketing*, *13*(1), 29–51. doi:10.1016/0167-8116(95)00027-5

Davis, F. D. (1989). Perceived usefulness, perceived ease of use, and user acceptance of Information Technology. *Management Information Systems Quarterly*, (September): 319–340. doi:10.2307/249008

Hole, K., Erlend, D., & Thorsheim, P. (2005). Securing Wi-Fi networks. *Computer*, (July): 27–34.

Kaufmann, H. R. (2004). Approaching the very core of the client: CRM and identity in private banking. In Geberl, S., Kaufmann, H. R., Menichetti, M., & Wiesner, D. (Eds.), *Aktuelle Entwicklungen im Finanzdienstleistungsbereich*. Heidelberg, Germany: Physica-Verlag.

Kitios, A., & Kaufmann, H. R. (2009). Status orientation and symbolic consumption of Cypriot consumers. *International Journal of Management Cases*, *11*(2).

Kotler, P., & Armstrong, G. (1999). *Principles of marketing* (8th ed.). London, UK: Prentice Hall International.

Kotler, P., Wong, V., Saunders, J., & Armstrong, G. (2005). *Principles of marketing*. London, UK: Pearson Hall.

Labuschagne, L., & Eloff, J. (2000). Electronic commerce: The information-security challenge. *Information Management & Computer Security*, *8*(3), 154. doi:10.1108/09685220010372582

Liao, C.-H., & Tsou, C.-W. (2007). Factors influencing the usage of 3G mobile services in Taiwan. *Online Information Review*, *31*(6), 759–774. doi:10.1108/14684520710841757

MisMe corporation. (2007). *About WiFi*. Retrieved November 2009 from http://www.misme.net/about_wifi.html

Moore, G. C., & Benbasat, I. (1991). Development of an instrument to measure the perceptions of adopting an Information Technology innovation. *Information Systems Research, 2*(3), 192–222. doi:10.1287/isre.2.3.192

Nabi, F. (2005). Secure business application logic for e-commerce systems. *Computers & Security, 24*, 208–217. doi:10.1016/j.cose.2004.08.008

O'Cass, A., & Fenech, T. (2003). Web retailing adoption: Exploring the nature of Internet users Web retailing behaviour. *Journal of Retailing and Consumer Services, 10*, 81–94. doi:10.1016/S0969-6989(02)00004-8

Pallant, J. (2006). *SPSS survival manual* (2nd ed.). Berkshire, UK: Open University Press.

Perreault, W., McCarthy, E. J., Parkinson, S., & Steward, K. (2000). *Basic marketing* (European edition). London, UK: McGraw Hill.

Porter, C., & Donthu, N. (2006). Using the technology acceptance model to explain how attitudes determine Internet usage: The role of perceived access barriers and demographics. *Journal of Business Research, 59*, 99–1007. doi:10.1016/j.jbusres.2006.06.003

Potter, B. (2006). Wireless hotspots: Petri dish of wireless security. *Hacking and Innovation, 49*(6), 50–56.

Saade, R. G., Nebebe, F., & Tan, W. (2007). Viability of the technology acceptance model in multimedia learning environments: A comparative study. *Interdisciplinary Journal of Knowledge and Learning Objects, 3*, 176–184.

Salisbury, D, Pearson, R., Pearson, A., & Miller, D. (2001). Perceived security and World Wide Web purchase intention industrial management + data systems. *ABI/INFORM Global, 101*(3/4), 165.

Schiffman, L., & Kanuk, L. (2004). *Consumer behavior*. New Jersey, USA: Pearson Prentice Hall.

Siano, A., Kaufmann, H. R., & Basile, G. (2009). The evolution of new consumer roles in branding. 2nd EMRBI Conference at the University of Salerno, Italy. 26-28. In D. Vrontis, Y. Weber, R. Kaufmann, & S. Tarba (Eds.), *Managerial and entrepreneurial developments in the Mediterranean area*. Cyprus: EuroMed Press. ISBN: 978-9963-634-76-7

Stuart, A. (2003). *Technofile: Why Wi-Fi?* Retrieved from http://www.inc.com/ articles/ 2003/ 09/ wifi.html

Thompson, R., Higgins, C., & Howell, J. (1991). Personal computing: Toward a conceptual model of utilization. *Management Information Systems Quarterly*, (March): 125. doi:10.2307/249443

Venkatesh, V., Morris, M. G., Davis, G. B., & Davis, F. D. (2003). User acceptance of information technology: Toward a unified view. *Management Information Systems Quarterly, 27*(3), 425–478.

Vijayasarathy, L. R. (2004). Predicting consumer intentions to use on-line shopping: The case for an augmented technology acceptance model. *Information & Management, 41*(6), 747–762. doi:10.1016/j.im.2003.08.011

Wright, K. (n.d.). *Best articles online, new wireless technology – Wi-Fi*. Retrieved 20 October, 2009, from http://www.bestarticlesonline.com/.../new-wireless-technology-wi-fi-.html

KEY TERMS AND DEFINITIONS

WiFi – Wireless Fidelity: the IEEE 802.11 standard for wireless data transmissions.

Marketing: business of promoting and selling products or services, that comprises market research and advertising.

Wireless Hotspots: a place where users can have wireless access to the internet.

Wireless Security: the prevention of unauthorized access to computers in a using wireless environment.

Consumer Behavior: the study that examines the behavior of people when of buying a product TAM-Technology Acceptance Model: an information systems theory that models the perceived usefulness and perceived ease of use determining an individual's intention to accept and use a system.

Compilation of References

Aaker, J. L., & Lee, A. Y. (2001). I seek pleasures, we avoid pains: The role of self regulatory goals in information processing and persuasion. *The Journal of Consumer Research*, *28*, 33–49. doi:10.1086/321946

Aaker, D. A., Kumar, V., & Day, G. S. (2001). *Marketing research* (7th ed.). New York, NY: John Wiley and Sons.

Aaker, D., Kumar, V., & Day, G. (2004). *Marketing research* (8th ed.). New Jersey: John Wiley & Son, Inc.

Abernathy, F. H., Dunlop, J. T., Hammond, J. H., & Weil, D. (2000). Retailing and supply chains in the information Age. *Technology in Society*, *22*, 5–31. doi:10.1016/S0160-791X(99)00039-1

Acquisti, A., & Varian, H. R. (2005). Conditioning prices on purchase history. *Marketing Science*, *24*(3), 367–381. doi:10.1287/mksc.1040.0103

Adamo, A., Bertacchini, P. A., Bilotta, E., Pantano, P., & Tavernise, A. (2010). Connecting art and science for education: Learning through an advanced virtual theater with "Talking Heads". *Leonardo*, *43*(5), 442–448. doi:10.1162/LEON_a_00036

Adams, D. A., Nelson, R. R., & Todd, P. A. (1992). Perceived usefulness, ease of use, and usage of information technology: A replication. *Management Information Systems Quarterly*, *16*(2), 227–247. doi:10.2307/249577

Addis, M., & Holbrook, M. B. (2001). On the conceptual link between mass customization and experiential consumption: An explosion of subjectivity. *Journal of Consumer Behaviour*, *1*(1), 50–66. doi:10.1002/cb.53

Ahn, H. J. (2010). Evaluating customer aid functions of online stores with agent-based models of customer behavior and evolution strategy. *Information Science*, *180*, 1555–1570. doi:10.1016/j.ins.2009.12.029

Aiello, A., & Cacia, C. (2009, August). Tourism International marketing strategies: the role of quality brand in the valorisations process. In *Proceeding of the 10th Vaasa Conference on International Business*, Vaasa, Finland.

Ajzen, I. (1991). The theory of planned behaviour. *Organizational Behavior and Human Decision Processes*, *50*, 179–211. doi:10.1016/0749-5978(91)90020-T

Ajzen, I., & Fishbein, M. (1980). *Understanding attitudes and predicting social behaviour*. Englewood Cliffs, NJ: Prentice-Hall.

Ajzen, I. (1985). From intentions to actions: A theory of planned behavior. In Kuhl, J., & Beckmann, J. (Eds.), *Action control: From cognition to behaviour* (pp. 11–39). Berlin, Germany: Springer.

Akkermans, H. (2001). Intelligent e-business: From technology to value. *IEEE Intelligent Systems*, *16*(4), 8–11. doi:10.1109/5254.941352

Aksoy, L., Bloom, P. N., Lurie, N. H., & Cooil, B. (2006). Should recommendation agents think like people? *Journal of Service Research*, *8*(4), 297–315. doi:10.1177/1094670506286326

Aladwani, A. M., & Palvia, P. (2002). Developing and validating an instrument for measuring user-perceived Web quality. *Information & Management*, *39*(6), 467–476. doi:10.1016/S0378-7206(01)00113-6

Alavi, M., & Leidner, D. E. (2001). Knowledge management and knowledge management systems: Conceptual foundations and research issues. *Management Information Systems Quarterly*, *25*(1), 107–136. doi:10.2307/3250961

Alexa - The Web Information Company. (2008). *Top sites by country*. Retrieved August 20, 2008, from http://www.alexa.com/site/ds/top%5C_500

Allen, T. (1977). *Managing the flow of technology*. Cambridge, MA: MIT Press.

Anderson, R. E., & Srinivasan, S. (2003). E-satisfaction and e-loyalty: A contingency framework. *Psychology and Marketing*, *20*(2), 123–138. doi:10.1002/mar.10063

Andrade, E. B., Kaltcheva, V., & Weitz, B. (2002). Self-disclosure on the Web: The impact of privacy policy, reward and company reputation. *Advances in Consumer Research. Association for Consumer Research (U. S.)*, *29*(1), 350–353.

Andrew, O., Clore, G. L., & Collins, A. (1990). *The cognitive structure of emotions*. Cambridge University Press.

Ang, S. H., Leong, S. M., & Lim, J. (1997). The mediating influence of pleasure and arousal on layout and signage effects. *Journal of Retailing and Consumer Services*, *4*(1), 13–24. doi:10.1016/0969-6989(95)00091-7

Anolli, L. (2005). *Psicologia generale*. Bologna, Italia: il Mulino.

Ansari, A., Essegaier, S., & Kohli, R. (2000). Internet recommendation systems. *JMR, Journal of Marketing Research*, *37*(3), 363–375. doi:10.1509/jmkr.37.3.363.18779

Ansari, A., & Mela, C. F. (2003). E-customization. *JMR, Journal of Marketing Research*, *40*(2), 131–145. doi:10.1509/jmkr.40.2.131.19224

Apple Inc. (2007). *Apple human interface guidelines*. Retrieved on September 8, 2008, from developer.apple.com/documentation/UserExperience/Conceptual/AppleHIGuidelines/index.html

Apte, C., Liu, B., Pednault, E., & Smyth, P. (2005). Business applications of data mining. *Communications of the ACM*, *45*(8), 49–53.

Arbuckle, J. L. (2006). *Amos (version 7)*. Chicago, IL: SPSS.

Areni, C. S., Duhan, D. F., & Kiecker, P. (1999). Point-of-purchase displays, product organization, and brand purchase likelihood. *Journal of the Academy of Marketing Science*, *27*(4), 428–441. doi:10.1177/0092070399274003

Arentze, T. A., Opewal, H., & Timmermans, H. J. P. (2005). A multipurpose shopping trip model to assess retail agglomeration effects. *JMR, Journal of Marketing Research*, *42*(1), 109. doi:10.1509/jmkr.42.1.109.56884

Ariely, D., Lynch, J. G. Jr, & Aparicio, M. IV. (2004). Learning by collaborative and individual-based recommendation agents. *Journal of Consumer Psychology*, *14*(1/2), 81–95. doi:10.1207/s15327663jcp1401&2_10

Armstrong, A., & Hagel, J. (1996). The real value of online communities. *Harvard Business Review*, (May–June): 134–141.

Arussy, L. (2009). Is customer experience relevant in a recession? *Customer Relationship Management*, *13*(4), 12.

Aubert-Gamet, V., & Cova, B. (1999). Servicescapes: From modern non-places to postmodern common places. *Journal of Business Research*, *44*(1), 37–45. doi:10.1016/S0148-2963(97)00176-8

Axhausen, K. W. (2003). *Social networks and travel: Some hypotheses. Arbeitsbericht Verkehrsund Raumplanung, 197, Institut für Verkehrsplanung und Transportsysteme*. Zürich: ETH Zürich.

Ayling, J. (2006, August 3). RFID revolution only just beginning. *just-food.com*.

Babin, B. J., & Darden, W. R. (1995). Consumer self-regulation in a retail environment. *Journal of Retailing*, *71*(1), 47–70. doi:10.1016/0022-4359(95)90012-8

Babin, B. J., Hardesty, D. A., & Suter, T. (2003). Color and shopping intentions: The intervening effect of price fairness and perceived affect. *Journal of Business Research*, *56*(7), 541–551. doi:10.1016/S0148-2963(01)00246-6

Backes, N. (1997). Reading the shopping mall city. *Journal of Popular Culture*, *31*(3), 1–17. doi:10.1111/j.0022-3840.1997.3103_1.x

Bäckström, K., & Johansson, U. (2006). Creating and consuming experiences in retail store environments: Comparing retailer and consumer perspectives. *Journal of Retailing and Consumer Services*, *13*(6), 417–430. doi:10.1016/j.jretconser.2006.02.005

Bagozzi, R. P. (2000). On the concept of intentional social action in consumer research. *The Journal of Consumer Research*, *27*, 388–396. doi:10.1086/317593

Bagozzi, R. P., & Dholakia, U. (1999). Goal setting and goal striving in consumer behavior. *Journal of Marketing*, *63*, 19–32. doi:10.2307/1252098

Bagozzi, R. P. (2007). The legacy of the technology acceptance model and a proposal for a paradigm shift. *Journal of the Association for Information Systems*, *8*(4), 244–254.

Baier, D., & Stüber, E. (2010). Acceptance of recommendations to buy in online retailing. *Journal of Retailing and Consumer Services*, *17*(3), 173–180. doi:10.1016/j.jretconser.2010.03.005

Bainbridge, W. S. (2004). *Berkshire encyclopedia of human-computer interaction*. USA: Berkshire Publishing Group.

Baker, J., Grewal, D., & Parasuraman, A. (1994). The influence of store environment on quality inferences and store image. *Journal of the Academy of Marketing Science*, *22*(4), 328–339. doi:10.1177/0092070394224002

Baker, J., Parasuraman, A., Grewal, D., & Voss, G. B. (2002). The influence of multiple store environment cues on perceived merchandise value and patronage intentions. *Journal of Marketing*, *66*(2), 120–141. doi:10.1509/jmkg.66.2.120.18470

Baker, J. (1998). Examining the informational value of store environments. In Sherry, J. F. Jr., (Ed.), *Servicescapes: The concepts of place in contemporary markets* (pp. 55–79). Chicago, IL: NTC Business Books.

Bakos, J. Y. (1998). The emerging role of electronic marketplaces on the Internet. *Communications of the ACM*, *41*(8), 35–42. doi:10.1145/280324.280330

Balconi, M., & Antonietti, A. (Eds.). (2009). *Scegliere, comprare. Dinamiche di acquisto in psicologia e neuroscienze*. Milano, Italia: Springer-Verlag.

Bansal, H. S., & Voyer, P. A. (2000). Word-of-mouth processes within a service purchase decision context. *Journal of Service Research*, *3*, 166–177. doi:10.1177/109467050032005

Barclay, D., Thompson, R., & Higgins, C. (1995). The Partial Least Squares (PLS) approach to causal modeling: Personal computer adoption and use as an illustration. *Technology Studies*, *2*(2), 285–309.

Barlow, A. K. J., Siddiqui, N. Q., & Mannion, M. (2004). Development in Information and Communication Technologies for retail marketing channels. *International Journal of Retail and Distribution Management*, *32*(3), 157–163. doi:10.1108/09590550410524948

Barney, J. B. (1991). Firm resources and sustained competitive advantage. *Journal of Management*, *17*(1), 99–120. doi:10.1177/014920639101700108

Baron, R. M., & Kenny, D. (1986). The moderator-mediator distinction in social psychology research: conceptual, strategic and statistical considerations. *Journal of Personality and Social Psychology*, *51*, 1173–1182. doi:10.1037/0022-3514.51.6.1173

Baron, S., Patterson, A., & Harris, K. (2006). Beyond technology acceptance: Understanding consumer practice. *International Journal of Service Industry Management*, *17*(2), 111–135. doi:10.1108/09564230610656962

Baroudi, J. J., & Orlikowski, W. J. (1989). The problem of statistical power in MIS research. *Management Information Systems Quarterly*, *13*(1), 7–106. doi:10.2307/248704

Barr, V. (2001). Self-scanning checkout picks up speed in mass retail markets. *Stores*, August, 66-70.

Bart, Y., Shankar, V., Sultan, F., & Urban, G. L. (2005). Are the drivers and role of online trust the same for all websites and consumers? A large-scale exploratory empirical study. *Journal of Marketing*, *69*(4), 133–152. doi:10.1509/jmkg.2005.69.4.133

Bartlett, F. C. (1932). *Remembering: A study in experimental and social psychology*. Cambridge, UK: Cambridge University Press.

Bartneck, C., & Reichenbach, J. (2005). Subtle emotional expressions of synthetic characters. *International Journal of Human-Computer Studies*, *62*, 179–192. doi:10.1016/j.ijhcs.2004.11.006

Barton, B., Byciuk, S., Harris, C., Schumack, D., & Webster, K. (2005). The emerging cyber risks of biometrics. *Risk Management*, *52*(10), 26–31.

Bass, B. M. (1990). *Handbook of leadership. Theory, research a managerial application*. New York, NY: The Free Press.

Bauer, H.-J., & Hallier, B. (1999). *Kultur und Geschichte des Handels*. Cologne, Germany: EHI.

Beckett, H., & Kavanagh, J. (2005). How it works in the real world (use of radio frequency identification tags for the improvement in the efficiency of the staffs). *Computer Weekly*, March 8, 50.

Bedi, H., & Yang, L. (2009). Fair electronic exchange based on fingerprint biometrics. *International Journal of Information Security and Privacy*, (July-September): 76–109. doi:10.4018/jisp.2009100705

Behrends, C. (2001). An der Schnittstelle zwischen Theorie und Praxis. In Hallier, B. (Ed.), *Praxisorientierte Handelsforschung*. Cologne, Germany, Germany: EHI.

Bellman, S., Lohse, G. L., & Johnson, E. J. (1999). Predictors of online buying behaviours. *Communications of the ACM*, *42*(12), 32–38. doi:10.1145/322796.322805

Ben-Ari, M., & Yeshno, T. (2006). Conceptual models of software artifacts. *Interacting with Computers*, *18*(6), 1336–1350. doi:10.1016/j.intcom.2006.03.005

Benbasat, I., & Weber, R. (1996). Research commentary: Rethinking diversity in Information System research. *Information Systems Research*, *7*(4), 389–399. doi:10.1287/isre.7.4.389

Benjamin, R., & Wigand, R. (1995). Electronic markets and virtual value chains on the information highway. *Sloan Management Review*, Winter.

Bennett, A., & Strydom, J. W. (2001). *Introduction to travel and tourism marketing*. JW Strydom.

Benyon, D., Turner, P., & Turner, S. (2005). *Designing interactive systems: People, activities, contexts, technologies*. Boston, MA: Addison-Wesley.

Berkowitz, L. (1993). Towards a general theory of anger and emotional aggression: Implications of the cognitive-neoassociationistic perspective for the analysis of anger and other emotions. In Wyer, R. S., & Srull, T. K. (Eds.), *Advances in social cognition, 6* (pp. 1–46). Hillsdale, NJ: Eribaum.

Bernard, S. (2005). *An introduction to enterprise architecture* (2nd ed.). Authors House.

Bernard, M. (2001). *User expectations for the location of web objects*. Paper presented at the CHI - Interactive Posters.

Bernard, M. (2002). *Examining user expectations for the location of common e-commerce web objects*. Retrieved on August 21, 2008, from http://www.surl.org/usabilitynews/41/web_object-ecom.asp

Bernstein, D. (1988). *Company image*. Guerini e Associati.

Berry, L. (1983). Relationship marketing. In L. Berry, G, L. Shostack, & G. D. Upah (Eds.), *Emerging perspectives on services marketing*. Chicago, IL: American Marketing Association.

Bertacchini, P. A., Bilotta, E., Pantano, P., Battiato, S., Cronin, M. D., Di Blasi, G., et al. Tavernise A. (2007). Modelling and animation of theatrical Greek masks in an authoring system. In R. De Amicis & G. Conti (Eds.), *Proceedings of the 5th Eurographics Italian Chapter (EG-IT-07)* (pp. 191-197). Germany: Eurographics Association.

Berthon, P., Pitt, L., & Watson, R. T. (1996). The World Wide Web as an advertising medium. *Journal of Advertising Research*, (Jan/Feb): 43–55.

Bettman, J. R., Luce, M. F., & Payne, J. (1998). Constructive consumer choice processes. *The Journal of Consumer Research*, *25*(3), 187–217. doi:10.1086/209535

Beverland, M., Lim, E. A. C., Morrison, M., & Terziovski, M. (2006). In-store music and consumer–brand relationships: Relational transformation following experiences of (mis)fit. *Journal of Business Research*, *59*, 982–989. doi:10.1016/j.jbusres.2006.07.001

Bharadwaj, A. S. (2000). A resource-based perspective on information technology capability and firm performance: An empirical investigation. *Management Information Systems Quarterly*, *24*(1), 169–196. doi:10.2307/3250983

Bhatnagar, P. (2003, May 28). Stuck at checkout? DIY lane's open. *CNN/Money*.

Biederman, I., Mezzanotte, R. J., & Rabinowitz, J. C. (1982). Scene perception: Detecting and judging objects undergoing relational violations. *Cognitive Psychology*, *14*(2), 143–177. doi:10.1016/0010-0285(82)90007-X

Bilotta, E. (1996). *Interfacce multimodali e aspetti psicologici dell'interazione uomo computer*. Cosenza, Italy: Edizioni Bios.

Bitner, M. J. (1990). Evaluating service encounter: The effects of physical surrounding and employee responses. *Journal of Marketing*, *54*(2), 69–82. doi:10.2307/1251871

Bitner, M. J. (1992). Servicescapes: The impact of physical surroundings on customers and employees. *Journal of Marketing*, *56*, 57–71. doi:10.2307/1252042

Bitner, M. J. (2001). Service and technology: Opportunities and paradoxes. *Managing Service Quality*, *11*(6), 375–379. doi:10.1108/09604520110410584

Bitran, G. R., & Ferrer, J.-C. (2007). On pricing and composition of bundles. *Production and Operations Management*, *16*(1), 93–108. doi:10.1111/j.1937-5956.2007.tb00168.x

Blackwell, R. D., Miniard, P. W., & Engel, J. F. (2001). *Consumer behavior*. Fort Worth, TX: Harcourt College.

Blanca, H., Julio, J., & Martin, J. M. (2009). The impact of self-efficacy, ease of use and usefulness on e-purchasing: An analysis of experienced e-shoppers. *Interacting with Computers*, *1*(2), 146–156.

Bleicher, P. (2005). Biometrics comes of age: Despite accuracy and security concerns, biometrics are gaining in popularity. *Applied Clinical Trials*, *14*(12), 18–20.

Bloch, P. H., Ridgway, N. M., & Dawson, S. A. (1994). The shopping mall as consumer habitat. *Journal of Retailing*, *70*(1), 23–42. doi:10.1016/0022-4359(94)90026-4

Bloch, M., Pigneur, Y., & Steiner, T. (1996). *The IT-enabled extended enterprise, applications in the tourism industry*. Retrieved from http://haas.berkley.edu

Bo, X., & Benbasat, I. (2007). E-commerce product recommendation agents: Use, characteristics, and impact. *Management Information Systems Quarterly*, *31*(1), 137–209.

Boardman, M. (2005). *The language of websites*. London, UK & New York, NY: Routledge.

Bodapati, A. V. (2008). Recommendation systems with purchase data. *JMR, Journal of Marketing Research*, *45*(1), 77–93. doi:10.1509/jmkr.45.1.77

Bodet, G. (2008). Customer satisfaction and loyalty in service: Two concepts, four constructs, several relationships. *Journal of Retailing and Consumer Services*, *15*(3), 156–162. doi:10.1016/j.jretconser.2007.11.004

Boedeker, M. (1997). Relationship marketing and regular customer cards: Daily product retailing in Finland. *Marketing Intelligence & Planning*, *15*(6), 249–257. doi:10.1108/02634509710184820

Boehner, K., DePaula, R., Dourish, P., & Sengers, P. (2007). How emotion is made and measured. *International Journal of Human-Computer Studies*, *65*(4), 75–291. doi:10.1016/j.ijhcs.2006.11.016

Bolle, R. M., Connell, J. H., Pankanti, S., Ratha, N. K., & Senior, A. W. (2004). *Guide to biometrics*. New York, NY: Springer-Verlag.

Bollen, K. A. (1989). *Structural equations with latent variables*. New York, NY: John Wiley & Sons.

Bolton, R. N., & Drew, J. H. (1991). A multistage model of consumers' assessments of service quality and value. *The Journal of Consumer Research*, *17*, 375–384. doi:10.1086/208564

Bone, P. F. (1995). Word-of-mouth effects on short-term and long-term product judgments. *Journal of Business Research*, *32*(3), 213–223. doi:10.1016/0148-2963(94)00047-I

Bonn, M., & Furr, H. S. (1999). Predicting a behavioural profile for pleasure travellers on the basis of Internet use segmentation. *Journal of Travel Research*, *37*(4), 330–340. doi:10.1177/004728759903700403

Booth, M. E., & Philip, G. (1998). Technology, competencies, and competitiveness: The case for reconfigurable and flexible strategies. *Journal of Business Research*, *41*(1), 29–40. doi:10.1016/S0148-2963(97)00009-X

Bowerson, D. J., & Daugherty, P. J. (1995). Logistics paradigms: The impact of information technology. *Journal of Business Logistics*, *16*(1), 65–80.

Boyd, D., & Ellison, N. (2007). Social network sites: Definition, history, and scholarship. *Journal of Computer-Mediated Communication*, *13*, 1–9.

Braun, J., & Sagi, D. (1990). Vision outside the focus of attention. *Perception & Psychophysics, 48*(1), 45–58. doi:10.3758/BF03205010

Bressler, S. L., & Kelso, J. A. S. (2001). Cortical coordination dynamics and cognition. *Trends in Cognitive Sciences, 5*, 26–36. doi:10.1016/S1364-6613(00)01564-3

Briley, D. A., & Wyer, R. S. (2002). The effect of group membership salience on the avoidance of negative outcomes: Implications for social and consumer decisions. *The Journal of Consumer Research, 29*(3), 400–415. doi:10.1086/344426

Brohman, M. K., Watson, R. T., Piccoli, G., & Parasuraman, A. (2003). Data completeness: A key to effective Net-based customer service systems. *Communications of the ACM, 46*(6), 47–51. doi:10.1145/777313.777339

Brooke, J. (1996). SUS: a quick and dirty usability scale. In Jordan, P., Thomas, B., Weerdmeester, B., & McClelland, A. (Eds.), *Usability evaluation in industry*. London, UK: Taylor and Francis.

Brown, S. R., & Melamed, L. E. (1990). *Experimental design and analysis (quantitative applications in the social sciences)*. London, UK: Sage Publishers.

Broydrick, S. C. (1998). Seven laws of customer value. *Executive Excellence, 15*, 15.

Brun, A., & Castelli, C. (2008). Supply chain strategy in the fashion industry: A model on product, retail channel and brand. *International Journal of Production Economics, 116*, 169–181. doi:10.1016/j.ijpe.2008.09.011

Bruner, G. C., & Kumar, A. (2005). Explaining consumer acceptance of handheld internet devices. *Journal of Business Research, 58*(5), 553–558. doi:10.1016/j.jbusres.2003.08.002

Brunetti, G., & Servidio, R. (2009). Conceptual design scheme for virtual characters. In F. Lehmann-Grube, & J. Sablatnig (Eds.), *Proceedings of the First International Conference, FaVE* (pp. 135-150). Germany: Springer.

Buchanan, R., & Gilles, C. (1990). Value managed relationship: The key to customer retention and profitability. *European Management Journal, 8*(4). doi:10.1016/0263-2373(90)90115-M

Buhalis, D. (1998). Strategic use of information technologies in the tourism industry. *Tourism Management, 19*(5), 409–421. doi:10.1016/S0261-5177(98)00038-7

Buhalis, D. (1999). Marketing the competitive destination for the future. *Tourism Management, 21*(1), 97–116. doi:10.1016/S0261-5177(99)00095-3

Buhalis, D., & Law, R. (2008). Progress in information technology and tourism management: 20 year on and 10 years after the internet, the state of e-tourism research. *Tourism Management, 29*(4), 609–623. doi:10.1016/j.tourman.2008.01.005

Buhalis, D., Tjoa, A. M., & Jafari, J. (Eds.). (1998). *Information and communication technologies in tourism: IT and the dynamic tourism marketplace*. Vienna, Austria: Springer-Verlag.

Buhalis, D. (1996). Information and telecommunication technologies as a strategic tool for tourism enhancement at destination regions. In S. Klein, B. Schmid, A. Tjoa & H. Werthner (Eds.). *Information and communication technologies in tourism, Conference Proceeding, ENTER'96,* (pp. 131-142). 17-19 January, Innsbruck. Spinger-Verlag, Vienna.

Burger, F. (Ed.). (1997). TSI@Web-databased supported tourist information on the web. In A. M. Tjoa (Ed.), *Information and communication technologies in tourism,* (1987). New York, NY: Springer-Wien.

Burke, R. R. (2002). Technology and the customer interface: What consumers want in the physical and virtual store. *Journal of the Academy of Marketing Science, 30*(4), 411–432. doi:10.1177/009207002236914

Burke, R. (2002). Hybrid recommender systems: Survey and experiments. *User Modeling and User-Adapted Interaction, 12*(4), 331–370. doi:10.1023/A:1021240730564

Burton-Jones, A., & Hubona, G. S. (2005). Individual differences and usage behavior: Revisiting a technology acceptance model assumption. *The Data Base for Advances in Information Systems, 36*(2), 58–77.

Bush, A. J., & Hair, J. F. (1985). An assessment of the mall intercept as a data collection method. *JMR, Journal of Marketing Research, 22*(2), 158–167. doi:10.2307/3151361

Bushman, R. M., & Smith, A. J. (2001). Financial accounting information and corporate governance. *Journal of Accounting and Economics*, *32*, 237–333. doi:10.1016/S0165-4101(01)00027-1

Buxmann, P., & Gebauer, J. (1999). *Evaluating the use of Information Technology in inter-organizational relationships*. Hawaii Conference on Systems Sciences.

Campbell, A. (2003). Creating customer knowledge competence: Managing customer relationship management programs strategically. *Industrial Marketing Management*, *32*, 375–383. doi:10.1016/S0019-8501(03)00011-7

Carpenter, J. M., & Moore, M. (2009). Utilitarian and hedonic shopping value in the US discount sector. *Journal of Retailing and Consumer Services*, *16*(1), 68–74. doi:10.1016/j.jretconser.2008.10.002

Carrasco, J., Hogan, B., Wellman, B., & Miller, E. J. (2006*). Collecting social network data to study social activity-travel behaviour: An egocentric approach*. Paper presented at the 85th Transportation Research Board Meeting, Washington DC, January 22-26.

Carroll, J. (2000). *Making use: Scenario-based design of Human-Computer Interactions*. Cambridge, MA: MIT Press.

Cartwright, D., & Zander, A. (1968). *Group dynamics: Research and theory*. New York, NY: Harper & Row.

Casalò, L., Flaviàn, C., & Guinalìu, M. (2008). The role of perceived usability, reputation, satisfaction and consumer familiarity on the website loyalty information process. *Computers in Human Behavior*, *24*(2), 325–345. doi:10.1016/j.chb.2007.01.017

Cassell, J., Bickmore, T., & Campbell, L. Vilhja´ lmsson, H., & Yan, H. (2000). Human conversation as a system framework: Designing embodied conversational agents. In J. Cassell, J. Sullivan, S. Prevost, & E. Churchill, (Eds.), *Embodied conversational agents* (pp. 29-63). Cambridge, MA: MIT Press.

Cecere, L. (2007, May 2). Three years of retail RFID pilots. *Forbes*.

Cespedes, F. V., & Smith, H. J. (1993). Database marketing: New rules for policy and practice. *Sloan Management Review*, *34*(Summer), 8–12.

Cha, H. S., Pingry, D. E., & Thatcher, M. E. (2009). What determines IT spending priorities? *Communications of the ACM*, *52*(8), 105–110. doi:10.1145/1536616.1536644

Chaffey, D. (2009). *E-business and e-commerce management strategy. Implementation and practice*. Financial Times Prentice Hall.

Chain Drug Review (2006, November 6). *RFID's day still coming, just later than expected.* (p. 37).

Chain Drug Review (2010, April 26). *RFID may yet prove to be next big thing*. (p. 88).

ChainStore Age (2005, February). *Fast growth seen in ESLs, RFID*, (p. 50).

Chan, S. H. (2009). The roles of user motivation to perform a task and decision support system (DSS) effectiveness and efficiency in DSS use. *Computers in Human Behavior*, *25*(1), 217–228. doi:10.1016/j.chb.2008.09.002

Chandra, C., & Kumar, S. (2001). Enterprise architectural framework for supply-chain integration. *Industrial Management & Data Systems*, *101*, 209–304. doi:10.1108/EUM0000000005578

Chang, C.-A., & Burke, R. R. (2007). Consumer choice of retail shopping aids. *Journal of Retailing and Consumer Services*, *14*, 339–346. doi:10.1016/j.jretconser.2006.12.003

Chang, H. H., & Wang, I. C. (2008). An investigation of user communication behaviour in computer mediated environments. *Computers in Human Behavior*, *24*(5), 2336–2356. doi:10.1016/j.chb.2008.01.001

Chang, C. A., & Burke, R. R. (2007). Consumer choice of retail shopping aids. *Journal of Retailing and Consumer Services*, *14*, 339–346. doi:10.1016/j.jretconser.2006.12.003

Chatterjee, P. (2001). Online reviews: Do consumer use them? In M. C. Gilly & J. Myers-Levy (Eds.), *Proceeding of the ACR* (pp. 129-134). Provo, UT: Association for Consumer Research

Chau, P. Y. K., Au, G., & Tam, K. Y. (2000). Impact of information presentation modes on online shopping: An empirical evaluation of a broadband interactive shopping service. *Journal of Organizational Computing and Electronic Commerce*, *10*(1), 1–22.

Chaudhuri, A. (2006). *Emotion and reason in consumer behavior*. Oxford, UK: Elsevier.

Chebat, J.-C. (1999). Effects of time of day on shopping behavior. *Perceptual and Motor Skills*, *88*(2), 421–422. doi:10.2466/PMS.88.2.421-422

Chebat, J.-C., & Michon, R. (2003). Impact of ambient odors on mall shoppers' emotions, cognition and spending: A test of competitive causal theories. *Journal of Business Research*, *56*, 529–539. doi:10.1016/S0148-2963(01)00247-8

Chebat, J.-C., & Morrin, M. (2007). Colors and cultures: Exploring the effects of mall décor on consumer perceptions. *Journal of Business Research*, *60*, 189–196. doi:10.1016/j.jbusres.2006.11.003

Chebat, J.-C., Sirgy, M. J., & St-James, V. (2006). Upscale image transfer from malls to stores: A self-image congruence explanation. *Journal of Business Research*, *59*(12), 1288–1296. doi:10.1016/j.jbusres.2006.09.007

Chebat, J.-C., & Michon, R. (2003). Impact of ambient odors on mall shoppers' emotions, cognition, and spending. A test of competitive causal theories. *Journal of Business Research*, *56*(7), 529–539. doi:10.1016/S0148-2963(01)00247-8

Chebat, J.-C., & Slusarczyk, W. (2005). How emotions mediate the effects of perceived justice on acceptance in service recovery situations: An empirical study. *Journal of Business Research*, *58*(5), 664–673. doi:10.1016/j.jbusres.2003.09.005

Chen, D., Doumeingts, G., & Vernadat, F. (2008). Architectures for enterprise integration and interoperability. *Computers in Industry*, *59*, 647–659. doi:10.1016/j.compind.2007.12.016

Chen, L., Gillenson, M. L., & Sherrell, D. (2002). Enticing online consumers: The technology acceptance perspective. *Information & Management*, *39*(8), 705–719. doi:10.1016/S0378-7206(01)00127-6

Chen, M., & Xie, J. (2008). Online consumer review: Word-of-mouth as a new element of marketing communication mix. *Management Science*, *54*(3), 477–491. doi:10.1287/mnsc.1070.0810

Chen, M., Zhang, D., & Zhou, L. (2005). Providing Web services to mobile users: the architecture design of an m-service portal. *International Journal of Mobile Communications*, *3*(1), 1–18. doi:10.1504/IJMC.2005.005870

Chen, Z., & Dubinsky, A. J. (2003). A conceptual model of perceived customer value in e-commerce: A preliminary investigation. *Psychology and Marketing*, *20*(4), 323–347. doi:10.1002/mar.10076

Chen, Y., & Xie, J. (2005). Third-party product review and firm marketing strategy. *Marketing Science*, *24*(2), 218–240. doi:10.1287/mksc.1040.0089

Chen, J., Geyer, W., Dugan, C., Muller, M., & Guy, I. (2009). Make new friends, but keep the old: Recommending people on social networking sites. In *CHI '09: Proceedings of the 27th international conference on Human factors in computing systems*, New York, NY, USA, pp. 201-210, ACM.

Chen, P. Y., Wu, S. Y., & Yoon, J. (2004). *The impact of online recommendations and consumer feedback on sales*. Paper presented at the 25th International Conference on Information Systems (ICIS 2004).

Chi, Y. L. (2009). A consumer-centric design approach to develop comprehensive knowledge-based systems for keyword discovery. *Science Direct. Expert Systems with Applications*, *36*(2), 2481–2493. doi:10.1016/j.eswa.2007.12.061

Childers, T. L., Carr, C. L., Peck, J., & Carson, S. (2001). Hedonic and utilitarian motivations for online retail shopping behavior. *Journal of Retailing*, *77*(4), 511–535. doi:10.1016/S0022-4359(01)00056-2

Chin, W. W. (1998). Issues and opinion on structural equation modeling. *Management Information Systems Quarterly*, *22*(1), vii–xvi.

Chin, W. W., & Newsted, P. R. (1999). Structural equation modeling analysis with small samples using partial least squares. In Hoyle, R. H. (Ed.), *Statistical strategies for small sample research* (pp. 307–341). Thousand Oaks, CA: Sage.

Chitturi, R., Raghunathan, R., & Mahajan, V. (2008). Delight by design: The role of hedonic vs. utilitarian benefits. *Journal of Marketing*, *72*, 48–63. doi:10.1509/jmkg.72.3.48

Chiu, Y.-T. H., Fang, S.-C., & Tseng, C.-C. (2010). Early versus potential adopters: Exploring the antecedents of use intention in the context of retail service innovations. *International Journal of Retail & Distribution Management, 38*(6), 443–459. doi:10.1108/09590551011045357

Choudhary, V., Ghose, A., Mukhopadhyay, T., & Rajan, U. (2005). Personalized pricing and quality differentiation. *Management Science, 51*(7), 1120–1130. doi:10.1287/mnsc.1050.0383

Churchill, G. A. (1979). A paradigm for developing better measures of marketing constructs. *JMR, Journal of Marketing Research, 16*(1), 64–73. doi:10.2307/3150876

Churchill, E., Cook, L., Hodgson, P., Prevost, S., & Sullivan, J. (2000). 'May I help you?': Designing embodied conversational agent allies. In Cassell, J., Sullivan, J., Prevost, S., & Churchill, E. (Eds.), *Embodied conversational agents* (pp. 64–94). Cambridge, MA: MIT Press.

Chuttur, M. Y. (2009). *Overview of the technology acceptance model: Origins, developments and future directions.* Indiana University, USA. Sprouts: Working Papers on Information Systems, 9 (37).

Cisco. (2008). *Connected retail architecture: Reuse network services to boost efficiency and profits.* Retrieved September 1, 2010, from https://www. cisco.com/ /ea Retail.pdf

Citrin, A. V., Stem, D. E. Jr, Spangenberg, E. R., & Clark, M. J. (2003). Consumer need for tactile input: An internet retailing challenge. *Journal of Business Research, 56*(11), 915. doi:10.1016/S0148-2963(01)00278-8

Clark, K. (2004). Biometrics technology touches down. *Chain Store Age, 80*(5), 20A.

Clark, K. (2005, February). Pointing to savings, *Chain Store Age*, p. 140.

Clarke, R. (2003). Question: When is a retailer not a retailer? Answer: When it's a media owner. *The European Retail Digest, 38*, 20–23.

Clemons, E. K., & Row, M. C. (1991). Sustaining IT advantage: The role of structural differences. *Management Information Systems Quarterly, 15*(3), 275–292. doi:10.2307/249639

Cliff, N. (1983). Some cautions regarding the application of causal modelling methods. *Multivariate Behavioral Research, 18*, 115–126. doi:10.1207/s15327906mbr1801_7

Clodfelter, R. (1998). Pricing accuracy at grocery stores and other retail stores using scanners. *International Journal of Retail and Distribution Management, 26*(11), 412–420. doi:10.1108/09590559810246368

Clodfelter, R. (2010). Biometric technology in retailing: Will consumers accept fingerprint authentication? *Journal of Retailing and Consumer Services, 17*(3), 181–188. doi:10.1016/j.jretconser.2010.03.007

Cohen, J. (1977). *Statistical power analysis for the behavioral sciences* (revised ed.). New York, NY: Academic Press.

Cohn, J. F., Reed, L. I., Moriyama, T., Xiao, J., Schmidt, K. L., & Ambadar, Z. (2004). *Multimodal coordination of facial action, head rotation, and eye motion.* Sixth IEEE International Conference on Automatic Face and Gesture Recognition (pp. 645-650). Korea: IEEE Computer Series.

Collins, J. (2004, August 13). Self-checkout gets RFID upgrade. *RFID Journal.*

Coltman, T. (2007). Why build a customer relationship management capability? *The Journal of Strategic Information Systems, 16*, 301–320. doi:10.1016/j.jsis.2007.05.001

Comegys, C., Hannula, M., & Väisänen, J. (2006). Longitudinal comparison of Finnish and US online shopping behaviour among university students: The five-stage buying decision process. *Journal of Targeting. Measurement and Analysis for Marketing, 14*(4), 336–356. doi:10.1057/palgrave.jt.5740193

Compeau, D., & Higgins, C. (1995). Application of social cognitive theory to training for computer skills. *Information Systems Research, 6*(2), 118. doi:10.1287/isre.6.2.118

Compete, Inc. (2006). *Embracing consumer buzz creates measurement challenges for marketers.* Retrieved 05 June, 2010, from: http://www.cymfony.com/

ComputerWeekly.com. (2002, November 21). *Stores look to self-scanning to cut checkout queues.*

Computerworld.com. (2008, 16 January). *Cisco enters digital signage market.* Retrieved on August 6, 2010, from www.computerworld.com

comScore. (2007). *Social networking goes global*. Reston, VA. Retrieved September 9, 2007 from http://www.comscore.com/press/release.asp?press=1555

Constantinide, E., & Fountain, S. J. (2008). Web 2.0: Conceptual foundations and marketing issues. *Journal of Direct. Data and Digital Marketing Practice*, *9*, 231–244. doi:10.1057/palgrave.dddmp.4350098

Cooper, C. (2008). *Tourism: Principles and practice. Pearson Education*. England: FT Prentice Hall.

Coretto, T. (2010a). *Customer intelligence = profits*. Retrieved June 15, 2010, from http://nyreport.com/articles/73044/customer_intelligence_profits

Coretto, T. (2010b). *7 characteristics of successful customer intelligence*. Retrieved June 15, 2010, from http://blog.pntmarketingservices.com

Coretto, T. (2010c). Six things customer intelligence is not. Retrieved June 15, 2010, from http://www.nyreport.com/articles/74949/six_things_customer_intelligence_is_not?page=0,1

Coretto, T. (2010d). *Does customer intelligence include prospecting?* Retrieved June 15, 2010, from http://altairdata.wordpress.com/2010/06/08/does-customer-intelligence-include-prospecting/

Corporate IT Update *(2010, January11)*. Retail Anywhere integrates DigitalPersona's fingerprint biometrics.

Corvi, E., & Bonera, M. (2005). *Web orientation and value chain evolution in the tourism industry*. Paper n. 51, Dipartimento di economia aziendale, Università degli studi di Brescia, Italy.

Cova, B., & Pace, S. (2006). Brand community of convenience products: New forms of custodies empowerment. *European Journal of Marketing*, *40*(9-10), 1087–1105. doi:10.1108/03090560610681023

Cox, R., & Brittain, P. (2004). *Retailing – an introduction* (5th ed.). Prentice Hall / Pearson.

Craig, A. B., Sherman, W. R., & Will, J. D. (2009). *Developing virtual reality applications. Foundations of effective design*. Oxford, UK: Elsevier.

Craik, K. (1943). *The nature of explanation*. Cambridge, UK: Cambridge University Press.

Cranor, L. F., Reagle, J., & Ackerman, M. S. (1999). Beyond concern: Understanding Net users' attitudes about online privacy. *Proceedings of the 1st ACM Conference of Electronic Commerce* (pp.1-8). New York, NY: ACM.

Creasey, S. (2009, May 9). Technology that may seem like science fiction today could become commonplace in the stores of tomorrow. *Grocer*, 46-50.

Cristopher, M., Payne, A., & Ballantyne, D. (1991). *Relationship marketing*. Butterworth-Henemann.

Cronbach, L. J. (1951). Coefficient alpha and the internal structure of tests. *Psychometrica*, *16*(3), 297–334. doi:10.1007/BF02310555

Cronin, J. J., & Taylor, S. A. (1992). Measuring service quality: A re-examination and extension. *Journal of Marketing*, *56*, 55–68. doi:10.2307/1252296

Cronin, J. J., Brady, M. K., & Hult, G. T. M. (2000). Assessing the effects of quality, value, and customer satisfaction on consumer behavioral intentions in service environments. *Journal of Retailing*, *76*(2), 193–218. doi:10.1016/S0022-4359(00)00028-2

Cross, R., & Sproull, L. (2004). More than an answer: Information relationships for actionable knowledge. *Organization Science*, *15*(4), 446–462. doi:10.1287/orsc.1040.0075

Culnan, M. J., & Armstrong, P. K. (1999). Information privacy concerns, procedural fairness and impersonal trust: An empirical investigation. *Organization Science*, *10*, 104–115. doi:10.1287/orsc.10.1.104

Culnan, M. J. (1993). How did they get my name? An exploratory investigation of consumer attitudes toward secondary information use. *Management Information Systems Quarterly*, *17*(3), 341–363. doi:10.2307/249775

Cummings, J. N. (2004). Work groups, structural diversity, and knowledge sharing in a global organization. *Management Science*, *50*(3), 352–364. doi:10.1287/mnsc.1030.0134

Cunningham, S. J., & Jones, M. (2005). Autoethnography: A tool for practice and education. In *Proceedings of the 6th ACM International Conference on Computer-human interaction*.

Curiel, J. (2006, June 25). The last days of privacy. *SFGate.com*, 1-6.

Currall, S. C., & Judge, T. A. (1995). Measuring trust between organizational boundary role persons. *Organizational Behavior and Human Decision Processes, 64*, 151–170. doi:10.1006/obhd.1995.1097

Curran, J. M., & Meuter, M. L. (2005). Self-service technology adoption: Comparing three technologies. *Journal of Services Marketing, 19*(2), 103–113. doi:10.1108/08876040510591411

Curry, A., & Penman, S. (2004). The relative importance of technology in enhancing customer relationships in banking – a Scottish perspective. *Managing Service Quality, 14*(4), 331–341. doi:10.1108/09604520410546860

Dabholkar, P. A., Bobbitt, L. M., & Lee, E.-J. (2003). Understanding consumer motivation and behavior related to self-scanning in retailing. Implications for strategy and research on technology-based self-service. *International Journal of Service Industry Management, 14*(1), 59–96. doi:10.1108/09564230310465994

Dabholkar, P. A. (1996). Consumer evaluations of new technology-based self-service options: An investigation of alternative models of service quality. *International Journal of Research in Marketing, 13*(1), 29–51. doi:10.1016/0167-8116(95)00027-5

Dabholkar, P. A., & Bagozzi, R. P. (2002). An attitudinal model of technology-based self-service: Moderating effects of consumer traits and situational factors. *Journal of the Academy of Marketing Science, 30*(3), 184–201.

Dahlén, M., & Lange, F. (2002). Real consumers in the virtual store. *Scandinavian Journal of Management, 18*(3), 341–363. doi:10.1016/S0956-5221(01)00012-4

Dalgety, D. (2002). *Online travel continues to grow*. Retrieved from Visitearth.com

Davenport, T. H. (2005). *Thinking for a living: How to get better performance and results from knowledge workers*. Boston, MA: Harvard Business School Press.

Davenport, T. H., Harris, J. G., & Kohli, A. K. (2001). How do you know your customers so well? *Sloan Management Review, 42*, 63–73.

Davenport, J. L., & Potter, M. C. (2004). Scene consistency in object and background perception. *Psychological Science, 15*(8), 559–564. doi:10.1111/j.0956-7976.2004.00719.x

Davenport, H. (1999). Groups and teams. In Brook, S. I. (Ed.), *Organisational behaviour*. London, UK: Financial Times Pitman publishing.

Davis, D. W., & Silver, B. D. (2004). Civil liberties vs. security. *American Journal of Political Science, 48*(1), 28–46. doi:10.1111/j.0092-5853.2004.00054.x

Davis, F. D., Bagozzi, R. P., & Warshaw, P. R. (1989). User acceptance of computer technology: a comparison of two theoretical models. *Management Science, 35*(3), 982–1003. doi:10.1287/mnsc.35.8.982

Davis, F. D. (1989). Perceived usefulness, perceived ease of use, and user acceptance of information technology. *Management Information Systems Quarterly, 13*(3), 319–340. doi:10.2307/249008

Davis, F. D. (1989). Perceived usefulness, perceived ease of use and user acceptance of information technology. *Management Information Systems Quarterly, 13*(3), 319–339. doi:10.2307/249008

Davis, J. M., Kettinger, W. J., & Kuney, D. G. (2009). When users are IT experts too. *European Journal of Information Systems, 18*, 1.

Davis, F. D. (1986). *A technology acceptance model for empirically testing new end-user systems: Theory and results*. Cambridge, MA: MIT.

Day, G. S. (1994). The capabilities of market-driven organizations. *Journal of Marketing, 58*(4), 37–52. doi:10.2307/1251915

De Dreu, C. K. W., & Van Vianen, A. E. M. (2001). Managing relationship conflict and effectiveness of organizational teams. *Journal of Organizational Behavior, 22*, 309–328. doi:10.1002/job.71

De Moerloose, C., Antioco, M., Lindgreen, A., & Palmer, R. (2005). Information kiosks: The case of the Belgian retail sector. *International Journal of Retail & Distribution Management, 33*(6), 472–490. doi:10.1108/09590550510603651 de Vita, P., Mercurio, R., & Testa, F. (2007). *Organizzazione aziendale: Assetto e meccanismi di relazione*. Torino, Italy: G. Giappichelli Editore.

Dean, D. H. (2008). Shopper age and the use of self-service technologies. *Managing Service Quality*, *18*(3), 225–238. doi:10.1108/09604520810871856

Dean, J. W., & Sharfman, M. P. (1996). Does decision process matter? A study of strategic decision-making effectiveness. *Academy of Management Journal*, *39*(2), 368–396. doi:10.2307/256784

DeCew, J. (1997). *In pursuit of privacy. Law, ethics, and the rise of technology*. Ithaca, NY: Cornell University Press.

Deck, S. (2001). *What is CRM?* Retrieved June 15, 2010, from www.cio.com/research/crm/edit/crmabc.html

Decker, R., & Höppner, M. (2006). Strategic planning and customer intelligence in academic libraries. [from http://www.emeraldinsight.com/Insight/ViewContentServlet?Filename=Publisher]. *Library Hi Tech*, *24*(4), 504–514. Retrieved June 15, 2010. doi:10.1108/07378830610715374

Dellaert, G. C., & Arentze, T, A., Bierlaire, M., Borgers, A. W. J., & Timmermans, H. J. P. (1998). Investigating consumers' tendency to combine multiple shopping purposes and destinations. *JMR, Journal of Marketing Research*, *35*(2), 177–189. doi:10.2307/3151846

Dellarocas, C. (2003). The digitalization of word of mouth: Promise and challenges of online feedback mechanisms. *Management Science*, *49*(10), 1407–1424. doi:10.1287/mnsc.49.10.1407.17308

DeLone, W. H., & McLean, E. R. (1992). Information systems success: The quest for the dependent variable. *Information Systems Research*, *3*(1), 60–95. doi:10.1287/isre.3.1.60

DeLone, W. H., & McLean, E. R. (2003). The DeLone and Mc Lean model of Information Systems success: A ten-year update. *Journal of Management Information Systems*, *19*(4), 9–30.

Denham, P. (1993). Nine- to fourteen-year-old children's conception of computers using drawings. *Behaviour & Information Technology*, *12*(6), 346–358. doi:10.1080/01449299308924399

Denis, J. L., Heber, Y., Langley, A., & Trottier, L. H. (2002). Explaining diffusion patterns for complex health care innovations. *Health Care Management Review*, *27*(3), 60–73.

Dennis, C., Murphy, J., Marsland, D., Cockett, W., & Patel, T. (2002). Measuring image: Shopping centre case studies. *International Review of Retail, Distribution and Consumer Research*, *12*(4), 353–373.

Dennis, C., Newman, A., Michon, R., Brakus, J., & Wright, L. T. (2010). The mediating effects of perception and emotion: Digital signage in mall atmospherics. *Journal of Retailing and Consumer Services*, *17*(3), 205–215. doi:10.1016/j.jretconser.2010.03.009

DeSanctis, G., & Fulk, J. (1999). *Shaping organizational forms*. Thousand Oaks, CA: Sage.

DeSanctis, G., & Scott Poole, M. (1994). Capturing the complexity in advanced technology use: Adaptive structuration theory. *Organization Science*, *5*(2), 121–147. doi:10.1287/orsc.5.2.121

Deshpande, R. (1983). Paradigms lost: On theory and method in research in marketing. *Journal of Marketing*, *47*, 101–110. doi:10.2307/1251403

Detlor, B., Sproule, S., & Gupta, C. (2003). Pre-purchase online information seeking: Search versus browse. *Journal of Electronic Commerce Research*, *4*(2), 72–84.

Devaraj, S., & Kohli, R. (2003). Performance impacts of Information Technology: Is actual usage the missing link? *Management Science*, *49*(3), 273–289. doi:10.1287/mnsc.49.3.273.12736

DHI. (1991). *Anforderungen des Handels an Verpackungen*. Cologne, Germany: DHI.

DHI. (1991). *Verpackungsverordnung und Entsorgung im Handel*. Cologne, Germany: DHI.

DHI. (1991). *Verpackung und Umwelt im Handel*. Cologne, Germany: DHI.

DHI. (1992). *Auf dem Weg zur: Direkten Produkt-Rentabilität*. Cologne, Germany: DHI.

Diep, V. C. S., & Sweeney, J. C. (2008). Shopping trip value: Do stores and products matter? *Journal of Retailing and Consumer Services*, *15*, 399–409. doi:10.1016/j.jretconser.2007.10.002

Digital Transaction News. (2007, November 15). Why Piggly Wiggly decided to roll out biometric payments (pp. 1-3).

Dijksterhuis, A., Smith, P. K., van Baaren, R. B., & Wigboldus, D. H. J. (2005). The unconscious consumer: Effects of environment on consumer behaviour. *Journal of Consumer Psychology*, *15*(3), 193–202. doi:10.1207/s15327663jcp1503_3

Dimmock, B. (2003, June). Will RFID replace the UPC? *Stores*, 88-90.

Dirks, K. T., & Ferrin, D. L. (2001). The role of trust in organizational settings. *Organization Science*, *12*(4), 450–467. doi:10.1287/orsc.12.4.450.10640

Donio, J., Massari, P., & Passiante, G. (2006). Customer satisfaction and loyalty in a digital environment: An empirical test. *Journal of Consumer Marketing*, *23*(7), 445–457. doi:10.1108/07363760610712993

Donovan, R. J., & Rossiter, J. R. (1982). Store atmosphere: An environmental psychology approach. *Journal of Retailing*, *58*, 34–57.

Donovan, R. J., Rossiter, J. R., Marcoolyn, G., & Nesdale, A. (1994). Store atmosphere and purchasing behaviour. *Journal of Retailing*, *70*(3), 283–294. doi:10.1016/0022-4359(94)90037-X

Doucet, G., Gøtze, J., Saha, P., & Bernard, S. (2009). *Coherency management – architecting the enterprise for alignment, agility and assurance*. AuthorsHouse.

Drennan, J., & McColl-Kennedy, J. R. (2003). The relationship between Internet use and perceived performance in retail and professional service firms. *Journal of Services Marketing*, *17*(3), 295–311. doi:10.1108/08876040310474837

Dubé, L., & Morin, S. (2001). Background music pleasure and store evaluation: intensity effects and psychological mechanisms. *Journal of Business Research*, *54*(2), 107–113. doi:10.1016/S0148-2963(99)00092-2

Dulude, L. (2002). Automated telephone answering systems and aging. *Behaviour & Information Technology*, *21*(3), 171–184. doi:10.1080/0144929021000013482

Dwyer, C., Hiltz, S. R., & Passerini, K. (2007). Trust and privacy concern within social networking sites: A comparison of Facebook and MySpace. *Proceedings of AMCIS 2007*, Keystone, CO. Retrieved September 21, 2007 from http://csis.pace.edu/~dwyer/research/DwyerAMCIS2007.pdf

Ebster, C., Wagner, U., & Valis, S. (2006). The effectiveness of verbal prompts on sales. *Journal of Retailing and Consumer Services*, *13*(3), 169–176. doi:10.1016/j.jretconser.2005.08.003

Edwards, S., & Shackley, M. (1992). Measuring the effectiveness of retail window display as an element of the marketing mix. *International Journal of Advertising*, *11*, 193–202.

EHI. (1993). *Mehrwegverpackungen*. Cologne, Germany: EHI.

EHI. (1994). *Scannersysteme- Neue Impulse für Organisation und Marketing*. Cologne, Germany: EHI.

EHI. (1997). *Elektronische Identifikation und Rückverfolgbarkeit von Tieren*. Cologne, Germany: EHI.

EHI. (1999). *Eurep GAP, Verification 2000*. Cologne, Germany: EHI.

EHI. (2000). *Eurepgap, Implementation*. Cologne, Germany: EHI.

EHI. (2001). *Transparenz in der Wertschöpfungskette: Fleisch*. Cologne, Germany: EHI.

EHI. (2005). *Tracing and tracking*. Cologne, Germany: EHI.

Ekman, P. Friesen, & W. V., & Hager, J. C. (2002). *The facial action coding system*. London, UK: Weidenfeld & Nicolson.

Elder, J. (2005, June 15). Ready to whip out your finger to pay? *Charlotte Observer*, p. 1A. *eWeek*. (2006, January 30). *Consumers resist retail biometrics*.

Ellen, P. S., & Bone, P. F. (1998). Does it matter if it smells? Olfactory stimuli as advertising executional cues. *Journal of Advertising*, *27*(4), 29–39.

Ellram, L. M., La Londe, B. J., & Weber, M. M. (1999). Retail logistics. *International Journal of Physical Distribution & Logistics Management*, *29*(7/8), 477–494. doi:10.1108/09600039910291993

Elmaghraby, W., & Keskinocak, P. (2003). Dynamic pricing in the presence of inventory considerations: Research overview, current practices, and future directions. *Management Science*, *49*(10), 1287–1309. doi:10.1287/mnsc.49.10.1287.17315

Epstein, S. (1993). Emotion and self-theory. In Lewis, M., & Haviland, J. M. (Eds.), *Handbook of emotions*. New York, NY: McGraw Hill.

Esslimani, I., Brun, A., & Bayer, A. (2008). Behavioral similarities for collaborative recommendations. *Journal of Digital Information Management*, *6*(6), 442–448.

Fabris, G. (2008). *Societing*. Milano, Italia: Egea.

Fairnet, A., & Plonchard, E. (2002). *Customer relationship management*. Milano, Italy: Etas.

Falkman, M. A. (2005, July). Future store shows off tomorrow's technology. *Packaging Digest*, 40-43.

Farris, J. S., Jones, K. S., & Elgin, P. D. (2002). Users' schemata of hypermedia: What is so 'spatial' about a website? *Interacting with Computers*, *14*(5), 487–502. doi:10.1016/S0953-5438(02)00011-5

Faul, F., Erdfelder, E., Lang, A.-G., & Buchner, A. (2007). G*Power 3: A flexible statistical power analysis program for the social, behavioral, and biomedical sciences. *Behavior Research Methods*, *39*(2), 175–191. doi:10.3758/BF03193146

Featherman, M. S., & Pavlou, P. A. (2003). Predicting e-services adoption: A perceived risk facets perspective. *Journal of Human-Computer Studies*, *59*(4), 451–474. doi:10.1016/S1071-5819(03)00111-3

Federal Trade Commission. (1996). *A report by the staff of the Federal Trade Commission, Technology Services of the National Institute of Standards and Technology, the States of Florida, Michigan, Tennessee, Vermont, and Wisconsin and the Commonwealth of Massachusetts*. Washington, DC: Federal Trade Commission.

Federal Trade Commission. (1998). *Price check II: A follow-up report on the accuracy of checkout scanner prices*. Washington, DC: Federal Trade Commission.

Fedorikhin, A., & Cole, C. A. (2004). Mood effects on attitudes, perceived risk and choice: Moderators and mediators. *Journal of Consumer Psychology*, *14*(1&2), 2–12. doi:10.1207/s15327663jcp1401&2_2

Feng, C., & Yuang, C. (2007). Application of collaborative transportation management to global logistics: An interview case study. *International Journal of Management*, *24*(4), 623–636.

Ferri, M. A., & Celant, A. (2009). I canali della distribuzione. In *L'Italia. Il declino economico e la forza del turismo*. Italy: Marchesi.

Ferri, M. A., Basile, G., & Aiello, L. (2010, April). *The relationship between integrated cultural product brand and individual: A conceptual framework*. Paper presented at the 7th CIRCLE Conference for Consumer Behavior and Retailing Research, Estoril Higher Institute for Tourism and Hotel Studies (ESHTE) Estoril, Portugal.

Finn, A., & Louvière, J. J. (1996). Mall perception, consideration and choice: Anchor store contribution. *Journal of Business Research*, *35*, 241–251. doi:10.1016/0148-2963(95)00129-8

Fishbein, M., & Ajzen, I. (1975). *Belief, attitude, intention and behavior: An introduction to theory and research*. Reading, MA: Addison-Wesley.

Fisher, S. G., Hunter, A., & Macrosson, W. D. K. (2001). A validation study of Belbin's team roles. *European Journal of Work and Organizational Psychology*, *10*(2), 121–144. doi:10.1080/13594320143000591

Fittkau and Maaß. (2010). *World Wide Web-Benutzer-Analyse (W3B), 29.* Fittkau and Maaß. Retrieved from www.w3b.org

Fitzgerald, L., Ferlie, E., Wood, M., & Hawkins, C. (2002). Interlocking interactions, the diffusion of innovations in health care. *Human Relations*, *55*(12), 1429–1449. doi:10.1177/001872602128782213

Fitzsimons, G. J., & Lehmann, D. R. (2004). Reactance to recommendations: When unsolicited advice yields contrary responses. *Marketing Science*, *23*(1), 82–94. doi:10.1287/mksc.1030.0033

Fleder, D., & Hosanagar, K. (2009). Blockbuster culture's next rise or fall: The impact of recommender systems on sales diversity. *Management Science*, *55*(5), 697–712. doi:10.1287/mnsc.1080.0974

Fleisher, C. S., Wright, S., & Allard, H. T. (2008). The role of insight teams in integrating diverse marketing information management techniques. *European Journal of Marketing*, *42*(7/8), 836–851. doi:10.1108/03090560810877187

FMI SuperMarket Research. (2003). Check it out—getting the best of the front end. *FMI SuperMarket Research, 5*(1).

Fodness, D., & Murray, B. (1997). Tourist information search. *Annals of Tourism Research*, *24*, 503–523. doi:10.1016/S0160-7383(97)00009-1

Ford, D., Gadde, L.-E., Håkansson, H., & Snehota, I. (2003). *Managing business relationships*. Wiley.

Ford, D. (Ed.). (1990). *Understeing business markets: Interaction, relationship e network*. London, UK: Academic Press.

Fornell, C., & Larcker, D. F. (1981). Evaluating structural equation models with unobservable variables and measurement error. *JMR, Journal of Marketing Research*, *18*(1), 39–50. doi:10.2307/3151312

Fox, B. (2002). Electronic shelf labels draw new interest as prices fall. *Stores*, January.

Foxall, G. R., & Soriano, M. Y. (2005). Situational influences on consumer's attitudes and behaviour. *Journal of Business Research*, *58*(4), 518–525. doi:10.1016/S0148-2963(03)00142-5

France, F., & Sclormcy, D. (2009). Biometrics improving financial accessibility. *Biometric Technology Today*, July-August, 10-12.

Franco, M. (1991). *I piccoli gruppi strutturati. Impatto sui modelli organizzativi aziendali*. Torino, Italy: G. Giappichelli Editore.

Franco, M. (2007). I livelli dell'attore organizzativo: gruppo. In de Vita, P., Mercurio, R., & Testa, F. (Eds.), *Organizzazione aziendale: Assetto e meccanismi di relazione*. Torino, Italy: G. Giappichelli Editore.

Franco, M., Di Virgilio, F., & Di Pietro, L. (2006, April). *A qualitative analysis of conflict types: An intragroup conflict*. Paper presented at the meeting of 10th International Workshop on Team working, EIASM, University of Groningen, Groningen, The Netherlands.

Franco, M., Di Virgilio, F., & Di Pietro, L. (2007, April). *Diversity in work groups: Effects on the level of conflict*. Paper presented at the meeting Workshop on Organisational Design, European Institute for Advanced Studies in Management, Bruxelles, Belgium.

Frasquet, M., Cervera, A., & Gil, I. (2008). The impact of IT and customer orientation on building trust and commitment in the supply chain. *International Review of Retail, Distribution and Consumer Research*, *18*(3), 343–359. doi:10.1080/09593960802114164

Friedman, B., Kahn, P. H., & Howe, D. C. (2000). Trust online. *Communications of the ACM*, *43*(12), 34–40. doi:10.1145/355112.355120

Fugate, D. L. (2007). Neuromarketing: A layman's look at neuroscience and its potential application to marketing practice. *Journal of Consumer Marketing*, *24*(7), 385–394. doi:10.1108/07363760710834807

Gaeth, G. J., Levin, I. P., Chakraborty, G., & Levin, A. M. (1991). Consumer evaluation of multi-product bundles: An information integration analysis. *Marketing Letters*, *2*(1), 47–57. doi:10.1007/BF00435195

Gale, B. T., & Chapman, R. W. (1994). *Managing customer value: Creating quality and service that customers can see*. New York, NY: Free Press.

Gallego, G., & van Ryzin, G. J. (1994). Optimal dynamic pricing of inventories with stochastic demand over finite horizon. *Management Science*, *40*(8), 999–1020. doi:10.1287/mnsc.40.8.999

Garbarino, E., & Strahilevitz, M. (2004). Gender difference in the perceived risk of buying online and the effects of receiving a site recommendation. *Journal of Business Research*, *57*(7), 768–775. doi:10.1016/S0148-2963(02)00363-6

García-Crespo, A., Chamizo, J., Rivera, J., Mencke, M., Colomo-Palacios, R., & Gómez-Berbís, J. M. (2008). SPERA: Social pervasive e-tourism advisor. *Telematics and Informatics*, *26*(3), 306–315. doi:10.1016/j.tele.2008.11.008

Garland, R. (1992). Pricing errors in the supermarket: Who pays? *International Journal of Retail and Distribution Management*, *20*(1), 25–30.

Garlin, F. V., & Owen, K. (2006). Setting the tone with the tune, a meta-analytic review of the effects of background music in retail settings. *Journal of Business Research*, *59*, 755–764. doi:10.1016/j.jbusres.2006.01.013

Gartner (2008). *IBM expands retail SOA foundation to enterprise level*. Retrieved September 1, 2010, from ftp://public.dhe.ibm.com/REL03003USEN.PDF

Gartner. (2010). *Dataquest alert: Lead IT spending growth in challenging* 2009. Retrieved September 1, 2010, from http://www.lesechos.fr/medias/2009/0227//300332929.pdf

Gathright, A. (2001, October 30). Biometric technology raises hopes, fears and skepticism. *San Francisco Chronicle*, A1.

Gefen, D., & Straub, D. (2000). The relative importance of perceived ease of use in IS adoption: A study of e-commerce adoption. *Journal of the Association for Information Systems*, *1*, 1–30.

Gefen, D. (2000). E-commerce: The role of familiarity and trust. *The International Journal of Management Science*, *28*(6), 725–737.

Gefen, D., Karahanna, E., & Straub, D. W. (2003). Trust and TAM in online shopping: An integrated model. *Management Information Systems Quarterly*, *27*(1), 51–90.

Gefen, D., Straub, D., & Boudreau, M.-C. (2000). Structural equation modeling techniques and regression: Guidelines for research practice. *Communications of AIS*, *7*(7), 1–78.

Gemperlein, J. (2006, June 18). Self-checkout? Just you wait. *Washington Post*.

Gentner, D., & Stevens, A. L. (1983). *Mental models*. Hillsdale, NJ: Lawrence Erlbaum Associates.

Ghosh, A. K. (1998). *E-commerce security: Weak links, best defenses*. John Wiley and Sons.

Giacomazzi, F., & Camisani-Calzolari, M. (2008). *Impresa 4.0. Marketing e comunicazione digitale a 4 direzioni*. Prentice Hall, Financial Times.

Gibson, J. J. (1986). *The ecological approach to visual perception*. Hillsdale, NJ: Lawrence Erlbaum Associates.

Gil, I., Cervera, A., & Frasquet, M. (2007). Empleo de ICT y efectos relacionales en la cadena logística. *Boletín Económico de ICE*, *2914*, 31–48.

Gil, I., Ruiz, E., & Servera, D. (2008a). Logistic service quality and technology: A comparison between supplier-retailer and retailer-consumer relationships. *International Review of Retail, Distribution and Consumer Research*, *18*(5), 495–510. doi:10.1080/09593960802573385

Gil, I., Servera, D., Berenguer, G., & Fuentes, M. (2008b). Logistics service quality: A new way to loyalty. *Industrial Management & Data Systems Journal*, *108*(5), 650–668. doi:10.1108/02635570810876778

Gilad, B., & Gilad, T. (1985). A system approach to business intelligence. *Business Horizons*, *8*(5), 65–70. doi:10.1016/0007-6813(85)90070-9

Gilbert, D. (2003). *Retail marketing management*. England: Pearson Education Limited.

Gillin, P. (2007). *The new influencers, a marketer's guide to the new social media*. Quill Driver Books\Word Dancer Press, Inc, CA, USA.

Gil-Saura, I., Berenguer-Congri, G., & Ruiz-Molina, M. (2009). Information and communication technology in retailing: A cross-industry comparison. *Journal of Retailing and Consumer Services*, *16*, 232–238. doi:10.1016/j.jretconser.2008.11.018

Go, F. M. (1992). The role of computerized reservation systems in the hospitality industry. *Tourism Management*, *13*(1), 22–26. doi:10.1016/0261-5177(92)90026-4

Golden, L. G., & Zimmerman, D. A. (1986). *Effective retailing*. Boston, MA: Houghton Mifflin.

Golden, L. L., & Johnson, K. A. (1983). The impact of sensory preferences and thinking vs. feeling appeals on advertising effectiveness. In Bagozzi, R. P., & Tybout, A. M. (Eds.), *Advances in consumer research, 10* (pp. 203–208). Ann Arbor, MI: Association for Consumer Research.

Golinelli, G. M. (2010). *Viable systems approach (VSA), governing business dynamics* (pp. 28–396). Padova, Italy: Cedam.

Goodstein, R. C., & Escalas, J. E. (1995). Improving pricing accuracy at the supermarket: Electronic shelving systems and public policy. *Journal of Public Policy & Marketing, 14*(2), 216–225.

Goodwin, C. (1991). Privacy: Recognition of a consumer right. *Journal of Public Policy & Marketing, 10*(1), 149–166.

Gordon, I. (1999). *Relationship marketing: New strategies, techniques and technologies to win the customers you want and keep them forever*. John Wiley and Sons Publishers.

Gorn, G. J., Goldberg, M. E., & Basu, K. (1993). Mood, awareness and product evaluation. *Journal of Consumer Psychology, 2*, 237–256. doi:10.1016/S1057-7408(08)80016-2

Grabner-Kraeuter, S. (2002). The role of consumers' trust in online-shopping. *Journal of Business Ethics, 39*(1/2), 43–50. doi:10.1023/A:1016323815802

Graham, E. (1988, May 13). The call of the mall. *Wall Street Journal*, 7R.

Granbois, D. H. (1968). Improving the study of customer in-store behavior. *Journal of Marketing, 32*, 28–32. doi:10.2307/1249334

Grant, D., Hall, R., Wailes, N., & Wright, C. (2006). The false promise of technological determinism. *New Technology, Work and Employment*, 21.

Gratzer, M., & Winiwarter, W. (2003). A framework for competitive advantage in e-tourism. In *Proceedings of the 10th International Conference on Information Technology and Travel & Tourism*, Springer-Verlag.

Green, A. J. K. (1994). Interacting cognitive subsystems: A framework for considering the relationship between performance and knowledge representation. *Interacting with Computers, 6*(1), 61–85. doi:10.1016/0953-5438(94)90005-1

Greenfield, E. (1992). OMR and barcode technologies: empowering education. *Technological Horizons in Education, 19*(7), 10.

Greenhalgh, T., Robert, G., Macfarlane, F., Bate, P., & Kyriakidou, O. (2004). Diffusion of innovation in service organizations: Systematic review and recommendations. *The Milbank Quarterly, 82*(4), 581–629. doi:10.1111/j.0887-378X.2004.00325.x

Gregor, S., & Benbasat, I. (1999). Explanations from intelligent systems: Theoretical foundations and implications for practice. *Management Information Systems Quarterly, 23*(4), 497–530. doi:10.2307/249487

Gretzel, U. (2006). Consumer generated content: Trends and implication of consumer behaviour. In Buhalis, D., & Costa, C. (Eds.), *Tourism business frontiers* (pp. 9–18). Butterworth-Heinernann. doi:10.1016/B978-0-7506-6377-9.50009-2

Grier, S. A., & Deshpandé, R. (2001). Social dimensions of consumer distinctiveness: The influence of social status on group identity and advertising persuasion. *JMR, Journal of Marketing Research, 38*(2), 216–224. doi:10.1509/jmkr.38.2.216.18843

Grimes, W. (2004, April 7). When the cashier is you. *New York Times*, F1.

Grocer. (2007, March 3). Self-checkout tills win for Henderson. *Grocer*, 16.

Groeppel, A., & Bloch, B. (1990). An investigation of experience-oriented consumers in retailing. *International Review of Retail, Distribution and Consumer Research, 1*(1), 101. doi:10.1080/09593969000000008

Gronroos, C. (1997). From marketing mix to relationship marketing: Towards a paradigm shift in marketing. *Management Decision, 35*(4), 322–339. doi:10.1108/00251749710169729

Grover, V., & Davenport, T. H. (2001). General perspectives on knowledge management: Fostering a research agenda. *Journal of Management Information Systems*, *18*(1), 5–21.

Gruenfeld, D. H., Mannix, E. A., Williams, K. Y., & Neale, M. A. (1996). Group composition and decision making: How member familiarity and information distribution affect process and performance. *Organizational Behavior and Human Decision Processes*, *67*(1), 1–15. doi:10.1006/obhd.1996.0061

GSI. (2010). *Danish retailer powers North American expansion with retail*, 1. Retrieved September 1, 2010, from http://www.gsi.ca/panda/files/Bestseller_web.pdf

Guatri, L., & Massari, M. (1992). *La diffusione del valore*. Milano, Italy: Egea.

Gulz, A. (2004). Benefits of virtual characters in computer based learning environments: Claims and evidence. *International Journal of Artificial Intelligence in Education*, *14*, 313–334.

Gulz, A., & Haakeb, M. (2006). Design of animated pedagogical agents-a look at their look. *International Journal of Human-Computer Studies*, *64*, 322–339. doi:10.1016/j.ijhcs.2005.08.006

Gummesson, E. (1994). Marketing relationship marketing operational. *International Journal of Service Industry Management*, *5*(5), 5–20. doi:10.1108/09564239410074349

Gummesson, E. (2008). *Total relationship marketing*. Oxford, UK: Butterworth-Heinemann, Elsevier Ltd.

Gurevych, I., & Mühlhäuser, M. (2007). Natural language processing for ambient intelligence. *Künstliche Intelligenz*, *2*, 10–16.

Guttentag, D. A. (2010). Virtual reality: Applications and implications for tourism. *Tourism Management*, *31*(5), 637–651. doi:10.1016/j.tourman.2009.07.003

Gwizdka, J., & Spence, I. (2007). Implicit measures of lostness and success in Web navigation. *Interacting with Computers*, *19*(3), 357–369. doi:10.1016/j.intcom.2007.01.001

Ha, S., & Stoel, L. (2009). Consumer e-shopping acceptance: Antecedents in a technology acceptance model. *Journal of Business Research*, *62*(5), 565–571. doi:10.1016/j.jbusres.2008.06.016

Ha, S., & Stoel, L. (2009). Consumer e-shopping acceptance: Antecedents in a technology acceptance mode. *Journal of Business Research*, *62*(5), 565–571. doi:10.1016/j.jbusres.2008.06.016

Hackett, P. M. W., & Foxall, G. R. (1994). A factor analytic study of consumers' location specific values, a traditional high street and a modern shopping mall. *Journal of Marketing Management*, *10*, 163–178. doi:10.1080/0267257X.1994.9964266

Hallier, B. (1987). Sich ins Regal hineinrechnen. *Absatzwirtschaft*, *10*, 87.

Hallier, B. (2001). *Praxisorientierte Handelsforschung*. Cologne, Germany: EHI.

Hallier, B. (2004). *EuroShop*. Cologne, Germany: EHI.

Hallier, B. (2009). *Modern stores*. Moscow, Russia: Interexpert.

Hallier, B. (2010). Systemkonkurrenz zwischen stationären Handel und Versandhandel. In Mattmüller, R. (Ed.), *Versand-Handels-Marketing*. Frankfurt, Germany: Deutscher Fachverlag.

Hallier, B. (1999). Wird ECR zum Club der Großen? In von der Heydt, A. (Ed.), *Handbuch Efficient Consumer Response*. Munich, Germany.

Hallier, B. (1995). Der Handel auf dem Weg zur Marketingführerschaft. *Absatzwirtschaft, 3*.

Hallier, B. (1997). Wal Mart-Mythos führt zur falschen ECR-Positionierung. In *Dynamik im Handel, 4*. Cologne, Germany.

Hangland, A., & Cimbalo, R. S. (1997). Human ethology, age and sex differences in mall walking. *Perceptual and Motor Skills*, *85*(3), 845–846.

Hannah, J. (2005). Privacy concerns, expense keep biometrics out of U.S. ATMs. *Information Week* October 12.

Hansen, T., & Møller-Jensen, J. (2009). Shopping orientation and online clothing purchases: The role of gender and purchase situation. *European Journal of Marketing, 43*(9/10), 1154–1170. doi:10.1108/03090560910976410

Hansen, H. R., Knotzer, N., & Madlberger, M. (2007). Empfehlungssysteme zur Verkaufsberatung im Internet. State-of-the-Art und Konsumentenakzeptanz. *Wirtschaftsinformatik, 49*(Sonderheft), 50-61.

Harris, L. C., & Goode, M. M. H. (2004). The four levels of loyalty and the pivotal role of trust: A study of online service dynamics. *Journal of Retailing, 80*(2), 139–158. doi:10.1016/j.jretai.2004.04.002

Harrison, J. V., & Andrusiewicz, A. (2004). A virtual marketplace for advertising narrowcast over digital signage networks. *Electronic Commerce Research and Applications, 3*(2), 163–175. doi:10.1016/j.elerap.2004.02.004

Hasan, B., & Ahmed, M. U. (2007). Effects of interface style on user perceptions and behavioral intention to use computer systems. *Computers in Human Behavior, 23*(6), 3025–3037. doi:10.1016/j.chb.2006.08.016

Haubl, G., & Trifts, V. (2000). Consumer decision making in online shopping environment: The effects of interactive decision aids. *Marketing Science, 19*(1), 4–21. doi:10.1287/mksc.19.1.4.15178

Hausman, A. V., & Siekpe, J. S. (2009). The effect of Web interface features on consumer online purchase intentions. *Journal of Business Research, 62*(1), 5–13. doi:10.1016/j.jbusres.2008.01.018

Hausman, A., & Stock, J. R. (2003). Adoption and implementation of technological innovations within long-term relationships. *Journal of Business Research, 56*(8), 681–686. doi:10.1016/S0148-2963(01)00313-7

Hawkins, G. E. (2003). *Customer intelligence*. New York, NY: Breezy Heights Publishing.

Hawkins, D., & Schuchter, A. (1995, June). *Travel marketing on the Internet*. Paper presented at the biannual Conference of the International Academy for the Study of Tourism, 24-30 June, Cairo, Egypt.

Hee-Woomg, K., Xu, Y., & Koh, J. (2004). A comparison of online trust building factors between potential customers and repeat customers. *Journal of the Association for Information Systems, 5*(10), 392–420.

Heidel, B. (1990). *Scannerdaten im Einzelhandelsmarketing*. Wiesbaden, Germany: Gabler Verlag.

Heiman, A., & Muller, E. (1996). Using demonstrations to increase new product acceptance: Controlling demonstration time. *JMR, Journal of Marketing Research, 33*, 422–430. doi:10.2307/3152213

Hendrix, P. E. (1999). Build it, and they will come. *Marketing Management, 8*, 31–35.

Henseler, J., Ringle, C. M., & Sinkovics, R. R. (2009). The use of partial least squares path modeling in international marketing. *Advances in International Marketing, 20*, 277–319.

Herlocker, J. L., Konstan, J. A., & Riedl, J. T. (2002). An empirical analysis of design choices in neighborhood-based collaborative filtering algorithms. *Information Retrieval, 5*(4), 287–310. doi:10.1023/A:1020443909834

Herrington, J. D., & Capella, L. M. (1996). Effects of music in service environments, a field study. *Journal of Services Marketing, 10*(2), 26–41. doi:10.1108/08876049610114249

Hevner, A. R., March, S. T., Park, J., & Ram, S. (2004). Design science in Information Systems research. *Management Information Systems Quarterly, 28*(1), 75–105.

Hewett, T. T., Baecker, R., Card, S., Carey, T., Gasen, J., Mantei, M., et al. Verplank, W. (2009). *ACM SIGCHI curricula for Human-Computer Interaction*. Retrieved June 18, 2010, from http://old.sigchi.org/cdg/index.html

Hildebrandt, L. (1988). Store perception and the prediction of performance in retailing. *Journal of Business Research, 17*, 91–100. doi:10.1016/0148-2963(88)90026-4

Hirschowitz, A. (2001). Closing the CRM loop: The 21st century marketer's challenge: Transforming customer insight into customer value. *Journal of Database Marketing, 10*(2), 168–178.

Hitt, L., & Brynjolfsson, E. (1996). Productivity, business profitability and consumer surplus: Three different measures of Information Technology value. *Management Information Systems Quarterly, 20*(2), 121–142. doi:10.2307/249475

Hoffman, D. L., & Novak, T. P. (1996). Marketing in hypermedia computer-mediated environments: Conceptual foundations. *Journal of Marketing*, *60*(3). doi:10.2307/1251841

Hoffman, D., Novak, T. P., & Peralta, M. A. (1999). Building consumer trust online environment: The case for information privacy. *Communications of the ACM*, *42*(4), 80–85. doi:10.1145/299157.299175

Hole, K., Erlend, D., & Thorsheim, P. (2005). Securing Wi-Fi networks. *Computer*, (July): 27–34.

Holland, J., & Baker, S. M. (2001). Customer participation in site brand loyalty. *Journal of Interactive Marketing*, *15*, 34–45. doi:10.1002/dir.1021

Holzwarth, M., Janiszewski, C., & Neumann, M. M. (2006). The influence of avatars on consumer shopping behavior. *Journal of Marketing*, *70*(4), 19–36. doi:10.1509/jmkg.70.4.19

Hopper, L. (1990). Rattling SABRE new ways to compete on information. *Harvard Business Review*, *68*(3), 118–125.

Hoque, F. (2000). *E-enterprise: Business models, architecture and components*. Cambridge, UK: Cambridge University Press. doi:10.1017/CBO9780511584923

Hornbæk, K. (2006). Current practice in measuring usability: Challenges to usability studies and research. *International Journal of Human-Computer Studies*, *64*(2), 79–102. doi:10.1016/j.ijhcs.2005.06.002

Howard, E. (1992). Evaluating the success of retail out of town shopping centres. *International Review of Retail, Distribution and Consumer Research*, *2*(1), 59–80. doi:10.1080/09593969200000004

Howard, T. (2005). Viral advertising spreads through marketing plans. *USA Today*.

Hsiao, M.-H. (2009). Shopping mode choice: Physical store shopping versus e-shopping. *Transportation Research Part E, Logistics and Transportation Review*, *45*, 86–95. doi:10.1016/j.tre.2008.06.002

Hsu, Y. (2006). The effects of metaphors on novice and expert learners' performance and mental-model development. *Interacting with Computers*, *18*(4), 770–792. doi:10.1016/j.intcom.2005.10.008

Hui, M., Dubé, L., & Chebat, J.-C. (1997). The impact of music on consumers' reaction to waiting for services. *Journal of Retailing*, *73*(1), 87–104. doi:10.1016/S0022-4359(97)90016-6

Humphrey, D. (2010). Retail payments: New contributions, empirical results, and unanswered questions. *Journal of Banking & Finance*, *34*, 1729–1737. doi:10.1016/j.jbankfin.2010.04.008

Hunt, S. D., & Morgan, R. M. (1995). The comparative advantage theory of competition. *Journal of Marketing*, *59*(2), 1–15. doi:10.2307/1252069

Iacobucci, D., Saldanha, N., & Deng, X. (2007). A meditation on mediation, evidence that structural equations models perform better than regressions. *Journal of Consumer Psychology*, *17*(2), 139–153. doi:10.1016/S1057-7408(07)70020-7

Iacobucci, D., Arabie, P., & Bodapati, A. (2000). Recommendation agents on the internet. *Journal of Interactive Marketing*, *14*(3), 2–11. doi:10.1002/1520-6653(200022)14:3<2::AID-DIR1>3.0.CO;2-J

IBM. (2004). Satisfy customer demand for self-service and reap the benefits of self checkout. *IBM Retail Store Solutions*, 1-8.

IBM. (2008). *Design principles*. Retrieved August 19, 2008, from https://www-01.ibm.com/software/ucd/designconcepts/designbasics.html

IBM. (2009). *The IBM retail industry framework. Integration, optimization, analytics*. Retrieved September 1, 2010, from http://www-01.ibm.com/software/industry/retail/

Infosys. (2008). *Enterprise architecture organization transformation enables IT governance and shared vision*. Retrieved September 1, 2010, from http://www.infosys.com

International Organization for Standardization. (1998). *9241-11 Ergonomic requirements for office work with visual display terminals (VDT)s - part 11 guidance on usability*. International Standards for Business, Government and Society.

Investment Weekly News (2009, November 28). Unisys research shows growing global acceptance of biometrics among consumers for protecting identities and personal information. (p. 129).

ISB. (1980). *Physischer Warenfluss*. Cologne, Germany: ISB.

ISB. (1988). *50 Jahre Selbstbedienung, Sonderausgabe*. Cologne, Germany: ISB.

ISB. (1989). *Logistikgerechte Versandverpackungen und Ladungsträger*. Cologne, Germany: ISB.

Ivory, M. Y., & Megraw, R. (2005). Evolution of web site design patterns. *ACM Transactions on Office Information Systems*, *23*(4), 463–497. doi:10.1145/1095872.1095876

Jacoby, J. (2002). Stimulus-organism-response reconsidered: An evolutionary step in modelling (consumer) behaviour. *Journal of Consumer Psychology*, *12*(1), 51–57. doi:10.1207/S15327663JCP1201_05

Jacoby, J., Szybillo, G. S., & Busato-Schach, J. (1977). Information acquisition behavior in brand choice situations. *The Journal of Consumer Research*, *3*, 209–216. doi:10.1086/208669

Jarvenpaa, S. L., & Todd, P. A. (1997). *Is there a future for retailing on the Internet? Electronic marketing and the consumer*. Thousand Oaks, CA: Sage.

Jarvenpaa, S. L., Tractinsky, J., & Vitale, M. (2000). Consumer trust in an Internet store. *Information Technology Management*, *1*(1/2), 45–71. doi:10.1023/A:1019104520776

Jehn, K., Rupert, J., & Nauta, A. (2006). The effects of conflict asymmetry on mediation outcomes. *The International Journal of Conflict Management*, *17*(2), 96–109. doi:10.1108/10444060610736594

Jehn, K. A., & Mannix, E. A. (2001). The dynamic nature of conflict: Longitudinal study of intragroup conflict and group performance. *Academy of Management Journal*, *44*, 238–251. doi:10.2307/3069453

Jin, S.-A. A. (2009). The roles of modality richness and involvement in shopping behaviour in 3D virtual stores. *Journal of Interactive Marketing*, *23*(3), 234–246. doi:10.1016/j.intmar.2009.04.005

Jobson, J. D. (1991). Applied multivariate data analysis: *Vol. 1. Regression and experimental design*. New York, NY: Springer.

Joch, A. (2005). *Capturing customer intelligence*. Retrieved June 15, 2010, from http://www.oracle.com/oramag/profit/05-may/p25feature.html

Johansen, J. I., & Monthelie, C. (1996). *Lojalitetsredovisning*. Gothenburg, Sweden: InfoNet Scandinavia.

Johnson-Laird, P. (1983). *Mental models: Toward a cognitive science of language, inference and consciousness*. Cambridge, MA: Harvard University Press.

Jones, P., Williams, P., Hillier, D., & Comfort, D. (2007). Biometrics in retailing. *International Journal of Retail & Distribution Management*, *35*(3), 217–222. doi:10.1108/09590550710735077

Jones, M. A. (1999). Entertaining shopping experiences, an exploratory investigation. *Journal of Retailing and Consumer Services*, *6*, 129–139. doi:10.1016/S0969-6989(98)00028-9

Jones, P., Clarke-Hill, C., Hillier, D., & Comfort, D. (2005). The benefits, challenges and impacts of radio frequency identification technology (RFID) for retailers in the UK. *Marketing Intelligence & Planning*, *23*(4), 395–402. doi:10.1108/02634500510603492

Jones, P., Comfort, D., & Hillier, D. (2009). *Marketing sustainable consumption within stores: A case study of the UK's leading food retailers*. Sustainability 2009. Retrieved from www.mdpi.com/journal/sustainability

Joshi, G. (2009). *Information Technology for retail*. Oxford Univ. Press.

Jugger, S. (1999). The power of packaging. *Admap Magazine*, October.

Junglas, I. A., & Watson, R. T. (2006). The U-Constructs: Four information drives. *Communications of AIS*, *17*, 2–43.

just-food.com. (2008, April 21). Woolworths launches self-service checkouts.

just-food.com. (2009, October 22). Tesco on defensive over self-serve stores.

Kamaruizaman, Y. (2007). Adoption of travel e-shopping in the UK. *International Journal of Retail and Distribution Management*, *35*(9), 703–719. doi:10.1108/09590550710773255

Kamis, A., Koufaris, M., & Stern, T. (2008). Using an attribute-based decision support system for user-customized products online: An experimental investigation. *Management Information Systems Quarterly*, *32*(1), 159–177.

Kaplan, R. S., & Norton, D. P. (2001). *The strategy-focused organization: How balanced scorecard companies thrive in the new business environment*. Boston, MA: Harvard Business Review.

Karaian, J. (2008, June 2). Back to basics. *CFO Europe Magazine*.

Karakostas, B., Kardaras, D., & Papthanassiou, E. (2005). The state of CRM adoption by the financial services in the UK: An empirical investigation. *Information & Management*, *42*(4), 853–863. doi:10.1016/j.im.2004.08.006

Katsikitis, M. (1997). The classification of facial expressions of emotion: A multidimensional-scaling approach. *Perception*, *26*, 613–626. doi:10.1068/p260613

Kaufmann, H. R. (2004). Approaching the very core of the client: CRM and identity in private banking. In Geberl, S., Kaufmann, H. R., Menichetti, M., & Wiesner, D. (Eds.), *Aktuelle Entwicklungen im Finanzdienstleistungsbereich*. Heidelberg, Germany: Physica-Verlag.

Keegan, S., O'Hare, G. M. P., & O'Grady, M. J. (2008). EasiShop: Ambient intelligence assists everyday shopping. *Journal of Information Science*, *178*(3), 588–611. doi:10.1016/j.ins.2007.08.027

Keil, M., Tan, B. C., Wei, K.-K., Saarinen, T., Tuunainen, V., & Wassenaar, A. (2000). A cross-cultural study on escalation of commitment behaviour in software projects. *Management Information Systems Quarterly*, *24*(2), 299–325. doi:10.2307/3250940

Keller, K. L., & Staelin, R. (1987). Effects of quality and quantity of information and decision effectiveness. *The Journal of Consumer Research*, *14*, 200–213. doi:10.1086/209106

Kerlinger, F. N., & Lee, H. B. (2000). *Foundations of behavioral research*. New York, NY: Harcourt College Publishers.

Kerr, S. (1990). Wayfinding in an electronic database: The relative importance of navigational cues vs. mental models. *Information Processing & Management*, *26*(4), 511–523. doi:10.1016/0306-4573(90)90071-9

Kim, H. Y., & Kim, Y. K. (2008). Shopping enjoyment and store shopping modes: the moderating influence of chronic time pressure. *Journal of Retailing and Consumer Services*, *15*, 410–419. doi:10.1016/j.jretconser.2007.10.003

Kim, H., & Niehm, L. S. (2009). The impact of website quality on information quality, value, and loyalty intentions in apparel retailing. *Journal of Interactive Marketing*, *23*(3), 221–233. doi:10.1016/j.intmar.2009.04.009

Kim, J., Fiore, A. M., & Lee, H.-H. (2007). Influences of online store perception, shopping enjoyment, and shopping involvement on consumer patronage behaviour towards an online retailer. *Journal of Retailing and Consumer Services*, *14*, 95–107. doi:10.1016/j.jretconser.2006.05.001

Kim, B.-D., & Kim, S.-O. (2001). A new recommender system to combine content-based and collaborative filtering systems. *Journal of Database Marketing*, *8*(3), 244–252. doi:10.1057/palgrave.jdm.3240040

Kitios, A., & Kaufmann, H. R. (2009). Status orientation and symbolic consumption of Cypriot consumers. *International Journal of Management Cases*, *11*(2).

Kiviat, B. (2008, March 24). The end of customer service. *Time*, 11.

Kleijnen, M., de Ruyter, K., & Wetzels, M. (2007). An assessment of value creation in mobile service delivery and the moderating role of time consciousness. *Journal of Retailing*, *83*(1), 33–46. doi:10.1016/j.jretai.2006.10.004

Kleijnen, M., Wetzels, M. G. M., & de Ruyter, K. (2004). Consumer acceptance of wireless finance. *Journal of Financial Services Marketing*, *8*(3), 206–217. doi:10.1057/palgrave.fsm.4770120

Kline, R. B. (2005). *Principles and practice of structural equation modeling* (2nd ed.). New York, NY: The Guilford Press.

Knights, D., & Willmott, H. (2007). *Introducing organizational behaviour and management*. Thomson Learning.

Kohler, T., Matzler, K., & Fuller, J. (2009). Avatar-based innovation: Using virtual worlds for real-world innovation. *Technovation*, *29*, 395–407. doi:10.1016/j.technovation.2008.11.004

Komiak, S. Y. X., & Benbasat, I. (2006). The effects of personalization and familiarity on trust and adoption of recommendation agents. *Management Information Systems Quarterly*, *30*(4), 941–960.

Konana, P., & Ray, G. (2007). Physical product reengineering with embedded information technology. *Communications of the ACM*, *50*(10), 72–78. doi:10.1145/1290958.1290965

Kondratjew, N. (1926). Die langen Wellen der Konjunktur. *Archiv für Sozialwissenschaft und Sozialpolitik, 56.*

Königstorfer, J., & Gröppel-Klein, A. (2007). Experiences of failure and anger when using the mobile and wired Internet: The interference of acceptance-and resistance-driving factors. *Marketing Journal of Resource Management*, *29*(1e), 34–47.

Kooser, A. (2005). Touch and go: Fingerprint readers mean better security—and no more tricky passwords. *Entrepreneur*, *33*(3), 30.

Korolishin, J. (2004). ESL for the little guy. *Stores*, October, 71-72.

Kostova, T., & Roth, K. (2002). Adoption of an organizational practice by subsidiaries of multinational corporations: Institutional and relational effects. *Academy of Management Journal*, *45*(1), 215–233. doi:10.2307/3069293

Kotler, P., Armstrong, G., Saunders, J., & Wong, V. (1999). *Principles of marketing* (2nd ed.). Prentice Hall Europe.

Kotler, P., & Keller, K. L. (2006). *Marketing management* (12th ed.). Upper Saddle River, NJ: Pearson Education, Inc.

Kotler, P., Wong, V., Saunders, J., & Armstrong, G. (2005). *Principles of marketing*. London, UK: Pearson Hall.

Koufaris, M., & Hampton-Sosa, W. (2004). The development of initial trust in an online company by new customers. *Information & Management*, *41*(3), 377–397. doi:10.1016/j.im.2003.08.004

Koufaris, M., & Hampton-Sosa, W. (2002). *Customer trust online: Examining the role of the experience with the website*. New York, NY: Zicklin School of Business, Baruch College Department of Statistics and Computer Information Systems Working Paper Series.

Kourouthanassis, P., & Roussos, G. (2003). Developing consumer-friendly pervasive retail systems. *Pervasive Computing*, *2*(2), 32–39. doi:10.1109/MPRV.2003.1203751

Kowatsch, T., & Maass, W. (2011. (accepted for publication)). A Design Model for Knowledge-based Pricing Services in the Retail Industry. *International Journal of Web Engineering and Technology.*

Kowatsch, T., Maass, W., & Fleisch, E. (2011). The role of product reviews on mobile devices for in-store purchases: consumers' usage intentions, costs and store preferences. *International Journal of Internet Marketing and Advertising: Special Issue on Information Systems and Marketing*, *6*(3), 226–243. doi:10.1504/IJIMA.2011.038237

Kowatsch, T., Maass, W., Filler, A., & Janzen, S. (2008). *Knowledge-based bundling of smart products on a mobile recommendation agent.* Paper presented at the IEEE 7th International Conference on Mobile Business (ICMB).

Kowler, E., Anderson, E., Dosher, B., & Blaser, E. (1995). The role of attention in the programming of saccades. *Vision Research*, *35*(13), 1897–1916. doi:10.1016/0042-6989(94)00279-U

Kozinets, R. (1999). E-tribalized marketing? The strategic implications of virtual communities of consumption. *European Management Journal*, *17*(3), 252–264. doi:10.1016/S0263-2373(99)00004-3

Kriebel, C., & Kauffman, R. (1988). *Measuring and modelling the business value of Information Technologies*. ICIT Press.

Kroll, K. M. (2006). Signs of progress. *Stores*, February, 58.

Krug, S. (2005). *Don't make me thin: A common sense approach to Web usability* (2nd ed.). Berkeley, CA: New Riders Publishing.

Kukar-Kinney, M., & Close, A. G. (2010). The determinants of consumers' online shopping cart abandonment. *Journal of the Academy of Marketing Science*, *38*(2), 240–250. doi:10.1007/s11747-009-0141-5

Kulviwat, S., Bruner, G. C. II, & Al-Shuridah, O. (2009). The role of social influence on adoption of high tech innovations: The moderating effect of public/private consumption. *Journal of Business Research*, *62*(7), 706–712. doi:10.1016/j.jbusres.2007.04.014

Kulviwat, S., Bruner, G. C. II, Kumar, A., Nasco, S. A., & Clark, T. (2007). Toward a unified theory of customer acceptance technology. *Psychology and Marketing*, *24*(12), 1059–1084. doi:10.1002/mar.20196

Kuo, C.-F., & Chu, C.-H. (2005). An online ergonomic evaluator for 3D product design. *Computers in Industry*, *56*(5), 91–112. doi:10.1016/j.compind.2005.02.002

Kurata, H., & Yue, X. (2008). Trade promotion mode choice and information sharing in fashion retail supply chains. *International Journal of Production Economics*, *114*, 507–519. doi:10.1016/j.ijpe.2007.05.021

Kurkovsky, S., & Harihar, K. (2006). Using ubiquitous computing in interactive mobile marketing. *Personal and Ubiquitous Computing*, *10*(4), 227–240. doi:10.1007/s00779-005-0044-5

Kurtz, D. L. (2008). *Contemporary marketing* (p. 322). Mason, OH: South-Western Cengage Learning.

Kveraga, K., Ghuman, A. S., & Bar, M. (2007). Top-down predictions in the cognitive brain. *Brain and Cognition*, *65*(2), 145–168. doi:10.1016/j.bandc.2007.06.007

Labuschagne, L., & Eloff, J. (2000). Electronic commerce: The information-security challenge. *Information Management & Computer Security*, *8*(3), 154. doi:10.1108/09685220010372582

Lamnek, S. (2005). *Qualitative Sozialforschung Lehrbuch, 4 auflage. Weihnhein, Germany & Basel*. Switzerland: Beltz Verlag.

Lang, A. (2000). The limited capacity model of mediated message processing. *The Journal of Communication*, *50*(1), 46–70. doi:10.1111/j.1460-2466.2000.tb02833.x

Lankhorst, M. M. (2009). *Enterprise architecture at work*. Springer Verlag. Lewis, N. (2005). *Enterprise integration: A practical guide for the retail solution architect*. Retrieved September 1, 2010, from http://msdn.microsoft.com/en-us/library/aa479078.aspx

Lavidge, R. J., & Steiner, G. A. (1961). A model for predictive measurements of advertising effectiveness. *Journal of Marketing*, *25*(6), 59–62. doi:10.2307/1248516

Law, R., Shansan, Q., & Buhalis, D. (2009). Progress in tourism management: A review of website evaluation in tourism research. *Tourism Management*, *31*, 297–313. doi:10.1016/j.tourman.2009.11.007

Lawrence, B. S. (1997). The black box of organizational demography. *Organization Science*, *8*, 1–22. doi:10.1287/orsc.8.1.1

Lazarus, R. S. (1991). *Emotion and adaptation*. New York, NY: Oxford University Press.

Leana, C. R. (1985). A Partial test of Janis, groupthink model: Effects of group cohesiveness and leader behavior on detective decision making. *Journal of Management*, *11*(1), 5–17. doi:10.1177/014920638501100102

Lee, K. C., & Chung, N. (2008). Empirical analysis of consumer reaction to the virtual reality shopping mall. *Computers in Human Behavior*, *24*, 88–104. doi:10.1016/j.chb.2007.01.018

Lee, M. K. O., & Turban, E. (2002). A trust model for consumer Internet shopping. *International Journal of Electronic Commerce*, *6*(1), 75–91.

Lee, H.-H., Fiore, A. M., & Kim, J. (2006). The role of the technology acceptance model in explaining effects of image interactivity technology on consumer responses. *International Journal of Retail and Distribution Management*, *34*(8), 621–644. doi:10.1108/09590550610675949

Lee, Y., Kozar, K. A., & Larsen, K. R. (2003). The technology acceptance model: Past, present, and future. *Communications of AIS*, *12*, 752–780.

Lee, N., Broderick, A. J., & Chamberlain, L. (2007). What is neuromarketing? A discussion and agenda for future research. *International Journal of Psychophysiology*, *63*, 199–204. doi:10.1016/j.ijpsycho.2006.03.007

Lee, H.-J., Fairhurst, A. E., & Lee, M.-Y. (2009). The importance of self-service kiosks in developing consumers' retail patronage intentions. *Managing Service Quality*, *19*(6), 687–701. doi:10.1108/09604520911005071

Lester, J., Stone, B., & Stelling, G. (1999). Lifelike pedagogical agents for mixed-initiative problem solving in constructivist learning environments. *User Modeling and User-Adapted Interaction*, *9*, 1–44. doi:10.1023/A:1008374607830

Leuthold, S., Bargas-Avila, J. A., & Opwis, K. (2008). Beyond Web content accessibility guidelines: Design of enhanced text user interfaces for blind internet users. *International Journal of Human-Computer Studies*, *66*(6), 257–270. doi:10.1016/j.ijhcs.2007.10.006

Levin, D. Z. (1999). *Transferring knowledge within the organization in the R&D arena*. Unpublished doctoral dissertation, North-western University, Evanston.

Levine, J. M., & Moreland, R. L. (1990). Progress in small group research. *Annual Review of Psychology*, *41*, 585–634. doi:10.1146/annurev.ps.41.020190.003101

Levy, S. J. (2006). How new, how dominant? In Lusch, R. F., & Vargo, S. L. (Eds.), *The service-dominant logic of marketing. Dialog, debate, and directions*. Armonk, NY: M.E. Sharpe.

Lewis, G. H. (1990). Community through exclusion and illusion, the creation of social worlds in an American shopping mall. *Journal of Popular Culture*, *24*(2), 121–136. doi:10.1111/j.0022-3840.1990.2402_121.x

Lewis, I. (2001). Logistics and electronic commerce: An interorganizational systems perspective. *Transportation Journal*, 5-13.

Lewis, L. (2007). A more able label? *Stores*, September, 54-57.

Li, C., & Hwang, M. (2010). An efficient biometrics-based remote user authentication scheme using smart cards. *Journal of Network and Computer Applications*, (January): 1–5. doi:10.1016/j.jnca.2009.08.001

Li, H., & Bukovac, J. L. (1999). Cognitive impact of banner ad characteristics, an experimental study. *Journalism & Mass Communication Quarterly*, *76*(2), 341–353.

Li, Y., Lu, L., & Li, X. (2005). A hybrid collaborative filtering method for multiple-interests and multiple-content recommendation in e-commerce. *Expert Systems with Applications*, *28*(1), 67–77. doi:10.1016/j.eswa.2004.08.013

Li, H., Daugherty, T., & Biocca, F. (2001). Characteristics of virtual experience in electronic commerce: A protocol analysis. *Journal of Interactive Marketing*, *15*(3), 13–29. doi:10.1002/dir.1013

Li, H., Daugherty, T., & Biocca, F. (2002). Impact of 3-D advertising on product knowledge, brand attitude, and purchase intention: the mediating role of presence. *Journal of Advertising*, *31*(3), 43–57.

Li, P., Huynh, T., Reformat, M., & Miller, J. (2007). A practical approach to testing GUI systems. *Empirical Software Engineering*, *12*(4), 331–357. doi:10.1007/s10664-006-9031-3

Li, H., Kuo, C., & Russell, M. G. (1999). The impact of perceived channel utilities, shopping orientation, and demographics on the consumer's online buying behaviour. *Journal of Computer Mediated Communication*, December.

Liang, T. P., & Lai, H. J. (2002). Effect of store design on consumer purchase: An empirical study of online bookstores. *Information & Management*, *39*(6), 431–444. doi:10.1016/S0378-7206(01)00129-X

Liao, C.-H., & Tsou, C.-W. (2007). Factors influencing the usage of 3G mobile services in Taiwan. *Online Information Review*, *31*(6), 759–774. doi:10.1108/14684520710841757

Licata, P. G. (2004, March 7). Checking yourself out, quite literally. *New York Times*, L13.

Lin, J. C., & Lu, H. (2000). Towards an understanding of the behavioural intention to use a website. *International Journal of Information Management*, *20*, 197–208. doi:10.1016/S0268-4012(00)00005-0

Lin, J.-S. C., & Hsieh, P.-L. (2006). The role of technology readiness in customers' perception and adoption of self-service technologies. *International Journal of Service Industry Management*, *17*(5), 497–517. doi:10.1108/09564230610689795

Lindeman, T. (2005, April 26). Self-scanning takes off. *Pittsburgh Post-Gazette*, E1.

Lindgaard, G., Fernandes, G., Dudek, C., & Brown, J. (2006). Attention Web designers: You have 50 milliseconds to make a good first impression! *Behaviour & Information Technology*, *25*(2), 115–126. doi:10.1080/01449290500330448

Linnhoff, S., & Langenderfer, J. (2004). Identity theft legislation: The fair and accurate credit transactions act of 2003 and the road not taken. *The Journal of Consumer Affairs*, *38*(2), 204–216. doi:10.1111/j.1745-6606.2004.tb00864.x

Liu, Y., & Zhang, Z. J. (2006). The benefits of personalized pricing in a channel. *Marketing Science*, *25*(1), 97–105. doi:10.1287/mksc.1050.0120

Lohse, G. L., & Spiller, P. (1998). Electronic shopping. *Communications of the ACM*, *14*, 81–87. doi:10.1145/278476.278491

Lonardi, G. (May 17, 2010). *Bravofly: è italiano il leader europeo nella vendita di biglietti aerei low cost.* Repubblica, sezione Affari & Finanza, (p. 15).

Lonnqvist, A., & Pirttimaki, V. (2006). The measurement of business intelligence. *Information Systems Management*, (Winter): 32–40. Retrieved from www.ism-journal.com. doi:10.1201/1078.10580530/45769.23.1.20061201/91770.4

Lorenzoni, G., & Lipparini, A. (1999). The leveraging of interfirm relationships as a distinctive organizational capability: A longitudinal study. *Strategic Management Journal*, *20*, 317–338. doi:10.1002/(SICI)1097-0266(199904)20:4<317::AID-SMJ28>3.0.CO;2-3

Lowson, R. H. (2001). Retail operational strategies in complex supply chains. *International Journal of Logistics Management*, *12*(1), 97–111. doi:10.1108/09574090110806253

Lowson, R. (2001). Analysing the effectiveness of European retail sourcing strategies. *European Management Journal*, *19*, 543–551. doi:10.1016/S0263-2373(01)00068-8

Lucas, P. (2005). Biometrics come into focus. *Collections & Credit Risk*, *10*(3), 18–21.

Luhmann, N. (1989). Familiarity, confidence and trust: Problems and alternatives. In Gambetta, D. (Ed.), *Trust: Making and breaking cooperative relations* (pp. 123–140). Oxford, UK: Basil Blackwell.

Luo, X. (2002). Trust production and privacy concerns on the Internet: A framework based on relationship marketing and social exchange theory. *Industrial Marketing Management*, *31*(2), 111–118. doi:10.1016/S0019-8501(01)00182-1

Lynch, J. C., & Ariely, D. (2000). Wine online: Search costs and competition on price, quality, and distribution. *Marketing Science*, *19*, 83–105. doi:10.1287/mksc.19.1.83.15183

Maass, W., & Varshney, U. (2008). Preface to the focus theme section: Smart products. *Electronic Markets*, *18*(3), 211–215. doi:10.1080/10196780802265645

Maass, W., & Filler, A. (2006). Towards an infrastructure for semantically annotated physical products. In Hochberger, C., & Liskowsky, R. (Eds.), *Informatik 2006* (pp. 544–549). Berlin, Germany: Springer.

Maass, W., & Janzen, S. (2007). *Dynamic product interfaces: A key element for ambient shopping environments*. Paper presented at the 20th Bled eConference. Retrieved from http://domino.fov.uni-mb.si/proceedings.nsf/2007

Maass, W., & Kowatsch, T. (2008a). *Adoption of dynamic product information: An empirical investigation of supporting purchase decisions on product bundles*. Paper presented at the 16th European Conference on Information Systems (ECIS).

Maass, W., & Kowatsch, T. (2008b). *Mobile decision support vs. interpersonal sales communication: predictors of buying intentions for price bundles*. Paper presented at the 21st Bled eConference.

Mach, Q. H., Hunter, M. D., & Grewal, R. S. (2010). Neurophysiological correlates in interface design: An HCI perspective. *Computers in Human Behavior*, *26*(3), 371–376. doi:10.1016/j.chb.2009.11.008

Machleit, K. A., Kellaris, J. J., & Eroglu, S. A. (1994). Human versus spatial dimensions of crowding perceptions in retail environments, a note on the measurement and effect on shopper satisfaction. *Marketing Letters*, *5*(2), 183–194. doi:10.1007/BF00994108

Machleit, K. A., & Mantel, S. P. (2001). Emotional response and shopping satisfaction, moderating effects of shopper attributions. *Journal of Business Research*, *54*, 97–106. doi:10.1016/S0148-2963(99)00093-4

MacKay, K., McVetty, D., & Vogt, C. (2005). Web-based information search and use: Is it new tourism reality? A preliminary examination of visitors to Canada's Four Mountain National Parks. In A. Rice & M. Conlin (Ed.), *Proceedings of the TTRA Canada 2005 conference. Ottawa, Canada, Travel and Tourism Research Association Canada Chapter*.

Maekawa, T., Itoh, Y., Kawai, N., Kitamura, Y., & Kishino, F. (2009). MADO interface: A window like a tangible user interface to look into the virtual world. In N. Villar, I. Shahram, M. Fraser, & S. Benford (Eds.), *Proceedings of the 3rd International Conference on Tangible and Embedded Interaction* (pp. 175-180). New York, NY: ACM.

Magalhães, R., & Silva, A. R. (2009). *Organizational design and engineering*. Working paper 1, INOV Centre for Organizational Design and Engineering (CODE). Retrieved October, 4, 2010, from http://is2.lse.ac.uk/wp/pdf/wp179.pdf

Maidel, V., Shoval, P., Shapira, B., & Taieb-Maimon, M. (2008). *Evaluation of an ontology-content based filtering method for a personalized newspaper*. Paper presented at the ACM Conference On Recommender Systems.

Malone, R. (2006, December 7). Can RFID invade your privacy? *Forbes*.

Malone, R. (2006, October 24). Smart Store. *Forbes*.

Management, M. H. (2009). University study finds proof of REID benefits. *Materials Handling Management*, *64*(12), 7–8.

Mandel, N., & Johnson, E. J. (2002). When Web pages influence choice, effects of visual primes on experts and novices. *The Journal of Consumer Research*, *29*(2), 235–245. doi:10.1086/341573

Mander, G. (2004). Self-service schemes must meet real needs to win shopper buy-in. *Gartner*, *G2*(June), 1–7.

Marcati, A., Guido, G., & Peluso, A. (2007). *What is marketing for SME entrepreneurs? The need to market the marketing approach*. Paper presented at the 6th International Conference in Marketing Trends, Paris, Venice. 26th-27th of January.

Marcus, B. W., & Wallace, S. L. (1997). *New dimensions in investor relations: Competing for capital in the 21st century. (Wiley Frontiers in Finance)*. John Wiley & Sons.

Marcussen, C. H. (2001). *Trends in European Internet distribution of travel and tourism services*. Research Centre of Bornholm, Denmark, 27 March.

Mari, C. (2008). *Consumer motivation. Foundations for a theory of consumption*. Naples, Italy: Edizioni Scientifiche Italiane.

Market Research Society. (2007). *Qualifications for market researchers*. Retrieved June 15, 2010, from www.mrs.org.uk/qualifications/index.htm

Marks, L. J., & Olson, J. C. (1981). Toward a cognitive structure Conceptualization of product familiarity. In Kent, M., & Arbor, A. (Eds.), *Advances in consumer research*. MI: Association for Consumer Research.

Marsh, R. (2005). Drowning in dirty data? It's time to sink or swim: A four stage methodology for total data quality management. *Journal of Database Marketing and Customer Strategy Management*, *12*(2), 105–112. doi:10.1057/palgrave.dbm.3240247

Marshall, G. W., Goebel, D. J., & Moncrief, W. C. (2003). Hiring for success at the buyer–seller interface. *Journal of Business Research*, *56*, 247–255. doi:10.1016/S0148-2963(02)00435-6

Marzocchi, G. L., & Zammit, A. (2006). Self-scanning technologies in retail: Determinants of adoption. *The Service Industries Journal*, *26*(6), 651–669. doi:10.1080/02642060600850790

Mathiassen, L., & Nielsen, P. A. (2008). Engaged scholarship in IS research. *Scandinavian Journal of Information Systems*, *20*, 2.

Mattila, A. S., & Wirtz, J. (2001). Congruency of scent and music as a driver of in-store evaluations and behaviour. *Journal of Retailing*, *77*(2), 273–289. doi:10.1016/S0022-4359(01)00042-2

Mattmüller, R. (2010). *Versand-Handels-Marketing*. Frankfurt: Deutscher Fachverlag.

McAfee, A. P. (2006). Enterprise 2.0: The dawn of emergent collaboration. *MIT Sloan Management Review*, *47*(3), 21–28.

McCabe, D. B., & Nowlis, S. M. (2003). The effect of examining actual products or product descriptions on consumer preference. *Journal of Consumer Psychology*, *13*(4), 431–439. doi:10.1207/S15327663JCP1304_10

McDonald, J. B., & Smith, K. (2004). The effects of technology-mediated communication on industrial buyer behavior. *Industrial Marketing Management*, *33*(2), 107–116. doi:10.1016/S0019-8501(03)00033-6

McGoldrick, P. J., & Thompson, M. G. (1992). The role of perception in the attraction of the out of town mall. *International Review of Retail, Distribution and Consumer Research*, *2*(1), 81–98. doi:10.1080/09593969200000005

McInnis, D. J. (2004). Where have all the papers gone? Reflections on the decline of conceptual articles. *ACR News*, 1-3.

McIntosh, D. (2009). Biometrics—a fad or the future? *Biometric Technology Today*, May, 9-12.

McIntyre, C. (2009). Diminishing varieties of active and creative retail experience. *Journal of Retailing and Consumer Services*, *16*, 466–476. doi:10.1016/j.jretconser.2009.08.001

McKechnie, S., Winkihofer, H., & Ennew, C. (2006). Applying the technology acceptance model to the online retailing of financial services. *International Journal of Retail and Distribution Management*, *34*(4/5), 388–410. doi:10.1108/09590550610660297

McKinley, E. (2006). From pallets to cartons. *Stores*, July, 74-76.

McKinney, M. (2008, April 26). Pay By Touch's backers claim fingers burned. *StarTribune.com*, 1.

McKinsey. (2007). How business are using Web 2.0: A McKinsey global survey. *The McKinsey Quarterly*. Retrieved August 7, 2010, from http://www.mckinsey-quarterly.com/Marketing /How_businesses_are_using_Web_20_A_McKinsey_Global_Survey_1913_abstract

McKnight, D. H., Choudhury, V., & Kacmar, C. (2002). The impact of initial consumer trust on intention to transact with a website: A trust building model. *The Journal of Strategic Information Systems*, *13*(4), 297–323. doi:10.1016/S0963-8687(02)00020-3

McKnight, D. H., Kacmar, C., & Choudhury, V. (2004). Shifting factors and the ineffectiveness of third party assurance seals: A two-stage model of initial trust in a Web business. *Electronic Markets*, *14*(3), 252–266. doi:10.1080/10196780420000245263

Mehrabian, A., & Russell, J. A. (1974). *An approach to environmental psychology*. Cambridge, MA: MIT Press.

Mentzer, J. T., & Williams, L. R. (2001). The role of logistics leverage in marketing strategy. *Journal of Marketing Channels*, *8*, 29–48. doi:10.1300/J049v08n03_03

Merten, K. (1997). *Kommunikation: Eine Begriffs- und prozessanalyse*. Opladen, Germany: Westdeutscher Verlag.

Messinger, P. R., & Narasimhan, C. (1997). A model of retail formats based on consumers' economizing on shopping time. *Marketing Science*, *16*(1), 1–23. doi:10.1287/mksc.16.1.1

Methlie, L. B. (2000). A business model for electronic commerce. *Teletronikk*, *96*(2), 8–19.

Meuter, M. L., Bitner, M. J., Ostrom, A. L., & Brown, S. W. (2005). Choosing among alternative service delivery modes: An investigation of customer trial of self-service technologies. *Journal of Marketing*, *69*, 61–83. doi:10.1509/jmkg.69.2.61.60759

Meuter, M. L., Ostrom, A. L., Roundtree, R. I., & Bitner, M. J. (2000). Self-service technologies: Understanding customer satisfaction with technology-based service encounters. *Journal of Marketing*, *64*, 50–64. doi:10.1509/jmkg.64.3.50.18024

Miceli, N., Ricotta, F., & Costabile, M. (2007). Customizing customatization: A conceptual framework for interactive personalization. *Journal of Interactive Marketing*, *21*(2), 6–25. doi:10.1002/dir.20076

Michael, R. W., & Attaway, J. S. (2003). At the interface: The nature of buyer-seller interactions and relationships. *Journal of Business Research*, *56*, 243–246. doi:10.1016/S0148-2963(02)00434-4

Michon, R., Chebat, J.-C., & Turley, L. W. (2005). Mall atmospherics: The interaction effects of the mall environment on shopping behaviour. *Journal of Business Research*, *58*(5), 576–583. doi:10.1016/j.jbusres.2003.07.004

Mick, D. G., & Fournier, S. (1998). Paradoxes of technology: Consumer cognisance, emotions and coping strategies. *The Journal of Consumer Research*, *25*, 123–143. doi:10.1086/209531

Migliarese, P., & Corvello, V. (in press). Organizational relations in organizational design and engineering. *International Journal of Organizational Design and Engineering*.

Mihal, W. L., Sorce, P. A., & Comte, E. T. (1984). A process model of individual career decision making. *Academy of Management Review*, *9*(1), 95–103. doi:10.2307/258236

Mild, A., & Reutterer, T. (2003). An improved collaborative filtering approach for predicting cross-category purchases based on binary market basket data. *Journal of Retailing and Consumer Services*, *10*(3), 123–133. doi:10.1016/S0969-6989(03)00003-1

Milne, G. R. (2000). Privacy and ethical issues in database/interactive marketing and public policy: A research framework and overview of the special issue. *Journal of Public Policy & Marketing*, *19*(1), 1–6. doi:10.1509/jppm.19.1.1.16934

Milne, G. R., & Boza, M. E. (1999). Trust and concern in consumers' perceptions of marketing information management practices. *Journal of Interactive Marketing*, *13*(1), 5–24. doi:10.1002/(SICI)1520-6653(199924)13:1<5::AID-DIR2>3.0.CO;2-9

Milne, G. R., & Gordon, M. E. (1993). Direct mail privacy-efficiency trade-offs within an implied social contract framework. *Journal of Public Policy & Marketing*, *12*(2), 206–215.

MisMe corporation. (2007). *About WiFi*. Retrieved November 2009 from http://www.misme.net/about_wifi.html

Money, B. R., & Crotts, J. C. (2003). The effect of uncertainty avoidance on information search, planning and purchases of international travel vacations. *Tourism Management*, *24*(2), 263–277. doi:10.1016/S0261-5177(02)00057-2

Monger, J., & Keen, C. (2004). The research proves it. We cannot wait to measure the customer experience. *DM Review Magazine*, June 2004. Retrieved June 15, 2010, from http://www.crm2day.com/content/t6_librarynews_1.php?id=

Montoya-Weiss, M. M., Voss, G. B., & Grewal, D. (2003). Determinants of online channel use and overall satisfaction with a relational, multichannel service provider. *Journal of the Academy of Marketing Science*, *31*(4), 448–458. doi:10.1177/0092070303254408

Moon, J., & Kim, Y. (2001). Extending the TAM for a World-Wide-Web context. *Information & Management*, *38*(4), 217–230. doi:10.1016/S0378-7206(00)00061-6

Moon, B.-J. (2004). Consumer adoption of the internet as an information search and product purchase channel: some research hypotheses. *International Journal of Internet Marketing and Advertising*, *1*(1), 104–118. doi:10.1504/IJIMA.2004.003692

Moore, G. C., & Benbasat, I. (1991). Development of an instrument to measure the perceptions of adopting an information technology innovation. *Information Systems Research*, *2*, 173–191. doi:10.1287/isre.2.3.192

Moore, G. C., & Benbasat, I. (1991). Development of an instrument to measure perceptions of adopting an information technology innovation. *Information Systems Research*, *2*(3), 192–222. doi:10.1287/isre.2.3.192

Morin, S., Dubé, L., & Chebat, J.-C. (2007). The role of pleasant music and servicescapes: A test of the dual model of environmental perception. *Journal of Retailing*, *1*, 115–130. doi:10.1016/j.jretai.2006.10.006

Moshagen, M., & Thielsch, M. T. (in press). Facets of visual aesthetics. *International Journal of Human-Computer Studies*.

Mulhern, F. J. (1999). Customer profitability analysis: Measurement, concentration, and research directions. *Journal of Interactive Marketing*, *13*(1), 25–40. doi:10.1002/(SICI)1520-6653(199924)13:1<25::AID-DIR3>3.0.CO;2-L

Müller-Lankenau, C., Wehmeyer, K., & Klein, S. (2005). Multi-channel strategies: Capturing and exploring diversity in the European retail grocery industry. *International Journal of Electronic Commerce*, *10*(2), 85–122. doi:10.2753/JEC1086-4415100204

Muller-Seitz, G., Dautzenberg, K., Creusen, U., & Stromereder, C. (2009). Customer acceptance of RFID technology: Evidence from the German electronic retail sector. *Journal of Retailing and Consumer Services*, *16*, 31–39. doi:10.1016/j.jretconser.2008.08.002

Mulpuru, S. (2007). *Which personalization tools work for e-commerce - and why*. Forrester Research.

MultichannelMerchant. (2010, June 1). *Bakers labels barcoding a success*.

Muramatsu, R., & Hanoch, Y. (2005). Emotions as a mechanism for boundedly rational agents: The fast and frugal way. *Journal of Economic Psychology*, *26*(2), 201–221. doi:10.1016/j.joep.2004.03.001

Murray, K. B., & Häubul, G. (2009). Personalization without interrogation: Towards more effective interactions between consumers and feature-based recommendation agents. *Journal of Interactive Advertising*, *23*(2), 138–146. doi:10.1016/j.intmar.2009.02.009

Murray, K. B. (1991). A test of service marketing theory: Consumer information acquisition activities. *Journal of Marketing*, *55*, 10–15. doi:10.2307/1252200

Muthén, L. K., & Muthén, B. O. (2007). *Mplus user's guide* (4th ed.). Los Angeles, CA: Muthén & Muthén.

Myers, J. G., Greyser, S. A., & Massy, W. F. (1979). The effectiveness of marketing's R&D for marketing management: An assessment. *Journal of Marketing*, *43*(1), 17–29. doi:10.2307/1250754

Myron, D. (2009). Social media spawns a new era in customer intelligence. *CRM Magazine*, *13*(6), 6–6.

Nabi, F. (2005). Secure business application logic for e-commerce systems. *Computers & Security*, *24*, 208–217. doi:10.1016/j.cose.2004.08.008

NACSInsight. (2009, June 9). Albert Heijn invests in customer-driven IT.

Nambisan, S. (2002). Designing virtual customer environments for new product development: Toward a theory. *Academy of Management Review*, *27*(3), 392–413. doi:10.2307/4134386

National Conference on Weights and Measures. (1995). *Examination procedure for price verification*. Gaithersburg, MD: National Institute of Standards and Technology.

National Petroleum News. (2006). NCR creates "mini" self-checkout solution. *National Petroleum News* March, 27.

Nault, B. R., & Dexter, A. S. (1995). Added value and pricing with information technology. *Management Information Systems Quarterly*, 449–464. doi:10.2307/249628

NCR. (2010). *NCR corporate website*. Retrieved September 1, 2010, from www.ncr.com

Nelson, S. (2005). Finger on the solution. *Stores*, July, 69-70.

Netemeyer, R., Bentler, P., Bagozzi, R., Cudeck, R., Cote, J., & Lehmann, D. (2001). Structural equation modelling. *Journal of Consumer Psychology*, *10*(1 & 2), 83–100. doi:10.1207/S15327663JCP1001&2_08

Newholm, T., McGoldrick, P., Keeling, K., Macaulay, L., & Doherty, J. (2004). Multi-story trust and online retailer strategies. *International Review of Retail, Distribution and Consumer Research*, *14*(4), 437–456. doi:10.1080/0959396042000260889

Newman, A. J. (2007). Uncovering Dimensionality in the Servicescape: Towards Legibility. *Service Industries Journal.*, *27*(1), 15–28. doi:10.1080/02642060601038601

Newman, J. W., & Werbel, R. A. (1973). Multivariate analysis of brand loyalty for major household appliances. *JMR, Journal of Marketing Research*, *10*, 404–409. doi:10.2307/3149388

Newman, A. J., & Cullen, P. (2002). *Retailing: Environment & operations*. London, UK: Thomson Learning.

Newman, A., Dennis, C. E., & Zaman, S. (2006). Marketing images and consumers' experiences in selling environments. *Marketing Management Journal*, Fall, 515-599.

Newsom, M. K., Collier, D. A., & Olsen, E. O. (2009). Using biztainment to gain competitive advantage. *Business Horizons*, *52*(2), 167–176. doi:10.1016/j.bushor.2008.10.005

Nielsen, J. (1993). *Usability engineering*. London, UK: Academic Press.

Nielsen, J. (2000). *Designing Web usability: The practice of simplicity*. Indianapolis, IN: New Riders Press.

Nielsen, J., & Norman D. A. (2000). Usability on the Web isn't a luxury. *InformationWeek, 65–73*.

Norberg, P. A., Horne, D. R., & Horne, D. A. (2007). The privacy paradox: Personal information disclosure intentions versus behaviors. *The Journal of Consumer Affairs, 41*(1), 100–126. doi:10.1111/j.1745-6606.2006.00070.x

Norman, D. A. (2004). *Emotional design*. New York, NY: Basic Books.

Norman, D. A. (1983). Some observations on mental models. In Gentner, D., & Stevens, A. L. (Eds.), *Mental models* (pp. 7–14). Hillsdale, NJ: Lawrence Erlbaum Associates.

North, A. C., Hargreaves, D. J., & McKendrick, J. (1999). The influence of in-store music on wine selections. *The Journal of Applied Psychology, 2*, 271–276. doi:10.1037/0021-9010.84.2.271

Novak, T. P., Hoffman, D. L., & Yung, Y. F. (2000). Measuring the customer experience in online environments: A structural modelling approach. *Marketing Science, 19*, 22–42. doi:10.1287/mksc.19.1.22.15184

Novo, J. (2001). *Drilling down: Turning customer data into profits with a spreadsheet*. Booklocker.com

Nowell, C., & Stanley, L. R. (1991). Length-biased sampling in mall intercept surveys. *JMR, Journal of Marketing Research, 28*(4), 475–480. doi:10.2307/3172787

Nunnally, J. C. (1967). *Psychometric theory*. New York, NY: McGraw-Hill.

Nunnally, J. C., & Bernstein, I. H. (1994). *Psychometric theory* (3rd ed.). New York, NY: McGraw-Hill Education.

Nysveen, H., Leif, B. M., & Pedersen, P. E. (2003). Tourism websites and value-added services: The gap between customer preferences and websites offering. *Information Technology & Tourism, 5*, 165–174. doi:10.3727/109830503108751117

Nysveen, H., & Lexhagen, M. (2002). Reduction of perceived risk through on-line value-added services offered on tourism businesses websites. In *Proceedings of the 9th International Conference on Information and Communication Technologies in Tourism, Innsbruck* (pp. 488–496). Vienna, Austria: Springer-Verlag.

O'Cass, A., & Fenech, T. (2003). Web retailing adoption: Exploring the nature of Internet users Web retailing behaviour. *Journal of Retailing and Consumer Services, 10*, 81–94. doi:10.1016/S0969-6989(02)00004-8

O'Mahoney, M., Hurley, N., Kushmerick, N., & Silvestre, G. (2004). Collaborative recommendation: A robustness analysis. *ACM Transactions on Internet Technology, 4*(4), 344–377.

OAAA. (2009). *Digital billboards, outdoor advertising's glowing future*. Outdoor Advertising Association of America. Retrieved on December 19, 2009 from, http,// www.oaaa.org/marketingresources/digital.aspx

Oatley, K. (1992). *Best laid schemes: The psychology of emotions*. New York, NY: Cambridge University Press.

Observatorio de las Telecomunicaciones y la Sociedad de la Información - Entidad Pública Empresarial Red. (2006). *Diagnóstico tecnológico del sector de comercio minorista*. Retrieved on September 3, 2007, from observatorio.red.es/estudios/documentos/diag_cminoristav1.3.pdf

Ochi, P., Rao, S., Takayama, L., & Nass, C. (2010). Predictors of user perceptions of Web recommender systems: How the basis for generating experience and search product recommendations affects user responses. *International Journal of Human-Computer Studies, 68*(8), 472–482. doi:10.1016/j.ijhcs.2009.10.005

Odekerken-Schröder, G., De Wulf, K., Kasper, H., Kleijnen, M., Hoekstra, J., & Commandeur, H. (2001). The impact of quality on store loyalty: A contingency approach. *Total Quality Management, 12*(3), 307–322. doi:10.1080/09544120120034474

OECD. (2000). *Knowledge management in the learning society*. Paris, France: OECD.

Oehlsen, N. (2005). Putting a finger on biometrics. *Credit Card Management,* May, 46-50.

Okhuysen, G. A., & Eisenhardt, K. M. (2002). Integrating knowledge in groups: How formal interventions enable flexibility. *Organization Science*, *13*(4), 370–386. doi:10.1287/orsc.13.4.370.2947

Opara-Nadi, G. (2005). Electronic self-checkout system vs cashier operated system: A performance based comparative analysis. *Capella University*, June.

Oracle. (2008). *Oracle on demand for markdown optimization*. Retrieved from http://www.oracle.com / markdown-optimization-data-sheet.pdf

Oracle. (2010). *Oracle and retail*. Retrieved from http:// www.oracle.com /us/industries/retail/022189.htm

Orlikowski, W. J. (2000). Using technology and constituting structures: A practice Lens for studying technology in organizations. *Organization Science*, *11*(4), 404–428. doi:10.1287/orsc.11.4.404.14600

Orlikowski, W. J. (2007). Sociomaterial practices: Exploring technology at work. *Organization Studies*, *28*(9), 1435–1448. doi:10.1177/0170840607081138

Orman, L. V. (2007). Consumer support systems. *Communications of the ACM*, *50*(4), 49–54. doi:10.1145/1232743.1232747

Otter, M., & Johnson, H. (2000). Lost in hyperspace: Metrics and mental models. *Interacting with Computers*, *13*(1), 1–40. doi:10.1016/S0953-5438(00)00030-8

Otto, J. R., & Chung, Q. B. (2000). A framework for cyber-enhanced retailing: Integrating e-commerce retailing with brick-and-mortar retailing. *Electronic Markets*, *10*(3), 185–191. doi:10.1080/10196780050177099

Oulasvirta, A. (2004). Task demands and memory in web interaction: A levels of processing approach. *Interacting with Computers*, *16*(2), 217–241. doi:10.1016/j.intcom.2003.12.004

Oulasvirta, A., Kärkkäinen, L., & Laarni, J. (2005). Expectations and memory in link search. *Computers in Human Behavior*, *21*(5), 773–789.

Özgener, S., & Iraz, R. (2006). Customer relationship management in small-medium enterprises: The case of Turkish tourism industry. *Tourism Management*, *27*, 1356–1363. doi:10.1016/j.tourman.2005.06.011

Page, T. J., Thorson, E., & Heide, M. P. (1990). The memory impact of commercials varying in emotional appeal and product involvement. In Agres, S. J., Edell, J. A., & Dubitsky, T. M. (Eds.), *Emotion in advertising* (pp. 255–268). New York, NY: Quorum Books.

Pallant, J. (2006). *SPSS survival manual* (2nd ed.). Berkshire, UK: Open University Press.

Palmatier, R. W., Dant, R. P., Grewal, D., & Evans, K. E. (2006). Factors influencing the effectiveness of relationship marketing: A meta-analysis. *Journal of Marketing*, *70*, 136–153. doi:10.1509/jmkg.70.4.136

Palmer, J. W., & Markus, M. L. (2000). The performance impacts of quick response and strategic alignment in specialty retailing. *Information Systems Research*, *11*(3), 241–259. doi:10.1287/isre.11.3.241.12203

Palmer, J. W., Bailey, J. P., & Faraj, S. (1999). The role of intermediaries in the development of trust on the WWW: The use and prominence of trusted thirds parties and privacy statements. *Journal of Computer-Mediated Communication*, 9.

Palmer, S. E. (1975). The effects of contextual scenes on the identification of objects. *Memory & Cognition*, *3*(5), 519–526. doi:10.3758/BF03197524

Panian, Ž., & Ćurko, K. (2010). *Poslovni informacijski sustavi*. Zagreb, Croatia: Element.

Pantano, E. (2010). New technologies and retailing: Trends and directions. *Journal of Retailing and Consumer Services*, *17*, 171–172. doi:10.1016/j.jretconser.2010.03.004

Pantano, E., & Naccarato, G. (2010). Entertainment in retailing: The influences of advanced technologies. *Journal of Retailing and Consumer Services*, *17*(3), 200–204. doi:10.1016/j.jretconser.2010.03.010

Pantano, E., & Tavernise, A. (2009). Learning cultural heritage through Information and Communication Technologies: A case study. In Chang, M., & Kuo, C.-W. (Eds.), *Learning culture and language through ICTs: Methods for enhanced instruction* (pp. 103–119). Hershey, PA: IGI Global. doi:10.4018/9781605661667.ch007

Pantano, E., & Servidio, R. (2009). Pervasive environments for promotion of tourist destinations. In Vrontis, D., Weber, Y., Kaufmann, R., & Tarba, S. (Eds.), *Managerial and entrepreneurial developments in the Mediterranean area* (pp. 1213–1335). Cyprus: EuroMed Press.

Pantano, E., Taversine, A., & Viassone, M. (2010). Consumer perception of computer-mediated communication in a social network. In *Proceedings of the 4th International Conference on New Trends in Information Science and Service Science* 2010, IEEE Computer series.

Parasuraman, A., & Colby, C. L. (2001). *Techno-ready marketing: How and why your customers adopt technology*. New York, NY: Free Press.

Parasuraman, A., Berry, L. L., & Zeithaml, V. (1988). SERVQUAL: A multiple-item scale for measuring consumer perceptions of service quality. *Journal of Retailing, 64*(1), 12–40.

Park, D. H., Lee, J., & Han, J. (2007). The effect of online consumer reviews on consumer purchasing intention: the moderating role of involvement. *International Journal of Electronic Commerce, 11*(4), 125–148. doi:10.2753/JEC1086-4415110405

Park, J., Stoel, L., & Lennon, J. S. (2008). Cognitive, affective and conative responses to visual simulation: The effects of rotation in online product presentation. *Journal of Consumer Behaviour, 7*(1), 72–87. doi:10.1002/cb.237

Parks, L. (2007). Tracking jeans in Mexico. *Stores*, March, 35-37.

Pavlou, P. A. (2003). Consumer acceptance of electronic commerce - integrating trust and risk with the technology acceptance model. *International Journal of Electronic Commerce, 7*(3), 69–103.

Pavlou, P. A., & Fygension, M. (2006). Understanding and predicting electronic commerce adoption: An extension of the theory of planned behavior. *Management Information Systems Quarterly, 30*(1), 115–143.

Payne, J., Bettman, J. R., & Johnson, E. J. (1991). Consumer decision making. In Kassarjian, H. H., & Robertson, T. S. (Eds.), *Handbook of consumer behavior* (pp. 50–84). Prentice-Hall.

Payne, S. J. (2007). Mental models in human–computer interaction. In Sears, A., & Jacko, J. A. (Eds.), *The human-computer interaction gandbook: Fundamentals, evolving technologies, and emerging applications* (pp. 63–76). New York, NY: Lawrence Erlbaum Associates. doi:10.1201/9781410615862.ch3

Peachey, A., Gillen, J., Livingstone, D., & Smith-Robbins, S. (2010). *Researching learning in virtual worlds*. London, UK: Springer.

Pelz, J. B., & Canosa, R. (2001). Oculomotor behavior and perceptual strategies in complex tasks. *Vision Research, 41*(25-26), 3587–3596. doi:10.1016/S0042-6989(01)00245-0

Peppard, J. (Ed.). (1993). *IT strategy, firm business*. London, UK: Pitman.

Peppers, D., & Rogers, M. (1997). *Enterprise one to one. Tools for competing in the interactive age*. New York, NY: Currency Doubleday.

Pereira, R. E. (2001). Influence of query-based decision aids on consumer decision making in electronic commerce. *Information Resources Management Journal, 14*(1), 31. doi:10.4018/irmj.2001010104

Perlin, K. (1997). *Layered compositing of facial expression*. ACM SIGGRAPH Technical Sketch. Retrived June 25, 2010, from http://mrl.nyu.edu/~perlin/experiments/facedemo

Perreault, W., McCarthy, E. J., Parkinson, S., & Steward, K. (2000). *Basic marketing* (European edition). London, UK: McGraw Hill.

Peter, P. J., & Olson, J. C. (1983). Is science marketing? *Journal of Marketing, 47*, 111–125. doi:10.2307/1251404

Peterson, R., & William, R. (1992). Measuring customer satisfaction: Fact and artefact. *Journal of the Academy of Marketing Science, 20*(Winter).

Petty, R., & Cacioppo, J. (1981). *Attitudes and persuasion: Classic and contemporary approaches*. Dubuque, IA: Brown.

Petty, R., & Cacioppo, J. (1986). The elaboration likelihood model of persuasion. *Advances in Experimental Social Psychology, 19*, 123–205. doi:10.1016/S0065-2601(08)60214-2

Pew Internet & American Life Project. (2006). *Internet evolution, Internet penetration and impact*. Retrieved 21 May, 2010, from http://www.pewinternet.org/PPF/r/182/report_display.asp

Pfeffer, J. (1983). Organizational demography. *Research in Organizational Behavior*, *5*, 299–357.

Pfeffer, J. (1993). Barriers to the advance of organizational science: Paradigm development as a dependent variable. *Academy of Management Review*, *18*(4), 599–620. doi:10.2307/258592

Pham, M. T. (2004). The logic of feeling. *Journal of Consumer Psychology*, *14*(4), 360–369. doi:10.1207/s15327663jcp1404_5

Phan, D. D., & Vogel, D. R. (2010). A model of customer relationship management and business intelligence systems for catalogue and online retailers. [from www.elsevier.com]. *Information & Management*, *47*, 69–77. Retrieved June 15, 2010. doi:10.1016/j.im.2009.09.001

Phang, C., Sutanto, J., Kankanhalli, A., Li, Y., Tan, B., & Teo, H. (2006). Senior citizens' acceptance of information systems: A study in the context of e-government services. *IEEE Transactions on Engineering Management*, *53*(4), 555–569. doi:10.1109/TEM.2006.883710

Phelps, J., Nowak, G. J., & Ferrell, E. (2000). Privacy concerns and consumer willingness to provide personal information. *Journal of Public Policy & Marketing*, *19*(1), 27–4. doi:10.1509/jppm.19.1.27.16941

Phelps, J. E., D'Souza, G., & Nowak, G. J. (2001). Antecedents and consequences of consumer privacy concerns: An empirical investigation. *Journal of Interactive Marketing*, *15*(4), 2–17. doi:10.1002/dir.1019

Pollock, N., & Williams, R. (2009). *Software and organisations: The biography of the enterprise-wide system or how SAP conquered the world*. Routledge.

Pollock, A. (1995). The impact of information technology on destination marketing. *EIU Travel and Tourism Analyst*, *3*, 66–83.

Ponder, N., Lueg, J. E., & Williams, Z. (2006). Salesperson assistance versus self-service in retailing: Are they both a matter of convenience? *Marketing Management Journal*, *16*(1), 72–87.

Porter, M. E. (1997). Creating advantages. *Executive Excellence*, *14*(2), 17–18.

Porter, M. (1985). Technology and competitive advantage. *The Journal of Business Strategy*, (Winter): 60–70. doi:10.1108/eb039075

Porter, M. E. (2001). Strategy and the Internet. *Harvard Business Review*, 63–78.

Porter, M. E. (1980). *Competitive strategy: Techniques for analyzing industries and competitors*. New York, NY: The Free Press.

Porter, C., & Donthu, N. (2006). Using the technology acceptance model to explain how attitudes determine Internet usage: The role of perceived access barriers and demographics. *Journal of Business Research*, *59*, 99–1007. doi:10.1016/j.jbusres.2006.06.003

Postmes, T., Spears, R., & Cihangir, S. (2001). Quality of decision making and group norms. *Journal of Personality and Social Psychology*, *80*(6), 81–99. doi:10.1037/0022-3514.80.6.918

Poston, R. S., & Speier, C. (2005). Effective use of knowledge management systems: A process model of content ratings and credibility indicators. *Management Information Systems Quarterly*, *29*(2), 221–244.

Potter, B. (2006). Wireless hotspots: Petri dish of wireless security. *Hacking and Innovation*, *49*(6), 50–56.

Powell, T. C. (1992). Organizational alignment as competitive advantage. *Strategic Management Journal*, *13*(2), 119–134. doi:10.1002/smj.4250130204

Prabhakar, S., Pankanti, S., & Jain, A. K. (2003). Biometric recognition: Security and privacy concerns. *IEEE Security & Privacy*, *1*(March/April), 33–42. doi:10.1109/MSECP.2003.1193209

Premazzi, K., Castaldo, S., Grosso, M., & Hofacker, C. (in press). Supporting retailers to exploit online settings for internationalization: The different role of trust and compensation. *Journal of Retailing and Consumer Services*.

Pride, W., & Ferrell, O. C. (2005). *Marketing*. Milano, Italia: EGEA.

Prideaux, B. (2002). Airline distribution systems: The challenge and opportunity of the Internet. In Buhalis, D., & Laws, E. (Eds.), *Tourism distribution channels: Practices, issues and transformations*. London, UK: Continuum.

Pritchard, S. (2010). Good analysis and use of data can make all the difference. *Financial Times World Retailing, 21*.

Professional Services Close-Up (2010, May 31). Global options group: Bode Technology releases RFID system.

Puccinelli, N. M., Goodstein, R. C., Grewal, D., Price, R., Raghubir, P., & Stewart, D. (2009). Customer experience management in retailing: Understanding the buying process. *Journal of Retailing, 85*(1), 15–30. doi:10.1016/j.jretai.2008.11.003

Puccinelli, N. M. (2006). Putting your best face forward, the impact of customer mood on salesperson evaluation. *Journal of Consumer Psychology, 16*(2), 156–162. doi:10.1207/s15327663jcp1602_6

Puccinelli, N. M., Goodstein, R. C., Grewal, D., Price, R., Raghubir, P., & Stewart, D. (2009). Customer experience management in retailing: Understanding the buying process. *Journal of Retailing, 85*(1), 15–30. doi:10.1016/j.jretai.2008.11.003

Puto, C. P., & Wells, W. D. (1984). Informational and transformational advertising, the differential effects of time. In Kinnear, T. C. (Ed.), *Advances in consumer research, 11* (pp. 572–576). Proto, UT: Association for Consumer Research.

Qiu, L., Li, Y., & Wu, X. (2008). Protecting business intelligence and customer privacy while outsourcing data mining tasks. [from www.springerlink.com]. *Knowledge and Information Systems, 17*, 99–120. Retrieved June 15, 2010. doi:10.1007/s10115-007-0113-3

Quelch, J. A., & Kein, L. R. (1996). The Internet and international marketing. *Sloan Management Review*, (Spring): 60–75.

Ram, S., & Jung, H.-S. (1991). Forced adoption of innovation in organizations: Consequences and implications. *Journal of Product Innovation Management, 8*(2), 117–126. doi:10.1016/0737-6782(91)90005-J

Raman, P., Brudvig, S., & Hofacker, C. F. (2006, October). *To give or not to give: Providing personal information to online firms.* Paper presented at the Direct Marketing Association Educators' Conference, San Francisco, CA.

Ranganathan, C., & Ganapathy, S. (2002). Key dimensions of business-to consumer websites. *Information & Management, 39*(6), 457–465. doi:10.1016/S0378-7206(01)00112-4

Rangarajan, T. S. (2010). Enhancing supply chain management using RFID. *Material Handling Management, 65*(5), 22–25.

Rao, B. P. (2000). Improving retail effectiveness through technology: A survey of analytical tools for physical and online retailers. *Technology in Society, 22*, 111–122. doi:10.1016/S0160-791X(99)00032-9

Ratchford, B. T. (2001). The economics of consumer knowledge. *The Journal of Consumer Research, 27*, 397–411. doi:10.1086/319617

Rebolledo, C., Ricard, L., & Prefontaine, L. (2005). The potential of Information Technology in facilitating relationship marketing: The case of large Canadian firms. *Journal of Relationship Marketing, 4*(1/2), 57–72. doi:10.1300/J366v04n01_05

Reda, S. (2005). What you don't know about RFID! *Stores*, September, 26-34.

Reda, S. (2006). Reading the RFID tea leaves. *Stores*, May, 46-50.

Reda, S. (2006). Stepping up the RFID effort. *Stores*, March, 26-30.

Reeves, B., & Nass, C. (1996). *How people treat computers, television and new media like real people and places*. Stanford, CA: Cambridge University Press.

Reicheild, F. F., & Sasser, E. (1990). *Zero defections: Quality comes to services. Harvard Business Review*. September-October.

Reinagel, P., & Zador, A. M. (1999). Natural scene statistics at the centre of gaze. *Network (Bristol, England), 10*(4), 341–350. doi:10.1088/0954-898X/10/4/304

Reinders, M. J., Dabholkar, P. A., & Frambach, R. T. (2008). Consequences of forcing consumers to use technology-based self-service. *Journal of Service Research*, *11*(2), 107–123. doi:10.1177/1094670508324297

Resatsch, F., Sandner, U., Leimeister, J. M., & Krcmar, H. (2008). Do point of sale RFID-based information services make a difference? Analyzing consumer perceptions for designing smart product information services in retail business. *Electronic Markets*, *18*(3), 216–231. doi:10.1080/10196780802265728

Retail, D. M. S. (2010). *E-tailing and multi-channel retailing*. Retrieved September 1, 2010, from http://www.dmsretail. com/etailing.htm

Retail, L. S. (2010). *LS Retail NAV*. Retrieved September 1, 2010, from http://www.lsretail.com

Reynolds, K. E., Ganesh, J., & Luckett, M. (2002). Traditional malls vs. factory outlets, comparing shopper typologies and implications for retail strategies. *Journal of Business Research*, *55*(9), 687–696. doi:10.1016/S0148-2963(00)00213-7

Richardson, P. S., Dick, A. S., & Jain, A. K. (1994). Extrinsic and intrinsic cue effects on perceptions of store brand quality. *Journal of Marketing*, *58*, 28–36. doi:10.2307/1251914

Richins, M. (1983). Negative word-of-mouth by dissatisfied consumers: A pilot study. *Journal of Marketing*, *47*(Winter), 68–78. doi:10.2307/3203428

Ringle, C. M., Wende, S., & Will, A. (2005). *SmartPLS 2.0 (M3)*. Hamburg, Germany: University of Hamburg.

Ringle, C. M., Wende, S., & Will, A. (2005). *SmartPLS* (release 2.0 M3). Hamburg, Germany: University of Hamburg. Retrieved from http://www.smartpls.de

Robertson, N., & Shaw, R. (2009). Predicting the likelihood of voiced complaints in the self-service technology context. *Journal of Service Research*, *12*(1), 100–116. doi:10.1177/1094670509333789

Robles, P. (2010). *Virgin America: Flying social media into the abyss*. Retrieved June 24, 2010, from http://econsultancy.com/blog/6133-virgin-america-flying-social-media-into-the-abyss

Rogers, E. M., & Kincaid, D. L. (1981). *Communication networks: Towards a paradigm for research*. New York, NY: Free Press.

Rogers, E. M. (2003). *Diffusion of innovations* (5th ed.). New York, NY: Free Press.

Rogers, E. M. (1962). *The diffusion of innovations*. New York, NY: The Free Press.

Ross, J. W., Weill, P., & Robertson, D. C. (2008). *Enterprise architecture as strategy*. Harvard Business School Press.

Roth, S. P., Schmutz, P., Pauwels, S. L., Bargas-Avila, J. A., & Opwis, K. (2010). Mental models for Web objects: Where do users expect to find the most frequent objects in online shops, news portals, and company web pages? *Interacting with Computers*, *22*(2), 140–152. doi:10.1016/j.intcom.2009.10.004

Roth, S. P., Tuch, A. N., Mekler, E. D. Z., Bargas-Avila, J. A., & Opwis, K. (in press). *Meet the users' expectations - Web objects are found faster on web pages when they are placed according to users' mental models*.

Rouse, W. B., & Morris, N. M. (1986). On looking into the black box: Prospects and limits in the search for mental models. *Psychological Bulletin*, *100*(3), 349–363. doi:10.1037/0033-2909.100.3.349

Roussel-Dupre, S. (2002, February 1). Self-service reality checkout. *Integrated Solutions for Retailers*.

Roussos, G., Kourouthanasis, P., & Moussouri, T. (2003). Designing appliances for mobile commerce and retailtainment. *Personal and Ubiquitous Computing*, *7*, 203–209. doi:10.1007/s00779-003-0221-3

Roussos, G., & Kostakos, V. (2009). RFID in pervasive computing: State-of-the-art and outlook. *Pervasive and Mobile Computing*, *5*(1), 110–131. doi:10.1016/j.pmcj.2008.11.004

Rowley, J. (1995). Multimedia kiosks in retailing. *International Journal of Retail & Distribution Management*, *23*(5), 32–40. doi:10.1108/09590559510089212

Rowley, J. (2000). Loyalty kiosks: Making loyalty cards work. *British Food Journal*, *102*(5/6), 390–397. doi:10.1108/00070700010329236

Rowley, J., & Slack, F. (2003). Kiosks in retailing: The quiet revolution. *International Journal of Retail & Distribution Management, 31*(6), 329–339. doi:10.1108/09590550310476060

Rowley, J., & Slack, F. (2007). Information kiosks: A taxonomy. *The Journal of Documentation, 63*(6), 879–897. doi:10.1108/00220410710836402

Roy, A. (1994). Correlates of mall visit frequency. *Journal of Retailing, 70*(2), 139–161. doi:10.1016/0022-4359(94)90012-4

Rubin, R., & Leigh, W. E. (2001). Mining customer intelligence: A practical framework for the development of a database marketing system for the small business. *Proceedings of the Second Annual USASBE/SBIDA Joint National Conference Proceedings* 2001: An Entrepreneurial Odyssey, February 7-10, 2001, Orlando, Florida. Retrieved June 15, 2010, from https://usasbe.org/knowledge/proceedings/proceedingsDocs/USASBE2001proceedings-071.pdf

Rulke, D. L., & Galaskiewicz, J. (2000). Distribution of knowledge, group network structure, and group performance. *Management Science, 46*(5), 612–625. doi:10.1287/mnsc.46.5.612.12052

Rundle-Thiele, S. (2005). Elaborating customer loyalty: Exploring loyalty to wine retailers. *Journal of Retailing and Consumer Services, 12*(5), 333–344. doi:10.1016/j.jretconser.2004.11.002

Russo, J. E., & Johnson, E. J. (1980). What do consumers know about familiar products? In Olson, J. E., & Arbor, A. (Eds.), *Advances in consumer research*. MI: Association for Consumer Research.

Ryals, L. (2003). Making customers pay: Measuring and managing customer risk and returns. *Journal of Strategic Marketing, 11*, 165–175. doi:10.1080/0965254032000133476

Ryan, T. M., & Jacobs, C. A. (2005). *Using investor relations to maximize equity valuation*. John Wiley & Sons, Inc.

Ryssel, R., Ritter, T., & Gemunden, H. G. (2004). The impact of information technology deployment on trust, commitment and value creation in business relationships. *Journal of Business and Industrial Marketing, 19*(3), 197–207. doi:10.1108/08858620410531333

Saade, R. G., Nebebe, F., & Tan, W. (2007). Viability of the technology acceptance model in multimedia learning environments: A comparative study. *Interdisciplinary Journal of Knowledge and Learning Objects, 3*, 176–184.

Salisbury, D, Pearson, R., Pearson, A., & Miller, D. (2001). Perceived security and World Wide Web purchase intention industrial management + data systems. *ABI/INFORM Global, 101*(3/4), 165.

Sambamurthy, V., Bharadwaj, A., & Grover, V. (2003). Shaping agility through digital options: Reconceptualizing the role of Information Technology in contemporary firms. *Management Information Systems Quarterly, 27*(2), 237–263.

Sanfety, A. G., Loewenstein, G., McClure, S. M., & Cohen, J. D. (2006). Neuroeconomics: Cross-currents in research on decision-making. *Trends in Cognitive Sciences, 10*, 108–116. doi:10.1016/j.tics.2006.01.009

Santa-Maria, L., & Dyson, M. C. (2008). *The effect of violating visual conventions of a website on user performance and disorientation. How bad can it be?* Paper presented at the SIGDOC.

SAP. (2010). SAP for retail – comprehensive solutions for your business. Retrieved September 1, 2010, from http://www.sap.com/industries/retail/index.epx

Savage, T. (2002). Future trends in retail IT systems. *European Retail Digest, 33*, 17–19.

Savitt, R. (1984). The wheel of retailing and retail product management. *European Journal of Marketing, 18*(6/7).

Schekkermann, J. (2004). *How to survive the jungle of EA Frameworks*. Trafford Publishing.

Schepers, J., & Wetzels, M. (2007). A meta-analysis of the technology acceptance model: Investigating subjective norm and moderation effects. *Information & Management, 44*(1), 90–103. doi:10.1016/j.im.2006.10.007

Schermerhorn, J. R. Jr, Hunt, J. G., & Osborn, R. N. (2008). *Organizational behavior* (8th ed.). New York, NY: John Wiley E Sons Inc.

Schiffman, L. G., Kanuk, L., & Hansen, H. (2008). *Consumer behaviour: A European outlook*. Financial Times/Prentice Hall.

Schiffman, L., & Kanuk, L. (2004). *Consumer behavior*. New Jersey, USA: Pearson Prentice Hall.

Schipul, E. (2006). The Web's next generation: Web 2.0. *Tactics,* March, 23.

Schlosser, A. E. (1998). Applying the functional theory of attitudes to understanding the influence of store atmosphere on store inferences. *Journal of Consumer Psychology, 7*(4), 345–369. doi:10.1207/s15327663jcp0704_03

Schmitt, B. H. (1999). *Experiential marketing: How to get customers to sense, feel, think, act, relate*. Free Press.

Schmutz, P., Heinz, S., Métrailler, Y., & Opwis, K. (2009). Cognitive load in e-commerce applications - measurement and effects on user satisfaction. *Advances in Human-Computer Interaction, 2009*, 121494. doi:10.1155/2009/121494

Schoenbachler, D. D., & Gordon, G. L. (2002). Trust and customer willingness to provide information in database-driven relationship marketing. *Journal of Interactive Marketing, 16*(3), 2–16. doi:10.1002/dir.10033

Schulz, D. P. (2008). 10 things you may have missed. *Stores*, March, 20.

Schuman, E. (2004). Self-service need not be self-delusional. *Storefrontbacktalk.com*, August.

Schumpeter, J. (1961). *Konjunkturzyklen. Eine theoretische, historische und statistische Analyse des kapitalistischen Prozesses*. Göttingen, Germany: Vandenhoeck & Ruprecht.

Schurr, P. H., & Ozanne, J. L. (1985). Influences on exchange processes: Buyer preconceptions of a seller's trustworthiness and bargaining toughness. *The Journal of Consumer Research, 11*(4), 939–953. doi:10.1086/209028

Schwarz, N. (2004). Metacognitive experiences in consumer judgment and decision making. *Journal of Consumer Psychology, 14*(4), 332–348. doi:10.1207/s15327663jcp1404_2

Scoresby, K. (2004). *Seven deadly excuses for poor design*. Retrieved October 2010, from http://www.zdnet.com/news/seven-deadly-excuses-for-poor-design/138101

Scott, D. (2007). *The new rules of marketing and PR: How to use news releases, blogs, podcasting, viral marketing, & online media to reach buyers directly*. New York, NY: John Wiley & Sons, Inc.

Sears, A., & Jacko, A. J. (Eds.). (2009). *Human-Computer Interaction. Development process*. London, UK: Taylor & Francis.

Seeman, E. D., & O'Hara, M. (2006). Customer relationship management in higher education: using information systems to improve the student-school relationship. *Campus-Wide Information Systems, 23*(1), 24–34. doi:10.1108/10650740610639714

Seffah, A., Donyaee, D., Kline, R. B., & Padda, H. R. (2006). Usability measurement and metrics: A consolidated model. *Software Quality Journal, 14*(2), 159–178. doi:10.1007/s11219-006-7600-8

Senecal, S., & Nantel, J. (2004). The influence of online product recommendations on consumers' online choices. *Journal of Retailing, 80*(2), 159–169. doi:10.1016/j.jretai.2004.04.001

Serif, T., & Ghinea, G. (2008). Mobile information access in the real world: A story of three wireless devices. *Computers in Human Behavior, 24*(4), 1385–1403. doi:10.1016/j.chb.2007.07.012

Servidio, R., Feraco, A., & Pantano, E. (2009). Usability evaluation of a Calabrian tourist visitor information center website. *Information Technology & Tourism, 11*(2), 95–110. doi:10.3727/109830509789994766

Sethuraman, R., & Parasuraman, A. (2005). Succeeding in the big middle through technology. *Journal of Retailing, 81*(2), 107–111. doi:10.1016/j.jretai.2005.03.003

Shaik, A. D., & Lenz, K. (2006). *Where's the search? Re-examining user expectations of web objects*. Retrieved August 21, 2008, from http://www.surl.org/usabilitynews/81/webobjects.asp

Shankar, V., Urban, G. L., & Sultan, F. (2002). Online trust: A stakeholder perspective, concepts, implications and future directions. *The Journal of Strategic Information Systems, 11*(4), 325–344. doi:10.1016/S0963-8687(02)00022-7

Sharkar, V., & Malthouse, C. (2007). The growth of interactions and dialogs in interactive marketing. *Journal of Interactive Marketing, 21*(2), 2–4. doi:10.1002/dir.20080

Sharma, G., Anantaram, C., & Hiranmay, G. (2009). The role of semantics in next-generation online virtual world-based retail store. In F. Lehmann-Grube, & J. Sablatnig (Eds.), *Proceedings of the First International Conference, FaVE* (pp. 91-105). Germany: Springer.

Shaw, J. B., & Barret-Power, E. (1998). The effects of diversity on small work group processes and performance. *Human Relations, 51*, 1307–1325. doi:10.1177/001872679805101005

Shaw, M. E. (1981). *Group dynamics: The psychology of small group behaviour*. New York, NY: McGraw-Hill.

Sheehan, K. B., & Hoy, M. G. (1999). Flaming, complaining, abstaining: How online users respond to privacy concerns. *Journal of Advertising, 28*(3), 37–51.

Sheehan, K. B., & Hoy, M. G. (2000). Dimensions of privacy concern among online consumers. *Journal of Public Policy & Marketing*, (Spring): 62–73. doi:10.1509/jppm.19.1.62.16949

Sheehan, K. B. (2005). Public opinions of online privacy: Definitions, assessment and implications for industry and public policy. In Krishnamurthy, S. (Ed.), *Contemporary research in e-marketing* (pp. 186–199). Hershey, PA: Idea Group Publishing.

Sheldon, P. J. (1997). *Tourism Information Technology*. UK: CAB International.

Sheldon, J., & Buzek, G. (2006, July 20). *Market study: 2006 North American self-checkout systems*. IHL Consulting Group.

Sherman, E., Mathur, A., & Smith, R. B. (1997). Store environment and consumer purchase behavior, mediating role of consumer emotions. *Psychology and Marketing, 14*(4), 361–378. doi:10.1002/(SICI)1520-6793(199707)14:4<361::AID-MAR4>3.0.CO;2-7

Sherman, E., & Smith, R. B. (1987). Mood states of shoppers and store perception, promising interactions and possible behavioral effect. *Advances in Consumer Research. Association for Consumer Research (U. S.), 14*(1), 251–254.

Shim, S., & Eastlick, M. A. (1998). The hierarchical influence of personal values on mall shopping attitude and behavior. *Journal of Retailing, 74*(1), 139–160. doi:10.1016/S0022-4359(99)80091-8

Shinoda, H., Hayhoe, M. M., & Shrivastava, A. (2001). What controls attention in natural environments? *Vision Research, 41*(25-26), 3535–3545. doi:10.1016/S0042-6989(01)00199-7

Shiv, B., & Fedorikhin, A. (1999). Heart and mind in conflict, the interplay of affect and cognition in consumer decision making. *The Journal of Consumer Research, 26*, 278–292. doi:10.1086/209563

Shneiderman, B. (2000). Designing trust into online experiences. *Communications of the ACM, 43*(12), 57–59. doi:10.1145/355112.355124

Shneiderman, B. (1997). *Designing the user interface: Strategies for effective human-computer interactions*. Addison Wesley Longman.

Siano, A. (2001). *Competenze e comunicazione del sistema d'impresa. Il vantaggio competitivo tra ambiguità e trasparenza*. Milano, Italy: Giuffrè.

Siano, A., Kaufmann, H. R., & Basile, G. (2009). The evolution of new consumer roles in branding. 2nd EMRBI Conference at the University of Salerno, Italy. 26-28. In D. Vrontis, Y. Weber, R. Kaufmann, & S. Tarba (Eds.), *Managerial and entrepreneurial developments in the Mediterranean area*. Cyprus: EuroMed Press. ISBN: 978-9963-634-76-7

Simon, F., & Usunier, J. (2007). Cognitive, demographic, and situational determinants of service customer preference for personnel-in-contact over self-service technology. *International Journal of Research in Marketing, 24*(2), 163–173. doi:10.1016/j.ijresmar.2006.11.004

Singer, P., Ferri, A., Aiello, L., & Cacia, C. (in press). -2010). Internet as a "point of synergy" between communication and distribution: Hypothesis of model applied to tourism. *Journal of Digital Content and Technological Application.*

Singer, P. (2002). *Economia digitale, dinamica evolutiva e valutazione delle imprese internet. Elementi di teoria ed evidenze empiriche.* Padova, Italy: Cedam.

Singer, P., & Cacia, C. (2009, October). *The role of Web investor relations for mitigating and manage stock exchange liquidity and enterprise risks*. Paper presented at the 9th Global Conference on Business & Economics 2009 Program, Cambridge University, UK.

Singh, J., & Sirdeshmukh, D. (2000). Agency and trust mechanisms in consumer satisfaction and loyalty judgments. *Journal of the Academy of Marketing Science, 28*, 150–167. doi:10.1177/0092070300281014

Sit, J., Merrilees, W., & Birch, D. (2003). Entertainment-seeking mall patrons, the missing segments. *International Journal of Retail and Distribution Management, 31*(2), 80–94. doi:10.1108/09590550310461985

Slack, F., & Rowley, J. (2002a). Kiosks 21: A new role for information kiosks? *International Journal of Information Management, 22*(1), 67–83. doi:10.1016/S0268-4012(01)00041-X

Slack, F., & Rowley, J. (2002b). Online kiosks: The alternative to mobile technologies for mobile users. *Internet Research: Electronic Networking Applications and Policy, 12*(3), 248–257. doi:10.1108/10662240210430928

Smalltree, H. (2006). *The best practices in customer intelligence*. Retrieved June 15, 2010, from www.SearchDataManagement.com

Smith, P., & Burns, D. J. (1996). Atmospherics and retail environments, the case of the power aisle. *International Journal of Retail and Distribution Management, 24*(1), 7–14. doi:10.1108/09590559610107076

Smith, R. B., & Sherman, E. (1993). Effects of store perception and mood on consumer behavior, a theoretical and empirical analysis. *Advances in Consumer Research. Association for Consumer Research (U. S.), 20*(1), 631.

Smith, H. J., Milberg, S. J., & Burke, S. J. (1996). Information privacy: Measuring individuals' concerns about organizational practices. *Management Information Systems Quarterly, 20*(2), 167–196. doi:10.2307/249477

Smith, M., Bailey, J., & Brynjolfsson, E. (2000). Understanding digital markets: Review and assessment. In Bryonjolfsson, E., & Kahim, B. (Eds.), *Understanding the digital economy*. Cambridge, MA: MIT Press.

Snellman, K., & Vihtkari, T. (2003). Customer complaining behaviour in technology-based service encounters. *International Journal of Service Industry Management, 14*(2), 217–231. doi:10.1108/09564230310474174

Söderlund, M., & Julander, C. R. (2009). Physical attractiveness of the service worker in the moment of truth and its effects on customer satisfaction. *Journal of Retailing and Consumer Services, 16*, 216–226. doi:10.1016/j.jretconser.2008.11.008

Southwell, M. (2002, July 2). Beyond the POS. *Business Trends*.

Spangenberg, E. R., Sprott, D. E., Grohmann, B., & Tracy, D. L. (2006). Gender-congruent ambient scent influences on approach and avoidance behaviors in a retail store. *Journal of Business Research, 59*, 1281–1287. doi:10.1016/j.jbusres.2006.08.006

Speier, C., & Venkatesh, V. (2002). The hidden minefields in the adoption of sales force automation technologies. *Journal of Marketing, 66*(3), 98–111. doi:10.1509/jmkg.66.3.98.18510

Spies, K., Hesse, F., & Loesch, K. (1997). Store atmosphere, mood and purchasing behavior. *International Journal of Research in Marketing, 14*, 1–17. doi:10.1016/S0167-8116(96)00015-8

Spool, J. M. (2008). *The wheres and whens of users' expectations*. Retrieved August 19, 2008, from http://www.uie.com/articles/user_expectations/

Sraeel, H. (2005). Hot on the ID theft trail: Will the real Jane Doe stand up? *Bank Technology News, 18*(2), 8.

Stahl, F., & Freudenschuss, M. (2006). *Mobile business through RFID technology: About the potential of RFID technology for distributing free and paid content on mobile devices in retail business*. Paper presented at the Americas Conference on Information Systems.

Stamboulis, Y., & Skayannis, P. (2003). Innovation strategies and technology for experience based tourism. *Tourism Management, 24*(1), 35–43. doi:10.1016/S0261-5177(02)00047-X

Stasser, G., Vaughan, S. I., & Stewart, D. D. (2000). Pooling unshared information: The benefits of knowing how access to information is distributed among group members. *Organizational Behavior and Human Decision Processes*, *82*(1), 102–116. doi:10.1006/obhd.2000.2890

Stecher, P. (1993). Building business and applications systems with the retail application architecture. *IBM Systems Journal*, *32*, 278–306. doi:10.1147/sj.322.0278

Steinfield, C. (2004). The missing link: Connecting physical and virtual channels through click & mortar electronic commerce. In Stanoevska-Slabeva, K. (Ed.), *Digital economy* (pp. 141–157). Heidelberg, Germany: Anspruch und Wirklichkeit.

Stern, B. B., Royne, M. B., Stafford, T. F., & Bienstock, C. C. (2008). Consumer acceptance of online auctions: An extension and revision of the TAM. *Psychology and Marketing*, *25*(7), 619–636. doi:10.1002/mar.20228

Stewart, G. L., Manz, C. C., & Sims, H. P. (1999). *Team work and group dynamics*. USA: John Wiley & Sons.

Stiglitz, J., & Weiss, A. (1981). Credit rationing in markets with imperfect information. *The American Economic Review*, *71*(3), 93–410.

Stipanuk, D. M. (1993). Tourism and technology. Interactions and implications. *Tourism Management*, August.

Stone, M., Fos, B., & Henderson, I. (2003). The quality of customer information management in customer life cycle management. *Journal of Database Marketing and Customer Strategy Management*, *10*(3), 240–254. doi:10.1057/palgrave.jdm.3240113

Strassmann, P. (1990). *The business ~due of computers: An executive guide*. Connecticut: The Information Economics Press.

Stuart, A. (2003). *Technofile: Why Wi-Fi?* Retrieved from http://www.inc.com/articles/2003/09/wifi.html

Stüber, E. (2011). *Kaufempfehlungen als Personalisierungsansatz im Internethandel – Eine empirische Untersuchung der Akzeptanz am Beispiel der Bekleidungsbranche*. Gabler.

Su, X. (2007). Intertemporal pricing with strategic customer behavior. *Management Science*, *53*(5), 726–741. doi:10.1287/mnsc.1060.0667

Summers, T. A., & Hebert, P. R. (2001). Shedding some light on store atmospherics. Influence of illumination on consumer behavior. *Journal of Business Research*, *54*, 145–150. doi:10.1016/S0148-2963(99)00082-X

Summers, J. O. (2001). Guidelines for conducting research and publishing in marketing: From conceptualization though the review process. *Journal of the Academy of Marketing Science*, *29*(4), 405–415. doi:10.1177/03079450094243

Sun, T., Tai, Z., & Tsai, K. C. (2009). The role of interdependent self-construal in consumers' susceptibility to retail sales persons' influence: A hierarchical approach. *Journal of Retailing and Consumer Services*, *16*(5), 360–366. doi:10.1016/j.jretconser.2009.04.002

Swaminathan, V., Lepkowska-White, E., & Rao, B. P. (1999). Browsers or buyers in cyberspace? An investigation of factors influencing electronic exchange. *Journal of Computer-Mediated Communication*, *5*(2), 1–24.

Sward, D. (2006). *Measuring the business value of information technology*. Intel Press.

Swoboda, B. S. (1998). Conditions of consumer information seeking: Theoretical foundations and empirical results of using interactive multimedia systems. *International Review of Retail, Distribution and Consumer Research*, *8*(4), 361–381. doi:10.1080/095939698342715

Szmigin, I., & Bourne, H. (1998). Consumer equity in relationship marketing. *Journal of Consumer Marketing*, *15*(6), 544–557. doi:10.1108/07363769810240545

Takács, G., Pilászy, I., Németh, B., & Tikk, D. (2009). Scalable collaborative filtering approaches for large recommender systems. *Journal of Machine Learning Research*, *10*, 623–656.

Tambo, T. (2010b). *IT Retail i et praktisk perspektiv (Lecture notes for retail managers training)*. TEKO-Center.

Tambo, T. (2010a). Bespoke ERP – a systems view on enterprise management. In *Proceedings of the 19th International Conference on Management of Technology*, Cairo

Tambo, T., & Koch, C. (2010). Enterprise architecture in the supply chain. *In Proceedings of Workshop on Enterprise Architecture Challenges and Responses.* Yonezawa.

Tanriverdi, H. (2005). Information technology relatedness, knowledge management capability, and performance of multibusiness firms. *Management Information Systems Quarterly*, *29*(2), 311–334.

Tatnall, A. (Ed.). (2010). *Web technologies: Concepts, methodologies, tools, and applications*. Hershey, PA: IGI Global.

Taylor, S. E., & Thompson, S. C. (1982). Stalking the elusive vividness effect. *Psychological Review*, *89*(2), 155–181. doi:10.1037/0033-295X.89.2.155

Teas, K. R., & Palan, K. M. (1997). The realms of scientific meaning framework for constructing theoretically meaningful nominal definitions of marketing concepts. *Journal of Marketing*, *61*(4), 52–67. doi:10.2307/1251830

Teas, R. K., & Agrawal, S. (2000). The effect of extrinsic product cues on consumers' perceptions of quality, sacrifice, and value. *Journal of the Academy of Marketing Science*, *28*, 278–290. doi:10.1177/0092070300282008

Tellis, G. J., & Gaeth, G. J. (1990). Best value, price-seeking, and price aversion: The impact of information and learning on consumer choices. *Journal of Marketing*, *54*(2), 34–45. doi:10.2307/1251868

Tenenhaus, M., Esposito Vinzi, V., Chatelin, Y.-M., & Lauro, C. (2005). PLS path modeling. *Computational Statistics & Data Analysis*, *48*(1), 159–205. doi:10.1016/j.csda.2004.03.005

Textilwirtschaft. (2010). *Die größten Bekleidungslieferanten in Europa* 2008. Retrieved September 1, 2010, from http://www.textilwirtschaft.de/business/markt/marktfakten/pdfs/64_org.pdf

Thatcher, A., & Greyling, M. (1998). Mental models of the Internet. *International Journal of Industrial Ergonomics*, *22*, 299–305. doi:10.1016/S0169-8141(97)00081-4

The Economist. (2006, March 11). Signs of the times. *The Economist, 378*(8468).

The State. (2008, March 27). Piggly Wiggly ends Pay By Touch service. *The State*, B7.

Theodoridis, P. K., & Chatzipanagiotou, K. C. (2009). Store image attributes and customer satisfaction across different customer profiles within the supermarket sector in Greece. *European Journal of Marketing*, *43*(5/6), 708–734. doi:10.1108/03090560910947016

Thibaut, J., & Kelley, H. H. (1959). *The social psychology of groups*. New York, NY: John Wiley.

Thomas-Hunt, M. C., Ogden, T. Y., & Neale, M. A. (2003). Who's really sharing? Effects of social and expert status on knowledge exchange within groups. *Management Science*, *49*(4), 464–477. doi:10.1287/mnsc.49.4.464.14425

Thompson, R., Higgins, C., & Howell, J. (1991). Personal computing: Toward a conceptual model of utilization. *Management Information Systems Quarterly*, (March): 125. doi:10.2307/249443

Thong, J. Y. L., Hong, S., & Tama, K. Y. (2006). The effects of post-adoption beliefs on the expectation-confirmation model for information technology continuance. *International Journal of Human-Computer Studies*, *64*, 799–810. doi:10.1016/j.ijhcs.2006.05.001

Thorbjørnsen, M., Supphellen, M., Nysveen, H., & Pedersen, P. E. (2002). Building brand relationship online: A comparison of two interactive applications. *Journal of Interactive Marketing*, *16*(3), 17–34. doi:10.1002/dir.10034

Timmers, P. (1998). Business models for electronic markets. *Electronic Markets*, *8*, 3–8. doi:10.1080/10196789800000016

Timmor, Y., & Rymon, T. (2007). To do or not to do: The dilemma of technology-based service improvement. *Journal of Services Marketing*, *21*(2), 99–111. doi:10.1108/08876040710737868

Todd, P., & Benbasat, I. (1999). Evaluating the impact of DSS, cognitive effort, and incentives on strategy selection. *Information Systems Research*, *10*(4), 356–374. doi:10.1287/isre.10.4.356

Toppo, G. (2006, February 23). Eye scans: A high-tech hall pass? *USA Today*, 12B.

Tosi, H. L., Mero, N. P., & Rizzo, J. R. (2000). *Managing organizational behavior*. Blackwell Business.

Towers, N., & Chen, R. (2007). Employing the participative paradigm as a valid approach of supply chain. *International Journal of Retail & Distribution, 36*, 627–637. doi:10.1108/09590550810883478

Trim, P. R. J., & Lee, Y.-I. (2006). The role of marketing intelligence officers in strategy formulation and implementation. In *Handbook of business strategy*, (pp. 125-130).

Tuch, A. N., Bargas-Avila, J. A., Opwis, K., & Wilhelm, F. H. (2009). Visual complexity of websites: Effects on users' experience, physiology, performance, and memory. *International Journal of Human-Computer Studies, 67*(9), 703–715. doi:10.1016/j.ijhcs.2009.04.002

Tung, L. L., & Tan, J. H. (1998). A model for the classification of information kiosks in Singapore. *International Journal of Information Management, 18*(4), 255–264. doi:10.1016/S0268-4012(98)00014-0

Turley, L. W., & Milliman, R. E. (2000). Atmospheric effects on shopping behavior, a review of the experimental evidence. *Journal of Business Research, 49*(2), 193–211. doi:10.1016/S0148-2963(99)00010-7

Ullman, S. (1984). Early processing of visual information. In Gazzaniga, M. S. (Ed.), *Handbook of cognitive neuroscience*. New York, NY: Plenum.

Underhill, P. (2004). *Call of the mall*. New York, NY: Simon & Schuster.

Urban, G. L., Sultan, F., & Qualls, W. J. (2000). Placing trust at the center of your internet strategy. *Sloan Management Review, 42*(1), 39–48.

Urban, G. (2003). *Customer advocacy: Is it for you?* MIT Sloan School of Management. Center for E-Business, Paper 175.

Ustundag, A., & Tanyas, M. (2009). The impacts of radio frequency identification technology on supply chain costs. *Transportation Research Part E, Logistics and Transportation Review, 45*(1), 29–39. doi:10.1016/j.tre.2008.09.001

van den Berg, P. E. W., Arentze, T. A., & Timmermans, H. J. P. (2009). ICT and social interaction: Modelling communication mode choice and its effect on travel behavior. *Proceedings of the IATBR Conference*. Jaipur, India, in press.

van der Heijden, H. (2006). Mobile decision support for in-store purchase decisions. *Decision Support Systems, 42*(2), 656–663. doi:10.1016/j.dss.2005.03.006

Van der Heijden, H. (2004). User acceptance of hedonic information systems. *Management Information Systems Quarterly, 28*(4), 695–704.

Van Raaij, E. M. (2005). The strategic value of customer profitability analysis. *Marketing Intelligence & Planning, 23*(4), 372–381. doi:10.1108/02634500510603474

van Schaik, P., & Ling, J. (2009). The role of context in perceptions of the aesthetics of web pages over time. *International Journal of Human-Computer Studies, 67*(1), 79–89. doi:10.1016/j.ijhcs.2008.09.012

Varki, S., & Colgate, M. (2001). The role of price perceptions in an integrated model of behavioral intentions. *Journal of Service Research, 3*(3), 232–240. doi:10.1177/109467050133004

Vaughan, M. W., & Dillon, A. (2006). Why structure and genre matter for users of digital information: A longitudinal experiment with readers of a web-based newspaper. *International Journal of Human-Computer Studies, 64*, 502–526. doi:10.1016/j.ijhcs.2005.11.002

Vedamani, G. G. (2006). *Retail management – functional principles and practices*. Jaico Publishing House.

Veletsianos, G., & Miller, C. (2008). Conversing with pedagogical agents: A phenomenological exploration of interacting with digital entities. *British Journal of Educational Technology, 39*(6), 969–986. doi:10.1111/j.1467-8535.2007.00797.x

Venkatesan, R., Kumar, V., & Ravishankar, N. (2007). Multichannel shopping: Causes and consequences. *Journal of Marketing, 71*(2), 114–132. doi:10.1509/jmkg.71.2.114

Venkatesh, V., Morris, M. G., Davis, G. B., & Davis, F. D. (2003). User acceptance of information technology: towards a unified view. *Management Information Systems Quarterly, 27*(3), 425–478.

Venkatesh, V., & Davis, F. D. (2000). A theoretical extension of the technology acceptance model: Four longitudinal field studies. *Management Science, 46*(2), 186–204. doi:10.1287/mnsc.46.2.186.11926

Verity, J., & Hoff, R. (1994, 14 November). The internet: How it will change the way you do business. *Business Week*.

Vesanen, J., & Raulas, M. (2006). Building bridges for personalization: A process model for marketing. *Journal of Interactive Marketing*, *20*(1), 5–20. doi:10.1002/dir.20052

Vickery, G., & Wunsch-Vincent, S. (2007). *Participative Web and user-created content. Web 2.0, wikis and social networking*. OECD.

Vignali, C., Vignali, G., Vranesevic, T., & Vrontis, D. (2006). *Basic selling skills*. UK: Foxwell & Davies.

Vijayasarathy, L. R. (2004). Predicting consumer intentions to use on-line shopping: The case for an augmented technology acceptance model. *Information & Management*, *41*(6), 747–762. doi:10.1016/j.im.2003.08.011

Visich, J., Li, S., Khumawala, B., & Reyes, P. (2009). Empirical evidence of RFID impacts on supply chain performance. *International Journal of Operations & Production Management*, *29*(12), 1290–1315. doi:10.1108/01443570911006009

Visual Retailing. (2010). *Mockshop – see what you can do*. Retrieved September 1, 2010, from http://www.visualretailing.com

Viswanathan, S. (2005). Competing across technology-differentiated channels: The impact of network externalities and switching costs. *Management Science*, *51*(3), 483–496. doi:10.1287/mnsc.1040.0338

von Reischach, F., Guinard, D., Michahelles, F., & Fleisch, E. (2009). *A mobile product recommendation system interacting with tagged products*. Paper presented at the Seventh Annual IEEE International Conference on Pervasive Computing and Communications (PerCom 2009).

Wakefield, K. L., & Baker, J. (1998). Excitement at the mall, determinants and effects on shopping response. *Journal of Retailing*, *74*(4), 515. doi:10.1016/S0022-4359(99)80106-7

Walczuch, R., & Lundgren, H. (2004). Psychological antecedents of institution-based consumer trust in e-retailing. *Information & Management*, *42*, 159–177.

Walker, R. H., Craig-Lees, M., Hecker, R., & Francis, H. (2002). Technology-enabled service delivery. An investigation of reasons affecting customer adoption and rejection. *International Journal of Service Industry Management*, *13*(1), 91–106. doi:10.1108/09564230210421173

Wang, H.-C., & Doong, H.-S. (in press). Argument from a spokeperson type: The recommendation strategy of virtual salespersons. *International Journal of Information Management*.

Wang, Y. (2008). Examining the level of sophistication and success of destination marketing systems: Impacts of organizational factors. *Journal of Travel & Tourism Marketing*, *24*(1), 82–98. doi:10.1300/J073v24n01_06

Wang, Y., Hernandez, M., & Minor, M. (2010). Web aesthetic effects on perceived online service quality and satisfaction in an e-tail environment: The moderating role of purchase task. *Journal of Business Research*, *63*(9-10), 935–942. doi:10.1016/j.jbusres.2009.01.016

Wang, W., & Benbasat, I. (2007). Recommendation agents for electronic commerce: Effects of explanation facilities on trusting beliefs. *Journal of Management Information Systems*, *23*(4), 217–246. doi:10.2753/MIS0742-1222230410

Wang, Y., Wang, Y., Lin, H., & Tang, T. (2003). Determinants of user acceptance of internet banking: An empirical study. *International Journal of Service Industry Management*, *14*(5), 501–519. doi:10.1108/09564230310500192

Wasko, M. M., & Faraj, S. (2005). Why should I share? Examining social capital and knowledge contribution in electronic networks of practice. *Management Information Systems Quarterly*, *29*(1), 35–57.

Watson, R. T., Pitt, L. F., Berthon, P., & Zinkhan, G. M. (2002). U-commerce: Expanding the universe of marketing. *Journal of the Academy of Marketing Science*, *30*(4), 333–347. doi:10.1177/009207002236909

Watzalawick, P., Beavin, J., & Jackson, D. (1967). *Pragmatics of human communication. A study of interactional patterns, pathologies, and paradoxes*. New York, NY: W.W. Norton e Co., Inc.

Webb, W. (2002, November 14). On-the-shelf electronic price labels. *EDN*, 36-40.

Weber, M. M., & Kantamneni, S. P. (2002). POS and EDI in retailing: An examination of underlying benefits and barriers. *Supply Chain Management: International Journal, 7*, 311–317. doi:10.1108/13598540210447755

Webster, F. E. (1996). The changing role of marketing in the corporation. *Journal of Marketing, 56,* October 1992, 1-17. In G. Cozzi, & G. Ferrero (Eds.) *Le frontiere del marketing*. Torino, Italy: Giappichelli.

Weijters, B., Rangarajan, D., Falk, T., & Schillewaert, N. (2007). Determinants and outcomes of customers' use of self-service technology in a retail setting. *Journal of Service Research, 10*(1), 3–21. doi:10.1177/1094670507302990

Weijters, B., Schillewaert, N., Rangarajan, D., & Falk, T. (2005). *Customers' usage of self service technology in a retail setting*. Vlerick Leuven Gent Management School Working Paper Series 2005/19.

Weinstein, A. (2002). Customer-specific strategies-customer retention: A usage segmentation and customer value approach. *Journal of Targeting. Measurement & Analysis for Marketing, 10*(3), 259–268. doi:10.1057/palgrave.jt.5740051

Werthener, H. (2001). *E-commerce and travel/tourism issues and challengers. EC3 – E-Commerce Competence Center*. Vienna, Austria: IFIT.

Werthner, H., & Klein, S. (1999). *Information Technology and tourism. A challenging relationship*. Vienna, Austria: Springer Computer Science.

West, M. A., Garrod, S., & Carletta, J. (1997). Group decision-making and effectiveness: Unexplored boundaries. In Cooper, C. L., & Jackson, S. E. (Eds.), *Creating tomorrow's organizations: A handbook for future research in organizational behavior*. John Wiley & Sons.

Westerman, S. J., Tuck, G. C., Booth, S. A., & Khakzar, K. (2007). Consumer decision support systems: Internet versus in-store application. *Computers in Human Behavior, 23*(6), 2928–2944. doi:10.1016/j.chb.2006.06.006

Westin-Harris. (1997). *Privacy and American business study, a survey*.

Wildstorm, S. H. (2005). New weapons to stop identity thieves. *Business Week, 3924*, 24.

Wilkie, W. L., & Morre, E. S. (2003). Scholarly research in marketing: Exploring the 4 eras of thought development. *Journal of Public Policy & Marketing, 22*(2), 116–146. doi:10.1509/jppm.22.2.116.17639

Wilson, M. (2005). Atmosphere key to well-being. *Chain Store Age, 81*(10), 94.

Wirtz, B. W. (2001). *Electronic business.* Wiesbaden,G ermany: Gabler Verlag.

Wixom, B. H., & Todd, P. A. (2005). A theoretical integration of user satisfaction and technology acceptance. *Information Systems Research, 16*(1), 85–102. doi:10.1287/isre.1050.0042

Wold, H. O. (1974). Causal flows with latent variables: Partings of the ways in the light of NIPALS modeling. *European Economic Review, 5*(1), 67–86. doi:10.1016/0014-2921(74)90008-7

Wold, H. O. (1982). Soft modeling: The basic design and some extensions. In Jöreskog, K. G., & Wold, H. O. (Eds.), *Systems under indirect observations, part II* (pp. 1–54). Amsterdam.

Wold, H. O. (1985). Partial least squares. In Kotz, S., & Johnson, N. L. (Eds.), *Encyclopedia of statistical sciences, 6* (pp. 581–591). New York.

Wolfinbarger, M., & Gilly, M. C. (2001). Shopping online for freedom, control, and fun. *California Management Review, 43*, 34–55.

Wong, C. Y., Arlbjørn, J. S., Hvolby, H., & Johansen, J. (2006). Assessing responsiveness of a volatile and seasonal supply chain: A case study. *International Journal of Production Economics, 104*, 709–721. doi:10.1016/j.ijpe.2004.12.021

Wong, J., & Law, R. (2005). Analysing the intention to purchase on hotel websites: A study of travellers to Hong Kong. *International Journal of Hospitality Management, 24*(3), 311–329. doi:10.1016/j.ijhm.2004.08.002

Woodruff, R. B. (1997). Customer value: The next source of competitive advantage. *Journal of the Academy of Marketing Science, 25*, 139–153. doi:10.1007/BF02894350

Woodside, A. G., Frey, L. L., & Daly, R. T. (1989). Linking service quality, customer satisfaction, and behavioral intention. *Journal of Health Care Marketing, 9*, 5–17.

Wright, L. T., Newman, A., & Dennis, C. E. (2006). Enhancing consumer empowerment. *European Journal of Marketing, 40*(9/10), 925–935. doi:10.1108/03090560610680934

Wright, K. (n.d.). *Best articles online, new wireless technology – Wi-Fi*. Retrieved 20 October, 2009, from http://www.bestarticlesonline.com/.../new-wireless-technology-wi-fi-.html

Wu, J. J., & Chang, Y. S. (2005). Towards understanding members' interactivity, trust, and flow in online travel community. *Industrial Management & Data Systems, 105*(7), 937–954. doi:10.1108/02635570510616120

Wu, F., Yeniyurt, S., Kim, D., & Cavusgil, S. T. (2006). The impact of information technology on supply chain capabilities and firm performance: A resource-based view. *Industrial Marketing Management, 35*, 493–504. doi:10.1016/j.indmarman.2005.05.003

Xiao, B., & Benbasat, I. (2007). E-commerce product recommendation agents: Use, characteristics, and impact. *Management Information Systems Quarterly, 31*(1), 137–209.

Xie, E., Hock-Hai, T., & Wen, W. (2006). Volunteering personal information on the Internet: Effects of reputation, privacy notices, and rewards on online consumer behavior. *Marketing Letters, 17*(1), 61–74. doi:10.1007/s11002-006-4147-1

Yalch, R., & Spangenberg, E. (1990). Effects of store music on shopping behavior. *Journal of Consumer Marketing, 7*, 55–63. doi:10.1108/EUM0000000002577

Yang, Y.-F. (2008). The roles of human resources, information technology, and marketing knowledge capabilities in performance: An extension of the resource-based theory perspective. *Social Behavior and Personality, 36*(9), 1269–1282. doi:10.2224/sbp.2008.36.9.1269

Yang, J.-M., Li, K. F., & Zhang, D.-F. (2009). Recommendation based on rational inferences in collaborative filtering. *Knowledge-Based Systems, 22*(1), 105–114. doi:10.1016/j.knosys.2008.07.004

Yin, R. K. (2008). Applications of case study research. *Applied Social Research Methods Series, 34.*

Yoo, C., Park, J., & MacInnis, D. J. (1998). Effects of store characteristics and in-store emotional experiences on store attitude. *Journal of Business Research, 42*(3), 253–263. doi:10.1016/S0148-2963(97)00122-7

Yoon, C., Cole, C. A., & Lee, M. P. (2009). Consumer decision making and aging: Current knowledge and future directions. *Journal of Consumer Psychology, 19*, 2–16. doi:10.1016/j.jcps.2008.12.002

Yoon, S. J. (2002). The antecedents and consequences of trust in online purchase decisions. *Journal of Interactive Marketing, 16*(2), 47–63. doi:10.1002/dir.10008

Youll, J., Morris, J., Krikorian, R., & Maes, P. (2000). *Impulse: Location-based agent assistance*. Paper presented at the Fourth International Conference on Autonomous Agents.

Zablah, A. R., Bellenger, D. N., & Johnston, W. J. (2004). Anticipating consumer behavior with analytics. *Industrial Marketing Management, 33,* 475-489. Retrieved June 15, 2010, from http://www.crm2day.com/content/t6_librarynews_1.php?id

Zahay, D. L., & Peltier, J. (2008). Interactive strategy formation: Organizational and entrepreneurial factors related to effective customer information systems practices in B2B firms. *Industrial Marketing Management, 37*, 191–205. doi:10.1016/j.indmarman.2006.10.004

Zajonc, R. B., & Markus, H. (1984). Affect and cognition, the hard interface. In Izard, C. E., Kagan, J., & Zajonc, R. B. (Eds.), *Emotions, cognition, and behavior*. Cambridge, UK: Cambridge University Press.

Zeithaml, V. A., Parasuraman, A., & Malhotra, A. (2002). Service quality delivery through websites: A critical review of extant knowledge. *Journal of the Academy of Marketing Science, 30*, 362–375. doi:10.1177/009207002236911

Zentes, J., Morschett, D., & Schramm-Klein, H. (2007). *Strategic retail management – text and international cases*. Gabler.

Zhang, L., Snavelvelyn, N., Curless, B., & Seitz, M. S. (2004). Spacetime faces: High resolution capture for modeling and animation. In [ACM Press.]. *Proceedings of, SIGGRAPH2004*, 548–558.

Zhang, X., & Li, Y. (2008). Use of collaborative recommendations for web search: An exploratory user study. *Journal of Information Science*, *34*(2), 145–161. doi:10.1177/0165551507080413

Zhenhui, J., & Benbasat, I. (2004). Virtual product experience: Effects of visual and functional control of products on perceived diagnosticity and flow in electronic shopping. *Journal of Management Information Systems*, *21*(3), 111–147.

Zielke, S. (2006). Measurement of retailers' price images with a multiple-item scale. *International Review of Retail, Distribution and Consumer Research*, *16*(3), 297–316. doi:10.1080/09593960600696990

Zielke, S. (2010). How price image dimensions influence shopping intentions for different store formats. *European Journal of Marketing*, *44*(6), 748–770.

Zielke, S., Lietke, B., Toporowski, W., & Boslau, M. (2008). Self-service and traditional check-outs in a retail setting: An analysis of perceptual differences and satisfaction drivers. *Marketing – Journal of Research and Management, 30*(1), 24-37.

About the Contributors

Eleonora Pantano is a Post doc research fellow at University of Calabria (Italy). She holds a PhD in "Psychology of Programming and Artificial Intelligence." Her research interests are related to the applications of advanced technologies to retailing and tourism, with emphasis on the investigation of consumer behaviour in pervasive environments. She has been Assistant teacher of Integrated Marketing Communication, Engineering Faculty, University of Calabria; visiting lecturer at College of Business, University of Illinois (USA); visiting lecturer at Master in Business and Administration (MBA) Marketing Module at the Faculty of Economics & Business, University of Zagreb (HR). Furthermore, she is member of the Editorial Board of numerous international journals, guest editor of the special issue of *Journal of Retailing and Consumer Services on Applications of New Technologies to Retailing*, 17(3); and of *International Journal of Digital Content Technology and its Applications* on Digital contents management for improving consumers experience, 4(7). She was the Highly Commended Award winner of the 2008/2009 Emerald/EMRBI Business Research Award for Young Researchers.

Harry Timmermans holds a PhD degree in Geography/Urban and Regional Planning. He studied at the Catholic University of Nijmegen, The Netherlands. His dissertation concerned a theory of the functional and spatial structure and the dynamics of central place systems. Since 1976 he is affiliated with the Faculty of Architecture, Building and Planning of the Eindhoven University of Technology, The Netherlands. First as an assistant professor of Quantitative and Urban Geography, later as an associate professor of Urban Planning Research. In 1986 he was appointed chaired professor of Urban Planning at the same institute. In 1992 he founded the European Institute of Retailing and Services Studies (EIRASS) in Eindhoven, the Netherlands (a sister-institute of the Canadian Institute of Retailing and Services Studies). His main research interests concern the study of human judgement and choice processes, mathematical modelling of urban systems and spatial interaction and choice pattens and the development of decision support and expert systems for application in urban planning. He has published several books and many articles in journals in the fields of marketing, urban planning, architecture and urban design, geography, environmental psychology, transportation research, urban and regional economics, urban sociology, leisure sciences and computer science.

* * *

Lucia Aiello was born in Vico Equense, Italy, in 1980. She currently holds the Tutor in Marketing and Communication, Studies and Research at University of Rome (Italy). She conducts training in marketing, communications, and visual merchandising at ENI, Italy. She holds a degree in Tourism

Management from "Parthenope" University, Italy (2003), and a Ph.D. in Marketing and Communication from University of Salerno, Italy (2010). Her main research interests are in: tourism, cultural activities, local development, Web communication, Web marketing, entrepreneurship and innovation, and technologies management for tourism. She has published in academic and professional books, journals, and conference proceedings. She has participated in sectorial research for the University Mercatorum, the G. Tagliacarne, ConfIndustria, and CUEIM.

Javier A. Bargas-Avila is a researcher at the Center of Cognitive Psychology and Methodology within the Department of Psychology at the University of Basel, Switzerland. He has a background in cognitive science and Human-Computer Interaction (HCI). His work focuses on HCI, user experience, user satisfaction, and usability.

Daniel Baier is the Professor of Marketing and Innovation Management at Brandenburg University of Technology Cottbus, Germany. He received his diploma degree in computer science and his doctorates in business administration and economics from the University of Karlsruhe, Germany. His research interests relate to marketing research for product and service innovations, to innovation management, to data and Web mining, operations research, and statistics, as well as to new technologies in online retailing and marketing research. His research has appeared in academic journals like Annals of Operations Research, Journal of Econometrics, Journal of Retailing and Consumer Services, Marketing ZFP, R&D Management, Zeitschrift für Betriebswirtschaft, and Zeitschrift für betriebswirtschaftliche Forschung.

Eleonora Bilotta is full Professor of Psychology at the University of Calabria. She has published more than 150 scientific papers and is co-author of various books. She is interested in the following research areas: complexity theory, dynamical systems, cognitive science and educational technology. She is director of the Laboratory of Psychology and Cognitive Science and coordinates the Ph.D. course in Science and Technologies of the Complex Systems. She has coordinated and coordinates many national and international research projects such as: PRIN "Mathematical Modelling of natural and artificial behaviour"; MASC (Music and Sound by Chaos), a project that involves the MacGill University of Montreal. Moreover she collaborated at the NETConnect project (Connecting European Culture through New Technology) – EU Culture 2000 European Programme; COMSON project (Coupled Multiscale Simulation and Optimization in Nanoelectronics) - a Marie Curie Research Training Network; PIA Innovation, a National Research project of research and development.

Josko Brakus (BSc, MSc, PhD) is a Lecturer in Marketing at Brunel Business School, Brunel University, West London. Joško conducts experimental research in consumer behaviour. He studies experiential marketing and branding practices. In particular, he investigates brands as sources of compelling experiences for consumers. He also studies managerial and consumer judgement and decision processes. He focuses on selective and biased information processing and on how individual differences (e.g., self-regulation) affect consumers' adoption of new products and their responses to marketing stimuli in general. Joško's work has been published in leading academic marketing journals such as the Journal of Marketing Research and the Journal of Marketing. Joško joined Brunel in May 2009. Previously he was an Assistant Professor at Simon Graduate School of Business, University of Rochester, upstate New

York. He holds a PhD in Marketing from Columbia Business School, Columbia University. Recently, Joško was co-awarded a Knowledge Transfer Partnership research grant in the amount of £540,000.

Claudia Cacia was born in Catania, Italy, in 1982. She currently holds the positions of Post doc Research Fellow in the Department of Business Studies and Research at University of Salerno (Italy). She holds a degree in Business Administration from Bocconi University, Italy (2005), and a Ph.D. in Marketing and Communication from University of Salerno, Italy (2009). Her main research interests are devoted to: economics and corporate management, financial communication, corporate strategy, entrepreneurship and innovation, technologies management for tourism, family firms, and medium businesses. She received the award for the second best paper on Entrepreneurship and Family Businesses by The European STEP Project, presented at the 6th Workshop on Family Firms Management Research, Culture and values, EIASM (Barcelona, ESADE Business School-Spain- 2010). She has published in academic and professional journals, books, and conference proceedings.

Angelo Camillo is an Assistant Professor of Strategic Management at the School of Business at Woodbury University in Burbank California, USA. He teaches course in Strategy and Business Policy, Global Enterprise Management, Business Ethics, and Organizational Behavior. He has also been Visiting Professor at various institutions in Europe and China where he taught Entrepreneurship and International Management. Professor Camillo conducts research in the area of competitive advantage, international management, entrepreneurship management, strategic marketing, and innovation. His international industry experience as an expatriate executive spans over 35 years in 11 countries and four continents. Professor Camillo received an associate degree in accounting and hospitality management from the Italian State School in Sorrento, Italy and an undergraduate degree in Hospitality Business Economics from the School of Hotel Economics in Heidelberg, Germany. He holds a MBA degree from San Francisco State University and a Ph.D. from Oklahoma State University. He also received several diplomas in Executive Management Development from Cornell University.

Sandro Castaldo is Full Professor of the Marketing Department of the Bocconi University where he teaches marketing and channel management. He is also senior professor and was former Chairman of the Marketing Department of the SDA Bocconi School of Management. He has a Ph.D. in Business Administration and Management from Bocconi University. His current research interests are in trust, retailing, and channel relationships. His works have appeared in Journal of Business Ethics, International Journal of Electronic Commerce, Industrial Marketing Management, Journal of Service Management and other outlets. He recently co-edited the book "Coopetition Winning Strategies for the 21st Century" (edited by S. Castaldo, G. Dagnino, F. Le Roy, S. Yami) Edward Elgar, Cheltenham, UK, 2010 and published the book "Trust in Market Relationship," Edward Elgar, Cheltenham, UK, 2007.

Richard Clodfelter is a professor in the Department of Retailing at the University of South Carolina (USA). He has taught courses in retail buying, Internet retailing, merchandising, and entrepreneurship, and his primary research interests include retail technology, e-commerce, and retail buying. Dr. Clodfelter has received the Distinguished Researcher and Scholar of the Year Award from the College of Hospitality, Retail, and Sport Management. In addition, he was the recipient of the Lester Sanders Award for Best Research Paper at the National Marketing Education Research Conference and the

Award of Excellence for the most outstanding paper to appear in The International Journal of Retail and Distribution Management during 1998. He also has been honored as Collegiate Teacher of the Year by the Southern Business Education Association. Dr. Clodfelter has published in the "Journal of Retailing and Consumer Services," "Journal of Product and Brand Management," "Frontiers in Direct and Interactive Marketing Research," "Journal of Fashion Marketing and Management," "International Journal of Retail and Distribution Management," "Journal of Shopping Center Research," "Marketing Educators' Journal," and "International Review of Retail, Distribution and Consumer Research." He is also the author of numerous textbooks, including "Retail Buying" and "Making Buying Decisions, Using the Computer as a Tool."

Vincenzo Corvello, PhD, is a researcher at the Department of Business Science, University of Calabria. He obtained his PhD in management engineering in 2007 from the Department of Management Engineering at the University Federico II of Naples. He holds courses of Management at the Faculty of Engineering of the University of Calabria. His research interests are in the fields of innovation processes organization and management; innovation in networks of SMEs; organization design; inter-organizational information systems. He published papers in International and Italian journals and chapters in international books like: International Journal of Production Economics; International Journal of Organizational Design and Engineering, Journal of Systemic, Cybernetics and Informatics; Lecture notes in informatics; Sistemi e Impresa. His research has been supported by national research grants.

Barry Davies joined the University after some years in the Department of Retailing at Manchester Metropolitan University. There, he was actively involved (as MFI Professor of Retail Marketing) in the development of pioneering programmes in retail management. He has been active in doctoral education, and in particular the development of professional doctorates. His research interests centre on service environments and interactions within them. He has a particular concern with decision support systems, especially in relation to space allocation problems. Barry's interests in these areas were kindled by his experiences in retail management. He has published more than 50 papers in marketing, management, and retailing journals. With colleagues he has also authored and edited a number of books. He co-edits a journal and serves on the board of a number of others.

Charles Dennis is Associate Editor (Retailing) of the *European Journal of Marketing* and and Professor of Marketing and Retailing and Director of Research at Lincoln Business School, the University of Lincoln (UK). His teaching and research area is (e-)retail and consumer behaviour. Charles is a Chartered Marketer and has been elected a Fellow of the Chartered Institute of Marketing for helping to modernise the teaching of the discipline. Charles was awarded the Vice Chancellor's Award for Teaching Excellence for improving the interactive student learning experience. Charles has published in journals such as *Journal of Business Research, Journal of Marketing Management* and *European Journal of Marketing*. Books include *Marketing the e-Business*, (1st & 2nd editions) (joint-authored with Dr Lisa Harris); the research monograph (combined textbook) *e-Retailing* (Routledge), and research monograph *Objects of Desire: Consumer Behaviour in Shopping Centre Choice* (Palgrave). His research into shopping styles has received extensive coverage in the popular media, including TV appearances with Sir Trevor McDonald OBE and Adrian Edmondson.

Loredana Di Pietro is a Post doc in research and Ph.D. in Organization, Technology and Human Resource fellow at University of Molise. She earned her MBA at University of Pisa, and Master Degree in Business Economics at University of Molise. Loredana Di Pietro is an Assistant Professor of Organizational Studies at Faculty of Business and at Faculty of Human and Social Sciences, University of Molise. She has been involved in researching organizational consumer behavior, business organization, team-based organization, and organizational conflict resolution. She is Editor for the International Journal of Digital Content Technology and its Applications. Over the years the results of her research activities have been enriched with the collaboration of peers, both nationally and internationally, with research articles and publications. She has presented research papers at seminars and conventions at the University of Groningen Research School in Holland, at the EIASM (European Institute for Advanced Study in Management) in Belgium, and at the University of Madrid Institute of Research in Spain.

Francesca di Virgilio is an Assistant Professor of Organizational Studies and Organizational Behavior at the Faculty of Human and Social Sciences, University of Molise, Italy. She received her Ph.D. degree in organization, technology, and development of human resources from Faculty of Economics, department of management, University of Molise, Italy. She was Visiting Researcher at Research Center of Industrial Relations & Organizational Behavior (IROB), Warwick Business School (UK). She is Editor for the International Journal of Digital Content Technology and its Applications. Her research interests include human knowledge, organizational structures, human resource management, and teambuilding. She is currently involved in international research project in Consumer's Group Knowledge, and in national research project in Group's Knowledge: well-being and distress in organizations.

Antonella Ferri was graduated with honors in Economics, from Rome University "La Sapienza." Since 1989 she practices the profession in Doctrines accountant. She earned a Ph.D. in economic theory and corporate governance processes at the Navale University of Naples (1994). Dr. Ferri is Researcher in Industrial Technology at the University La Sapienza, Rome, Faculty of Economics. From 1999 to 2002, she was an adjunct professor of marketing course at The University of Tuscia, Department of Economics. She was Associate Professor of Economics and Management of Cultural Heritage in November 2003 at the University of Naples Parthenope. Since 2000, she has been Professor of Tourism Marketing for the Master in Economics and Management of Tourism, Department of Economics, University La Sapienza. Since 2008, she is Professor of Economics and Business Management, Marketing and Web Communication at the Mercatorum Universitas, Rome. She has also been a researcher and consultant for Institute Tagliacarne, Chambers of Commerce, Industrial Union and CUEIM. She has published in academic and professional books, journals and conference proceedings.

Massimo Franco is Professor of Organization and Human Resources Management at University of Molise. He serves as the Chairman of the Business Management Studies within the Undergraduate and Graduate Programs. He is also Associate Director of the Department of Economics, Management and Social Sciences (SEGeS) of the University of Molise and a member of the Evaluation Board of the University of Sannio in Benevento. He is Associate Editor-in-Chief for the International Journal of Digital Content Technology and its Applications. His research interests and publications focus on new forms of organizations, team based organizations, teamwork and teambuilding, knowledge management, human resource management and competences, and well-being and distress in organizations.

Marta Frasquet is Associate Professor of Marketing at the University of Valencia, where she has been working since 1993. Her main teaching and research interests are in the field of channels of distribution, supply chain management, retailing, and consumer behaviour. She has taught these topics in undergraduate and postgraduate specialized courses. She has a significant number of contributions to international conferences and papers published in international journals. She has been invited professor at the University of Edinburgh (UK), at the Universitá Bocconi (Italy) and at the University of North Carolina Wilmington (USA).

Irene Gil Saura, Professor Dr., is a member of the Marketing Department of the University of Valencia since 1988 and has been a visiting scholar at several European universities. Her studies are published in several international journals, e.g. Tourism Management, Annals of Tourism Research, International Review of Retail, Distribution and Consumer Research, International Journal of Retail & Distribution Management, International Journal of Service Industry Management, Industrial Marketing Management, et cetera. She is currently interested in service marketing, consumer behavior, and retailing.

Monica Grosso is a Ph.D. Candidate in Business Adminstration and Management at the Bocconi University and SDA Bocconi School of Management. She collaborates as a researcher with the Retailing Observatory of the SDA Bocconi School of Management. Her research interests are in retailing and channel management, and in particular, in private label and collaborative relationships within distribution channels.

Bernd Hallier is Managing Director of the EHI Retail Institute (www.ehi.org) since 1985. He is also the President of EuroShop - worldwide exhibition for retail investment goods (www.euroshop.de), Chairman of the Board of the Orgainvent (www.orgainvent.de), founder of EUREPGAP (now GlobalGAP- Good Agricultural Practice) (www.globalgap.org). Prof. Hallier manages also the European Retail Academy (www.european-retail-academy.org) as an executive member of the Board of Trustees. Apart from his teaching activity in Germany he lectures in trade marketing and trade management at the Moscow academy of business at the Government of Moscow and in Philadelphia/USA. Additionally, he is monitoring for the EU in 2010 "new skills" for commerce to prepare international blended learning systems. His book "EuroShop" published in German/English in 2004 and in 2008 in Russian became a standard shopfitting literature.

Kevin Hapeshi is the Head of Department of Computing at the University of Gloucestershire. He is a Psychologist having completed his PhD in the development of reading skills at University of Wales, Swansea. He has worked as a Lecturer and Researcher in Cognitive Science and Human Factors at several UK institutions, including the University of Nottingham (reading skills); The City University, London (Knowledge Engineering); University of Cardiff (automatic speech recognition; speaker recognition; multimodal interfaces); University of West of England (sleep deprivation and pilot performance). He has over 60 publications in a variety of topics, with the focus of most being speech interfaces, multimodal human-computer interfaces, and information processing.

Hans-Ruediger Kaufmann, after extensive experience in German Bank Management, completed his sponsored PhD in 1997 and was as research assistant and then lecturer (p-t) at Manchester Metropolitan

University. Later, he worked in Budapest, first as Course Director Marketing for the Chartered Institute of Marketing and then as an Assistant Professor in Marketing at the International Management Centre Budapest as well as a contractual consultant. At the University of Applied Sciences Liechtenstein he was Academic Director Private Banking and, later, Head of the Competence Centre International Management. Since October 2006 he is an Associate Professor at the University of Nicosia. He has been a launching member and President (2007-2009) of the international research network on consumer behaviour, CIRCLE. He is currently Vice-President of the EuroMed Research Business Institute (EMBRI). He is member of the editorial board of a variety of journals and an Associate Editor of the World Review of Entrepreneurship, Management, and Sustainable Development.

Björn Kniza is Junior Marketing Manager at Fortis Deutschland Lebensversicherung AG and wrote his diploma thesis in cooperation with xplace, Germany. He received a diploma in business administration from the University of Göttingen in 2009.

Tobias Kowatsch is Research Associate and PhD candidate at the Institute of Technology Management, University of St. Gallen. His research interests in the management of Information Systems are the adoption and diffusion of innovations, consumer behaviour, mobile business, smart products, Internet of things, and recommendation agents. He studied Computer Science in Media at Hochschule Furtwangen University (HFU) in Germany and wrote his Diploma thesis about Semantic Web applications in 2005. After studying in Scotland and doing social volunteering in South Africa, he attended the Master's programme in Computer Science in Media at HFU in 2006 and passed with distinction. Then, he conducted empirical research on user acceptance of mobile Information Systems at the Research Center for Intelligent Media at HFU.

Despo Ktoridou holds a B.Sc., M.Sc. in Computer Engineering and a Ph.D. in the field of Expert Systems from Saint Petersburg Electrotechnical University in Russia. Her doctoral dissertation was "Distance Learning Educational Systems for Early Childhood Education: Methods, Facilities and Means." From 1993 until 1999, Dr. Ktoridou worked as an engineer for computer companies. In 1999, she was employed as an Assistant Professor of Educational Technology at Intercollege – Cyprus. Since 2007 she is an Associate Professor of MIS at the University of Nicosia. She is an ACM and IEEE member as well as an ETEK certified engineer. Her research and teaching interests concentrate in the areas of expert systems applications, e-learning, m-learning, social networking application in higher education, and managing IT innovations. She has presented papers in numerous refereed international conferences and has published many papers in refereed journals. Dr. Ktoridou participated in EU and local funded programs and was invited by foreign universities as a guest lecturer.

Christos Liassides graduated in 2005 in Information Technology and Communication Systems. Right after that, he took a job at Columbia Ship Management, whose headquarters are in Limassol, but the company is German. He worked there as a network administrator in the IT department. In 2007, he decided that an MBA would be a benefit for his career, so he began the University of Nicosia MBA program in Cyprus hoping for new opportunities in the future.

Wolfgang Maass leads the Smart Products team at the Institute of Technology Management, University of St. Gallen (ITEM-HSG). He is Lecturer at ITEM-HSG and Professor of Media and Computer Science at Hochschule Furtwangen University where he is also Head of the Research Center for Intelligent Media. His research interests in information management and Information Systems include ambient intelligence with a focus on smart products, Internet of things, electronic knowledge markets, and paid content. He is a member of the editorial board of "Electronic Markets" and a reviewer for the "European Journal of Information Systems," "The International Journal on Media Management," and the "Journal of Theoretical and Applied Electronic Commerce Research." He received his PhD from the University of Saarbrücken in Germany. At the end of 2007, he was awarded with a habilitation by the Department of Management at the University of St. Gallen.

Richard Michon is associate professor of marketing at the Ted Rogers School of Management, Ryerson University, Toronto. He currently teaches competitive industry analysis, business intelligence, and decision models to graduate and undergraduate students. Prior to joining TRSM, he taught retail research, merchandising, and operation management at the Ted Rogers School of Retailing. He also lectured graduate and undergraduate students at the University of Quebec in Montreal and HEC-Montreal. Dr. Michon's research and publication domains include psychometrics, structural equation modeling, hierarchical linear models, consumer values, retail fashion, and retail atmospherics. He is a member of the editorial review board of the Journal of Business Research (Elsevier), European Retail Research, and the Journal of Business Environment. He has published in the International Journal of Retail and Distribution Management, Journal of Business Research, Journal of Fashion Marketing and Management, Journal of Marketing Theory and Practice, Journal of Retailing and Consumer Services, Journal of Services Marketing, and the Journal of Applied Psychology. Richard Michon is a graduate of the Royal Military College, Kingston, Ontario, and l'École des Hautes Études Commerciales, Montreal. He has an honours degree in liberal arts, a MBA in finance, and a PhD in marketing and quantitative methods. He has acquired extensive professional experience in the natural resources, utilities, financial services, retail, food processing, telecommunications, nonprofit, and manufacturing industries. He has completed several mandates in competitive market analysis, consumer insight, data mining, and analytical CRM in Canada, the U.S., and in Europe.

Andrew Newman, professor, has established an international reputation for his work in several marketing and management fields, evidenced by his publications in prominent journals, international conferences, textbooks, and an acknowledgment by international peers in these related fields. This reputation extends to practitioners on both sides of the Atlantic and the media, who frequently seek Professor Newman's advice on global matters of consumer behaviour and strategic marketing. He has held a visiting chair at Missouri State University in the United States and has recently commenced a new visiting professorship at Mansoura University in Egypt. Professor Newman enjoys corporate access at the highest level to support his project research work and executive education initiatives. His track record of EPSRC, KTP funded research, and project sponsorship feeds productively into MBA teaching and executive education. Professor Newman created the Marketing and Retailing special interest group (SIG) for the British Academy of Management (BAM), and currently sits on BAM council.

Klaus Opwis is a full professor of Cognitive Psychology and Psychological Methodology in the Department of Psychology at the University of Basel, Switzerland. He has many years of research experience, especially within the areas of cognitive psychology (memory and learning, perception, thinking and problem solving) and Human-Computer Interaction. He has published over 70 research papers in peer-review journals and books.

Katia Premazzi has a PhD in Business Administration & Management from Bocconi University, Milano, and is currently Assistant Professor at the Marketing Department of the same university where she teaches marketing, channel management, and retailing courses at undergraduate and graduate level, both in Italian and English. She is also Professor at the Marketing Department of SDA Bocconi School of Management, where she teaches in the distance learning modules of the EMMS (Executive Master in Marketing and Sales), and in some courses of the executive education programs. She also collaborates as a researcher with the Retailing Observatory. Her current research interests are in retail innovation, shopping behaviour, and corporate social responsibility in retailing.

Sanda Renko is the associate professor in the Department of Trade in the Faculty of Economics and Business at the University of Zagreb, Croatia. She has carried out research on topics such as retailing and wholesaling, category management, logistics, channels of distribution, et cetera. She has presented papers at several conferences such as CIRCLE, EUROMED, International Retailing Conference, et cetera.

Sandra P. Roth is a PhD student at the Center of Cognitive Psychology and Methodology within the Department of Psychology at the University of Basel, Switzerland. She has a background in clinical psychology and Human-Computer Interaction (HCI). She is interested in the role of mental models and users' expectations in HCI, especially in the Internet.

María-Eugenia Ruiz-Molina is Assistant Professor in the Marketing Department of University of Valencia, where she earned her PhD in Business Administration and Management. Her studies are published in several international journals, e.g. Industrial Management and Data Systems, International Review of Retail, Distribution and Consumer Research, International Journal of Retail & Distribution Management, International Journal of Hospitality Management; International Journal of Contemporary Hospitality Management, et cetera. Dr. Ruiz-Molina's current research interests are consumer behavior, retailing, and ICT business solutions.

Emanuela Salvia is working on her PhD course on "Psychology of Programming and Artificial Intelligence" at University of Calabria (Italy). The dissertation will focus on linguistic analysis of a corpus of 100 criminal trials. Her main research interests focus on the cognitive aspects of language and complex systems.

Rocco Servidio is Assistant Professor in General Psychology at the Department of Linguistics, Humanistic Faculty, of the University of Calabria, Italy. His research interests include Web usability, Human- Computer Interaction (HCI), and e-learning. He works also in the field of Information and Communication Technologies applied to tourism, investigating how these technologies change tourism activities. Particularly, he is interested in the study of tourist websites usability, applying the HCI methodologies. For this last aspect, interface design and user interaction evaluation have been analyzed. The scientific effort focuses on finding the common denominator between HCI and tourism applications.

Pierpaolo Singer was born in Rome in 1961. He graduated in Economics at Luiss University with 110/110. From October 1985, he was Lecturer in the Department of Industrial and Commercial Technique of Rome University "La Sapienza." In June 1992, he became Researcher in the Department of Business Studies and Research at University of Salerno. During the year 1994 he taught at University of Latina. During the year 1995 he taught Corporate Finance and Production at Latina and Salerno Universities. From 1996 till 2000 he taught Financial Analysis at University of Salerno and Production in University of Latina. In 1999 became Associate Professor at Cassino University where he taught Corporate Finance and Management. In 2002 became Full Professor at Salerno University where he teaches Advanced Corporate Finance and General Management. From 2005 he also teaches General Management at Rome University "La Sapienza." He has published in academic and professional books, journals, and conference proceedings.

Eva Stüber is Assistant Professor and the Chair of Marketing and Innovation Management at Brandenburg University of Technology Cottbus, Germany. She received her diploma in business administration and economics from the University of Saarbrücken, Germany, and her doctoral degree from Brandenburg University of Technology Cottbus, Germany. The topic of her dissertation is the acceptance of personalization in online retailing. For her outstanding research on retailing innovations she has received the Wolfgang Wirichs award from the German retailing association. Her research has appeared in the Journal of Retailing and Consumer Services and other journals.

Assunta Tavernise is a PhD in Psychology of Programming and Artificial Intelligence and collaborates with the Laboratory of Psychology at University of Calabria. Her research interests concern various scientific topics from an interdisciplinary point of view and comprise the following areas: Human Computer Interaction, virtual agents, educational technology, virtual worlds/games, edutainment, and ICT for cultural heritage. At the moment, she is carrying out studies on non verbal communication of virtual agents for the realization of didactic tutors, guides, and virtual shopper assistants. Moreover, she is working on the HCI approach to learning virtual worlds, carrying out research on the design of personalized paths as well as on the evaluation of educational portals/software. She has worked in national and international projects, among which are "Virtual Museum Net of Magna Graecia" (ROP2000–06, www.virtualmg.net) and "Connecting European Culture through New Technology - NETConnect" (Culture 2000 Programme, http://www.netconnect-project.eu/, http://netconnect-project.eu/index.aspx).

Torben Tambo is Associate Professor of Aarhus University, Institute of Business and Technology, and has previously worked 17 years worked with management and Information Technology in manufacturing, supply chain, and retailing enterprises. His research interests include Information Technology, supply chain management, fashion, retailing, e-commerce, and business – technology interrelationships; his research has been published in Journal of Enterprise Architecture, Journal of Economic Dynamics and Control, International Association for Management of Technology, International Conference of Information Systems – Logistics – Supply Chain. Torben Tambo is MSc in IT and organisation from

Technical University of Denmark (1989), and holds a Graduate Diploma in Business Administration from Aarhus School of Business (1993).

Waldemar Toporowski is Professor and Chair of Retailing at Georg-August-Universität Göttingen. He received his PhD in 1995 from the University of Cologne, before he joined the Institute of Marketing and Retailing at the University of Göttingen in 2003. His research focuses on retailing and distribution, especially marketing channels, retail logistics, retail technologies, and store brands. He has published in European Retail Research, Marketing, Journal of Research and Management, and several leading German-language journals, such as Die Betriebswirtschaft (DBW) and Zeitschrift für betriebswirtschaftliche Forschung (ZfbF).

Alexandre N. Tuch is a PhD student at the Center of Cognitive Psychology and Methodology within the Department of Psychology at the University of Basel, Switzerland. He has a background in cognitive science and Human-Computer Interaction (HCI). He is interested in the role of emotion in HCI, especially the way aesthetics affects the user's emotional state during interaction.

Len Tiu Wright is Professor of Marketing and Research Professor at De Montfort University, Leicester. She has held full time appointments at the universities of Keele, Birmingham, and Loughborough, and has visiting lecturing and examining experiences with other UK universities, e.g. Oxbridge, and with institutions that are overseas. Len Tiu has consultancy and industrial experience and has researched in the Far East, Europe and North America. Her writings have appeared in books, in American and European academic journals, and at conferences, where some have gained overall best paper conference awards. She is on the editorial boards of a number of leading marketing journals. She is the Founding Editor of *Qualitative Market Research – An International Journal.*

Stephan Zielke is Associate Professor based at Rouen Business School, France, and Georg-August-Universität Göttingen, Germany. He received his PhD in 2001 from the University of Cologne and worked for an international marketing and strategy consultancy, before he joined the Institute of Marketing and Retailing at the University of Göttingen in 2004. Since 2010, he is also Associate Professor at Rouen Business School. His research focuses on retail marketing, especially in-store marketing, retail technologies, retail pricing, and store brands. He has published his research in Advances in Consumer Research, European Journal of Marketing, European Retail Research, Journal of Product and Brand Management, Journal of Retailing and Consumer Services, Marketing, Journal of Research and Management, Psychology & Marketing, The International Review of Retail, Distribution and Consumer Research, and several leading German-language journals.

Index

D

E

F

G

H

I

K

L

M

N

O

P

R

S

T

U

V

W

CPSIA information can be obtained at www.ICGtesting.com
263891BV00005B/80/P

9 781609 607388